MW01625076

A TRIUMPH OF GENIUS

EDWIN LAND, POLAROID, AND THE KODAK PATENT WAR

BY RONALD K. FIERSTEIN

Cover design by Amanda Fry/Ankerwycke.

Printed in the United States of America

19 18 17 16 15 5 4 3 2 1

Library of Congress Cataloging-in-Publication Data

Fierstein, Ronald K., 1950– author.
A triumph of genius / Ronald K. Fierstein. — First edition.
pages cm
Includes bibliographical references and index.
ISBN 978-1-62722-769-8
1. Land, Edwin Herbert, 1909–1991. 2. Inventors—United States—Biography. 3. Polaroid Land camera—History. 4. Instant photography. 5. Polaroid Corporation. I. Title.
TR140.L28F54 2014
771.3—dc23

2014034317

TABLE OF CONTENTS

PROLOGUE vii

LANDISMS ix

CHAPTER 1

A Very Special Young Man 1

CHAPTER 2

Polaroid Is Born 19

CHAPTER 3

One-Step Photography 39

CHAPTER 4

A Consumer Curiosity? 57

CHAPTER 5

Color 79

CHAPTER 6

Absolute One-Step Photography 105

CHAPTER 7

The Big Push—The Bigger Divide 121

CHAPTER 8

Aladdin Emerges 147

CHAPTER 9
The Coming Storm 171

CHAPTER 10
A Fight for Survival 197

CHAPTER 11
Kodak's View 219

CHAPTER 12
The Battle Is Joined 235

CHAPTER 13
Early Skirmishes in Discovery 251

CHAPTER 14
Searching for the Facts 265

CHAPTER 15
Kodak's Big Assault 275

CHAPTER 16
Land Joins the Fray 289

CHAPTER 17
The Long, Slow Road to the Courthouse 303

CHAPTER 18
The Final Turn 319

CHAPTER 19
Over the Precipice 333

CHAPTER 20
Land's Day in Court 347

CHAPTER 21
The Cross-Examination 369

CHAPTER 22
After the Main Event 389

CHAPTER 23
Kodak's Defense 403

CHAPTER 24
The Battle of the Experts 421

CHAPTER 25
The Waiting Game 441

CHAPTER 26
Victory at Last 465

CHAPTER 27
The Hammer Falls Harder 479

CHAPTER 28
Aftermath 501

CHAPTER 29
Epilogue 515

ACKNOWLEDGMENTS 535

DEDICATION 539

ENDNOTES 541

INDEX 613

PROLOGUE

It's always about the light. Any photographer will tell you that. It was a beautiful December late afternoon in the desert town of Santa Fe, New Mexico, a place renowned for its unique light. The rocky landscape, faintly dusted with snow, would have been brightly lit in dramatic burnt orange and rose rays refracted from an afternoon sun sinking in the sky toward sunset. It is that magical light that illuminates the beauty in anything it reaches. It is the photographer's best friend and secret weapon.

The little girl was excited as her father snapped away, taking image after image of her with his spiffy, high-tech camera. The father watched his daughter with delight and shared her joy. The wonders of light had enthralled him since his youth. They had already brought him great wealth and success. He was all of thirty-four.

But the year was 1943, and the "high-tech" camera was a Rolleiflex box model filled with Kodak roll film.[1] His daughter was disappointed when she learned that it would be weeks before she would be able to see the pictures her father had taken. "Why can't I see these pictures right now? I don't want to wait."[2]

In those days, the film would have to be unloaded from the camera, taken to a pharmacy, and shipped off to a laboratory to be developed into negatives. That process would involve a series of baths in various chemicals, all done in the dark at controlled temperatures. Once dry, each negative would then have a precise amount of light projected through it, the image being focused onto photographic paper held at just the right distance so that the image was sharp. That paper, or positive, would next be developed to bring out the image, washed, and then processed further to stabilize the print. Finally, after drying, it could be shipped back for the little girl to see, long after the enchantment of the moment had faded away.

When confronted with an upset child asking silly questions seemingly impossible to answer, most fathers would respond with a comforting, if exasperated, "Because," or "That's just the way it is," or even "I don't know." But not this father. As a colleague acknowledged many years later, this man "never had an ordinary reaction to anything."[3] Instead of dismissively laughing off his daughter's frustration, he shared it, he embraced it, and on the spot, at that moment, he committed himself intellectually to solving his daughter's dilemma. He would figure out a way to create a photographic system—a revolutionary camera and film combination—that would allow images to be viewed immediately after being taken. He would invent "one-step photography."

LANDISMS

"If you are able to state a problem . . . then the problem can be solved."

—*Life*, October 27, 1972

"Every creative act is a sudden cessation of stupidity."

—*Forbes*, June 1, 1975

"Optimism is a moral duty."

—*Fortune*, January 1974, p. 83

"The only thing that keeps us alive is our brilliance. The only thing protecting our brilliance is our patents."

—*Newsweek*, May 10, 1976, p. 86

"There's a rule they don't teach you at Harvard Business School: if anything is worth doing, it's worth doing to excess."

—The *New York Times*, April 27, 1977

"Do not undertake a program unless the goal is manifestly important and its achievement is nearly impossible."

—*Land's Polaroid*, Peter C. Wensberg, p. 170 (Houghton Mifflin 1987)

"If you are right, the facts at the moment don't matter because in due course the facts will catch up with you."

—*Popular Photography*, October 1977

A worthwhile invention "must be startling, unexpected and must come into a world that is not prepared for it."

—The *New York Times*, May 15, 1983

"What the physical sciences teach the social sciences is how to fail without a sense of guilt."

—*Boston Globe*, October 18, 1976

CHAPTER 1

A VERY SPECIAL YOUNG MAN

Life was getting progressively tougher for Avram Solomonovitch, his wife, Ella, and their three sons. They lived near Kiev, in what is today Ukraine.[1] During the reign of Tsar Alexander III in the late nineteenth century, violent anti-Jewish pogroms had led to a full assault by the Russian government against its Jewish people. It was the world of Sholem Aleichem and *Fiddler on the Roof*. As conditions continued to deteriorate for the Jewish population in the early 1880s, Avram decided to take advantage of a program that helped Russian Jews escape to America.[2] It was called the Jewish Colonization Association and had been organized by a wealthy European banker, Maurice "Baron" Hirsch. With its help, Avram and his young family sailed from Odessa to freedom. When they arrived in New York City, they were processed at the Castle Garden immigration facility, where Avram's name was anglicized to Abraham Land. The boys were renamed Samuel, Harry, and Louis.

After their arrival, the Lands were sent to an agricultural colony in Colchester, Connecticut. It was part of Baron Hirsch's organization. Abraham worked on the farm as they all acclimated to their new country. By the end of the decade, they had moved to New Jersey, where they became naturalized citizens in 1888. Abraham got into the scrap metal business and eventually moved the family back to Connecticut, settling in New Haven. In the years after their arrival in America, Abraham and Ella had five more children—two sons and three daughters. It was a large family, but their father's focus on education led all of the children into successful lives. Two of the younger boys attended Yale and went on to become lawyers. Two of the daughters became teachers. The others were successful in business.

Harry, Abraham's second oldest, had been born in Russia in 1880. He followed his father into the scrap metal business. Just after the turn of the century, Harry married Mattie Goldfaden, a fellow immigrant from Russia whose family had settled in Cleveland. They settled in Bridgeport, Connecticut, where, in 1904, their first child, a daughter they named Helen, was born. Five years later, on May 7, 1909, Mattie gave birth to their second child, a boy. He was named Edwin Herbert Land.

As a child, Helen apparently had a hard time pronouncing her younger brother's name and called him Din, a nickname that would be used by his closest friends and colleagues throughout his life.[3] The family moved to Norwich, Connecticut, around 1919, where they settled in a Jewish area. Eddie, as he was known around the neighborhood,[4] joined the Boy Scouts and was bar mitzvahed. Harry's scrap metal business prospered. He developed a valuable client in Electric Boat, a company that built submarines for the U.S. Navy. Harry also invested heavily and successfully in real estate, although much of the family fortune would later be lost in the 1937 recession.[5]

In the early years of the twentieth century, most of the rest of the Land family moved to Brooklyn, New York, where they remained close and observant of their faith, gathering to celebrate Jewish holidays in groups as large as forty cousins and uncles and aunts. Conspicuously absent from these festivities were Harry and his family. A niece described them as being "not unfriendly but cool and distant." Mattie was "very, very aristocratic [and] very well educated." Whatever the reason, Harry's aloofness seemed eventually to infect Edwin, who by young adulthood had distanced himself from his larger family, as well as from his Jewish ancestry. He told relatives that he had given up the Jewish faith for business reasons. "I'm trying to live down even the honorable part of my past," he once told a journalist.[6] In later life, Edwin developed a penchant for privacy and secrecy and was known to resist any and all attempts to have a biography written about him.[7] As noted by Victor McElheny, a journalist who covered, worked for, and eventually managed to write a book about him, Land did his best to "keep . . . his origins obscure . . . [and to] cultivate a mystery [leaving] few clues to his boyhood."[8] At his father Harry's funeral, a nephew asked Uncle Edwin why he didn't participate more in family gatherings. His answer was "my work is my life."[9]

While some aspects of Edwin's personality may have been enigmatic, the one thing that became unquestionably clear early on was that Edwin Land was blessed with a gifted intellect and an inquisitive mind. When Edwin was about five, Harry brought home a new Gramophone, then the state of the art for phonographs. Edwin was too curious about the device to resist disassembling it. Dressed in Dr. Dentons, one-piece pajamas with a flap in the back, he proceeded to lay out all the pieces of the machine on the living room rug. Unfortunately, the young boy did not leave sufficient time for the reassembly, and when Harry came home, the back flap of the pajamas came down, and a proper spanking ensued. The episode had a long-standing effect on the budding scientist. "From then on . . . nothing or nobody could stop me from carrying through the execution of an experiment,"[10] a promise that clearly led to what would become his famous propensity for nonstop and intense experimental work.[11]

Edwin Land's fascination with light started as a young boy, when he became interested in stereopticons, kaleidoscopes, and stereoscopes, three early nineteenth-century devices that manipulate light.[12] The stereopticon, introduced in 1850 by William and Frederick Langenheim and also known as a "magic lantern," was a slide projector with two lenses that produced realistic-looking scenes from superimposed images on glass photographic slides. The kaleidoscope, invented in England by Sir David Brewster in 1816, uses multiple mirrors mounted within a tube at various angles to create duplicate images of objects viewed through it. By filling the tube with small colored objects, beautiful patterns of color are created by the light coming through the tube. Land later described the device as "the color television of the 1850s," noting that "no respectable home would be without a kaleidoscope in the middle of the library."[13] Sir Charles Wheatstone invented the stereoscope, a device that converted two side-by-side images into a three-dimensional composite. In the 1830s, Brewster and others produced many models of stereoscopes in England and France. Beginning in 1862, popular hand-held stereoscopes, first designed by Supreme Court Justice Oliver Wendell Holmes and his business partner, Joseph Bates, were widely marketed. These were early forms of entertainment in the days before moving pictures were invented. Land later called himself the undisputed "king of the stereopaths."[14]

Young Edwin's favorite three-dimensional images were those of caves. The local library in Norwich had a stereoscope for use by the public. Recalling its great popularity, Land described how it "transported the child through the interplay of stalagmites and stalactites into the distant depths of the caves, having converted the two slightly faded sepia flat dull photographs into a vivid reality in which you could hear the dripping water, smell the dampness, fear the darkness as you sat with your legs crossed under you on the chair in the dear old library."[15] The effect and experience these visual devices could create was a profound discovery for young Land, one that would affect him deeply. Reflecting back years later on the impact of the stereoscope, Land observed: "Our Western race, hopelessly immersed in all the philosophical intricacies from Plato to Wittgenstein, was not ready to notice that this presumed toy was a device in which the child and the three-dimensional space he rejoiced in seeing comprised one single union of mind and matter, of soul and body, of man and nature, a single union of what in fact had never been divided and did not need reuniting."[16]

Inspired by these devices, Land, still a preteen, began to read about optical science and discovered the textbook *Physical Optics* by Robert Wood, a professor of physics at Johns Hopkins University who was well known as "an experimenter of great ingenuity in the areas of optics, light, electricity and photography."[17] Land slept with Wood's book under his pillow and read it "nightly in the way that our forefathers read the Bible."[18] It was an intellectual awakening that would shape his life. In Wood's text, Land discovered "daring theories and daring adventures and colorful experiments, a feeling for life as seen through science that captivated me from then through the rest of my years."[19]

The next phenomenon of light to catch Land's eye occurred when he was thirteen. His curiosity was stirred at a summer camp, Mooween, which he attended near his home in Connecticut. Barney "Cap" Girden, who had a wide-ranging interest in science and nature, ran the camp. Land and his friends were all amazed when Girden used a chunk of mineral crystal known as Iceland spar to eliminate the glare reflecting off of a shiny tabletop.[20] A Danish physicist, Erasmus Bartholin, had observed this effect as early as the mid-seventeenth century. A normal beam of light has within it rays—oscillations of the electric and magnetic fields—moving in all directions. This random and unorganized action produces glare. The crystal is able to filter out all light rays except those that are traveling in a single straight plane. The result, visually, is the elimination of glare that is present in normal diffused light. In the late eighteenth century, the English physicist Thomas Young explained the effect by describing how the light rays were "polarized" at right angles to each other. When Land later consulted *Physical Optics*, he discovered that Wood had described the effect with "deceptive simplicity." "Rays of light exist . . . which possess a one-sidedness and behave differently when differently orientated. For example, it is possible to obtain light that a glass or water surface refuses to reflect at a certain angle of incidence. Such light is said to be polarized."[21] (See Fig. 1-1.)

Later that summer, Land took a nighttime automobile drive with a camp counselor. In those days, vehicles were equipped with only very faint headlights. When the car almost crashed into a team of horses hitched to a farmer's cart, Land concluded that brighter headlights would be more useful but had an inherent drawback—they would create glare for oncoming traffic, and perhaps pose an even greater safety hazard. Polarization, Land thought, like that produced by the Iceland spar crystal, might somehow offer a solution. This possibility quickly became a subject for discussion by Land, Girden, and the other campers for the rest of that summer, and for summers to come.[22]

Land attended high school at Norwich Free Academy, where he continued to demonstrate his superior intellect, and some athletic ability as well. He had an impressive high school career, excelling on the debating team and the track team. But what really set him apart was his work in the physics lab, where he began investigating the phenomenon of polarized light. His teacher, Raymond Case, recalled that by Land's senior year, "he was already working at a level where I couldn't help him."[23] The inscription under Land's graduation photograph in the 1926 Norwich Free Academy yearbook reads, "Ed is some star in his studies and we are sure that he will make a name for himself and *Alma Mater* in college."[24] Only seventeen years old, Land graduated with "near-perfect marks" and enrolled that fall at Harvard University. His stay at Harvard, however, was to be brief.

Immediately upon arrival, Land resumed the experiments with polarized light that he had been conducting at home while in high school.[25] It would not be overstating it to say that Land had become completely preoccupied with the subject. Years later he would recall "the full vividness of my own need at the age of seventeen to do something scientifically significant and tangibly demonstrable. At the age in which each week seems like a year . . . I decided that the great opportunity was polarized light."[26] More specifically, Land was obsessed with finding a solution to the automobile headlight dilemma. Brighter light bulbs could be made and a car's electrical system could power them, but the glare would dazzle drivers of cars going in the opposite direction. Prisms made of the crystals like those Land had encountered at summer camp could theoretically do the job, but they were impractical because of their bulk. They were also very expensive. "There are not enough of . . . [those] great and precious [crystals] to cover the headlights of the cars in a single city block, and the headlights of these cars would be more expensive than the entire remainder of the cars," Land lamented.[27] Instead, Land set out, as he later described it, to determine "whether or not one could make a material in the form of a synthetic sheet which could polarize light."[28]

At the end of a single semester, it had become apparent to him that the pursuit of a practical polarizer was a more pressing problem and worthwhile objective than his education, and so he "stunned family and friends" by taking a leave of absence to concentrate on his research.[29] As Thomas Edison had done decades earlier, Land moved to New York and immersed himself in a study of the scientific literature on optics in the Great Reading Room of the New York Public Library on Fifth Avenue.[30] Harry Land was not pleased with Edwin's decision to leave Harvard, and the son was still dependent on his father for financial support. Accordingly, they struck a bargain in which Edwin enrolled for the spring term at New York University, and Harry continued to provide his college allowance.[31]

Land, however, did not spend much, if any, time at NYU's Washington Square campus. Instead, he put in long days of research at the midtown library, immersing himself in optics and light polarization. This consuming absorption would remain at the heart of Land's work over the rest of his career. A lifelong motto was: "If anything is worth doing, it's worth doing to excess."[32] Land had learned early on that total engrossment was the best way for him to work. He strongly believed that this kind of concentrated focus could also produce extraordinary results for others.

Late in his career, Land recalled that his "whole life has been spent trying to teach people that intense concentration for hour after hour can bring out in people resources they didn't know they had."[33] He once vividly described the intensity of his scientific focus in a speech he gave for an audience of student inventors at MIT's Junior Science Symposium, as he articulated his philosophy of intellectual investigation and how it impacted the other elements of his life.

> My own recollection of your age is a curious alternation which, I think, goes on through life for the scientist. It is alternation between the one mood and attitude of feeling part of the race as a whole, part of the family, part of the neighborhood; the mood of being in love with friends, women, men, people all over the world; the mood of being in love with what is great in music and art—all that on the one hand and then, quite suddenly, a separateness that comes

> during the preoccupation with a particular scientific task. There is a need, a transient need, a violent need for being just yourself, restating, recreating, talking in your own terms about what you have learned from all the cultures, scientific and non-scientific, before you and around you. You want to be almost alone, with just a few friends. You want to be undisturbed. You want to be free to think, not for an hour at a time, or three hours at a time, but for two days or two weeks, if possible, without interruption. You don't want to drive the family car or go to parties. You wish people would just go away and leave you alone while you get something straight. Then you get it straight and you embody it, and during that period of embodiment you have a feeling of almost divine guidance. Then it is done, and, suddenly, you are alone and you have to go back to your friends and the world around you and to all history, to be refreshed, to feel alive and human once again. It is this interplay between all that is richly human and this special, concentrated, uninterrupted mental effort that seems to me to be the source, not only of science, but also of everything that is worthwhile in life.[34]

Settled in New York City, Land was soon conducting experiments in search of a solution to what had been the long-standing obstacle to realizing the practical benefits of polarized light. He needed a way to duplicate the effect achieved with the large block of crystal in a material that would be of more practical use. "Land has very little respect for the forces of nature," a colleague noted. "He wants to rearrange the forces of nature, and just goes ahead and does it."[35]

The effort continued through 1927 and 1928 in a series of rented rooms in Manhattan; his first laboratory was located in a room he rented just a few blocks north of famed Times Square in midtown Manhattan, on West 55th Street, off Broadway. This was the pre-Depression-era New York of Jimmy "Beau James" Walker, the legendary bon vivant mayor. The "Murderers Row" lineup of New York Yankees, featuring Babe Ruth and Lou Gehrig, was in the process of winning back-to-back World Series by twice sweeping its National League opponents four games to none. In the spring of 1927, Land moved uptown to a small house on Riverside Drive and 104th Street, with a basement room in which he could work. This location offered the resources of Columbia University, in terms of its physics laboratories and the availability of students to help him with his experiments. Early in 1928, Land moved to another rental on 106th

Street and Amsterdam Avenue, again with a basement in which he could set up shop.[36]

Around this time, Land met a young woman who was interested in his work, and who became his primary research assistant.[37] Her name was Helen Maislen, but she was known as Terre. She read with him in the library and then became his laboratory aide in an era when, according to Professor F.W. Campbell of Britain's Royal Society, "women were not normally allowed in Physics Laboratories."[38] Years later, Land admitted that many of his early experiments were conducted at Columbia clandestinely. He and Terre would climb out a hallway window and then along a ledge on the outside of the building. They would climb back into a physics laboratory window and then proceed to remove the hinges from a locked cabinet to get at the necessary equipment.[39] Much later, after he demonstrated the results of some experiments to the professor in charge, Land was finally given a key to the facility.[40]

During this period, Land tried several approaches to finding a practical way to harness the magic of polarized light. One of the first, as he later explained, was to make polarizers by reflecting light from metallic plates. For a time he worked with other kinds of materials. Often Land would read up on a predecessor's work in the field and then duplicate that experimental work in his own laboratory, trying to learn whatever he could from those unsuccessful attempts to solve the problem. As he reviewed the physics literature for more ideas on how to pursue a polarizer that would have practical applications, Land eventually came upon the work of William Herapath. In 1852, Herapath had discovered that when quinine was added to the urine of a sick dog, tiny needle-shaped crystals formed.[41] In a random distribution, these crystallites appeared black when overlapped and otherwise clear. This property was attributed to polarization.[42] The crystalline material was later designated herapathite.

Unfortunately, neither Herapath nor other scientists who had tried over the next seven decades had been able to find a way to make a useful material out of herapathite. The problem appeared to be that the tiny, fragile crystallites flew apart when touched.[43] No one had been able to come up with a method of fixing them in place, in the uniform orientation necessary to realize their unique polarizing properties, let alone in a material that could be adapted for various applications. For Land, this was just the challenge he was looking for. Initially, he began to work with larger crystals that he thought might be easier to handle than the ones Herapath had discovered. But when he duplicated Herapath's work, he concluded that his smaller crystallites "were far and away the best optically. . . . It was at once apparent that in Herapath's extraordinary product there resided the clue to the development of satisfactory polarizers and to the utilization of polarized light on a large scale in an industrial way."[44] And so he set out to succeed where Herapath and others had failed.

It was now the spring of 1928, and Land believed that he was finally on the right track. With his college allowance in some jeopardy, given the fact that he was no longer in school, Land went home to visit his father. He told him what he had been doing and then took Harry out for a drive at night. According to Land, as they drove along with headlights shining in their eyes, he told his father, "I think I know how to take the glare out of car headlights. . . . Can I borrow five thousand dollars to get a business going?" Harry was supportive and agreed to give Edwin the money. But he was concerned that his young son might not be able to fend for himself in the business world. Land recalled his father warning him to "watch out. Once you develop it, the big companies will take it away from you."[45] It was advice that Land took to heart, and filed away for future use.

Land decided to continue his work in the kitchen of his parents' summer house on Long Island Sound, near New London, Connecticut. He hired Joseph Friedman, an organic chemist, to assist him. Together they searched for a way of fixing an array of the small crystals on a clear, flexible material in the proper orientation. Finding the right medium proved to be the easy part. Early on, they had success employing the kind of lacquer used in spray paint, replacing the colored particles that tinted the paint with the tiny crystallites.

At the end of the summer, Land and Friedman moved their operation to a one-room laboratory in a nearby office building Harry owned. There, he and Friedman made a breakthrough discovery on how to orient the crystallite particles. Normally, the lacquer would be opaque, but Land found that "when placed in . . . [an] electric field . . . [it] would become transparent and clear." This was the effect they were looking for. A momentous experiment, it was, as Land described it, "the first time that truly polarized particles had been brought to a homogenous orientation and held there by an electric field."[46]

Unfortunately, it turned out that using an electric field for this purpose was dangerous and impractical. Instead, Land decided to try exposing the suspension of crystallites in lacquer to a magnetic field. That, however, would require sophisticated equipment. So he returned to New York City, and received permission to use the large electromagnet in the Columbia physics lab. Land devised what would be a crucial experiment to duplicate the phenomenon he had observed earlier when he used an electric field to orient the crystals. A vessel containing a suspension of herapathite crystals in a lacquer medium was placed within the gap of the electromagnet. "Then came the most exciting single event in my life," Land later recalled:

> Before the magnet was turned on, the . . . particles . . . [were] oriented randomly so that the liquid was opaque and reddish black in color. When the [magnetic] field was turned on—and this was the big moment—slowly and somewhat sluggishly the cell became lighter and quite transparent. . . . This first polarizer experiment was a success.[47]

The magnetic field was apparently working to orient the crystals. Land had finally found the basis for a safe process that might lead to the commercial production of useful polarizers. In further experiments, he found that if he dipped a piece of plastic into the lacquer and exposed it to the magnetic field, the crystals would adhere to and coat the plastic. Once the lacquer dried, the magnet could be turned off, but the crystals would be stuck in place, properly aligned. The now-dry piece of coated plastic sheet was a polarizer.[48] Land had done it. As described by Harvard University professor Karel Svoboda, Land had "solved the problem of orienting tiny crystallites in a lacquer to create the first optical polarizers . . . [using] electrostatic and magnetostatic forces to achieve the alignment of the crystallites."[49] Years later, an article in *Scientific American* explained how Land's sheet polarizer worked:

> [Land's] solution was to imbed vast numbers of tiny crystals parallel to one another in a cellophane-like sheet, so that they would govern light like one huge crystal. Light waves, which normally vibrate helter-skelter like a Fourth of July sparkler, are combed out as they pass through the sheet's invisible "slots," and proceed in orderly parallel rows.[50]

It was September 1928. Edwin Land had solved a problem that had eluded some of the greatest minds in the world of physics for almost a century. He was 19 years old. Many years later, holding up a piece of his plastic polarizer for some students, he remarked how "it seems so plausible now for these to exist. At that time, it was the most exciting fantasy in the domain of physics that an adolescent could have."[51] He ruminated about the nature of scientific investigation in terms that anyone who tackles the Sunday crossword puzzle would readily understand.

> It is a curious property of research activity that after a problem has been solved the solution usually seems obvious. This is true not only for those who have not previously been acquainted with the problem, but also for those who have worked over it for years. As they regard their finished work they cannot help wonder why a simple, rational process that can be performed in a day took them, rational people, ten years to develop. In research, as in the whole civilizing process, why does it take so long to learn so little?[52]

Although Land had surmounted a huge obstacle, he knew that a great deal of additional research would be necessary to turn his laboratory success into a usable product that could be manufactured on a large-scale basis. Given his grand ideas for the ultimate commercial potential of his invention, Land recalled his father's admonition about protecting himself. He turned to his acquaintance from camp, Cap Girden. Girden advised Land to seek the help of his former camp counselor, Julius Silver, who by this time had graduated from law school and had opened a law practice. It turned out that Silver did not do patent work, but he referred Land to someone he knew who did, Donald Brown.[53] In retrospect, this was to be a pivotal moment in Land's life and career. Together, Silver and Brown would provide Land with the kind of diligent and savvy guidance someone of Land's creativity so desperately needs. Brown would go on to serve Land as his principal patent attorney for nearly four decades. Silver would become Land's principal legal and business advisor for the rest of Land's life. These were to become two of his most enduring and critical relationships.

On September 19, 1928, Land brought to Brown's New York City office a sample of his plastic sheet polarizer and demonstrated how it worked. Brown was impressed, describing the effect as "black magic."[54] Brown was still a young lawyer in his late twenties, having graduated from New York University Law School in 1922. Together with an associate, William W. Fraser, Brown immediately began work on a patent application for his teenage client. Land continued his work that winter to improve and perfect his invention. On April 26, 1929, still short of Land's twentieth birthday, the first patent application on his sheet polarizer was filed with the U.S. Patent Office in Washington, DC.

By that summer, Land believed that he had "carried the polarizer program far enough so that I could simply consult and advise a manufacturer."[55] His patent attorney offered to make his first introduction for him to a potential industrial customer. On August 14, 1929, Donald Brown wrote a letter of introduction for Land to Dr. C.E. Kenneth Mees, the head of research at the photography giant Eastman Kodak Company of Rochester, New York.[56] It would actually take a number of years for this connection to blossom into a business relationship. In part, this was because Land's assessment of the status of his project may have been a bit optimistic. Successful experiments in a laboratory setting are one thing, but they turned out to be a far cry from having the process or the product ready for industrial use. Nonetheless, this introduction would prove to be another pivotal moment in Land's career.

With progress on the polarizer front having been made, Land decided that it was time to resume his education, and so he returned to Harvard for the fall semester of 1929. By this time, Land and Terre Maislen had become inseparable, and she followed him back to Cambridge. They were married on November 10, 1929.[57] Interestingly, these events bookended one of the most dramatic days in U.S. history, the "Black Thursday" stock market crash on October 24, 1929 that led to the Great Depression. There is no evidence that this calamity, which affected almost every American household, had any impact on Land or his pursuit of his mission to perfect his polarizer. The marriage would last more than sixty years and would produce two daughters, Jennifer and Valerie, although Land's family life was surely forever compromised by his overarching dedication to his laboratory life.

During that period, Land was described as having "a shock of black hair, dark piercing eyes, a jerky manner, and a sophisticated but incurable enthusiasm about almost everything in the world but especially about . . . [his] light polarizer."[58] Although back at Harvard, his main focus remained on the task of continuing his developmental work. The experiments he had done at Columbia had demonstrated some useful principles, but unfortunately, he had also learned that the same techniques could not be used to mass-produce sheet polarizers for industrial use.

That first year back at Harvard, Land befriended a physics graduate student who was his lab instructor in one of the courses he was taking. George Wheelwright III came from a wealthy family—he was the oldest of two sons of a prosperous paper manufacturer in Ware,

Massachusetts—and had accumulated a sizable bank account of his own from a successful boys' camp he had created with his brother, and some smart investments in the stock market.[59] Luckily, he had turned his stocks into cash over the summer of 1929, just prior to the market crash. Just five years Land's senior, Wheelwright took a special interest in his student's work. Wheelwright engaged the help of two professors, Theodore Lyman and John McCloud, to help secure laboratory space in a new physics building for his protégé.[60]

Once in his own laboratory, Land worked on several aspects of his invention. He developed methods for making smaller crystals, so that there would be trillions of crystals in just a square inch of plastic. Even more importantly, he searched for a better technique for orienting the crystals in the necessary pattern. According to an account published in the *Christian Science Monitor* a few years later, Land came up with the concept for a potential solution one day while playing with a sheet of rubber that had a pencil sitting on it.[61] He found that when he stretched the rubber sheet, the pencil would eventually line itself up in the direction of the stretch. The same thing occurred when he put some toothpicks on a rubber sheet; when the rubber was stretched, the toothpicks would fall into line, parallel to the direction of the stretch.[62] Land immediately embarked on a long series of experiments to see if the same principle could be applied to the crystals in his sheet polarizer.

By the summer of 1930, he had his answer. Land concluded that it was, indeed, possible. The only complication was that, according to Land, "extremely high elongation" of the plastic sheet was necessary to have the crystals line up in the proper orientation.[63] Eventually, he found that the same effect could be achieved by extruding a suspension of crystals "between long narrow slits."[64] With this breakthrough, Land was well on his way to being able to produce a useful product. The extruded plastic material had each of the requisite characteristics Land had been seeking. As described admiringly years later by a distinguished fellow scientist, Land's "exciting new polarizer had many advantages: it was inexpensive, thin and could be cut easily to any size and shape to fit the application."[65]

Although Land was making progress—staying up nights and working through weekends in his Harvard lab—his studies were not going as well. He had trouble handing in lab reports on time, causing Wheelwright to appeal to his young wife for help. Exasperated, Wheelwright told Terre that Land rapidly lost interest in his lessons. He would work on a lab assignment "as long as he doesn't understand it, but as soon as he understands it he wants somebody else to do it."[66] For Land, the intellectual exercise was done at that point, and he was on to the next problem. Writing up the results was, for him, a waste of time. Terre agreed to help complete the lab reports, and Wheelwright often biked over to Land's house to pick them up. Some of Land's professors would stop by his lab to admire his work, but, nonetheless, they docked him when he was late to class. Despite Land's status as a bit of a maverick student, there was a clear recognition in the Harvard physics department that he was doing important work. As a result, he was asked to make a presentation to the department's regular colloquium. On February 8, 1932, Edwin Land became the first—and, for sixty-seven years thereafter, the only—undergraduate in the history of the Harvard physics department to do so.[67] His talk was entitled "A New Polarizer for Light in the Form of an Extensive Synthetic Sheet."

Not all of Land's fellow students were impressed by the presentation. Wheelwright, who by this time had become a valuable lab assistant in Land's work, was disappointed that the event did not make his protégé better understood by his peers. Wheelwright began to become disenchanted with the "intellectual snobs" who failed to recognize the significance of what Land was up to. It eroded his belief in and his commitment to the academic path he was pursuing. Wheelwright began to wonder whether it was worth working for his doctorate, instead of going out into the real world and actually doing something. Wheelwright and Land often took long evening walks together. On one of those occasions Wheelwright asked Land whether he might be interested in leaving Harvard to pursue his work. They would set up their own lab together. Perhaps more importantly, Wheelwright said that he had the money to carry out the plan. Land did not immediately respond, as he pondered the notion. By this point, Land was within a semester of graduation. He finally turned to Wheelwright and said, "An education without a degree?" Wheelwright told him that he didn't think the degree mattered. Land gradually concurred. He was certainly anxious to complete the work on his polarizer. And so, at the end of the spring term in 1932, Land and Wheelwright both left Harvard without their degrees, and set up shop in a rented room in Cambridge. The company was to be known as Land-Wheelwright Laboratories. Wheelwright put Land on a salary of $2,000 per year.[68] (See Fig. 1-2.)

There is no evidence that Land ever regretted his decision not to complete his formal education and get his degree. Yet it's clear that he realized later in life that his situation was unique and that a normal and formal education was important. Land would go on to support the cause of education throughout his career, particularly in the sciences, becoming a major benefactor of various academic institutions and one of the people most responsible for the genesis of public radio and television.

As Land continued to pursue a commercial version of his polarizer, he also began to follow up on the business leads that he and his attorneys had initiated. Julius Silver had written to J.H. Hunt, the chairman of the Committee on Investigation of New Devices in the research department at General Motors. Hunt had authored an article several years earlier reporting on efforts to combat the effects of headlight glare. Silver wrote to Hunt suggesting that he might be interested in the work of "a client who has developed a practical method of solving the problem by means of polarized light."[69] Hunt finally responded in March of 1932 and began corresponding with Silver, seeking information about his client's work.

That summer, with GM's interest sufficiently piqued, Silver and Land traveled to Detroit to visit the company's research center. They brought with them samples of Land's polarizer, which the GM engineers put through a series of tests. In a memo Land wrote on his return from the trip, he noted that the GM engineers "were very much impressed" with this polarizer.[70] During their day in Detroit, Land and Silver got to meet the legendary Charles Kettering, the inventor of the automobile self-starter and, by this time, the head of GM's research division. According to Land's memo, Kettering "was satisfied that polarization was an ideal solution" to the problem of glare.[71] In October of 1932, Land once again visited Detroit and conducted additional demonstrations with improved materials. According to a telegraph Land sent to Silver, "Hunt [was] very anxious to see [the] demonstration and talk about [a] royalty."[72] It was all very promising.

During this early period, Land and Wheelwright worked in two locations. First, they set up in an empty dairy barn in rural Weston, Massachusetts. When this proved too remote, they moved in mid-1933 to a dusty basement at 168 Dartmouth Street in Boston. This location was ideal because of its proximity to the libraries they needed, including the facilities just across the Charles River at MIT. Even though business prospects seemed to be blossoming, learning how to make the polarizer in sufficient quantities for commercial use was proving to be more difficult than Land had imagined. Land continued to work on the problem, as he simultaneously and confidently pursued other markets for the product he knew he would, at some point, perfect. On June 13, 1933, Land received a telegram from his patent lawyer, Donald Brown.[73] The immediate news was good. Land had been granted his first patent on his polarizer. It was U.S.

Patent 1,918,848, Polarizing Refracting Bodies. However, in the process of getting that application and others that followed through the U.S. Patent Office, Brown had uncovered some other, more worrisome, information.

Land was not alone in his pursuit of a solution to automobile headlight glare through the use of polarizers. Lewis Warrington Chubb, the director of research at the Westinghouse Electric Company, had filed an application for a patent broadly covering the concept of using anti-glare polarizers for automobile headlights. In 1919, Chubb's wife had been killed in an automobile accident when the glare from oncoming headlights blinded the driver of the car she was riding in.[74] Chubb had devoted himself to finding a way to prevent such accidents. He had come upon the properties of polarized light in 1920, and had proposed in a patent application that it be used to eliminate headlight glare. As of 1933, however, no patent had issued on Chubb's application.

The main problem was that Chubb only had an idea, generally an insufficient basis for obtaining a patent. He was unable to disclose any practical material for carrying out his concept. Land's patent, and his other pending applications, did. Although perfecting a large-scale manufacturing process would remain an ongoing challenge for Land, by 1933, he had produced small quantities of an effective polarizer in a plastic sheet. This gave Land an edge on the patent front, but Chubb had some advantages as well, particularly with respect to the goal of commercializing the concept. Several years earlier, Chubb had joined with some other rival inventors working on the headlight problem and had formed an alliance they called Polarized Lights, Inc.[75] This group began challenging competing patent applications in the field, and also began contacting car manufacturers and other potential customers about taking licenses under whatever patents they might be issued, even though at that point they had neither patents to license nor any usable product to sell. But, given Chubb's status in the industrial world, Polarized Lights had more relationships and easier access to key business partners than Land's fledgling company.

Besides the automobile industry, Chubb's operation vied with Land for the attention of Eastman Kodak. Kodak's director of research, Kenneth Mees, had followed up on the original introduction to Land made by Donald Brown years earlier. By early 1934, Mees and Land had met, and Kodak had become fully familiar with Land's work, recognizing several potential applications for his sheet polarizer material. First and foremost, Kodak was interested in using the polarizers in filters for photographic lenses. Mees also had other ideas, such as using them in automobile windshields in a joint effort with Libby-Owens-Ford, a company that manufactured glass for those windshields.[76] Kodak also shared Land's vision about the possibility of using the material to create a three-dimensional movie system.[77] Kodak's patent attorneys conducted a review of Land's patents as compared with the still-pending applications of Chubb in the early spring of 1934. Their conclusion was clear. Land was in "a very formidable patent situation," wrote one Kodak attorney. Another wrote, "We are strongly of the opinion that Chubb is not entitled to claims dominating the polarizing sheet material itself, and accordingly we do not believe that Chubb can become a dominant factor in the field which the Kodak Company contemplates entering."[78]

Based upon these opinions, as well as all of the satisfactory technical testing that had been conducted by Kodak on Land's polarizer material, Kodak decided to move forward. In late April 1934, Land began to discuss the parameters of a deal with Fordyce Tuttle of Kodak's research department. Land-Wheelwright would purchase cellulose acetate sheeting from Kodak, convert it into polarizer material, and sell it back to Kodak for use as photographic filters. Land-Wheelwright would also issue a license under Land's patents for Kodak to use the polarizer material for that purpose. Before a formal agreement could be entered, however, Land had to prove that his company could deliver its polarizer sheeting in sufficient quantity. Over that summer, Land was still able to produce only relatively small pieces at a time, the largest being only eight inches by eleven inches. Nonetheless, the quality of the sample material delivered to Kodak proved to be satisfactory and, on November 30, a contract was finalized for Kodak's purchase of polarizer material for $25,000 from Land-Wheelwright. In the first few days of December, Kodak sent Land-Wheelwright a down payment of $2,500.[79] Eastman Kodak had become Land's very first significant customer.

One problem remained, however. As Land later retold the story:

> We took the order before we had made anything we could sell to Eastman. . . . After all, what was the use of making much if we hadn't sold it yet? So we had the order, and we had to go to work. . . . We had air mattresses on the floor in the lab, and had food sent in, and after working for 20 or 40 hours a man would fall down and we would slip a mattress under him as he fell. . . .
>
> We had to build the machines and get the rolls going, and then turn out an awful lot of sheeting, out of which we were able to

> cut some . . . and send to Eastman. . . . We packed the whole thing in a small box which we carefully wrapped around in black tape . . . sent [it] to Eastman Kodak and sent a bill for $5,000.[80]

This first shipment left Boston on December 27, 1934. On January 3, 1935, Tuttle wrote that he had "received your precious shipment . . . [and that] the material came through in good shape."[81] Land was elated. It was a landmark moment. "The company was in business. We delivered in time," he later recalled. Land had successfully taken his invention from the laboratory to the world of commerce. His company had an actual product to sell and had sold it. Land's friend, Clarence Kennedy, an art professor at Smith College, had an idea for a name. Land liked the suggestion, and called his new product: polaroid.[82]

CHAPTER 2

POLAROID IS BORN

While Land was determined to show that polaroid would be a solution to headlight glare, he knew that it had many other potential applications. One of Land's friends took a scrap of the material with him when he went fishing.[1] He told Land that his casting had been more accurate than ever before; peering through the polarizing plastic had eliminated the reflection of the sky on the surface of the stream, enabling him to see the fish under water. Land's polarizer could thus transform the way sunglasses functioned—and the obvious customer for this application was American Optical, a leading manufacturer located outside Boston.

Eastman Kodak was also interested in the potential for polaroid's use in sunglasses. Kodak's Fordyce Tuttle, with whom Land had been working on the photographic filter, by this time known as Polascreen, expected to go with Land to discuss the idea with American Optical. Even before the deal for polarized camera filters with Eastman Kodak was finalized, Julius Silver and Tuttle had discussed prices and quantities for sunglass production with Kodak. But Wheelwright had contacts at American Optical that would enable Land to approach the company without any help from Kodak, and so Land opted to go it alone.[2]

The eventual meeting with American Optical provided Land with a chance to showcase his evolving ability to make dramatic and persuasive demonstrations, a talent that, later in life, would lead to his being described as "part thespian and part Barnum."[3] Described as "a dark-haired young man of twenty-five who looked as though he might be interested in tennis and dancing,"[4] Land arranged to meet a representative of American Optical at the Copley Plaza Hotel in Boston in late July 1934. He rented a room on the sunny side of the hotel and arrived carrying nothing but a bowl full of goldfish. When the American Optical representative came to the room, Land showed him the bowl, which he had placed on a windowsill in the full glare of sunlight. "Can you see any fish?" Land asked. Of course, the man could not—until Land held a piece of polarized material in front of the fishbowl. The American Optical executive, who had seen every version of sunglasses on the market, exclaimed that "he had never seen anything like this."[5]

Samples of the material were sent to American Optical for extended tests, and, on November 5, 1935, the company signed a contract with Land-Wheelwright for producing material required for the lenses in Polaroid Day Glasses.[6] The laminating of polaroid to lenses proved to be more difficult than Land had imagined, but eventually the new and revolutionary sunglasses hit the market in time for the holiday buying season in 1936. Advertisements touted them as "a necessity for sportsmen, motorists [or] anyone who is outdoors during winter or summer. . . . [With] Polaroid Sun Glasses you will see no glare when looking at water, sand, snow, highly polished automobile bodies, macadam roads or any shiny surface."[7] A pair sold for $3.75 ($64.30 in 2014 dollars).

Land also came up with a more fanciful version—adjustable sunglasses that allowed the wearer to regulate the degree of glare screened out by using a pair of polarized lenses, one movable and rotating on top of another fixed lens. While it proved to be unsuccessful in the sunglasses market, this configuration has proven to be very useful in photography to this day. (See Figs. 2-1 and 2-2.)

By 1935, with the Kodak and American Optical deals in place, Land-Wheelwright was up and running. Land was now confident enough in the performance of polaroid to be able to "realize one of . . . [his] fantasies." He sent a sample of the material to Robert Wood, the scientist whose early work had so inspired him.[8] Wood conducted experiments on it that confirmed its polarizing properties, a triumphant validation for Land.

Although Land continued to pursue other potential markets for polarizing material, including its use to make brighter and clearer advertising displays, he remained focused on getting the automobile industry to adopt the technology. He faced formidable competition from the Chubb group, Polarized Lights, as they both aggressively pursued deals with Ford, General Motors, General Electric, Pittsburgh Plate Glass, Chrysler, and others. Kodak, still interested in exploring the automobile market in concert with a glass manufacturer, continued to talk to Chubb as well as to Land. Chubb himself called Kodak's Kenneth Mees in March 1935 to find out about the scope of its arrangement with Land-Wheelwright. Mees assured him that it was limited to the photographic field. In June, Kodak even supplied Chubb with some samples of the photographic filter it was making under license from Land's company.[9]

Despite these efforts, the automobile companies were in no hurry to make a deal with either of the contenders. They knew about the competition between Land and Chubb that extended to the U.S. Patent Office but remained unconvinced about the need for or practicality of adopting polarizer technology. In order for every motorist to get the benefit of the polarizers, every car on the road would have to be equipped. This meant that every windshield and every headlight on every automobile would have to be outfitted, a daunting and expensive task. The alternative of requiring every driver to wear special polarized glasses seemed unrealistic. Putting a polarizer overlay on headlights raised issues with regard to the stability of the material when exposed to sun and dust and rain and snow.[10] Accordingly, neither company's prospects were encouraging.

In early July 1935, attorneys for Chubb's company proposed to Brown and Silver the possibility of a merger between Land-Wheelwright and Polarized Lights. Chubb was aware that, although he had independently come up with the idea of using polarizers in automobile headlights, it was Land's company that had developed and patented the actual polarizing material. Thus, even if Chubb became successful in wresting a patent on the concept out of the U.S. Patent Office, any interested manufacturer knew that it would need both companies to proceed. However, Chubb's proposal was not met with enthusiasm. As far as Land and his counselors were concerned, Chubb's attempts to patent the concept were never going to be successful. Land was already well on his way to having an income stream from Kodak and American Optical on the two successful applications of polaroid. Land-Wheelwright didn't really need Chubb or his company.[11]

Land decided to take a new approach to influence Detroit. He thought that if he publicized the potential benefits of the technology, perhaps he could put pressure from the public on the automobile industry, which in time might lead to governmental intervention with either states or, more likely, the federal government requiring the adoption of this safety feature. Serendipity struck. In early January 1936, Robert Sparks was an electrical engineer who was teaching science and math in a Hartford, Connecticut, high school. He had heard about the polarizer proposal and had contacted several journalists to promote the idea that polarizers should be adopted for use in all automobiles. Sparks even conducted a demonstration of the polarizer effect for James Spearing, the automotive columnist for the *New York Times*.[12] When a news item concerning Sparks' efforts appeared in the *Hartford Daily Times*, it caught Land's eye.[13]

Land jumped on Sparks' story and issued his own press release fully describing his idea for automotive safety and how he proposed it would work. The release, which touted some of the many other potential uses of polaroid—including its use in creating three-dimensional movies, one-way glass for privacy, sunglasses, and as an aid for detecting structural defects in aircrafts, bridges, and other forms of construction—announced a demonstration at the Waldorf-Astoria on January 30, 1936.

The now twenty-six-year-old Land had the opportunity to indulge his inner showman for the second time in his young career. As the press release distributed at the event trumpeted, "Polaroid, a light-polarizing glass that solves a problem as old as science, was introduced to the public for the first time today."[14] Land took the podium and explained:

> The idea of utilizing polarized light is not new, but previous means of accomplishing this involved the use of complicated and expensive material. "Polaroid," on the other hand, is a material that possesses this exceeding rare quality of permitting the passage of light vibrating in a single plane, while light in all other planes is checked.
>
> Light from the sun, for instance, is a composite of a great many waves vibrating in all possible planes at right angles to the direction of travel of the beam. Such light is said to be "unpolarized." In effect, the waves of the sun are infinitely tangled together. Polarization can be visualized as a complete untangling or "combing out" of these light waves so that the wave motions all lie in the same plane. "Polaroid" is the optical comb that accomplishes this effect.[15]

Land followed with a series of presentations to illustrate, as the *New York Times* described it, "the powers of the new substance called 'polaroid.'" He began with a dramatic demonstration of its ability to eliminate headlight glare. James Spearing was in attendance and issued a full report:

> At one end of a suite of rooms Mr. Land had set up two bright headlights on standards at the usual height and spacing of lamps on cars. At the other end of the suite there were two similar lights. When any one stood behind one pair of the lamps in a position

corresponding to that of the driver of an automobile he faced the blinding glare of the opposing lights. The lenses of both sets of lights had been polarized, but this was not noticeable to the unprotected eye. They seemed to shine with the same terrific brilliance that every driver of an automobile has to fight on the road at night.

When the man in the driver's position held a piece of polarized glass in front of his eyes, however, the opposing headlights virtually went out. There wasn't the slightest glare from them and the headlights of the hypothetical driver's car illuminated clearly all furniture and floor ahead of him. Standing there, easily seeing everything out front with no fierce attack of light on the eyes, one could not help wishing that he were on the road enjoying the pleasure of nocturnal driving that nowadays can only be imagined while one suffers the pain of the actual experience.[16]

As impressive as this demonstration was, "the thrill" of the event came later. George Wheelwright handled this part of the presentation and explained how the magic of three-dimensional movies could be achieved using their special material. According to an observer, "when polaroid was applied to the motion-picture . . . observers were ushered into a seemingly living fairyland of forms and colors. With two projectors operating in unison on a single screen, the pictures, when one viewed them through special spectacles, took on all the semblances of real life. Flesh color became realistic. The human form became rounded."[17] At the event, "it was announced that movies made and viewed by this system would soon be available through equipment supplied by one of the large photographic companies."[18] That company was identified as Eastman Kodak, but attendees were asked not to publish the name, presumably because a final deal had not been consummated.

Land's publicity offensive was successful. Spearing called Land's new substance "amazing" and reported on how it "thrilled the group of scientists and newspaper men" who were present.[19] Another *New York Times* reporter wrote: "All who witnessed the demonstration . . . regarded it as one of the most remarkable inventions in many years of optical development work."[20] The *St. Louis Post-Dispatch* reported back home: "Two young men who learned how to make 100,000 minute crystals stand on end on the point of a pin, exhibited the result [that] solves the problem of automobile headlight glare."[21] A scientist from a subsidiary of American Telephone & Telegraph (which had been working to find its own polarized material) told a reporter from the *New York Herald Tribune*, "It's everything [Land] says it is and more. There's no doubt that Land has solved a problem that every physicist working with light has struggled with for nearly a century."[22]

With two customers for its product, and an aggressive campaign under way to attract many more, Land-Wheelwright quickly grew from its initial two-man and one-woman operation. By early 1936, the company had twenty-five employees working at its Dartmouth Street facility. Land employed a unique strategy for identifying good employment prospects. "I don't care what the people know if they're willing to work hard," he once told Wheelwright, "and they consider it a pleasure to come here and work."[23] He was intuitive and impulsive about the people he interviewed, looking for candidates who had "a moderately appropriate educational background . . . [and a] sparkle in their eye."[24]

> When I meet someone for the first time, often I can tell right away whether he may be a potential scientist. In talking to this person, how much is he ahead of you? When you draw a breath to say the next thing, does he know what you are going to say before you say it? Does he delight in the construction you are making? Does he turn the conversation quite subtly because he perceives where it is going and wishes it to go somewhere else? Not all scientists are that alert. There are many scientists who, for all their marvelous training, are just plain dull. You sit with them and nothing is happening. They have been stultified somehow and the world is going by them.[25]

A few men Land hired during this early period would go on to play pivotal roles in Land's business, and thus his life, for the rest of his career. One was Howard Rogers. Like Land, Rogers was a Harvard dropout. He had left school to help with his family's finances, hoping to enroll instead at MIT once things settled down. It never happened. Rogers was working at a gas station earning $25 a week when he heard about Land-Wheelwright from his brother Nick. Nick had met Wheelwright on a ski trip, and had landed a job for himself in the rapidly growing concern.[26]

Rogers applied for a job. In early March 1936, he met Land at a soda fountain across the street from the Dartmouth Street lab.[27] It was a brief meeting. Rogers later recalled that Land had quickly sized him up, giving him the impression that Land "could see into my head. It was really a kind of interesting sensation of having your head briefly searched for content."[28] Rogers took a pay cut to $10 a week when he joined Land's operation, working in the dusty facility as a production assistant in manufacturing the company's plastic sheet polarizers.[29] He would work at Polaroid for the next half century. (See Fig. 2-3.)

Continuing the publicity effort, in May 1936 Wheelwright demonstrated the three-dimensional movie effect to a meeting of the Society of Motion Picture Engineers. The assembly was duly impressed, issuing "mingled gasps of astonishment and a mild patter of handclapping as the three-dimensional scenes unfolded," according to one report. "Seen through a pair of glasses with polaroid lenses . . . the spectators watched figures so real and life-like they seemed likely at times to pop out and shake hands with the audience."[30] However, this group of technicians wondered how enough pairs of specialized glasses could be produced, and how would the public react to wearing them?

In June, Land renewed his assault on the automotive industry, appearing before the semiannual meeting of the Society of Automotive Engineers to demonstrate that with his polarized material, "night driving would approach the comfort and safety of day driving."[31] Perhaps the most significant event in the public relations campaign was the creation of an exhibit, Polaroid on Parade, at the New York City Museum of Science and Industry located in Rockefeller Center. Its debut of the exhibit in December 1936 attracted the elite of the science press and many important industrial figures. The museum's vice president, Felix Warburg, the wealthy son of the man who had designed the U.S. Federal Reserve System, presided over the event. In the museum's press release, Professor Wood, who had studied the sample of polaroid Land had sent him the previous year, called it "the most significant invention in the field of optics, certainly within the last generation, probably in the last century."[32] Herbert Nichols, a correspondent for the *Christian Science Monitor*, reported that Land was introduced at the event as "the originator of an invention to be ranked with that of the electric light, lens systems, and color dyes."[33] There was no doubt that the twenty-seven-year-old Land's star was on the rise.

In addition to attracting interest from the worlds of industry and science, Land's high profile brought his fledgling company to the attention of the financial community as well. This was a welcome development, as continuing its technical research while simultaneously running a public relations campaign was an expensive proposition, taxing Land-Wheelwright's resources. There had been previous unsuccessful proposals for raising further capital, including a 1934 deal for Kodak to take over the company that came very close to being signed. But the two young entrepreneurs backed out when it appeared that they would have to move their operation, and themselves, to Rochester. Of even more concern was the impression they got that the more senior research scientists at Kodak intended to exercise a paternal control over them, taking away their research independence and assigning them to projects of Kodak's choosing.[34]

Wheelwright and a "scared stiff" Land had met with J.P. Morgan in their quest for capital investment. While Morgan was fascinated by the technology, he declined to get involved, suggesting he was too old for a venture of that nature. However, according to Wheelwright, Morgan agreed to introduce the pair to "people like Kuhn, Loeb, the banking firm that made a specialty of backing Jews with real talent."[35] Following up on the excitement generated by the Museum of Science exhibit, Silver was able to introduce Land to Bernard Baruch and James Warburg, a banker and the younger brother of the Museum's Felix. Warburg, in turn, introduced Land to Averell Harriman, the heir to the Union Pacific Railroad fortune, and others.[36]

With such major financiers expressing interest in Land's polarizer technology, a contract with a group of investors was signed on August 10, 1937, to form the Polaroid Corporation.[37] The investors would provide $750,000 in financing, half of which would be payable immediately. Land entered into a ten-year employment contract, and he and Wheelwright and Terre assigned to the new corporation all of their patents and patent applications. Although the investors would own about half of the new corporation, through a complicated stock voting arrangement devised by Silver, Land retained complete control of the company. When it came time to determine how shares would be split between Land and Wheelwright, Wheelwright insisted that Silver broker the deal. Silver was reluctant, telling Wheelwright that he was Land's attorney and that Wheelwright should seek counsel of his own. Wheelwright did that, but his advisors insisted that he get more than fifty percent of their shares, a nonstarter as far as Silver and Land were concerned. Ultimately, Wheelwright returned to Silver and, citing the fair and informal way the three of them had conducted their business over the years, insisted that Silver structure the deal. Silver offered Wheelwright ten percent.[38] Wheelwright accepted, and never complained thereafter.

Before the arrangement was completed, in a development that would soon have a significant impact on the new company, on July 24, 1937, the U.S. Patent Office granted Lewis Chubb a basic patent on the use of polarized light for eliminating headlight glare.[39] His application had been pending since 1920. Land's immediate reaction was to challenge the validity of Chubb's patent, and Donald Brown filed suit in September 1937 to do just that. But given the uncertainty of prevailing in its challenge, and the fact that Polaroid was finally on secure financial footing, it finally seemed to be the right moment to cut a deal with Chubb, thus enabling a unified effort to win over the automobile industry. Accordingly, one of Polaroid's investors, Lewis Strauss, embarked on a negotiation to buy Chubb's company, Polarized Lights.[40]

Silver, however, continued to believe Chubb's patent to be useless and no obstacle to Polaroid's further efforts in the field. But the new investors decided to be more proactive. Strauss had several good arguments to use in trying to convince Chubb's representatives to make a deal. If Chubb's patent was indeed valid, it was limited to a specific technique of *using* polarized light to reduce glare. Land's recent work had produced techniques that were outside the scope of Chubb's patent. Moreover, the original problem still existed—while Chubb might have a claim to the concept, it was only Land who had the patent on the actual material required for implementation. Eventually, Chubb and his colleagues signed an agreement on March 7, 1938, under which Polaroid acquired Polarized Lights for 7,000 shares of Polaroid stock and a cash payment of $25,000.[41] Lewis Chubb became a Polaroid employee.

In the meantime, Land's public relations campaign to bring pressure on the automobile industry resulted in a major feature in *Fortune* describing how glare "vanished" when Polaroid's system was used.[42] By late 1939, Rogers and Land had developed a new type of polarizer, called the "K Sheet," that could stand up to the environmental wear and tear on headlights.[43] With pressure mounting, in 1939 the Automobile Manufacturers Association reached a cease-fire agreement with Polaroid. The association would initiate an engineering study to evaluate fully the headlight glare proposal. Polaroid and General Electric would conduct that study jointly.[44] As it turned out, the entire effort was put on hold when World War II broke out, but at least there was now a formal framework for exploring the industry's adoption of the concept.

Another by-product of Land's publicity blitz was that more and more installations of his polarizer material appeared on the scene. And with each dramatic application, Land's renown only continued to soar. One outstanding example was the use of polarizers in the adjustable windows of an observation car on the Union Pacific's elite streamliner *City of Los*

Angeles. These were essentially twenty-nine large versions of Land's variable sunglasses, which could make the colors of the passing landscape more vivid and glare-free or screen out all outside light. *Fortune* forecasted that "gadget-loving passengers will never tire of playing" with these "wonderful windows."[45] (See Fig. 2-4.)

Land's polarizer technology reached a truly massive audience by "startling" attendees at the New York World's Fair of 1939–1940.[46] Five million fairgoers visited the Chrysler Corporation pavilion and saw a three-dimensional film entitled *In Tune with Tomorrow*. Using special cardboard and plastic polarized eyeglasses supplied by Polaroid, the viewers experienced "fascination and sometimes alarm" as they watched the construction of a Plymouth automobile with "wheels and springs and pistons [that] seemed to be moving in thin air, out beyond the screen, just out of reach."[47] It was a new experience for so many, rooted in Land's childhood fascination with three-dimensional imagery.

This burgeoning public interest was matched by recognition from the academic and scientific communities. The many awards and honors accorded to Land over his lifetime began to flow in this era. For example, the Franklin Institute, an organization and museum founded in 1824 to promote the mechanical arts, awarded him the Elliott Cresson Medal for 1938. It called his polarizer "a contribution of the first order in applied optics [that was] epoch-making in the field of illumination and color production."[48] An even more significant honor was his being named as one of eighteen Modern Pioneers on the Frontiers of Industry by the National Association of Manufacturers.[49] This one-time event in 1940 was organized to celebrate the 150th anniversary of the American patent system.

A committee of scientists, headed by Karl T. Compton, the president of MIT, headed the selection committee, and chose the recipients from an initial list of more than 1,000 nominees. In making the presentations, Compton described Edwin Land to a tee as he articulated the essential attributes of the pioneers being honored that night: "a state of mind that includes curiosity, an idealism which is dissatisfied with the restrictions and imperfections of the present, a great inward urge for discovery and an ability to translate this dissatisfaction and inward urge into constructive achievement."[50] Land shared this honor with Wilbur Wright, Henry Ford, and a collection of the most important industrial scientists of the day, including Lee DeForrest, the inventor of the vacuum tube, and Charles Kettering, the head of General Motors research with whom Land had met years earlier with regard to his proposal to eliminate headlight glare. The

Associated Press reported Land's ascendency: "Among the sage elders of science a young man . . . [of] thirty takes rank today as one . . . upon whom the title 'modern pioneer' has been bestowed in recognition of his inventive genius."[51] (See Fig. 2-5.)

With Land's growing acclaim, George Wheelwright was beginning to feel a profound change occurring in his role at Polaroid, and in his relationship with his former student and current partner. Although a major shareholder and a corporate vice president, Wheelwright was no longer involved in any of the company's technical research. Since joining the company, Howard Rogers had quickly become Land's primary scientific collaborator. Perhaps as a result of these shifting relationships, Wheelwright saw less and less of Land. Polaroid was clearly becoming Land's enterprise in every way. Finally, with some encouragement from his brother, Wheelwright approached Land. The two spent a "very emotional" few days talking over the situation. Although Land urged him to stay, Wheelwright concluded that his departure would be "for the good of the company."[52] Wheelwright apparently recognized that Polaroid was then, as it would be for decades to come, the personification of Edwin H. Land. Wheelwright left Polaroid in 1942.[53] After a stint as a Navy pilot in World War II, Wheelwright lived tranquilly on his ranch on the coast of Northern California until his death at the age of ninety-seven in 2001.[54]

Wheelwright's departure during this period was offset with the arrival of a self-described "hayseed from upstate New York."[55] William McCune had graduated with an engineering degree from MIT in 1937. He went to work for General Motors and was scheduled to transfer to its German operation, when Hitler went on the rampage in August 1939. McCune resigned and, while living in New York, consulted with some of the placement people at MIT about other job opportunities. Nat Sage, an alumni advisor, told him about Polaroid, and made an introduction.[56] McCune interviewed with Jack Latham, a fellow MIT alumnus working at Polaroid who was also a close friend of Land's. While McCune was "intrigued," sensing some "fascinating" things about the company, he was a bit frustrated because it was difficult to learn much about Polaroid, other than that it was busy making polarized lenses for sunglasses. When McCune returned to New York, a telegram from Sage was waiting for him: "Think twice before you turn down Polaroid."[57] McCune joined Polaroid in late 1939 and was immediately assigned to set up a quality control department for the company's polarizer production facility. It would be several years before his path would cross with Land's in any significant way.

Whatever progress was being made at Polaroid during these years, ongoing events in Europe seemed to overshadow them more and more. Although Land had historically been oblivious to other major occurrences beyond his immediate focus, like the stock market crash of 1929, the threat of war had his full attention. As the prospect of another world war loomed, some key leaders of the U.S. scientific community worried that America had not taken full advantage of its technical capabilities during World War I. With hostilities spreading in Europe in 1940, and with the probability of America's involvement increasing, Vannevar Bush, a former dean of engineering at MIT and the inventor of one of the early computers, took steps to put America on a fully effective wartime footing. In early 1940, he joined forces with James Bryant Conant and Karl Compton, then presidents of Harvard and MIT, respectively, to propose the formation of a group to coordinate a constructive program between America's military and its scientific community, within both industry and academia. When Holland and Belgium surrendered to Hitler's forces in May, and the German army slashed into France in early June, Bush accelerated his initiative.

Bush contacted his friend Harry Hopkins, Franklin Delano Roosevelt's close advisor, and requested a meeting. On June 12, 1940, Bush presented his proposal to President Roosevelt in four brief paragraphs typed on a single piece of paper. The meeting took just ten minutes. Roosevelt approved the idea—initialing the document "OK-F.D.R."—and the National Defense Research Committee (NDRC) was launched. Its mission was to enlist in the war effort "the support of scientific and educational institutions and organizations, and of individual scientists and of engineers throughout the country."[58] Bush's novel approach was to farm out the work on contracts to existing academic and industrial institutions, rather than trying to attract the talent into a centralized government effort, an approach that had been proven ineffective in the previous war.

Over the course of the next five years, Bush's organization, later expanded and renamed the Office of Scientific Research and Development (OSRD), accomplished its mission with brilliant effect. More than 6,000 engineers and scientists participated; according to one estimate, this included more than two-thirds of all American physicists. Perhaps most notably, Bush took responsibility for following up on a letter President Roosevelt had received from Albert Einstein in 1939 advising him that a powerful new type of bomb could be constructed from the fission of a uranium isotope. As the *New York Times* later reported, it was Bush who "shepherded the complex atomic bomb project to fruition, [and] who [later] explained the technical details of the bomb and outlined its awesome powers to President Harry S. Truman soon after Roosevelt's death."[59]

It wasn't long after the formation of the NDRC that Land was contacted by one of Bush's representatives. It was clear that there would be myriad applications for Land's new sheet polarizer in the war effort.[60] Later, OSRD would exploit Land's intellect and creativity by involving him in other projects well beyond Polaroid's original area of expertise. Land's first meeting with the NDRC occurred on November 1, 1940.[61] He became immediately engaged in the challenge, and the opportunity, presented to him.

As Christmas approached, Land gathered the entire Polaroid staff together for a meeting at their facility on Main Street in Cambridge. They numbered about seventy-five.[62] By this time, England was enduring constant bombing by the Germans, longing for America to intervene. But Roosevelt hoped to keep America out of another European war. Japan's attack on Pearl Harbor extinguished that hope just a year later. Land, however, was certain from the start that American involvement was inevitable. William McCune, who had attended that Christmas meeting, recalled that Land told his employees that he believed the war going on in Europe "was of much greater significance to the United States than most people felt."[63] Land predicted that the United States would be in the war within the next year. As a result, he had decided to make "a big change" in Polaroid's focus. From that moment on, the company would devote itself to one sole purpose, "to win this war."[64]

In his talk to Polaroid's employees, and in future comments, Land made it clear that he was not motivated by the chance to profit from the war effort. In fact, he said that he didn't expect to make much. His motivation, instead, was to combat "the disease that is spreading over the world," one that "goes on for generations . . . [and] does not stop when war stops."[65] Victor McElheny wrote that Land "did not make clear whether he thought the disease was intolerance or totalitarianism. He may have been thinking of the anti-Semitism that drove his grandparents from Russia and now, decades later, was fueling Hitler's 'final solution.'"[66] In any event, Land, as he explained years later, was ready "to give up flexibility and freedom to get some specific things done. In war you beat the plowshare into a sword." Land added, "whatever distaste a scientist may have for using science for military purposes is put aside when he is confronted by absolute evil."[67]

By the time the Japanese attacked Pearl Harbor in December 1941, Polaroid was making a variety of special filters for various naval instruments, and had devised a machine-gun trainer that displayed a three-dimensional image of an attacking plane and simulated tracer bullets. It also wrestled with the problem of overcoming the glare that would often blind a gunner, particularly when forced to look into or close to the sun. This was a challenge whether the gunner was on a naval vessel, on an airplane, or on the battlefield aiming at a tank or other enemy target.[68] Polaroid produced millions of what Land proudly called "the best damn goggles in the world."[69] None other than General George S. Patton appeared on the cover of *Newsweek* outfitted in a pair.[70] (See Fig. 2-6.)

As Polaroid's involvement in the war effort deepened, it designed and manufactured for the government a wide range of optical devices, including a rangefinder devised by a group of Polaroid researchers in which Lewis Chubb was "one of the key men."[71] Howard Rogers worked on a team assigned to the problem of finding a method by which high-quality lenses could be crafted from plastic to replace the precision glass lenses formerly produced by Germany, and thus no longer available.[72] Other devices included various types of polarizing and non-polarizing filters, gunsights, periscopes, and bombsights.[73] Based on the design of its variable sunglasses, Polaroid produced a variable-density goggle for use by aircraft gunners. When the Navy asked six companies to make proposals for a device to determine the elevation of aircraft, Polaroid sent in a working prototype before any of the other five had even acknowledged the request.[74] Land also led the effort to develop dark-adaptation glasses. Navy officers and pilots used the glasses when they had to adapt quickly from dark environments to the light, or vice versa. Two decades later, it was a pair of Polaroid night vision goggles that garnered publicity from being used in the surveillance conducted by Lebanese counterintelligence agents leading to the capture of notorious British spy Kim Philby.[75]

Throughout the war, Land and his team repeatedly proved their willingness and ability to tackle special assignments. In September 1944, the U.S. Navy sent an officer to Cambridge to request Polaroid's help in coping with Japanese "kamikaze" suicide attacks, pilots who were flying their planes directly into U.S. warships. These attacks had forced the Navy to use ultraviolet lighting for its aircraft carriers, allowing the ships to stay blacked out at night while still providing some illumination for those working on the flight deck and for pilots making night landings. Special goggles were required to make the system workable; they had to remain "stable under the heat, brilliant sunlight, humidity, salt air, and other harsh conditions of the tropical regions of the western Pacific."[76]

Land showed the officer a filter that he thought would solve the problem. When the officer asked for fifty pairs to cover the most critical landing situations in the battlefield, Land asked him how many pairs of goggles would be needed to supply all aviators then making landings in enemy waters. The officer thought 200 would do it. Land then suggested that, to save time, the officer stay around for a week so he could take the goggles with him. The officer was stunned, and asked Land if he was serious. Indeed he was. Land set to work with a young colleague, Louis Rosenblum. Nine days later, the officer left Polaroid with 200 sets of goggles, carried aboard his train for Washington, D.C., by "a retinue of porters." They were promptly airlifted to the dozens of aircraft carriers in the Pacific and were used in engagements thereafter in the Philippines, Iwo Jima, and Okinawa.[77]

The war effort also gave Polaroid a chance to find new applications for its three-dimensional technology. In 1938, Joseph Mahler, a refugee from the Nazi takeover of Czechoslovakia, joined Polaroid. While a teenager, he had, like Land, become fascinated with photography and stereo imaging.[78] Mahler showed Land another form of three-dimensional imaging using polarized light, known as an anaglyph.[79] The technique used eyeglasses in which each lens was a different color, generally red and blue. This system differed from Land's earlier three-dimensional system that, like a stereoscope, had blended two side-by-side images into one three-dimensional image using different specialized eyeglasses. In the anaglyph, two differently colored images are superimposed on each other but printed slightly askew. The special eyeglasses separate them, and then the brain does the rest, recombining them into a three-dimensional image. Out of this came a technology known as the Vectograph. Using the anaglyph principles, the Vectograph could produce three-dimensional images on a single piece of film or paper. There was no longer the need for two projectors to blend two synchronized images or a stereoscope card to display two separate side-by-side images. Only a special pair of polarized glasses was necessary to create the "dramatic . . . three-dimensional representation" from a single image.[80] Land and Mahler received their first patent on the Vectograph in June 1940.[81]

The first public demonstration of the Vectograph was to the Optical Society in Rochester, New York, in October 1940. It was during the Roosevelt-Willkie presidential election campaign. Land created an image for the audience to view and decided to have some fun with it. As he described it years later, "I had the glasses arranged so that if you closed one eye you saw the Democrat and if you closed the other one, you saw the Republican. I asked all the Democrats to close the left eye and the Republicans to close the right eye, and of course they all applauded."[82] According to MIT's Stephen A. Benton, one of the leading researchers in the field and the inventor of the "rainbow" hologram often seen today on credit cards, the Vectograph was "the highest technical development of polarized 3-D print and projection technology."[83]

Thanks to this technological breakthrough, pilots flying as fast as 350 miles per hour over heavily fortified enemy targets or strategic locations could capture three-dimensional images that were later printed on photographic paper or as transparencies.[84] These aerial photographs could reveal vital topographical details; they could "pierce such camouflage as false shadows and make-believe 'gardens' painted on roofs of war factories. Vividly they [could] show the contours of shorelines, cliffs, hills and valleys in enemy-held territory. Trees, ditches and bushes, important in providing cover, appear in full relief like an exact-scale model."[85] Polaroid provided to the military the means to print three-dimensional images of many strategic sites and battlefields throughout the war, including Guadalcanal and the Normandy invasion coast.[86] Because the images could be easily reproduced, bomber pilots and men going into ground action could carry their own personal copies, together with simple viewers. In all, Polaroid provided approximately $2 million in Vectograph materials during the war.[87]

Polaroid's total commitment to the war effort led to a whole range of research projects and, ultimately, new products for the company. Each had its own unique connection to Land's interests and ongoing work. The first synthesis of quinine was achieved under a Polaroid program set up by Land and conducted by a young Polaroid chemist named Robert Woodward, who would go on to win a Nobel Prize.[88] Quinine, a substance found in the bark of the cinchona tree, was vital for the treatment of overseas personnel suffering from malaria. The need for a synthetic substitute was acute: ninety-seven percent of the world's supply came from Dutch cinchona plantations on the Indonesian islands of Java and Sumatra, both occupied by Japan during the war.[89] Land had a jump-start on solving the problem because the crystals used in his early sheet polarizers were made of a compound that included quinine. When his supply dried up, Land set upon two simultaneous research programs: one to develop crystals for making sheet polarizers that didn't require quinine and another to learn how to synthesize quinine.[90] Eventually, both programs were successful, but the latter one became relevant to the war effort.

Ironically, the synthetic process that Woodward and a colleague, William Doering, developed turned out to be very expensive, compared with the cost of the natural material. Medical researchers soon discovered that another drug, Atabrine, could serve as an effective substitute for quinine when administered in higher doses than had been previously prescribed for it. Nonetheless, the work that Land had shepherded proved to be of great importance. It was praised as an achievement that "chemists the world over have attempted for almost 100 years."[91] James Conant called it "a discovery which I think will stand as one of the landmarks of pure science in the history of American organic chemistry."[92]

One of the most unusual projects undertaken at Polaroid was the work Land and some colleagues did to help the Navy find a way to accurately attack ships at sea. Aiming conventional naval artillery shells from a distance was difficult, with both the attacking ship and the target bobbing in the ocean. The Air Force wanted to bomb ships from high altitudes of 15,000 feet or more, but hitting even large ships from that altitude was problematic.[93] Vannevar Bush asked Land if Polaroid could tackle the problem. Land agreed, and the Navy issued a contract with Polaroid for $7 million to develop a 1,000 pound guided bomb.[94] Land posed a question to a top-secret team of six researchers that he sequestered from the rest of the company: "Why do bombs dropped from level flight at high altitude almost always miss the target even as large as a capital ship?"[95] After studying a number of ongoing, but as yet unsuccessful, research approaches, Land focused in on one in which the location of the target would be detected by its thermal radiation, that is, any heat it emitted. The problem remained in how to direct the bomb to that detected location. David Grey, a physicist who had worked on Land's optics projects, posited an approach using a gyroscope to keep the bomb pointed in the right direction. Grey wrote later, "Immediately, Din [Land] understood the principle."[96]

As Grey worked to refine his gyroscope mechanism, Land put together a team to create an integrated device that incorporated the heat-seeking apparatus with the directional components. It became known as the Dove Project. According to Grey, what Land came up with was "simple," although "only in comparison to what an ordinary mortal would envision."[97] Through sources in the NDRC, Land knew what other companies were working on and so was able to reach out and obtain the materials he needed, such as a thermal detector under development at Bell Labs, and subminiature vacuum tubes being developed at Eastman Kodak as part of their research on a proximity fuse for munitions.[98]

As the project continued, he brought in William McCune to serve as its project manager, and this became the first substantial work they would collaborate on. Using bombers flown by the Navy, testing was done on Monomoy Point, a spit of sand extending southward from the elbow of Cape Cod, in Chatham, Massachusetts.[99] The result of their efforts was the first heat-homing ordnance in history. Years later, the Air Force wanted to develop an air-to-air missile. Its original idea was to use radar. However, some military ordnance personnel familiar with Polaroid's previous work convinced the officers in charge that since flying aircraft were, as McCune later noted, "pretty darn hot targets," a heat-homing approach might be better suited.[100] The ultimate result was the now familiar Sidewinder missile.

The first completed device from the Dove Project was tested in the summer of 1945, before the atom bomb was dropped on Hiroshima on August 6, 1945. But it was too late to be used against the Japanese Navy. Just days later, when Japan unconditionally surrendered, Land's interest in the war effort vanished as quickly as it had begun. His stated goal of winning the war had been achieved. On August 15, 1945, the day after Japan laid down its arms, Land walked into McCune's office.[101] The Dove Project contract with the Navy had not yet been completed. Nonetheless, Land told McCune that he wanted to stop work on it and refocus his efforts and those of the rest of the company on a new frontier. The Dove Project was turned over to Eastman Kodak.[102]

Polaroid's involvement in the war effort was the source of many fascinating stories. Perhaps the most vivid chronicle of the use of a Polaroid product during the war appeared just after peace was finally achieved, written in an eyewitness account of a historic event that had occurred earlier in 1945: when the *Enola Gay* dropped the first atomic bomb on Hiroshima, its crew were wearing special Polaroid adjustable goggles.

> Back at the right waist window, Sgt. Bob Shumard, the assistant flight engineer, turned his polaroids to full intensity and prepared to take advantage of the fact that he had the best seat for the show. When the bomb went off it looked blue through his polaroids, but he noted that the interior of the plane lighted up as if flash bulbs had been set off inside the cabin. He adjusted his polaroids to mild

intensity and looked down at Hiroshima. A large white cloud was spreading rapidly over the whole area, obscuring everything and rising very rapidly. Shumard shouted into the intercom: "There it goes, and it's coming right back at us."[103]

In this dramatic fashion, the effective end of the deadliest conflict in human history was viewed firsthand through the invention Land had made while still a teenager, less than two decades earlier. Land had committed his and Polaroid's full resources to the war even before America was actually engaged; its awesome conclusion was viewed through pairs of Polaroid goggles.

Although making a profit had not been one of Land's motives, his company's contributions to the war effort had been good for his business and its future. When the war had started, Polaroid's gross annual revenues were about $1 million. By the time it ended, Polaroid's business had grown to about $15 million annually. The company had employed only about seventy-five people in 1940. By 1945, the number of Polaroid employees had grown to between 1,200 and 1,300.

More significant than its sheer size, however, was the composition of the work force Land had assembled. Decades later, Stephen Benton characterized Land's unique approach:

> Land had a penchant for bringing a variety of eclectic and unorthodox thinkers to Polaroid, giving them the basic equipment they needed for their research without much fretting about the short-term payoff, and just turning them loose for long periods of time. . . . Land generally left them alone, waiting for them to call him, or calling them in when he had a short term project that needed their special skills. This provided fertile ground for new ideas that might come to the fore.[104]

This kind of unique organization would be well positioned for the challenges Land would set before it in the years to come—challenges that were already being investigated in complete secrecy by a few close associates assigned by Land to the task. With the war over, the time had come to unleash the full creative vigor of the scientific talent Land had amassed on his next technological adventure.

CHAPTER 3

ONE-STEP PHOTOGRAPHY

Edwin Land turned thirty-five in May 1944, as the tide of World War II was about to turn following the D-day invasion of Normandy a few weeks later. His newborn company would earn over $13 million that year (approximately $176 million in 2014 dollars).[1] His company's contributions to the war effort were widely recognized. One of his colleagues in the Academy of Sciences characterized Land as "a commanding, an effulgent presence in the world that was emerging from the war against Hitler."[2]

Yet for all that success, Land was troubled about the future of his company. His polarizing technology, so successful in military applications, had not established a substantial beachhead in the civilian sector. The use of polarizers for sunglasses and optical instruments proved to be the most viable commercial applications. Polaroid would remain in that technological niche for many years. It was, however, clearly not enough. Land continued to work on making three-dimensional motion picture technology attractive to the major studios but did not make any progress until, seemingly out of nowhere, in the 1950s, Hollywood produced a slew of films in a short period using the technology. At one point, more than fifty productions were in the works, with an average of seven three-dimensional productions being filmed every week.[3] Polaroid was also reportedly manufacturing twelve million pairs of disposable polarizing eyeglasses a month and selling them to movie theaters nationwide. Total sales were around 100 million pairs.[4] This helped the company make a profit in 1953 of an estimated $26 million ($232 million in 2014 dollars).[5] Unfortunately, the demand for three-dimensional movies ceased as quickly as it had begun.

Land's goal of getting Detroit to adopt his anti-glare technology remained beyond his reach—the automobile companies considered it too expensive, too impractical, and, arguably, unnecessary. He did not give up easily, however. As Land later admitted, "I was too young, too romantic to stop."[6] After the war, in an attempt to renew his campaign to encourage the public to demand improved safety, Land participated in a demonstration of his system by the Automobile Manufacturers Association. According to a report in the *New York Times*, this test of the polarized headlights "indicated their potential worth."[7] However, General Motors staged its own elaborate thirteen-hour program to convince the public that the adoption of polarization was not necessary. No one from Polaroid was present at this event, and one journalist who did attend characterized it as General Motors' attempt to do "a snow job on [polarized headlights]."[8]

Despite General Motors' argument against the desirability of using this technology, some observers nevertheless thought "the Polaroid system passed all the extensive tests on General Motors proving grounds."[9] The industry, however, continued to resist and, in early December, General Motors claimed that the highway glare problem was overstated, urging that further surveys be conducted to assess the actual extent of the problem.[10] This was part of the industry's strategy to resist adopting polarizing technology by subjecting it to the deliberations of an unending series of boards and surveys and other bureaucratic dead-ends.

Later that month, Land made a presentation to the Highway Research Board, a quasi-governmental group organized in 1924 to provide independent advice to the president, Congress, and federal agencies on science and technology issues affecting transportation. He described the advantages of his system and estimated the cost to manufacturers as only seven dollars per vehicle. If manufacturers wanted to pass the cost along to the consumer, they could probably charge thirty to eighty dollars.[11] Land was convinced that the automobile industry exaggerated the cost and minimized the benefits of his system.

> If you will look back over the history of important new improvements that we now take for granted in our day-to-day life, you will find, I believe, that each was introduced for obvious and overwhelming reasons, but that at the same time each of these improvements presented a number of minor disadvantages which may have seemed significant at the time the change was proposed but which we have now come to disregard. You will also find that

> the minor disadvantages have been balanced by a number of minor advantages to which little attention was given when the improvement was introduced. If pressed, I am sure we could all point out a number of minor disadvantages of such overwhelming improvements as the telephone, the electric light, or even the automobile. All of us could at the same time point out a great many minor advantages of these improvements beyond the obvious ones. This, I believe, will prove true of the Polaroid headlight system.[12]

Despite Land's impassioned arguments, J.H. Hunt, a representative of the auto industry, characterized the figures Land had presented to the Highway Research Board as "wholly unrealistic." Although he acknowledged that a workable set of headlights had been developed in a partnership between General Electric and Polaroid, he stated unequivocally that the industry was "opposed to its introduction at this time due to serious questions as to its usefulness and the added risk for drivers of cars not so equipped."[13]

As Land ultimately recognized, the adoption of his system was fatally hampered by the fact that there was no competitive advantage for any car company in using it first. Since all cars needed to incorporate the technology as simultaneously as possible, it was either going to be all, either voluntarily or as directed by the government, or none. No state or federal governmental agency ever stepped in to direct the adoption of the technology in the way that seat belts would be required decades later. Herbert Nichols, a journalist with the *Christian Science Monitor* who had followed the story, believed that the industry killed the idea even though the demonstrations clearly showed that the system worked. According to Nichols, the industry concluded that it "just didn't need anything to sell automobiles. They realized they could sell all the automobiles they could make."[14] Thus, with no economic or competitive incentive, why bother with a system that clearly added costs and admittedly presented implementation issues?

After more than two decades, Land reluctantly gave up the fight. But he learned one very important lesson. "I knew then that I would never go into a commercial field that put a barrier between us and the customer."[15] Rather than deal with other companies as intermediaries, he would market his innovative products directly to the public. He believed "that the role of industry is to sense a deep human need, then bring science and technology to bear on filling that need. Any market already existing is inherently boring and dull."[16] Land, like Steve Jobs many decades later, believed that his company should "give people products they do not even know they want."[17] Fortunately, he already had such a product in mind. (See Fig. 3-1.)

Toward the end of 1943, Land joined his family in Santa Fe, New Mexico, for an early December vacation. As the war approached its conclusion, he knew that Polaroid could not rely on sales of sheet polarizer to sustain it, much less grow it. Land needed to find a new way to exploit his organization's intellectual capital and entrepreneurial energy. When he arrived in Santa Fe, there was none in sight.

Then came the epiphany; why, his daughter asked, did she have to wait weeks to see the photographs her father had taken of her? Land, during the course of a long solitary walk, decided on the challenge that would ignite Polaroid's creativity: the development of "a camera that will make a picture right away."[18] "Within the hour the camera, the film and the physical chemistry became so clear that with a great sense of excitement I hurried to the place where a friend was staying to describe to him in detail a dry camera which would give a picture immediately after exposure."[19]

Serendipitously, Donald Brown, also on vacation in Santa Fe, was the friend Land sought out. Years later, Land recalled that his conception "was so real that [he] spent several hours on this description."[20] As his patent attorney listened, Land "was able to construct the whole scheme and define all the variables. It was a delightful scientific problem."[21]

Brown's reaction was to make sure that Land's idea was immediately memorialized so as to establish a firm date of conception. He had long-ago convinced Land how important it was to protect intellectual property by filing a well-documented patent application, which is why in 1940 Polaroid established a department to keep tabs on all technological developments in the company. This collection of written disclosure documents was the foundation upon which a program was organized to patent whatever came out of Polaroid's ongoing research in a vast number of areas, including Land's special new venture. (See Fig. 3-2.)

Within days, Land returned to Cambridge. Brown asked his new associate, Charles Mikulka, who had joined Polaroid in 1942, to supervise the process. On December 16, Mikulka had Land disclose everything to him in detail. Mikulka recorded the event in longhand on a legal-size pad. The rudimentary disclosure has the basic elements of the camera and film process that would take Land and his team the next three years to perfect: a photosensitive negative, a container of developer reagent, a positive print, and a pair of rollers to release the reagent from its container and spread it between the negative and the positive. In his handwritten memorandum, Mikulka articulated the generic elements of the concept in more legalistic "patentese":

> The novel film comprises a photosensitive layer and material for developing the latent image produced in said layer upon exposure . . . [which] may comprise a silver halide in gelatin and the self-development of each exposed frame may be effected . . . by having one or more containers filled with developer secured to or formed with each film frame . . .
>
> Means are provided in the camera, e.g., pressure rollers, for releasing the developer after exposure of each frame . . .
>
> The developed photosensitized layer acts to form the positive print which is stripped from the remainder of the film preferably on a suitable backing sheet . . .[22]

Land signed the document on December 28, 1943. During the same period, Mikulka collected from two other Land colleagues evidence of the disclosures Land had made to them. On December 23, one of Land's laboratory associates, Frederick Binda, wrote a memorandum to Mikulka reporting that "on December 13, 1943 Mr. E.H. Land called me to one side and said he had a secret to tell me . . . about a new photographic camera in which you simply photograph a subject and from that same camera roll comes out a finished picture."[23] On December 27, Richard Kriebel, a Polaroid public relations man, executed a handwritten memorandum, witnessed by Mikulka, that described what Land had told him:

> On December 10, 1943, at his home, Edwin H. Land disclosed to me a novel self-developing film which when mounted in a camera of novel construction is adapted to produce a positive print shortly after exposure of the film within the camera. The print is of such character as to require no fixing or other treatment after removal from the camera. . . . The positive print is stripped from the remainder of the film within the camera or as it emerges from the camera.[24]

Land was, of course, eager to start work on the project as soon as he returned from Santa Fe. He had great confidence in what could be achieved by following his own personal version of the scientific method.

> If you are able to state a problem—any problem—and if it is important enough, then the problem can be solved. Long before he puts the problem into words, the scientist knows how to confine his questions to ones that he thinks are answerable. He wouldn't be able to formulate them otherwise. His taste, discernment, wisdom, shrewdness and experience have established within him an inner knowledge of what is feasible. However, you must pick a problem that is manifestly important. It must be important to you and your colleagues, and more important than anything else. You can't necessarily separate the important from the impossible. If the problem is clearly very important, then time dwindles and all sorts of resources which have evolved to help you handle complex situations seem to fall into place letting you solve problems you never dreamed you could solve.[25]

Experience, however, had taught Land that the path leading to reaching a goal was not always a direct one. The ideal system he envisioned during his walk in Santa Fe was simple and elegant but, at least initially, unachievable. "Expose the film, pass it through a pair of steel rollers, then when it came out . . . [of the camera] you'd have a simple, dry finished picture."[26] In what Land later called his 1944 "fantasy dream" system, "the integral film would require no timing and no peeling apart."[27] Many years later, William McCune, recalled his "vivid" memory of the "dream" system that Land described to him in 1944, during a moment's break from their work on the heat-homing missile: "[A] camera, small and light, that while technically complicated would be easy to use. The operator would compose the desired picture, press the button, and out would come a finished, dry picture in full color."[28] Notwithstanding Land's prodigious confidence and ability, it would be decades before such a system could be perfected.

The disclosures collected by Mikulka involved the stripping or peeling-apart of the positive print from the rest of the film, rather than having the prints emerge intact from the camera. This was a point of departure on a long road ahead. Land later acknowledged, "You always start with a fantasy. Part of the fantasy technique is to visualize something as perfect. Then with the experiments you work back from the fantasy to reality, hacking away at the components."[29] Not that accepting the unavoidability of this evolutionary process was easy for Land. As McCune confided: "Everything we've done has always been a compromise to him, which has been almost intolerable."[30]

Within days of returning to Cambridge from Santa Fe, Land assigned one of his laboratory assistants, Eudoxia Muller, to begin conducting experiments. Muller was sequestered in a separate section of Land's laboratory at 730 Main Street in Cambridge and put to work in complete secrecy on the project Land dubbed SX-70.[31] At the time, Land had been numbering all of his special research projects sequentially. When he began this investigation, seventy was the next number up.[32] Land decided to begin the investigation by working with aspects of the imaging process Polaroid had developed for its Vectograph system.[33] On December 18, Muller did some rudimentary experiments "according to Mr. Land's instructions" on transferring an image from a negative to a positive directly by pressing them together with the help of a squeegee. It would be the first minute step in what Muller later characterized as a research program of "four thousand, nine hundred and ninety" steps.[34] (See Fig. 3-3.)

Land undertook nothing less than a revolutionary advance in photography, dispensing with the many steps required to develop a conventional film negative and print a positive. Generally speaking, film is comprised of a plastic base on which is coated a layer of light-sensitive silver halide (generally, silver bromide) crystals suspended in gelatin. The scientific theory behind how silver halide crystals on a film's surface operate to form a photochemical image is quite complex, but a readily understandable version was provided to the readers of *Scientific American* in April 1947: "Each crystal . . . is equipped with a 'trigger' . . . which is 'set' for development when light falls upon it in taking the picture. When the film is exposed to the lights and the darks of the scene photographed, the triggers are set in the white areas but not in the black."[35] Land referred to these "triggers" as "active centers or sensitivity specks."[36] The pattern of light reacting with millions of crystals leaves a real, although at this point invisible, image within the photosensitive layer corresponding to the amount of radiation reaching each individual crystal. This image is called the "latent image" and controls the rest of the process of making a photograph.

The first step in making a picture from an exposed negative is to develop the latent image. To accomplish this, the film is put into a developer solution in which the crystals in the exposed areas are chemically converted into metallic silver, forming the black of the negative. Next in the process, a solvent bath gets rid of the unexposed crystals by dissolving them to leave areas of transparency in the plastic base. The result is the familiar black-and-white negative. Light is then passed through the negative in a precise amount onto a sheet of photographic paper. This transfers the image from the negative to the positive print. Once so exposed, the print then has to be subjected to a similar series of chemical baths and physical steps as were performed on the negative. As *Scientific American* put it, "After you snap the shutter, eleven separate operations must be performed before you can see the finished dry print."[37] In all of these operations, the choice of chemicals, their temperature and concentration, and the timing of the sequential steps are all critical to the quality of the finished print. The developing and printing process can take thirty minutes or more from start to finish.

Although in Land's system the negative would be exposed in a camera through a lens in the same way as in conventional photography, "from there on everything is different because all of the [other] steps . . . [would be] subsumed into . . . [a] single one step."[38] For this reason, Land later decided to call his process one-step photography.[39] His stated goal was to

> devise a camera and photographic process that would produce a finished positive print, directly from the camera, immediately after exposure. From the point of view of the user, the camera was to look essentially like an ordinary camera, the process was to be dry, the film was to be loaded in one of the usual ways, the positive print was to look essentially like a conventional paper print, and this print was to be completed within a minute or two after the picture was taken.[40]

Land broke the problem into discrete parts, assigning those parts to members of a small, but slowly expanding, group of associates working with him surreptitiously. Polaroid would continue to devote most of its resources to war-related work, but Land assigned a personal priority to solving an extremely complex problem.

Most essential would be the imaging system itself. From the outset, Land knew that he wanted to use a diffusion transfer system in which the negative image is developed in situ and a soluble complex formed of the unexposed silver halide, washed away during conventional processing, transfers into a receiving sheet to form a positive print. Diffusion transfer imaging was not a new technique in 1943. The earliest experiments in the field can be traced back to 1857, when a scientist named Belfield Lefevre transferred silver from an early photograph known as a daguerreotype to a gelatin layer coated on another support. Land had used diffusion transfer in his work on the Vectograph process. Work in diffusion transfer by both Agfa and Gevaert over the following decades was largely devoted to developing low-sensitivity imaging material used primarily in document copying.

In the years just prior to World War II, a pair of European scientists, Andre Rott of the German company Agfa and Edith Weyde of the Belgian firm Gevaert, did substantial work on silver image transfer. Rott's efforts resulted in the very first commercial office copier system commercialized by Gevaert in 1940 under the trade name Transargo. Weyde's work resulted in a competitive document copying system called Copyrapid in the late 1940s. However, until Land started his work in late 1943, little, if any, work had been done on diffusion transfer processes using the kind of high-sensitivity photographic materials suitable for making photographs. Weyde had indeed conducted some experiments using photographic emulsions, and even "obtained very good pictures with a soft gradation" that needed to be "well washed" to survive, but her work was never published.[41] She never pursued those experiments, switching her efforts instead to the document-copying field.

As work to find a suitable imaging chemistry progressed, several processes were tried in Land's lab, including two known as the Exhausted Developer Process and the Oxidized Developer Process. However, the one that worked the best and was ultimately adopted was known as the Soluble Silver Complex Process.[42] In this process, silver complexes in the negative that did not form the latent image diffused over to a receiving sheet just a few thousandths of an inch away and formed the positive image. These diffusing complexes of silver were created through the development of the silver halide photosensitive emulsion in the negative. They could form either a black-and-white or a sepia-toned image, depending on the chemicals in the image-receiving layer on the print paper. Land's initial experiments in early 1944 were conducted using standard-issue Kodak negatives.[43] Later that same year, Land approached Cyril "Sy" Staud, one of the Kodak scientists with whom he was involved in war activities.[44] He asked his colleague for a supply of standard Kodak photographic emulsion without divulging anything about the purpose he intended to use it for. Having seen from his early work that suitable images could be made using existing negative material, Land wanted to create negatives for his further experimentation that used photosensitive chemicals he knew would be readily available. Standard-issue Kodak chemicals were the perfect foundation upon which to build his process. Staud agreed to help his colleague, and Land had the materials he needed.

With a direction for the imaging process established, the team focused its research on developing the other necessary components for the system—composing the optimal processing solution, creating an image-receiving layer and print structure, and then developing the mechanical components necessary to make the process work. In that regard, one of the most important challenges was finding a method for storing the processing solution before it would be applied to the film. Several approaches were explored. Each of the techniques was based on Land's fundamental concept that after exposure, the film would be run through a pair of metal rollers that would be used to superimpose the negative and the positive on each other, while rupturing whatever container held the processing solution and spreading it in between. A frangible liquid-containing wall within the film was considered, as was a rupturable membrane covering a layer of the solution.

In September 1944, Frederick Binda apparently came up with the idea of using a "pod"—that is, a small packet that would contain either the processing solution or a solvent for the processing solution that would dissolve the pre-dispersed developer when it was burst by the rollers. The pod could be attached or positioned on the leading edge of either the negative or the positive. Land picked up on this concept and, before long, had further developed it into embodiments with multiple compartments, or ones that burst in two or more locations.[45]

It is impossible to overstate how many different ideas were postulated during this period and how many different approaches and techniques Land and his small crew experimented with. The problems were many, and each one would inevitably engender the proposal of several possible solutions. Every suggestion was dutifully recorded and carefully written up in hundreds of comprehensive disclosure reports collected by Donald Brown's patent department. This voluminous, day-by-day collection is a testament to the scientific credo of trial and error, as well as to the organized effort Brown established to protect Polaroid's intellectual property. Ultimately, "with a great deal of courage," as Land would later confess, the pod was adopted for use as the component that would contain the processing solution. It would be used in every single one-step photographic system ever produced. Land loved to joke about all of the "young whippersnappers who get out of MIT, and the first thing they try to do when they come to the company is [to] eliminate the pod."[46]

Indeed, one of Land's great strengths as a scientist, one that made him virtually indefatigable, was his understanding that to research necessarily meant to endure without discouragement the "error" part of the "trial and error" axiom. As he put it, "an essential aspect of creativity is not being afraid to fail. Scientists [pursue] a great invention by calling their activities hypotheses and experiments [and make] it permissible to fail repeatedly until in the end they [get] the results they want."[47] This process was an exciting one for Land, and for his associates. Land was effective in this endeavor not because he single-handedly came up with every idea and solved every problem. As he had already demonstrated in his early career, while he had many great and profound ideas of his own, he was also incredibly perceptive in identifying talent and in recognizing the germ of a great idea in others' work. He could assemble and motivate members of a team to great synergistic effect. Knowing instinctively which ideas to pursue, he could guide the process to incorporate those individual contributions into the overall effort. One of his associates during that period described "the spirit, the joy, the excitement of working with Land. . . . He was a charismatic leader. . . . He [could] choose and train people to do the work just the way he wanted it done, [and could] select people to fill in the voids in his own scientific background."[48]

As work progressed through 1944, Land increasingly focused on the SX-70 project. Although the members of Land's research team remained few in number, he began to monopolize the resources of the small, and apparently overworked, Polaroid patent department. In particular, Land conscripted Charles Mikulka to spend a great deal of time in the laboratory he and Muller were using, learning the technology and following their progress. Apparently, Donald Brown began to believe that Mikulka was spending too much time on SX-70 to the detriment of all the other ongoing projects at the company. He became unhappy that one of his key men had pretty much been hijacked to work on Land's pet project, with little to show for the time invested.[49] In March 1944, Mikulka had started work on the first patent application relating to the SX-70 project.[50] It was filed with the U.S. Patent Office in June.[51] By October, it remained the only application—or "case," as they referred to applications—to come out of Mikulka's sequestration. Brown was upset and wrote to Mikulka to address the situation. Mikulka shared Brown's note with Land, and the two collaborated on a response in which Land clearly, though diplomatically, exercised his authority and demonstrated once and for all his commitment to this new venture.

In a memorandum dated October 17, 1944, Mikulka responded to Brown under Land's direction:

> While it is true that we have filed only one case in SX-70, that case is a relatively broad and comprehensive one . . . cover[ing] about eight to ten different film structures and several processes. Mr. Land and I feel that the time I have spent at Miss Muller's laboratory . . . will give me a background that will prove very beneficial in the preparation and prosecution of future cases. Whether this time could have been better spent from Polaroid's point of view in preparing other cases, I do not know. Mr. Land feels that it was well worth spending [the] extra time.[52]

Mikulka tried to explain to Brown that his work on SX-70 was time-consuming because of the peculiarities of working with Land:

> 1. He is the only source of information.
> 2. He is constantly inventing, so that at a time when perhaps it would be more orderly to finish off a pending application, he will describe a new embodiment of the invention.
> 3. Generally speaking, his schedule is so full that he cannot with any degree of reliability fix a time for appointment. . . . I do not see how anybody working with Mr. Land can possibly circumvent a situation of this character.[53]

In the end, though, Land did not completely let his patent counsel's protest go unheeded. He agreed to limit Mikulka's time on the SX-70 project, but instead of communicating directly with Brown, Land again asked Mikulka to convey the message. "Mr. Land wants me to continue to spend half my time on SX-70," Mikulka wrote to his immediate boss.[54] More importantly, Mikulka also delivered two other significant messages from Land. First, that Land considered "SX-70 to have as much post-war commercial possibility as any other project in Polaroid."[55] Second, and perhaps most ironically, given what would transpire over the coming decades, that Land "thinks that the photographic art is one in which patents have always played a dominant part and that it is an art in which apparently as many, if not more, patent conflicts have arisen as in any other field."[56] That observation was a preview of things to come.

As 1944 turned into 1945, work on SX-70 continued in top secret. To virtually every Polaroid employee, as well as to the world at large, Polaroid's primary activity was its continuing production of an assortment of polarizing apparatus, mostly for the military. In fact, demand continued to be so high that the company was straining to keep up. In his report to shareholders in early 1945, Land announced that Polaroid "continues to accelerate its contribution to the war effort" but admitted that "as a result of substantially increased demands, unfilled orders for the latest types of goggles and filters are at a peak level."[57] He went on to talk about Polaroid's future activities in sunglasses and its intention to pursue entering into some sort of agreement with the automobile companies. Yet despite Land's prediction to Brown that one-step photography would be the key to Polaroid's future success, nothing was mentioned about the intense research being conducted in secret to perfect this revolutionary process.

Land's research group had produced clear photographic images through various diffusion transfer processes as early as March 1944, but by early 1946, the group's work had progressed tremendously. On January 11, 1946, one of the patent attorneys assigned to the project, who reviewed the project activities daily, was apparently impressed enough to write a memorandum to record what he must have thought was a landmark event: "I saw some pictures today in SX-70 which were most remarkable for the degree of contrast obtained."[58] Obviously, momentum toward achieving a satisfactory one-step photographic process was building. Correspondingly, so were the activities of the patent department, as Land, with the help of Brown and Mikulka, had intended. By this point, several additional applications had been filed, and there were more than forty-eight additional applications in the pipeline.[59] The breadth of the work being done had expanded, and the central concern was to protect it all, even though choices had not yet been made as to what would be used in a commercial product, should the project come to a successful conclusion.

In mid-1946, Land thought that finally he "had quite handsome looking pictures and [so] decided the time had come to tell Kodak, as friends, what we were doing in the photographic field that might, in the long run, turn out to be most significant."[60] He invited Kenneth Mees, Kodak's head of research, to come to Cambridge to see what he was up to. Mees, twenty-one years Land's senior, had been at Kodak since 1912, having been recruited by George Eastman himself. He was an English chemist who had written extensively about photographic science, which made him highly desirable as an employee when Eastman had decided it was time to build a proper research organization at Kodak's Rochester headquarters. At the time, Kodak had employed fewer than ten scientists. Eastman's ambitious and far-reaching job assignment for Mees was no less than "the future of photography."[61] Over the years, Mees had gone on to build one of the most respected research departments in industrial America, becoming one of the outstanding figures in photographic science and technology. But with the big corporation about to be let in on his latest big secret, Land's father's early admonition spurred him to action. Before Mees' visit, Land had one of his patent attorneys do a review of the patent applications that were in the process of being prepared, so that he could identify those that "should be filed prior to the arrival of Dr. Mees."[62]

Mees brought with him a few other Kodak executives. In his laboratory, Land demonstrated the new process. After the demonstration, Land was eager to find out what his visitors thought of what they had seen. Before Land could utter the "wonderful phrase [he had prepared] about what great chemistry and all we had," Mees held up a pod, and said, "Gentleman, this is the transcendent invention."[63]

It was beginning to look as though the SX-70 research project might come up with a marketable product after all. As with his polarizer invention, Land's fertile imagination had come up with a whole variety of possible applications for the new photographic process, even before its development had been completed. In October 1946, Land asked one of his patent attorneys to compile a list of the "various uses . . . which have been suggested" beyond the amateur photography market.[64] The result was a list of twelve disparate potential products, including x-rays, movies, television, metal or wood reproductions, meter-reading recorders, a race-track camera, and an automatic portrait studio.

After the successful demonstration for Kodak, Land knew that Polaroid's entry into the consumer photography business was inevitable. In fact, it was probably coming just in time to save the company. While Polaroid had earned approximately $17 million in 1945, with the war's end, revenues were already projected to be much lower in 1946, and would actually end up below $5 million for the year. As a result, Land wondered whether the time had come to crack the door of secrecy surrounding his new one-step photography system. But, competing forces buffeted him. On the one hand was his innate circumspection and secretive nature, compounded by his recognition that the project was still in the developmental stage and not fully protected by patents. On the other hand was his concern for the morale of his company's employees, as well as his innate boyish enthusiasm for wanting to show off his new technology and its potential for Polaroid after the company's long digression into the war effort. This latter impulse highlighted what was clearly becoming the trademark dichotomy of Land's personality. In counterpoint to his penchant for disappearing deep into his laboratory for great lengths of time was the enormous joy he derived from wowing an audience with the magic of his latest discovery.

So Land decided to let employees in on the big secret. But given his concerns, he felt that full disclosure would be too risky. In order to resolve his dilemma, Land decided to stage a most unusual event—one as full of drama as possible. On Christmas Eve of 1946, Land rented a movie theater in Harvard Square and invited all of Polaroid's employees to attend. Everyone was apparently in a party mood. One "happy busload of people started their Christmas party a little early," Land recalled, got lost, and had to be rescued by some "scouts" he sent out to find them.[65] The evening began with a trumpet voluntary by a Boston Symphony Orchestra musician. Then a magician took the stage and performed some tricks. The symbolism was not subtle: what was coming was a gift of magic from Land to his Polaroid community. Land took the stage and introduced a film entitled *The Horn Blows at Midnight.* It was about an angel, played by Jack Benny, on a mission to Earth. Land's description of the film was probably more intriguing than enlightening:

> [The film is] about a subject I cannot talk to you about . . . I'm not free, in other words, to tell you anything about SX-70 for a couple of months, but what happens in this movie happens in heaven, and I can't see why I am responsible for anything you learned from heaven. If you keep your wits about you, and if you're good detectives in the way our magician taught us to be, you will see near the end of this movie something that may probably suggest to you what SX-70 is.[66]

With that, the house went dark and the film began. At one point in the film, an angel needs a passport photo. A large studio camera is set up, the shutter is snapped and, miraculously, the passport photo emerges directly from the camera. Immediately, without warning, the film stopped, and the house lights came back on. Standing on the stage was Land, who announced in a "tone of euphoria . . . 'That's SX-70!'"[67] His audience left the theater more confused than informed. (See Fig. 3-4.)

Within a few weeks, however, Land decided to make his one-step photography process public. Land arranged a demonstration at the winter meeting of the Optical Society of America, to be held at the Hotel Pennsylvania in New York City on February 21, 1947. As part of that event, a paper on the technology had to be presented at the meeting. Land enlisted the help of Elizabeth McCune, William McCune's wife, to help put that document together. William McCune worked with Land and a few others to build some special cameras for the demonstration. One was a large format camera mounted on a tripod that would produce eight-by-ten-inch prints large enough for a group to see. This camera was motorized so that the components of the film were moved through the camera mechanically. They also built a few handheld cameras that could be used by some of the Polaroid people to take photographs of people at the meeting.[68]

The cameras built for the demonstration required a double loading operation. On one end, a roll of unexposed negative was loaded. This negative differed from conventional film in that it was coated on paper instead of a plastic base. This would allow the negative to be pulled from the camera without exposing it further. On the other end, where the exposed film would ordinarily be wound up and collected in a normal camera, a roll of photographic printing paper was loaded. That paper had pods of processing fluid mounted on it in predetermined intervals. Just above the paper roll was a pair of steel rollers, through which both the negative and the paper rolls were threaded. In the handheld models, after the photograph was taken, a knob was turned on the outside of the camera to start the processing. As the rollers began to turn, the negative and positive print paper would come together face-to-face and be fed between them. The pressure of the rollers would burst the pod at the leading end and spread the processing chemicals between a predetermined length of negative and paper, forming a sandwich. After a minute, the sandwich would be peeled apart to reveal the final dry print. The remainder of the negative, and the now empty pod, would be discarded. (See Fig. 3-5.)

The meeting was scheduled for a Friday. The day before, Land and his colleagues packed all of their equipment into a truck to be driven to New York. They left on the train. However, the entire Polaroid contingent was heading into a huge blizzard, the biggest in at least six years. Land and McCune arrived safely, but when they awoke on Friday morning they found, as McCune remembered it, that "New York was shut down. There was no traffic, no nothing."[69] New York was paralyzed, and Land and McCune were left wondering what had happened to their truck.

Fortunately, the truck somehow made it through, and the meeting went off as scheduled. The large format camera was unloaded and set up in the front of the room. It was a conventional camera with a special back built to house the one-step components. Land began his presentation, and explained how Polaroid's new camera "will make it possible for anyone to take pictures anywhere, without special equipment for developing and printing and without waiting for his films to be processed."[70] He then invited the president of the Optical Society to the front. Land had him pose and then took his photograph. He turned the crank and out came the sandwich of negative and positive. In a minute, Land peeled away the print, showed it to the audience, and handed it to the president. According to McCune, "It astonished everybody . . . everyone went wild."[71] Land, the showman, was once again working his magic. As photographers milled around photographing him and his camera, Land took pictures of them. When he immediately showed them the results, Land teasingly said, "Now, let me see your work."[72] Land also took a photographic self-portrait, which was widely featured in newspapers and magazines, including a full page "Picture of the Week" in *Life*.[73] Edwin Land was thirty-seven years old; his dream would soon become a reality for everyone who wanted to snap a picture and get a finished print within a minute after it was taken. (See Fig. 3-6.)

Even though no timetable for the actual commercial introduction of the new camera and film was announced at the event, reaction from the press to the demonstration was as ecstatic as Land could have hoped. "There is nothing like this in the history of photography," reported the *New York Times*.[74] "Sensational," read the story in the *Christian Science Monitor*.[75] "One of the greatest advances in the history of photography . . . Few modern inventions are so completely new," concluded the feature in *Scientific American*.[76] "A photographer's dream come true," announced *Business Week*.[77] Widely reprinted was the opinion of Arthur C. Hardy, professor of optics and photography at MIT, that Land's new process was "as revolutionary as the transition from wet plates to daylight-loading film."[78]

It was reported at the time that Eastman Kodak had assisted Land in his work on the project by supplying various photographic "papers and chemicals." In his formal paper to the Optical Society, Land expressed his appreciation for Kodak's help directly. "The writer also wished to thank the Eastman Kodak Company for its cooperation in supplying a variety of special emulsions for this investigation."[79] Several reporters noted that Land had achieved the full measure of the promise George Eastman made in 1888: "Kodak Cameras—You press the button, we do the rest."[80] Kodak had opened the window of photography to everyone, which guaranteed its tremendous growth and success. As a journalist for *Newsweek* observed after Land's demonstration, "the 'rest' to this day, has meant: develop, rinse, fix, wash, and dry the film; then expose the positive and repeat. Or, alternatively, take the roll to the drugstore and call again for the prints."[81] (See Fig. 3-7.)

But, as *Business Week* pointed out, the amateur photographer really did not want to rely on Kodak or anyone else to have to "do the rest." Instead, what "the amateur has really wanted [was] to press the button and do it all himself—right away."[82] Now, Land's one-step process was going to enable that amateur to "do the rest." Robert Lenzner, a reporter for the *Boston Globe* who, in later years, probably covered Land more thoroughly and knew him better than any other journalist, wrote that the one-step camera "appealed to Americans' innate love for instant gratification."[83] As Land expressed it, what he was delivering was "the realization of an impulse: see it, touch it, have it."[84]

CHAPTER 4

A CONSUMER CURIOSITY?

By 1947, Land had moved into his own laboratory and office separate from the rest of Polaroid's operations and located in a building just across the street from Polaroid's facility at 730 Main Street in Cambridge. It was an old brick two-story factory building at 2 Osborn Street, sandwiched between Harvard and MIT, and had an important place in the history of technology. There, Thomas Watson had received the very first two-way long distance telephone call from Alexander Graham Bell in 1876.

In the wake of the publicity generated by the New York City demonstration, anticipation for the public release of Polaroid's one-step system was high. Unfortunately, Land and his colleagues were not even close to being ready. The technology was close but not the final designs of a mass-producible version of the camera or the film. Getting there would take almost two years. (See Fig. 4-1.)

During this period, business for Polaroid continued to decline. The profitable military sales from the war years had dried up, as expected. Fortunately, a special tax provision provided the company with funds sufficient to carry on. During the war, the government had instituted an emergency tax on what were characterized as "excess profits" made during the war. Congress had first passed this kind of provision in March 1917 to limit excess profits made in World War I, as well as to raise revenue to fight the war. President Franklin Roosevelt brought it back in 1942 for World War II. However, once the war was over, in order to support the transition to normal economic activity, the tax code was amended and the taxes paid rebated to companies like Polaroid.[1]

Land's plan for his commercial one-step product was to secure supplies of the negative component from an outside source; his company would then manufacture the positive (or image-receiving sheet) and assemble it with the negative into the composite film for sale to the public. The positive was really the unique structure in the process, and that is where all of Polaroid's research efforts had been directed. All of the imaging experiments done by Muller, Land, and others had used a standard Kodak black-and-white photographic silver halide emulsion. So now Polaroid turned to Kodak to provide a negative containing the same standard emulsion for its commercial film, but this time coated on a paper base. Land had anticipated this arrangement when he made his 1946 presentation to Kodak's Kenneth Mees. Sure enough, Kodak agreed to sell negative to Polaroid for a fixed price.[2] In this, Land's second deal with Kodak, the vendor-customer relationship was reversed; instead of supplying Kodak, Polaroid became Kodak's customer, and by the late 1960s, it ranked as one of Kodak's top-three corporate accounts, alongside R.J. Reynolds and Phillip Morris, cigarette manufacturers for whom Kodak provided cellulose acetate filter tips.[3]

As a result of this new relationship, and to enable it to do the best job, Kodak was interested in learning more about the process that Polaroid was working on. A special arrangement was made whereby some Kodak scientists, led by Henry Yutzy, would come to Cambridge so Polaroid could show them what it was doing. In terms of intellectual property, the agreement provided that any inventions made by Kodak in the photographic diffusion transfer field as a result of the work it was doing for Polaroid would belong to Polaroid. However, Kodak would have the option of using any of Polaroid's technology in the document-copying field.[4] It was the start of a unique, "cordial," and cooperative relationship between the companies that would last for more than two decades.[5] While Kodak's management had no interest in getting directly into the field of one-step photography that Polaroid was pioneering, they were very supportive, mentoring Land and his colleagues.[6]

As development work at Polaroid on the positive part of the film sandwich continued, Polaroid decided that the images in the first consumer version of the film were going to be sepia toned, as had been the case in all of the early public demonstrations of the process. The sepia color of the image was due to the chemistry of the process. Development produced a brown-silver colloid that diffused to form the image, rather than a pure black compound that could produce a black-and-white image. In the early stages of the research, this chemistry had proven easier to adapt to the diffusion transfer process. Much more additional work would be necessary to modify the imaging chemistry to form the kind of silver compound necessary to achieve black-and-white images and to develop a different kind of positive sheet to receive it. The initial sepia film was marketed as Type 40 Polaroid Land film.

It also soon became apparent that the consumer version of the system was not going to be able to operate like the demonstration models. In those, after the shutter was tripped and the film exposed, the negative and the positive were brought together, and the resultant sandwich was advanced through the rollers and out of the camera to develop in the ambient light. But the paper used to make the commercial version of the film was not opaque enough to prevent unwanted additional light from reaching the photosensitive layer of the film after the film exited the camera but before processing was completed.[7] The film used in the demonstrations had been made by hand, and thick paper supports were possible.[8] The camera mechanism had also been custom designed to accommodate the thick supports of this demonstration film.

But a consumer version would require thinner paper film supports for both mechanical and economic reasons. If the paper material is too thick, it would be too stiff to rotate properly around the rollers, making it difficult to get the proper spread of processing reagent. Also, paper thick enough to be opaque was going to be costly. As a result, the first commercial camera, the Model 95 Polaroid Land Camera, was designed so that the negative and positive sandwich stayed inside the camera. After a minute, once the processing was done, the camera back would be opened and the print peeled off and lifted out. As it turned out, the kind of "idealized camera and film" system used in the demonstrations, which allowed for out-of-camera processing of peel-apart film, would not actually become practicable until the 1960s.[9] (See Fig. 4-2.)

The design for the Model 95 had interesting features beyond the special components necessary for the one-step process. It was the very first to employ drop-in loading of film in any camera that used roll film. Other cameras of the day required the user to thread the film through one end of the rear part of the camera and onto the take-up spool on the other. The Polaroid film was completely different because it comprised two spools, one carrying the rolled-up negative material and the other carrying the image-receiving sheet material with the pods positioned in between image lengths. A paper leader on the starting end connected these two lengths of positive and negative on their respective spools. To load the Model 95, one opened the back flap and simply dropped the two spools into place on opposite ends of the camera. Then the photographer slipped the leader through the slot to the outside, shut the camera, and then pulled the leader, advancing the film into position for taking the first picture.

Another notable attribute of the Model 95 was the weight of the camera, which was the result of pragmatic considerations. At the time, there was a fairly high federal excise tax of approximately twenty-five percent charged to purchasers of amateur cameras, "amateur" being defined as weighing less than four pounds.[10] Since it was already apparent to Land and McCune that their camera was going to be relatively heavy anyway, they made sure that the Model 95 weighed more than four pounds, so it could avoid the tax. Today, when one lifts one of these original cameras, one is struck by its massive weight as compared with cameras of more recent vintage. It was truly a solid, yet sophisticated, piece of equipment.

Land's initial philosophy was that Polaroid should manufacture as little as possible on its own.[11] Instead, it would rely on other companies, including one to build cameras to its specifications. Once the design for the Model 95 camera was complete, Polaroid contracted with an outside firm, Samson United, of Rochester, New York, to manufacture it. The firm had been recommended to Polaroid by one of its Kodak colleagues. This arrangement proved to be short-lived, as Samson United went bankrupt, forcing Polaroid to switch suppliers in 1952 to another company, U.S. Time, later known as Timex, of Middlebury, Connecticut.[12]

On the film side, Polaroid assembled the final product from negatives sourced from Kodak with other elements—the positive image-receiving sheet, pod, and processing composition—that it manufactured itself. Land soon learned that there were clear disadvantages to having outside companies involved. It made it more challenging for Polaroid to keep quality control to the very high standards Land required and to maintain control over the invaluable intellectual property that would inevitably be generated by the process of adapting Polaroid's products from the laboratory to mass production. Over the years, these issues and others led Polaroid eventually to manufacture most of its own critical components, but at this initial stage the company had neither the desire nor the resources to do so.

By early 1948, no date for the commercial introduction of the Polaroid one-step system had yet been announced. The system did make some news, however, when the first four patents covering the technology issued from the U.S. Patent Office on February 10, 1948.[13] "Four patents on cameras for making instantaneous pictures have been issued to Dr. Edwin H. Land, president and director of research, Polaroid Corp.," read a newspaper account.

"Cameras of the Land type turn out a finished picture one minute after the shutter is snapped."[14] It is significant that the article referred to Land, who had never returned to Harvard to earn even his bachelor's degree, as "Dr. Land." The previous year, Tufts College, which later became Tufts University in 1955, had given Land an honorary doctorate.[15] Over the course of his career, he would collect fourteen additional doctorates, including one from his alma mater, Harvard, in 1957. From this point forward, and for the rest of his life, out of admiration and respect, most people within and outside of Polaroid would refer to him as "Dr. Land."

The patent program Donald Brown had put into place was finally beginning to show results in the new photographic area of Polaroid's research. The initial patents came just in time for what was hoped to be the public introduction of the new system later that year. Over the next decade, more than 300 patents would be issued to Polaroid from the work done toward developing a one-step imaging system. By the end of 1959, Land would have a total of 245 patents to his name. In addition to 129 in the field of one-step photography, 101 related to light polarization, fifteen to optics, and the balance to a variety of his scientific pursuits.[16] In a speech the following year, Brown underscored the revolutionary nature of the new one-step photography system by noting that only one of the first 200 patent applications his department filed on one-step photography technology was rejected by the U.S. Patent Office on the ground that its subject matter had been previously disclosed in the scientific literature.[17]

Now the pressure was on for Polaroid to get its new system out in time for Christmas shopping in 1948. The company had actually posted a loss in 1947. The funds it had been receiving from reclaiming the excess profits taxes it had paid during the war were running out. The company had to introduce its new products before the end of the year or face dire fiscal consequences. In late February, Polaroid's annual report to its shareholders was released. In it, for the first time, the company announced that a "folding type camera using the Land one-step process" would be introduced by the end of the year.[18] No date or details were provided, but the press, including the *New York Times* in its Camera Notes column, picked up the news. To help build anticipation, while work on the commercial products continued, Land put together a series of events to keep his new camera and film in the public eye.

In May 1948, the big news in America was Citation's victory in the Kentucky Derby. The popular colt, with his legendary rider Eddie Arcaro in the stirrups, would go on to win horse racing's Triple Crown over the next few weeks, the last horse to accomplish that feat until Secretariat came along in 1973. Polaroid's new one-step process was about to start making its own headlines that month. Land was invited to attend the 167th annual meeting of the American Academy of Arts and Sciences. Honorees at the meeting included Enrico Fermi, the Nobel Prize winner who was a key collaborator in developing the atomic bomb, and General Dwight D. Eisenhower. Land was awarded the society's Rumford Medal for his work in polarized light and one-step photography. He was asked to make a presentation about his work and submitted a paper entitled *Theory and Application of Synthetic Polarizing Sheets*.[19] His demonstration, however, was on the fundamentals of his new one-step photography system, which made a big splash.

Later that November, just weeks before the Polaroid system was scheduled to go on sale, and the day after it was confirmed that Harry Truman had squeaked by in an upset victory over Thomas Dewey for his own term as president, Land made a high-profile presentation of the commercial version of the new one-step system to the Photographic Society of America at its annual convention in Cincinnati, Ohio. Fittingly, the other major order of business for the meeting was awarding the first Progress Medal of the Photographic Society of America to Kodak's Mees, who presented a lecture on the work of the Kodak research laboratories he had been so instrumental in building.[20] (See Fig. 4-3.)

The headlines coming out of the event were dominated by Land's demonstration. "Picture-in-a-minute photography emerged from the laboratory stage and became a reality in a demonstration last night before the Photographic Society of America," read the lead in the *Cincinnati Enquirer*.[21] The *New York Times* reported that the "introduction of the Polaroid Land Camera, which utilizes the one-step picture-making principle first announced last year, was the highlight of the [event] . . . [and] brought frequent applause during the course of the demonstration by Dr. Edwin H. Land, the inventor."[22] The photography columnist for the *Boston Globe*, George Green, had some fun with his report of the proceedings. Green, who followed the hometown Polaroid very closely, and who knew Land quite well, had known about the system for some time but had been embargoed from writing about it. In his monthly column, The Camera Eye, Green gave his readers the big news with apparent great relief:

> After more than a year of "off the record" contact with the "photographic A-bomb," I am now able to bring you the story you've been waiting for.

> Photography at your fingertips became an actuality on Friday, November 5, 1948 at the P.S.A. convention in Cincinnati when Edwin Land demonstrated his invention, the Polaroid camera.
>
> Remember—you read about the camera originally in the Boston Globe March 1947 issue of the Camera Eye . . . and since that time your reporter has heard all sorts of rumors . . . principally that it was a gadget and Polaroid would never produce it. If you've ever kept a secret for more than a year you can well imagine how much restraint I had to exercise to refrain from telling these rumor-mongers that they were out of focus.
>
> The Polaroid camera is not a gadget. It stands on the same plane as the comparison between a Model T Ford and a DC 6 airplane . . . both will get you to your destination but one does it much faster and just as efficiently.
>
> This camera with the built-in darkroom will open the windows of photography to those who never before evinced any desire to take pictures because either they didn't care to process the film themselves or want to wait until a commercial finisher did it.[23]

Despite the excitement generated in the press by the new technology, some retailers of cameras seemed to have lingering doubts about the forthcoming system. Was it more than a gadget? Was the price that Land had announced—"well under $100" for the camera and about $1.50 for a roll of eight pictures[24]—going to be prohibitively expensive? These reservations led to a wait-and-see attitude on the part of many dealers. But a reporter for *National Photo Dealer*, the trade publication of camera dealers, seemed to be impressed by what he saw in Cincinnati.

> I have seen the Land camera and been conquered. . . . Anybody who doubts that this four-pound, two-ounce gadget invented by Dr. Edwin H. Land . . . is going to light a bonfire in photography had better be prepared to be convinced at the earliest opportunity, or sooner.
>
> Friends, the Polaroid Land Camera is no flash-in-the-pan picture-making device. . . .
>
> A leading camera manufacturer whose name I shall not mention shook his head a bit sadly and said: "It's a good camera."
>
> . . .

> The Land camera is so simple to operate that even a child can be taught to handle it with ease.[25]

Unfortunately, this glowing review came too late to have any impact on Polaroid's plans for how it would market its cameras and film. Due to the industry's initial ambivalence, the decision had already been made not to sell through the normal distributors who dealt in photographic products. Instead, Polaroid would initially sell directly to a yet-to-be-determined retailer. This unusual marketing approach continued for some time because, as McCune later recalled, it "was only after several years . . . that the photographic guys began to realize that selling Polaroid film was a very lucrative business."[26]

The date for the public release had been set for the day after Thanksgiving, November 26, 1948. Polaroid, however, had only a very small amount of product to sell. Working with Samson United, only fifty cameras could be manufactured before the introduction date. The film, being made and assembled by hand at Polaroid's facility in Cambridge, was also in very short supply. Nonetheless, there was nothing to do but proceed. A local department store chain—Jordan Marsh—was chosen to receive the initial supply of cameras and as many rolls of film as Polaroid could produce.[27] One of McCune's engineers loaded up the first batch of cameras and film into the trunk of his car, and took them over to the Jordan Marsh department store in Boston's Downtown Crossing.[28] No one knew what was going to happen next.

The Jordan Marsh camera department was known for selling low-cost Brownie cameras and other Kodak equipment. Polaroid's products, however, were aimed at a completely different market. The initial price for the Polaroid Model 95 was set at $89.75 (the equivalent of over $886 in 2014). An eight-pack of film was put on sale for $1.75 (about $17 in 2014). At those prices, would any of the store's customers be interested? A demonstration platform was set up, with a sign that read: "May we take your photograph with the new Polaroid Land Camera?"[29] Any suspense over whether there would be interest in the new system was short-lived. A crowd grew quickly, and excitement spread "as people clamored to buy on the spot."[30] That initial shipment sold out in one day.[31] (See Fig. 4-4.)

This was good news indeed. Polaroid's principal public relations person at the time, Richard Kriebel, announced that demand for Polaroid's camera was double the company's top expectations. The industry took note of the successful launch. A "leading" Boston photographic dealer admitted "he had changed his mind about the camera being simply an interesting gadget and was [now] convinced it has an important and permanent place in the photographic industry."[32] Polaroid pressed ahead with increased production as best it could, building its own automated assembly plant for the film under the leadership of McCune.[33] By early the following year, about 10,000 Model 95 cameras were being manufactured every month.[34] They sold just about as quickly as they were delivered from the factory. Macy's sold over 4,000 cameras in the first week alone.[35] In fact, there were even instances of bitter competition over access to the limited supplies. (See Figs. 4-5 and 4-6.)

Polaroid was finally off and running in the photography business. Sales that first year totaled about $5 million.[36] Polaroid's red ink was gone and soon forgotten. The company had a hot product that was rapidly establishing its place in the popular culture of the time. For example, the Polaroid one-step system became an object of discussion in the sports world. This was an era well before instant replay could even be imagined. Jimmy Powers, the legendary sports columnist for the *New York Daily News*, recognized the potential of Polaroid's new system to remedy the problem of disputed calls in Major League Baseball.

> We all make mistakes. And most of us welcome suggestions on how to avoid making mistakes. . . . A new process has been developed by . . . Polaroid . . . by which prints may be made in about one minute. Under the Land process there is no doubt umpires would have the photos make the decision an accurate one, not a snap guess, and have the game resume in less time than it would ordinarily be halted.[37]

One-step photography was becoming widely accepted by the public; press reports described it as an alternative to normal photography, even pointing out that a Polaroid print "costs just a little more than you'd pay for drug-store processing of ordinary prints of this size."[38] Lowell Mason, a member of the Federal Trade Commission, presented newly reelected President Truman with a Polaroid camera. In a series of photos that were published in newspapers across the country, Truman can be seen smiling as he examined the photographs he had produced in just one minute.[39] (See Fig. 4-7.)

Kodak's public posture on the successful introduction of the Polaroid Model 95 was one of proud, if somewhat dismissive, benevolence. The press reported Kodak as characterizing the new Polaroid system as "little more than a gimmick," a "fad," or "an ingenious toy with no real commercial possibilities" or "appeal."[40] In a review in *Popular Photography* of photographic highlights for the year in 1947, written by Kodak's Walter Clark, Polaroid's one-step photography was buried at the bottom of a long list of Kodak achievements for the year, despite the widespread publicity that Land's demonstration had engendered. "The process is too young to indicate what applications it will have," read Clark's review.[41] An unnamed member of Kodak management, however, probably reflected more of the corporate mentality toward Polaroid in a telegram sent out to Kodak personnel. According to Land, who saw the telegram, it said: "Anything that is good for photography is good for Kodak. This is good for photography."[42]

By any objective standard, Land's was a remarkable achievement. He had moved from the basic technology to a demonstration prototype and finally was mass-producing and marketing a highly popular product. It was amazing enough to conceive of, and then to create the prototype for, an instant photographic system that eliminated all of the processing steps required to develop conventional film. But, as any research engineer will acknowledge, taking that kind of discovery from the laboratory, or the highly controlled environment of a demonstration, to a reliable, mass-produced commercial product, is another major undertaking altogether.

Through the inspiration and the direction of Land, all of this was accomplished with great success in a relatively short period of time. The journey from the moment in Santa Fe when his daughter asked her pivotal question to a credible demonstration for Eastman Kodak was achieved in just over two years, while the full commercialization occurred in less than five years. All the while Polaroid had ceaselessly delivered on its commitment to America's military in World War II. As an admirer observed at the time, "I would be willing to bet that 100 Ph.D.'s would not have been able to duplicate Land's feat in 10 years of interrupted work."[43] This revolutionary advance in photography was possible only because during this period Land had taken one of his characteristic leaps of faith and made a personal commitment to translate an idea into reality, aided by the best and brightest of his researchers. Only someone with Land's unique mind-set about the process of invention, and his innate confidence, could have undertaken such an endeavor.

From the outset, Land intended to move beyond rather old-fashioned sepia prints to black-and-white and, eventually, color. In the late 1940s, black-and-white was still the serious photographer's medium. Kodak had pursued the color film market since 1904, when George Eastman famously told one of his associates that "if we could achieve a practical color process, it might have quite a vogue."[44] A physicist and engineer named John Capstaff developed Kodak's first color film for professionals, called Kodachrome. It was introduced in 1914, and was capable of making beautiful portraits, but it was impractical as a consumer product because, among other reasons, it required the exposure of two negatives through specific filters, which then had to undergo an extremely cumbersome developing process. Twenty years later, the trade name Kodachrome was reutilized when Kodak introduced color film for the amateur photographer. Notwithstanding the high quality of these products, color film was expensive and, in its early versions, presented difficulties to the amateur photographer in challenging lighting situations, such as indoor photography. Accordingly, black-and-white film continued to dominate the market through the 1950s and into the 1960s.

Land pushed ahead with his usual intensity. Following the war, Howard Rogers moved from his work in the polarizer lab to join Land's burgeoning team developing film for the one-step process. During that period, Land kept a blackboard in the lab with a list of things to do. One day sometime in 1947, Rogers noticed the word "color" at the bottom of the list. He asked Land if he could work on that and Land agreed.[45] Thus began an investigation by Rogers that would consume the next fifteen years. To assist Rogers, Elkan Blout, a Harvard PhD chemist that Land and Robert Woodward had recruited to join Polaroid in 1943, was assigned to help synthesize the various compounds that Rogers would need for his work.[46] For the special color photosensitive emulsions required for Rogers' project, Land once again turned to his colleagues at Kodak. "By then, we were a very good customer [of Kodak's] as well as friends," recalled Land.[47] Cyril J. Staud, who by this time had become director of Kodak's Research Laboratories and would later succeed Mees as Kodak's head of research in 1955, came through as he had in 1944 and again provided Land with the emulsions Polaroid needed for its research.[48] The understanding with Staud was, should anything come out of Polaroid's investigations, a discussion would ensue about enlarging Kodak's charter so that it could manufacture components for Polaroid's color film as well.[49]

Initially, Rogers knew very little about photographic color imaging, but just as Land had done decades earlier to come up with a practical polarizing material, Rogers began his quest in the library, learning all about the imaging chemistry used in conventional color film.[50] It was known as "color coupler" chemistry and had been developed, as a popular saying in the photography world went, by two scientists known as "God and Man," Kodak researchers Leopold Godowsky Jr. and Leopold Mannes. Kenneth Mees was instrumental in guiding and supporting the work that led to this historic breakthrough at Kodak's Rochester research labs in the early 1930s. This chemistry was the basis for Kodachrome, Kodak's first color consumer film, released first in 1935 as movie film and then in 1936 as slide film. It was also the forerunner of the improved color coupler chemistries employed in later products like Kodacolor, released in 1941.

Rogers decided to begin his investigation by trying to adapt color coupler chemistry to the diffusion transfer imaging process used in one-step photography. Land had already done some work in the area, developing a color diffusion transfer chemistry known as the Exhausted Developer Process.[51] Rogers began by picking up where Land had left off, but the Exhausted Developer Process was never successfully perfected for commercial use. As Rogers later explained: "Initially . . . [Land] gave me a couple of ideas to start with, which I tried to carry out. As time went by, I found various roadblocks, so I had to change things a bit. But he was always very supportive—particularly if something worked."[52]

With Rogers at work on color one-step film, Land selected another of his closest associates to lead the effort of creating a film that could be used with existing Polaroid cameras but would yield black-and-white prints. Meroe Morse, described as "a cheerful, intelligent, energetic young woman," had joined Polaroid directly after graduating from Smith College in 1944.[53] An art student of Land's friend, Clarence Kennedy, she had absolutely no scientific training. Yet she apparently had the right stuff for Land and excelled in his laboratory under the tutelage and inspiration he provided. Howard Rogers once described how Morse was the perfect collaborator for Land because of the way she naturally embraced his scientific philosophy while matching his energy and compulsion. According to Rogers, Morse emulated Land's laboratory style, which was

> to propose the hypothesis, to test the hypothesis, to modify the hypothesis, to test with another experiment—[the investigation proceeding like] a sequential train moving at high speed, several hypotheses and experiments per hour.
>
> And yet the environment was calm. The results seemed more like the magical growing of a plant in time-lapse pictures.[54]

There was some skepticism about whether Land and Morse would be successful in creating a black-and-white one-step film. Knowing that this would not be an easy task, Kodak's Mees predicted, with warm scientific camaraderie, that Land would "kill himself" if he tried.[55] Later, Land noted how he had appreciated Mees' warning but that he had found the endeavor "a rather thrilling pursuit."[56] The challenge had to do with the nature of the silver compound that would be formed upon development to diffuse and create the image and the physical and chemical structure of the image layer itself. Land and Morse, after a period of intense work, eventually developed an imaging chemistry and structure for the image-receiving part of the film in which "the deposited image silver was restricted to the interstices of a matrix of colloidal silica."[57] That is, after transferring, the silver forming the image was arrayed within a very thin layer of a crystal-like material. The silver aggregates were much larger than those used in sepia film. As Land explained, "the aggregates would be of such size that the light throughout the spectrum would be absorbed uniformly and they would look black."[58] According to colleagues, "Land was fond of saying that the silver in this black-and-white film was 'a new form of matter.'"[59]

The additional work to complete the development of the first Polaroid black-and-white film took less than two years. On May 1, 1950, Land introduced the film in another demonstration to the Photographic Society of America. As one journalist reported, in his presentation, Land "conceded that many photographers prefer the brilliance and crispness of black-and-white prints to sepia."[60] As a result, Land announced that Polaroid had developed this new Type 41 film and would be making it available to photographic dealers later that month at the same price as its sepia film. The speed of the film was ASA 100, the same as Polaroid's sepia film and Kodak's popular black-and-white film of the era. Land publicly credited Morse for her valuable contribution to the work that had led to this new film.[61] In what was becoming a Polaroid tradition, when the first advertisement for Type 41 film appeared in *Life*, the film was not actually ready for delivery to retailers. But the public announcement created the kind of challenge the Polaroid workers were up for, and they delivered. As Land described it, the pressure created for him and his associates "a peculiarly poignant kind of thrill, a very special Polaroid kind of thrill. . . . We have never failed to deliver what we promised," Land quickly added.[62] Polaroid's first black-and-white film was finally available, but almost immediately a major problem emerged.

Polaroid had established a copy service through which customers could get additional copies of prints by sending them to Polaroid for duplication. From observing some of the early black-and-white prints sent in for this purpose, Land and Morse noticed that there were problems with the images. They looked as if they were beginning to degrade with time. It quickly became apparent that the black-and-white film had a stabilization problem. The technique used in sepia film, in which the stabilizing chemical was built into the paper film base, was not working for black and white. It was panic time on Osborn Street. Land embarked on what he later called a "violent . . . effort," and, eventually, with the help of his colleagues, a solution was found.[63] The black-and-white prints would have to be swabbed manually with a print coater to apply a solution that would stabilize the image and form a thin plastic coat over it for protection. Land was disappointed that the print coater had proven necessary since it added an extra step to the process. "My first ambition [for my one-step film was] neither to wash it nor to use a print coater," Land later confessed.[64] However, while inconvenient, the print coater turned out to be hugely effective, providing black-and-white prints that proved perfectly stable for decades. Customers accepted the procedure without complaint, and thus "the first and only real crisis in sixty-second photography" was averted.[65]

Polaroid's marketing studies were showing that its customers bought film at a higher rate than photographers using conventional cameras, perhaps three to five times as much.[66] Of course, some of this had to be attributed to the novelty of the new process, but the phenomenon was nonetheless very good for business. Indeed, through the first few years, Polaroid experienced significant problems in manufacturing enough film to keep up with demand. Given the immediate popularity of the product, and the fact that Polaroid was still building and debugging its facilities as it learned how to be a film manufacturer, such growing pains were to be expected. But with the help of talented and devoted Polaroid people like Morse and McCune and another Polaroid engineer of note, Otto Wolff—a man characterized by one of his colleagues as "a great originator of mechanical ideas and solutions"[67]—the problems would inevitably be overcome.

The results were apparent in the company's earnings, as well as in the critical acclaim heaped upon the film. Polaroid sales went from $1.5 million in 1948 to $9.3 million in 1951. By 1957, Polaroid sales would reach the level of $48 million (over $407 million in 2014 dollars), and by 1959, $90 million (over $737 million in 2014 dollars).[68] Polaroid's stock would split several times during this period, multiplying in value seven times from

1954 to 1958 alone. Land and all of his early collaborators suddenly found themselves enjoying the wealth that accrued from this success. In 1950, Polaroid had 429 employees, a number that would blossom to 2,873 over the next decade. On September 27, 1956, the one millionth Land camera, a Model 95A, came off the assembly line. It was sold on New Year's Eve at the Village Camera Shop in South Orange, New Jersey.[69]

Polaroid continued to improve and evolve its products. New cameras were developed and introduced with better focusing and exposure controls. A lower-priced model, the Highlander, was released in the mid-1950s, further broadening the popularity of one-step photography. Land and Morse continued to work on the black-and-white film to improve it in terms of film speed and image quality. Higher film speed equates to higher film sensitivity to light. Such films take better pictures in low-light situations that involve things like cloudy days, shadows, or indoor scenes. Through these efforts, Polaroid was able to introduce PolaPan black-and-white films with ASA film speeds of 200 and 400 in May 1955. Praise for the new film was enthusiastic. *Popular Photography* declared, "It is almost impossible to describe the new version of Polaroid PolaPan 200 without lapsing into superlatives . . . [and] only the most technically proficient darkroom workers are able to match the brilliance and gradation of prints that are being torn from Polaroid Land Cameras by even the rankest of beginners."[70] The photographic trade and technical press appreciated the rapid evolution of Polaroid's products over a relatively short period of time, recognizing the commitment of the company to the technology it was pioneering.

With acceptance of his one-step process firmly established, there was another aspect of the system that was important to Land, and that he did not overlook. One of Land's lifelong beliefs was in "the homogeneity and the continuum between science and art."[71] From his earliest publications, Land had stressed that there was an aesthetic utility to one-step photography.[72] In addition to its utility for amateur photographers, Land also felt that one-step photography could provide a new artistic medium that a wide range of people could use. With the advances in film quality, he even believed that the professional photographic artist might find the process of interest. To promote this idea, Land formed an early alliance with the soon-to-become-famous black-and-white landscape photographer Ansel Adams. They had met in 1948, introduced by Clarence Kennedy, who saw to it that Adams was invited to a party at Land's home to celebrate his recent announcement of the one-step process.[73] Land invited Adams to his laboratory the next day, where he took a picture of Adams with his prototype camera. Adams described the moment in his autobiography:

> [As the print] was peeled from its negative after just sixty seconds, the sepia-colored print had great clarity and luminosity. We were both beaming with the satisfaction of witnessing a photographic breakthrough come alive before our eyes. For Land it represented confirmation of a dream; for me it was a thrilling experience relating to the future of my craft and my first adventure with instant photography.

It was an instant match, according to Adams: "From our first meeting, I responded warmly to Land's intellect and personality; we seemed intuitively to understand each other. Land has an extraordinary curiosity about everything and the discipline to satisfy it." Land became interested in Adams' work and purchased some prints. In early 1949, Land wrote to Adams, telling him how much he and his family and friends were enjoying them. "My own admiration for your combination of aesthetic and technical competence is complete," Land wrote. With the letter, Land sent along one of the just-released commercial Polaroid cameras and a supply of film. Eventually, Land engaged Adams as a consultant to Polaroid. According to Adams, Land's "aim was to produce the most perfect picture-making process, and he felt that I, an exacting photographer, could provide important feedback. Since I balanced creative ideals with a practical approach, were I pleased with his product, so too might other creative and professional photographers."[74]

Land put Adams together with Morse, and the two consulted and collaborated regarding ongoing projects to improve the one-step process. Adams provided his creative input and direction to many technological advances being considered or pursued in Polaroid's laboratories. "As a Polaroid consultant," Adams explained, "I was primarily involved with the qualities and performance of materials in reference to my professional and creative approach." Adams truly appreciated the fact that someone like Land cared about the artistic aspects of his company's products, and was not only focused on producing and marketing a system aimed at the mass consumer. "It is unfortunate that most photographic manufacturers know or care little about creative photography. They have a vast knowledge of advertising and sales . . . but none of aesthetics. . . . Land believed that if the manufacturer includes objectives of highest quality in the social and aesthetic sense, all of his products should benefit."[75]

At first, Adams met resistance among his brethren. "In the early days of Polaroid," he later recalled, "I found that the majority of professional and creative photographers dismissed the process as a gimmick. I was considered by my colleagues a bit eccentric because of my enthusiasm and championing of what they considered a beguiling toy."[76] But in the long run, Adams' work with Polaroid over the years was a collaboration that resulted in many novel artistic applications of Land's imaging technology, including ultra-large-format enlargements of Adams' work that hang on the walls of museums around the world. Adams admitted that "many of . . . [my] most successful photographs from the 1950s onward have been made on Polaroid film."[77] Land also enlisted Adams to give courses in photography to Polaroid's employees, which became a very popular attraction. Over the course of time, as the photographer's fame soared, and the photographic improvements from Land's laboratory kept coming, Adams contributed mightily toward Land's process, helping one-step photography achieve, in Adams' words, its place as "a most important branch of the tree of photography."[78]

Adams' collaboration with Land and Polaroid lasted for decades. In a genuine tragedy, however, Meroe Morse's life was cut short by cancer in 1969, at the age of forty-five. She had been one of Land's most valuable and closest associates during the great leap of one-step photography from Land's mind, to the laboratory and into the marketplace. Morse had possessed the intellect and the temperament to work at Land's furious and intense pace. She once famously told Land: "A day is all too short. It always seems to me that we just really get warmed up to our problems and then it's time to quit."[79] At a service for Morse held in the Memorial Church at Harvard, Land could not bring himself to speak and instead asked his other closest compatriot, soft-spoken Howard Rogers, to deliver a eulogy. Rogers' words no doubt spoke for them both and for the many others at Polaroid who had had the privilege of working with and knowing Morse during her all-too-brief tenure.

> I know the difficulty of the problems she solved, and the scientific and artistic skill she brought to bear in solving them. Like many people of great effectiveness, she seemed to have more strength and energy than most people, and could work longer on more problems at once. She somehow could keep her charm and cheerfulness when there was a deadline to meet. The magic produced in her laboratory often seemed to go beyond what could reasonably be expected. . . .
>
> Everyone who came in contact with her felt uplifted by her presence and attention. . . . Hard problems were made easier, and impossible

> problems possible. . . . Admiration and love for Meroe are built into the deepest core of us, and she has lastingly improved the quality of all our lives.[80]

As one-step photography grew in popularity through the early 1950s, Land's fame and stature, particularly within the scientific community, continued to grow immensely. Still in his early 40s, he had already contributed two unique and utterly new technologies to the world. He had built a hugely successful company based on his talent and that of the gifted team he had assembled, protected by a carefully constructed patent portfolio. The awards and honors rolled in commensurate with this achievement. In May 1951, just two days after his forty-second birthday, Land was elected president of the American Academy of Arts and Sciences.[81] His position as a leading light in the world of entrepreneurial technology was thus forever assured.

In the late 1950s, some voices in academia began to express reservations about the usefulness of the U.S. patent system. In particular, Seymour Melman, a distinguished professor of industrial engineering at Columbia University, wrote a paper that was published in the summer of 1958 by the subcommittee on patents of the U.S. Senate Judiciary Committee.[82] Melman argued that the patent system was counterproductive. His thesis was that the allure of patents impeded basic research, particularly in the universities, because it diverted scientists away from pure research to applied research areas in which they had the hope of securing patents and thus making money from royalties. Melman contended that the current scheme did not reward the individual inventor because most inventions originated in group efforts in which it was impossible to ascertain who contributed what and how much. The professor's basic conclusion was that the patent system was unnecessary to the progress of research and development.

The patent community was extremely concerned about the potential impact of this paper to affect future legislation on patent laws. More immediately, it was also feared that the paper might have an adverse effect on the judiciary, making it more difficult to have patents upheld when challenged in court. In June 1958, Land was invited to speak at the annual dinner of the Boston Patent Law Association (BPLA) scheduled for the following spring. With some prodding from Donald Brown, he quickly accepted.[83] This would give Land an opportunity to be heard by federal judges, both from the district court and the circuit court of appeals, located in Boston.

In September, with the impact of Melman's paper spreading, Bob Thompson and Herbert Kenway, two officers of the BPLA, approached Donald Brown. Given the fact that Polaroid's success had been built on the foundation of hundreds of patents, they wondered whether Land might be interested in using his upcoming address to defend the patent system and to refute Melman. Brown reviewed Melman's paper and wrote a detailed memorandum to Land explaining the situation and making the case that he was uniquely suited to the task.

> It seems to me that the story of the development of the Land process and Land camera is one which should counter practically every conclusion reached by Melman. The initial stages of the work were solely yours and deliberately so; it was only after the basic inventions had been made that engineers and chemists were called in to assist. . . .
>
> I think it is fair to say also that the rapid growth of our photographic business has stimulated research in a fairly wide field among competitors and potential competitors. I think it is not incorrect to say that all of this work was greatly aided, if not made possible, by the fact that we could safely rely upon a strong patent structure as the principal means for recovering research costs and development charges and showing a satisfactory profit. I think it is questionable whether we would have embarked upon any such costly and extensive new product development if we could not have been reasonably assured of patent protection. I think it is also questionable whether we would have been as free in our public disclosures of new ideas and concepts and especially whether we would have been as free in our disclosures to our competitors of these new ideas and concepts if we could not have protected them by patents. I doubt whether there is any other recent comparable commercial development where patent coverage is so intimately tied in with extensive research and development.[84]

Land agreed. Fittingly, the dinner took place at the Museum of Science in Boston on April 2, 1959.[85] The BPLA did some advance publicity on a national scale, and, even before the event, requests for copies of Land's speech came in to them from legal organizations across the country. Brown reported this burgeoning interest to Land, noting that "your inventions and the growth of Polaroid Corporation's business under its patents have been repeatedly used as the shining example of the success of the patent system in the American economy."[86] When the day came, Land did not disappoint. It was a record turnout for the event, with thirteen federal judges as the guests of honor. The speech was a tour de force. Filled with passion, inspired by his inimitable comprehension of and commitment to the inventive process, and informed by his unique personal experience in turning technological dreams into the reality of a successful business enterprise, Land attacked each of Melman's theses with an articulate and intellectually compelling barrage:

> The . . . operation of the patent system in a good corporation enables the scientist to discover, invent, publish, participate in the scientific activity of the academic world unselfconsciously and naturally—knowing that the corporation which is supporting him will be protected, and that the scientific world in which his mind is rooted will be continuously repaid by the prompt publication, both in patents and in scientific journals, of all that he has learned. He does not need to wear one hat in the world of applied science and another in the world of pure science.
>
> The horrible, unthinkable alternative to all this is a cesspool of secrecy, an industrial environment where a true scientist would be embarrassed to participate because he could not talk freely about what he knows, and where he could not use freely what he had learned.
>
> Now, the attack on this position takes the following philosophical form: Science is a group effort. . . . Science is men using their minds together. Science is the community mastering knowledge. Invention, these attackers say, is the *obvious* consequence of this group-learning process, and invention, therefore, belongs to the group; invention can belong to no man. They maintain that no single mind burning with a hard gem-like flame can cut through stupidity and ignorance, but that the generalized heat, the great warm front of intellectual social advance, can consume the ignorance of the past.
>
> ***
>
> There is something warm and appealing and cozy about this picture of the human race marching forward, locked arm in arm and mind to mind; and there are insecure ages in life and insecure people in life to whom this vision of progress by phalanx

> brings comfort and strength. But I, for one, think this is nonsense socially and scientifically. . . . I think whether outside science or within science there is no such thing as *group* originality or *group* creativity or *group* perspicacity.
>
> I do believe wholeheartedly in the individual capacity for greatness. . . . Profundity and originality are attributes of single, if not singular, minds. Two minds may sometimes be better than one, provided that each of the two minds is working separately while the two are working together; yet three tend to become a crowd.
>
> ***
>
> Just as the great steps in scientific history are taken by the giants of the centuries when they slough off the tentacles of the group mind, so every significant step in each . . . single field, is taken by some individual who has freed himself from a way of thinking that is held by friends and associates who may be more intelligent, better educated, better disciplined, but who have not mastered the art of the fresh, clean look at the old, old knowledge.
>
> ***
>
> By very definition things which we care about most—the important breakthroughs—do not occur spontaneously in multiple because they are the result of a very special way of seeing, by a very special mind. . . .
>
> It should be the role of our patent system to bring encouragement, a sense of reward, and a stimulus to prompt publication to men in applied science. There are a thousand new fields ready to be opened. Only a handful of these will be explored by large corporations, leaving many areas untouched. Without the protection of the patent system, young scientific entrepreneurs cannot be counted on to develop the rest.[87]

Land's address was an instant sensation among the corporate patent community and bar. It was published in the *Harvard Business Review*, and many corporations disseminated it to their executives and lawyers. For example, Harry Mayer, general patent counsel for the General Electric Company, called it "the most impressive and significant statement on this subject that has been made in many years."[88] He had copies distributed to every lawyer in his company, along with a memorandum advising that "the high standing of Dr. Land as a scientist and industrial entrepreneur gives special interest to his views." Mayer proposed that General Electric print 10,000 copies of Land's address for distribution to "all Congressmen and to the presidents of all substantial corporations in the United States."[89] Brown made sure that Land was aware of this response.[90] Land had, from the time of his father's admonition to protect his inventions, availed himself of the patent system to its fullest. Now he became its most vocal public advocate.[91] As one of his long-time patent attorneys characterized it, he had become, indeed, "Edwin Land—Champion of Patents."[92]

CHAPTER 5

COLOR

Leading Polaroid's search to find a way to make a peel-apart color film that could be used in existing Polaroid cameras, Howard Rogers had tried to adapt the color coupler chemistry used by Kodak in all of its conventional color films to the diffusion transfer process employed in Polaroid's one-step film. Consistent with his philosophy of "turning his people loose [on scientific investigations] for long periods of time,"[1] Land had allowed Rogers to work without any interference for almost six years. Rogers, having failed to make this adaptation work, embarked on a new path.

Color photographs are made by using combinations of three different colored dyes familiar to anyone who buys printer cartridges today—cyan, magenta, and yellow. In color film, each of these dyes is associated with a photographic emulsion sensitive to the color opposite those dyes on the visual spectrum—that is, red, green, and blue, respectively. The process is known as subtractive color reproduction. Because white light is a combination of all colors, it is necessary to subtract out unwanted colors to reproduce a given color in a picture. For example, if the scene you are photographing has a red cape, the photosensitive emulsion in the negative sensitive to red light will react in the area in which the red cape is located. This, in turn will cause the associated dye color—cyan, or blue—to be subtracted from the image, leaving just magenta (red) and yellow in that area that, when combined, will create the red color of the cape. A huge range of subtle color combinations is required to reproduce what we see in a color photograph.

In conventional color photography, color photographs are made by first creating a negative that has all the *opposite* colors of the actual scene. Color coupler imaging chemistry is used. In that system, the dyes are actually created during the development process—the couplers are chemical compounds capable of forming a colored dye by joining together with molecules of developer that are oxidized in exposed regions of the film. In chemistry terms, a molecule is "oxidized" when it loses an electron, an event that primes it to react further. Only oxidized molecules of developer can "couple"—non-oxidized molecules are inert. The developer molecules are oxidized in exposed regions by reacting with the activated silver halide. In the case of a red cape, in areas of the negative where red light hits the photosensitive emulsion, the silver halide becomes activated. The color developer, which is contained in the processing solution applied to the exposed negative in the lab, becomes oxidized by the activated silver halide in those regions. This allows the oxidized developer molecules to "couple" with their associated color couplers to form cyan dye. Thus, in the negative, the red areas of the picture look cyan.

By exposing that negative onto color photographic paper in a subsequent operation, the colors are reversed back to the proper hues in the final image, using the same imaging chemistry, which is this time incorporated into the photographic paper. In this instance, when white light is passed through the negative, the red (magenta) component of the light is absorbed by the cyan (blue) in the negative and so does not reach the photographic paper. Only the green (yellow and blue combined) and blue components of the light reach the paper, so that when the print is developed, magenta and yellow dye will be formed by the color couplers in that positive area, resulting in a red image.

For Polaroid's one-step process, a method had to be found to get the correct dyes to diffuse simultaneously from the negative to the image-receiving layer in the film unit in order to produce a color photograph. After several years of failed attempts, Rogers came up with a novel idea. He began investigating the use of a preformed dye attached to a molecule of developer that could regulate its diffusion as a function of the latent image.[2] He called these new compounds "dye developers." Rogers knew instinctively that he was on the right track. "When an idea like this comes that you're sure is good, it spreads throughout your body," Rogers explained years later. "I felt intoxicated, but more 'all there' than usual—almost as if I were a giant. Then I went to draw my new molecule for Land."[3]

Rogers recorded the concept in his laboratory notebook in September 1953.[4] He then worked for the next two years to find the right dye developer compounds that could be employed to make a reliable diffusion transfer film. Elkan Blout provided the necessary chemical expertise in synthesizing those materials. In the process Rogers envisioned, the dye developer molecule would initially be mobile in the processing solution, but would become immobilized in exposed areas, leaving it unable to transfer to the image-receiving layer. Thus, only dye developers in unexposed areas could transfer. During the same period, Rogers also realized that a mirror image of this process was possible; that is, a dye developer process could also be created in which the dye developer molecules were initially immobile but would become mobilized when and where developed so they could transfer to form the image, usually by splitting off the dye portion from the dye developer in exposed regions of the film.

Rogers named these molecules "negative" dye developers.[5] The other species, in which the dye developers were initially mobile and were immobilized by development, were named "positive" dye developers. By 1955, Rogers had conducted successful image transfer experiments with positive dye developers and had observed the effects of the process using negative dye developers.[6] He patented both.

Moving forward with his concept, Rogers focused his work on the positive dye developer version because it was the most practical one for his immediate purposes. The emulsions needed to provide an image with correct colors using the positive dye developer process were the normal color silver halide emulsions. But reproducing the correct colors in a scene using negative dye developers required special photosensitive emulsions with quite different color sensitivities. In the mid-1950s, these "reversal emulsions," also known as "direct positive emulsions," did not have sufficient sensitivity for photographic usage and were employed only for document copying purposes.[7] Polaroid had received normal color emulsions from Kodak to conduct its experiments. As was the case during the research on the original one-step process in the 1940s, Rogers' work was premised on the assumption that such normal emulsions would be commercially available from Kodak at a reasonable cost when the time came to put together a commercial product. Accordingly, Rogers needed to perfect the positive dye developer process for a possible Polaroid one-step color film, an endeavor that consumed him for the next two years. (See Fig. 5-1.)

The basic negative structure Rogers conceived was to put dye developers of the three image-forming colors in three separate layers behind their three associated photosensitive silver halide layers in the film unit. After exposure, the processing composition in the pod would be spread within the film unit. The dye developer molecules would be soluble in the processing composition and thus would be initially mobile, able to diffuse to the image-receiving layer. In places where the photosensitive emulsion had been exposed, the dye developer would be oxidized and thus be rendered immobile. Everywhere else the dye developer molecules would be free to diffuse to the image-receiving layer to form the picture. Using the red cape as an example again, in areas where light from the red cape struck the negative, the silver halide emulsion in the photosensitive layer sensitive to red light would be exposed. As a result, the cyan dye developer in the adjoining layer would become immobilized so that no cyan dye diffused in that spot. Only yellow and magenta dyes would be free to diffuse to the image-receiving layer in that spot, where they would combine to create the red color in the picture. (See Fig. 5-2.)

By the fall of 1956, Rogers was producing color images with his dye developer chemistry that Land described as "impressive and promising."[8] Kodak, which had supplied the photographic emulsions and other materials used by Polaroid in its research, was certainly aware that Polaroid was attempting to produce color photographs with its one-step process. On November 30, Land and McCune traveled to Rochester to show key people at Kodak that Polaroid had succeeded. They met with Wren Gabel and Henry Yutzy, among others. Gabel was assistant general manager at Kodak and had been assigned to handle these discussions with Polaroid by Kodak president Albert Chapman, with whom he worked closely.[9] Yutzy already knew Land from some government defense research activities both were involved in.

According to McCune, at this initial meeting and at subsequent meetings, the Kodak representatives were shown "full-color prints which we produced by these techniques."[10] Land recalled that Yutzy was "very enthusiastic" when he saw a color print for the first time, perhaps even earlier than this Rochester meeting, at one of the government events the two attended together.[11] According to Land, Yutzy said, "That looks commercial."[12] In light of what they had been shown, as McCune and Land had hoped, Kodak expressed interest in manufacturing the new color negatives for Polaroid.

Notwithstanding Kodak's interest in the project, everyone involved on both sides knew that this was going to be a major undertaking. The negative that Kodak was supplying to Polaroid for its black-and-white film was essentially a conventional silver halide emulsion coated on a paper base as opposed to the transparent plastic base that Kodak used for other negative manufacturing. The proposed color negative for Polaroid was going to be a completely new product. It would have to incorporate the new imaging chemistry that had been developed by Howard Rogers. At the time of these discussions, Kodak executives knew nothing of the details of how it worked.

Thus, in order for Kodak to be able to accommodate Polaroid, it was clear that a tremendous amount of preliminary work would be necessary. Polaroid would have to disclose its new, proprietary imaging chemistry to Kodak so that its engineers could figure out a way to incorporate it into the multiple layers that would have to be coated onto the paper base of the negative. The difficulty of the task was a reality that Gabel and his colleagues at Kodak appreciated more than McCune, who initially assumed that Polaroid's "new color products can readily be made by . . . [Kodak] utilizing your existing equipment."[13] Kodak, however, realized that a great deal of research would be necessary to enable them to manufacture the negative on a large commercial scale and for reliable use in Polaroid's one-step process. This would take time, and require a cooperative venture between the two companies. Kodak would have to invest significant resources in terms of engineering manpower and capital in order to build new facilities that could produce the requisite special film coatings.

Accordingly McCune and Gabel began to explore an agreement under which work could commence. Kodak was certainly willing to talk, but it was unwilling to commit without a resolution of the details. McCune and Land traveled to Rochester on several occasions to, as Gabel put it, explore "our informal thoughts on possible areas of agreement . . . and in no way commit either of us . . . [in the] hope, of course, that these will assist us in coming to an understanding which would then be the basis for an agreement."[14] Each company had its own concerns. Polaroid was anxious to protect the valuable technical information it was going to have to disclose to Kodak. Some of this material was patented, but some of it comprised trade secrets that Polaroid would otherwise never have to disclose publicly. There was also the issue of how to treat any inventions made during the program—in the imaging chemistry, the operation of Polaroid's one-step process, or Kodak's film manufacturing techniques.

Given that Polaroid was eager to put its color film on the market "as promptly as possible considering all the circumstances involved,"[15] McCune sought assurances from Kodak that it would treat the development program with the necessary urgency. Gabel did not want to overpromise Polaroid in this regard, so he proposed that Kodak agree only to undertake "to attempt to develop" the Polaroid negative "consistent with other demands upon its personnel and facilities." "Kodak cannot guarantee the success of its attempts" and would accept "no legal obligation regarding" its efforts, he insisted. The one concession Gabel offered was that Kodak would be willing to agree that it would "treat this development [program] as if it were one of its own important color projects."[16]

Kodak faced a difficult situation. Given its lucrative business of supplying material for Polaroid's black-and-white film, Kodak wanted this new income stream but only if it would be rewarded financially for the significant investment of time, expertise, and resources required to help take Polaroid's color process from the laboratory to the marketplace. However, as the undisputed king of amateur photography, Kodak was not going to be satisfied merely playing the role of a component supplier to Polaroid, rather than participating directly in a new and potentially significant part of the industry.

As a result, Gabel made two additional proposals. First, Kodak wanted to be able to synthesize and sell to Polaroid the chemical compounds used in the imaging chemistry. Since Kodak had this capability and Polaroid had no other source of supply, this was acceptable. Second, and much more problematic, was a proposal that foreshadowed the dispute that would drive the two companies into litigation two decades later: Kodak wanted to market its own line of one-step photographic products. Gabel wanted Polaroid to agree to a license that would, at some future time to be determined, permit Kodak to enter Polaroid's one-step photography field using Polaroid's technology. This request was a major concern for Polaroid—it needed Kodak as a supplier and developmental partner, but its executives feared having to compete with Kodak products employing the same technology.

At this point, Polaroid was a fraction of the size of Kodak, the unchallenged dominant force of the worldwide photography industry. In 1956, Polaroid had net profits of $3.6 million on sales of just over $34 million.[17] In contrast, Kodak boasted record earnings of $94 million on sales of $762 million.[18] Kodak was twenty to thirty times the size of Polaroid and enjoyed unchallenged domination of the distribution channels for photographic products. The comparison was well beyond David and Goliath; it was more like Mighty Mouse versus Tyrannosaurus Rex. Given Polaroid's precarious position, even the hint of any possible competition had significant consequences for the company. For example, when a rumor appeared that Bell & Howell was going to release a competitive camera, Polaroid's stock was battered even though the rumor was absolutely baseless.[19]

After extensive deliberations with Polaroid's board of directors, McCune proposed that, with certain stringent conditions, licenses to

Kodak of Polaroid's technology, in applications other than Polaroid's color one-step photography process, might be possible. In fact, Kodak already had an option to license certain Polaroid technology for use exclusively in the document copy field under the agreement reached in 1947 relating to the supply of negative for Polaroid's sepia and black-and-white films. That option, which was never exercised by Kodak, did not extend to color and specifically excluded "all aspects" of photography. The applications that McCune now was willing to consider included photographic printing as well as document copying and enlarging using conventional processing techniques—that is, "wet processing in one or more baths." To make sure there would be no confusion that might allow Kodak to compete directly with Polaroid, specifically excepted from any licensing arrangements were Polaroid's patents on the use of a pod or a "viscous processing solution," the two mechanical photographic elements that made Polaroid's unique one-step process possible.[20] (See Fig. 5-3.)

Gabel raised new issues with McCune's proposal. Mindful of potential antitrust implications, Gabel insisted on certain terms.[21] First, Kodak was reluctant to have Polaroid completely dependent upon it for the supply of its negative. Accordingly, Gabel insisted "that Kodak will not supply all of Polaroid's requirements indefinitely and that Polaroid will develop other sources of supply than Kodak for color negative."[22] Kodak also demanded that any licenses that might later be granted to Kodak under the agreement be extended to other companies at the same time. In this regard, Kodak was anxious to avoid any appearance of attempting to create a monopoly, or duopoly, with Polaroid.

Because of its "commanding reach over the photographic marketplace," Kodak had been subject to periodic antitrust scrutiny from the early twentieth century onward. In 1902, it paid $3 million to settle a private suit against Goodwin Film & Camera Co.[23] "Trustbusters" of the U.S. government began to watch the company closely as early as 1911, scrutiny that George Eastman considered "damn foolishness."[24] Nonetheless, the government's pursuit of Kodak continued over the decades, forcing it to divest itself of various products and ancillary operations in 1916, and again in 1921. In December 1954, Kodak had settled a government antitrust suit against it by agreeing that it would no longer sell its film with the cost of developing and printing included in the price. Kodak had also been forced to share its previously secret process for developing Kodachrome film with any company interested in setting up a processing business.

Gabel refused to give up on Kodak's goal of getting a share of Polaroid's market. He therefore requested a provision whereby Polaroid would agree to open its one-step photography field to competition should "Polaroid's color business . . . become a substantial part of the total amateur color film business."[25] This request remained a nonstarter for Polaroid. McCune and his colleagues could not envision any circumstances under which Polaroid would be willing to license Kodak or anyone else to manufacture film for use in Polaroid cameras. As McCune later characterized it, "we just didn't know how to give them a license that would make sure that they just didn't swallow us up."[26] In looking back at the negotiation, Land was just as adamant. "We were not interested in making a supply arrangement with Kodak if through the course of doing that, our very future would be jeopardized because of the completely dominant position Kodak had in the photographic field and because of their unlimited resources."[27]

Polaroid was, however, still willing to move forward with the noncompetitive licenses to Kodak in the document and photographic copy fields, assuming that Kodak would accept the time limitations and other restrictive provisions Polaroid had imposed. It was not willing, however, to extend those licenses to other companies. Kodak may have wanted that provision to avoid antitrust problems, but Polaroid was only grudgingly agreeing to any licensing to Kodak at all because of its dependence on material supplied by Kodak. In May 1957, McCune wrote to Gabel to explain the reasons for Polaroid's position.

> During our discussions you have urged us to agree at this time to extend licenses in the copy [and conventional camera] field[s] . . . to others than Kodak at such times as Kodak's licenses in these areas become effective. We are mindful of the considerations that counsel the adoption of such a program, but we believe that, in the present state of development of these areas, any decision along these lines at this time would be premature and contrary to the best interests of our company and of its stockholders.
>
> We are in the process of creating a potential field of activity which, currently is, in the commercial sense, in early infancy with most of its production and marketing problems as yet unresolved. We do not deem it tenable to contend that an invention in this stage imposes on its owners any obligation to share the benefits of

> discovery with others at least until the efficacy of the contribution has been evaluated in the competitive market place.
>
> Against this background, the proposed grant to Kodak is justified by a special and unique motivation. Its purpose is to constitute part of the quid pro quo in an entire and indivisible exchange of undertakings that may result in expediting the commercialization of a broad field of special interest for our enterprise, namely the extension of instantaneous photography into the field of color. . . . In this sense, no other American company is in the position to grant us equivalent or comparable consideration for our licenses. . . .
>
> On the other hand, the proposed licenses to Kodak are nonexclusive and do not foreclose the possibility of further licensing at an appropriate time and on reasonable terms in the light of the circumstances as they may develop. . . . In our experience, indiscriminate licensing and premature licensing are capable of creating confusion and impotence.
>
> We believe that a policy which will permit us to view the landscape beyond the crest of the hill before we commit our resources is a policy which is in accord with the best interests of our company and at the same time it is in no sense contrary to the public interest.[28]

In short, there was simply no way at this point in its evolution that Polaroid would consider issuing licenses to its technology that would allow Kodak or any other company to compete directly with it in the field of one-step photography.

As McCune and Gabel continued to look for common ground on which to base an agreement, an event of worldwide significance occurred that lured Land away from the exigencies of Polaroid's business and back into an endeavor in which he had secretly become engaged following the end of World War II—America's military intelligence effort. Due to his contributions during the war effort, Land was on the short list for inclusion in any intelligence activities that emerged as the postwar period evolved into the Cold War. For example, in the summer of 1951, Land led a team of eminent scientists asked to make suggestions on how America could improve its air defenses. Known as Project Charles, the group was organized by the president of MIT, James Killian. It included Jerome Wiesner, later President Kennedy's science advisor, and the renowned mathematician and computer scientist John von Neumann, who had been one of the leading contributors to the Manhattan Project that had developed the atomic bomb.[29]

Almost immediately after the launch of Project Charles, the Air Force asked fifteen of these scientists to study means of improving its ability to conduct strategic reconnaissance behind the Iron Curtain in the Soviet Union and Eastern Europe. This secret project, code-named Beacon Hill, was again led by Killian and included Land among other prominent physicists, astronomers, and aeronautical engineers. These projects were conducted during the Truman administration, but they would prove to be the start of what would become Killian's long tenure in the role of organizing America's best and brightest minds in America's defense. Killian had become a close friend of Dwight Eisenhower's during the retired general's tenure as president of Columbia University following the war. Accordingly, when Eisenhower became America's thirty-fourth president in January 1953, he turned to Killian to organize the country's scientific community in support of the military.

In March 1953, the Science Advisory Committee (SAC) in the Office of Defense Mobilization, a group that included both Killian and Land, warned Eisenhower about America's "vulnerability to a surprise attack."[30] Since Pearl Harbor was still fresh in the minds of the war veterans who now served in government, Eisenhower took the matter very seriously. As 1953 unfolded, his concern was intensified by top-secret intelligence that the Soviets had developed an intercontinental bomber capable of delivering the atomic bomb, which had been detonated by the USSR in 1949. Eisenhower believed that "modern weapons . . . made it easier for a hostile nation with a closed society [like the USSR] to plan an attack in secrecy and thus gain an advantage denied to the nation with an open society."[31]

On March 27, 1954, Eisenhower challenged the committee to come up with a proposal to address this Soviet bomber threat. After some internal discussion over the following months, Killian proposed to Eisenhower that a task force of scientists be recruited to examine the issue. On July 26, Eisenhower authorized the formation of what became known as the Technological Capabilities Panel (TCP) of the SAC and granted it access to all of the nation's military and intelligence secrets.[32] This was a time when paranoia ran rampant in government circles in light of Senator Joseph McCarthy's ongoing investigations of "spying in high places."[33] Nevertheless, Eisenhower decided that to be effective, these scientists needed to know everything. Land was named as a member of the TCP's steering committee and was put in charge of perhaps the most significant of the three areas of investigation around which the effort was structured, that of U.S. intelligence capabilities.

At the time that the TCP was organized, Land was actually residing temporarily in Hollywood, where he was advising Alfred Hitchcock on the technology of three-dimensional movies. This was during that brief period when the phenomenon of three-dimensional movies was all the rage in American theaters. Land returned immediately to Cambridge to assume his duties. He quickly assembled a small but impressive group to tackle the assignment. His intelligence panel included James Baker, an expert in optics from Harvard; Edward Purcell, the Harvard Nobel laureate in nuclear physics; and Joseph Kennedy, the chemist from Washington University in St. Louis who had helped to isolate plutonium. Donald Welzenbach, the CIA historian, aptly described Land's approach: "As in every effort he undertook, Din Land was impetuous and demanding. He was determined that his country should have the best intelligence capabilities that science could provide and money could buy."[34] As he and his group investigated the condition of U.S. intelligence, Land became more and more distressed at the poor state of affairs that they found. "We would go in and interview generals and admirals in charge of intelligence and come away worried," Land admitted years later. "Here we were, five or six young men, asking questions that these high-ranking officers couldn't answer."[35]

After about six months of work, the TCP presented its report to the National Security Council at a meeting in the White House on February 14, 1955. It was entitled *Meeting the Threat of Surprise Attack*. Among the many proposals that emanated from the report, Land's section on intelligence was particularly notable. Killian later remarked that Land's group "had the leadership and brilliance to make lasting contributions to the technology of intelligence."[36] With respect to the CIA, Land's observations rang true to Eisenhower, especially as they were consistent with another study of the CIA he had initiated. Retired General James H. Doolittle, who had gained fame from leading the bombing raid on Tokyo in 1942, had led that inquiry. As a result, Eisenhower established a group to monitor the CIA and to consult with him on foreign intelligence activities. Land was appointed to this group, which continued through both Eisenhower terms and into the Kennedy administration that followed.

Perhaps more significantly, out of Land's small team came the high-flying aerial reconnaissance plane ultimately known as the U-2. Eisenhower had expressed his interest in getting a better idea of just what the Russians were up to. According to a perhaps apocryphal account, Land told Eisenhower, "Well, why don't we take a look and find out."[37] Land had heard General Doolittle predict that it would take ten years for the Air Force to get some kind of aerial surveillance over the Soviet Union.[38] The challenge inspired Land to prove Doolittle wrong. The germ for the U-2 idea had come from a CIA administrator who told Land about a plane capable of flying at high altitudes that had been under development by Lockheed but had been rejected by the Air Force.[39] When Land looked into the matter, he decided that this plane could be the basis for what Killian later called "a reconnaissance system of revolutionary power."[40]

Land personally marshaled all of the talent necessary to develop the various aspects of the plane, from the aeronautics to the engines, using his Osborn Street offices as a command center. He recruited James Baker, Kodak's Henry Yutzy, and Richard Perkin, president of the scientific instrument giant Perkin-Elmer Co., to contribute their expertise to the photographic system with which the reconnaissance plane would be outfitted. Land prodded and pushed the project—which involved the essential yet controversial concept of conducting reconnaissance overflights of the Soviet bloc during peacetime—through the necessary government channels, including then director of central intelligence Alan Dulles.[41] In early November 1954, Land advised Dulles of the plane's unprecedented potential: "A single mission in clear weather can photograph in revealing detail a strip of Russia two hundred miles wide and twenty-five hundred miles long and produce four thousand sharp pictures."[42] "This seems to us the kind of action and technique that is right for the contemporary version of the CIA," Land argued. "[It is] a modern and scientific way for an agency that is always supposed to be looking, to do its looking."[43]

The concept for the U-2 was deemed too highly classified for the TCP report given to the National Security Council, and so, with Dulles aboard, Land and Killian took it to Eisenhower directly on November 24, 1954. The president endorsed the idea and, in agreement with Land's suggestion, placed authority for the project not with the military but with a special group of scientists under the auspices of the CIA.[44] Land later told a colleague that he had been able to convince Eisenhower, an avid golfer, of the potential of the U-2 by using an unusual analogy. Normally, Land explained to the president, a golfer could see a golf ball up to about 150 yards away. With the photography system designed for the U-2, that golf ball could be seen clearly at 2,000 yards.[45]

Reconnaissance flights began in 1956, but despite its top-secret nature, the U-2 wound up in the headlines repeatedly in the years to come. On May 1, 1960, a U-2 flown by Francis Gary Powers was shot down over the Soviet Union as it gathered intelligence from its high altitude. Powers had failed to trigger the plane's self-destruct mechanism before parachuting to safety, and so the U-2's wreckage was captured by the Soviets. After an embarrassing propaganda-laced trial, Powers was convicted of espionage and spent two years in a Soviet prison before being exchanged for a Soviet spy, Col. Rudolf Abel, in February 1962. That same year, the U-2 once again came to prominence as the source of military intelligence showing that the Soviets were erecting missile launchers in Cuba, the discovery that led to the Cuban Missile Crisis.

The U-2 remains in service to this day. Land's small group had accomplished a feat of supreme importance for America's ongoing intelligence capability, one that would prove helpful to America's intelligence efforts repeatedly over the coming decades. In fact, the effectiveness of the group was such that the CIA turned it into a permanent fixture, as an advisory instrument for the president. The "Land Panel," as it came to be known, served every president for decades—Eisenhower, Kennedy, Johnson, and Nixon[46]—and continued to be the source of other creative ideas that ultimately turned into effective tools for the U.S. military. These included the Polaris missile-launching submarine, the supersonic Blackbird intelligence airplane, the Corona spy satellite, and the miniature camera on a stick used by the first man on the moon, Apollo 11 astronaut Neil Armstrong.

In the 1970s, Land convinced President Nixon to authorize the Keyhole 11 space satellite, one that used a space telescope aimed at earth to provide the finest intelligence pictures yet.[47] Richard Bissell, the CIA officer responsible for several of these efforts, credited Land with always "figuring out the leanest possible structures and shortest possible pathways to technical success."[48] Land also served on an elite panel, including Killian, General Doolittle, and Edward Purcell, which delineated a plan for a U.S. space program that eventually led to the formation of NASA.[49] Together, these efforts of Land over three decades of service represented a contribution to his country of inestimable value, all done in complete secrecy, without the fanfare that attended Land's more public achievements in polarized light and one-step photography.

On October 4, 1957, as McCune and Gabel continued to negotiate an agreement regarding a color negative for Polaroid, the Soviet Union launched into space the first man-made satellite with a powerful long-range rocket. The satellite was called *Sputnik*—it orbited the earth every ninety-six minutes, sending back radio signals from transmitters onboard. The event was met in the United States with a wide range of reaction, from a profound appreciation to virtual panic. The *New York Times*, taking the philosophical high road, noted that in addition to the political points scored by the Soviets through this accomplishment of its "'new socialist society,'" the event had a broader significance. "This was also an achievement of profound scientific significance for all mankind. It represented a step toward escape from man's imprisonment to earth . . . and opened new vistas of knowledge and travel in space."[50] Reaction on the other extreme was clearly influenced by the intensification of the Cold War. Less than a year earlier, Soviet Premier Nikita Khrushchev had issued to "capitalist states" his infamous threat that "we will bury you."

In fact, many more people saw the *Sputnik* launch as having sinister connotations. To them, it was a demonstration of Soviet technological and military supremacy.[51] First, *Sputnik* suggested that the Soviets were ahead of America in rocket technology and scientific know-how. One supposedly knowledgeable scientist from the Smithsonian Astrophysical Observatory was quoted as issuing the dire prediction that "no matter what we do now, the Russians will beat us to the moon . . . I would not be surprised if the Russians reached the moon within the week."[52] Edward Teller, the physicist responsible for the hydrogen bomb, appeared on television claiming that by falling behind the Soviets in technological capability, the United States had lost "a battle more important and greater than Pearl Harbor."[53]

Perhaps more threateningly, the successful rocket launch into outer space suggested that the Soviets had an intercontinental missile capable of carrying an atomic weapon to U.S. shores. President Eisenhower moved quickly to try to allay his countrymen's fears. He considered, but decided against, telling the American people about the still-secret U-2 spy plane that was providing the U.S. government with detailed intelligence on the Soviet's actual capabilities, information that tended to minimize the fear of imminent danger. His was a "calm estimate," meant to lower the level of hysteria found in some quarters.[54] Privately, Eisenhower was indeed concerned by the implications of the Soviet *Sputnik*, particularly the public's reaction to it, and again turned to the scientific community for consultation.

On October 15, he convened a meeting of the SAC in which he sought his advisors' assessments on the significance of the event and their suggestions on how to proceed.[55] Eisenhower admitted to the group that he couldn't "understand why the American people have got so worked up about this thing. It's certainly not going to drop on their heads."[56] At that meeting, Land made a truly significant contribution. In what was described by others as "an eloquent statement," he told Eisenhower that it was imperative for America to meet the Soviet challenge by fostering in our youth and educational system a renewed interest in the sciences.[57] "We [must] begin a rebirth of building, using the mind, enjoying the scientific adventure. . . . Otherwise Russian scientific culture will leave us behind as a decadent race."[58] Land encouraged Eisenhower that he was uniquely positioned to inspire this necessary commitment.

Land's suggestion resonated with Eisenhower.[59] Soon after, the *New York Times* reported that the "Administration is preparing to embark on a nation-wide educational campaign, led by President Eisenhower, to 'awaken' the American people to the need of scientific education and research."[60] Just days later, on November 7, Eisenhower delivered a radio and television address to the nation in which he began by calling for a renewed dedication to scientific education and basic research.[61] The following week, on November 13, Eisenhower, in a speech given in Oklahoma City, again called for a renewed commitment to science education.[62]

Eisenhower had addressed the country's reaction to *Sputnik* by providing a proactive agenda for people to focus on. James Killian credited Land with "a demonstration of intellectual leadership" that brought this strategy about.[63] Land took pride that he was able to, as Killian recalled, "convey to the president . . . something of the humanistic and aesthetic values of science."[64] Killian, who shouldered an immeasurable burden as the man responsible to decades of presidents for organizing America's scientific community in aid of our nation's military and intelligence agenda, provided a firsthand assessment of Land's contribution to his country: "fresh insights, a sense of adventure, and a 'vision of greatness.'"

> Land is an authentic genius. His powers of exposition, his facility in expressing complex ideas in novel, witty and clarifying ways, can lift a meeting or a report to a higher level of discourse. . . . In [his] assignments he pointed the way to the development of new intelligence-gathering technology . . . that have given unique powers, benign in their operation, to American intelligence agencies, undergirding policy decisions of immense consequence and saving the nation billions of dollars.
>
> In meetings with presidents his eloquence and lucid exposition incited their latent imagination and prompted them to make decisions and to undertake leadership roles that had been, until then, beyond their reach.[65]

Years later, in 1988, Judge William Webster, then serving as director of the CIA, praised Land's service to his country, noting how his "ingenuity and resourcefulness have allowed us to uncover some of the closely held secrets of our adversaries. The contributions Dr. Land has made to national security are innumerable and the influence he has had on our present intelligence capabilities is unequaled."[66]

Land's involvement in these activities changed his life forever. It created a new circle of colleagues among the most elite in the realms of science and technology, with whom he developed long and lasting relationships. It also influenced his personal behavior and lifestyle, leading him to have twenty-four-hour security for the duration of the Cold War. William Baker, the president of Bell Laboratories, and a fellow member of the Foreign Intelligence Advisory Board, warned that Land was "regarded by the Soviets as an absolutely prize package."[67] Given the opportunity, "they would grab him," Baker believed. These circumstances surely contributed to Land's evolving propensity for privacy, secrecy, and, what some described as, his reclusiveness.

The personal relationships that Land forged with Presidents Truman and Eisenhower, and later Kennedy and Johnson, demonstrated the respect each had for his abilities. Land seemed to appreciate each president for his own special character. Of Eisenhower, Land described "happy recollections . . . of a wonderful man."[68] Contrary to popular belief in some circles, Land found Eisenhower to be "remarkably literate. I've heard him give extemporaneous speeches that were exactly right."[69] Land appreciated the "sense of youthful hope" that John Kennedy brought to the country. He also appreciated his intellect. "Kennedy had a wonderful, retentive mind," Land recalled. "You could tell him something, then come back three or four months later, and he would pick up the sentence where you had stopped."[70] Land seemed to appreciate the honest, intellectual modesty of Lyndon Johnson. He once recalled a comment Johnson had made after a meeting that involved a highly technical subject. "We've been walking in the high cotton today," Johnson wryly remarked.[71] In recognition of his contributions, President Kennedy awarded Land the Presidential Medal of Freedom in 1963, and President Johnson conferred the National Science Medal on him in 1968.[72] (See Fig. 5-4.)

It was the presidency of Richard Nixon, however, that ended, at least temporarily, Land's service as a presidential advisor. During Nixon's second term in office, Land's name appeared on the list of 200 "enemies" of the Nixon administration, a fact disclosed by John Dean during his testimony to Congress during the Watergate scandal. According to a Dean memorandum released with the list, the purpose of compiling the names was to provide a target for the Nixon administration's effort "to use the available Federal machinery to screw our political enemies." Although Land joked to a colleague that he was proud to be on the list because "it was the only honor he had received without working for," Land could not have been amused.[73] He resigned as a presidential advisor shortly thereafter, returning to serve President Ford when he replaced Nixon.[74]

In the late fall of 1957, while Eisenhower was addressing the nation's collective anxiety over the implications of the *Sputnik* launch, representatives of Polaroid and Kodak continued to meet. They discussed the terms of an agreement to secure Kodak's help in developing and then manufacturing a color negative using Rogers' imaging chemistry. As had been the practice with Polaroid's sepia and black-and-white films, Kodak would manufacture the negative and ship it to Polaroid, where it would be assembled into film rolls with elements—positive, processing reagent and pod, etc.—that Polaroid produced itself. As the terms of an agreement began to coalesce, Charles Mikulka became involved with McCune as Polaroid's legal counsel.[75] Eventually, the two sides were able to come to an agreement and, on December 12, 1957, a contract was executed. It incorporated the terms that McCune, with continuing input from Land and others at Polaroid, had negotiated with Gabel over the many months of meetings and correspondence.

Without guaranteeing success, Kodak undertook a "research and development program attempting to develop a method for manufacturing, on a commercial basis," a color negative for use by Polaroid. Kodak had the right to discontinue the attempt anytime after five years. If successful, Kodak agreed to sell the negative material to Polaroid at a "commercially acceptable price," with a two-year notice requirement if Kodak decided to cease production. Provisions were included for the safeguarding of proprietary information each party would have to divulge to the other, as well as any inventions made during the development program. Polaroid received licenses under Kodak patents necessary for it to manufacture negative itself as a secondary source of supply. As had been discussed, a license of Polaroid technology to Kodak for the copying and conventional photographic printing and enlarging fields was included in the agreement, with Kodak agreeing to pay a royalty of five percent of Kodak's "net selling price" of any products incorporating the licensed technology.[76] This license provision was not all that Kodak wanted, but it was all that Polaroid was comfortable granting under the circumstances.

With the agreement finally in place, work began immediately. Representatives of both companies met the following week in Cambridge for two days to begin the exchange of information and to set up a plan to conduct the developmental program.[77] The Kodak contingent included three of its most prominent photographic scientists: Yutzy, Hanson, and Damschroder. George Petersen, an attorney from Kodak's patent department, was present as well. Land led the Polaroid team, which also included McCune, Rogers, Blout, and Mikulka. Land opened the meeting by describing the history of Polaroid's work with color diffusion transfer imaging, culminating in Rogers' discovery of the dye developer process. He also described the structure and elements of the three-color negatives that Polaroid had developed in its laboratory to conduct its experiments and to make the photographs that had previously been shown to Kodak. The task at hand was clear: transforming the hand-made negative into a version that could be manufactured en masse on Kodak's production equipment. Land also described issues that Polaroid was experiencing in its experiments with various dyes and other chemical components, and the Kodak group undertook to find improved substances.[78]

The joint meetings on the developmental program became a regular event for the next few years, occurring once a month, every month, alternating between Rochester and Cambridge.[79] Polaroid's Blout, the chemist who had worked so closely with Rogers to synthesize the chemicals he needed for his work on the dye developer process, called the joint effort a "splendid collaboration." Stanley Mervis, a Polaroid patent attorney who monitored the program for patentable developments, recalled how the "two teams worked closely and with mutual respect . . . [so that] progress was expedited."[80] It was an exciting time, and these were high-energy people working very hard to take Polaroid's one-step process to its next major milestone, color. The program had its special moments, including one recalled by Blout many years later.

> Although the Polaroid/Kodak meetings were intense, there was a camaraderie and lightheartedness that I have never forgotten. I remember one particularly grim evening in a drab hotel in Rochester. McCune and Land were both tense and were grousing about being obliged to travel to Rochester every two months

> when Land suggested to McCune, "let's try some bed-jumping." "Bed-jumping" turned out to be a contest to see who could jump the highest and farthest over the hotel's beds. The two spent the next hour jumping over the beds in our various rooms. Although Land was adept and competitive, McCune was six years younger and in much better physical condition, and he won. This sort of silliness, which we all enjoyed, provided some relief to the hard-driving, competitive work of the young group that gathered at Polaroid in the 1950s.[81]

Despite the hijinks, the work toward a Polaroid color negative progressed. At the Kodak annual shareholders meeting in April 1959, Kodak president Albert Chapman was asked about Kodak's relationship to Polaroid. Chapman described Polaroid as "both a fine competitor and a fine customer." He stressed that Kodak had "no plans for entering the picture-in-a-minute field" that Polaroid dominated.[82] Chapman acknowledged that Kodak was working with Polaroid to explore a new color process and noted that while Land had shown optimism about the process, there was still no indication of when rolls of color film would become commercially available. A year later, at Polaroid's 1960 annual shareholders meeting, Land got up in front of an audience of about 800 shareholders and, using one of Polaroid's commercial one-step cameras, took a photograph of two women wearing colorful Easter bonnets. After about two minutes, he opened the back of the camera and peeled out a full color photograph. To at least one observer, the picture was "disappointing both in contrast and color saturation."[83] Land warned his audience that this was only a "progress demonstration" and not a product announcement. He assured the onlookers that better results had been achieved in the laboratory work on the film. "I know we will not be ready to market the film by our next stockholder's meeting," Land explained, "but I believe that at our next meeting we will be free to tell you just where we stand on our time schedule."[84]

Sure enough, the following April, at the 1961 Polaroid annual shareholders meeting, Land announced that the joint development program had reached the point of producing test runs of color negative on Kodak's production equipment. "I am setting 1962 as a target date for introduction of color film," Land announced.[85] Typically optimistic, Land's goal turned out to be not quite achievable. In the summer of 1962, production runs of negative manufactured by Kodak began arriving at Polaroid's Waltham,

Massachusetts, plant for assembly, with the positive that Polaroid was producing, into the finished roll-film product. Disappointingly, early testing of the new color one-step film had shown that the finished image had a stability problem and would also require a coating operation to be performed by hand after the photograph was taken from the camera. Since Polaroid's very successful black-and-white film required this swabbing step, an operation its customers had been performing for a decade at this point, Land believed at first that "it didn't seem like too great a disappointment . . . [to] still have to use a coater."[86] Accordingly, materials for manufacturing the coaters were ordered, and the packaging for the new film was designed to include one in each box together with the necessary instructions.

While putting together the instructions for the packaging, it was determined that for best results, the coating operation had to occur within five seconds from the time the print was peeled away from the back of the camera. The time interval for black-and-white film was not as critical. Land realized that it would be "almost impossible to get a customer to coat in five seconds."[87] He also realized that "most customers simply wouldn't believe it," with the result that many pictures would suffer from color degradation because they had not been stabilized quickly enough. Land suddenly changed his mind and decided that the print coater was going to be a problem after all and that it was essential to avoid "terminating an elegant process with a clumsy afterstep."[88] To solve the problem, Land began an intense research effort that summer to eliminate the print coating operation. The task was complicated by the requirement that any solution he came up with would have to be built into the positive part of the film that Polaroid was producing, rather than the negative portion being manufactured by Kodak. It was simply too late in the process to reconfigure the production facilities already in operation in Rochester. Land's understanding that every different type of film required its own custom approach to stabilization also complicated the task. He later described the challenge.

> From the earliest period of conception in the 19th century, photography depended on two inextricably interwoven processes: making the image appear and keeping the image from disappearing, so the history of photography is strewn with the skeletons of inventors who did not take seriously, from their first concept, the ecology of permanence. As part and parcel of each generative cycle in photography, there must be a new concern for the

> permanence of the very specific chemical type of image associated with that cycle.[89]

To find a solution, Land moved up to Rogers' laboratory and, with a small group of people, "went at it day and night."[90] Early on the morning after Labor Day in 1962, Land called Mervis and asked him to come to the laboratory. When Mervis arrived, Land showed him a color print that had not been coated but that had true colors. Land had discovered a technique for incorporating into the image-receiving sheet a neutralizing layer of polymeric acid that stabilized the print. The experimental example was signed by Land and by his lab assistants with the notation "2 a.m., 9/3/62."[91] The new layer, thereafter known around Polaroid in recognition of its inventor as the "L-Coat," was incorporated into the new color film, and all color films produced by Polaroid for decades to come. The last-minute rescue of Polaroid's new color film—the elimination of the need for a print coater—was a clear example of Land's indefatigable attitude toward research. "Almost every day . . . we took on things which people might think would take a year or two," he said. "They weren't particularly hard. What was hard was believing they weren't hard."[92]

With this critical hurdle surmounted, the time had come to put the new color film on the market. Initially, consideration was given to calling it Polychrome, perhaps as homage to the contribution Kodak had made toward its development.[93] Although packaging was designed with that name, in the end the idea was dropped. On January 24, 1963, Polaroid announced that "the long awaited Polaroid Land Color Film which makes a finished color picture in 50 seconds goes on sale next week." According to Polaroid, "it was to be called 'Polacolor Film.'" Underlined for emphasis in the company's press release was the statement that "the print does not even require the coating that's needed with Polaroid Land black-and-white pictures." The Polaroid press release went on to describe the advances in technology that had made Polacolor possible:

> Dozens of new inventions, many new molecules and hundreds of new laboratory and manufacturing techniques are embodied in new Polacolor Film. Of these, two key concepts—one for the negative and one for the positive—gave special meaning to all the inventions and techniques developed in the entire program.
>
> The first of these concepts is the work of Howard G. Rogers, manager of Polaroid's Color Research. He proposed using a pre-formed

> complete dye linked to a developer in a single molecule so the molecule could control its own transfer from the negative to the positive. Rogers' idea for such [dye developer] molecules . . . is, in effect, the heart of the negative in the new Polacolor film.
>
> ***
>
> The second major concept . . . is the work of Dr. Edwin H. Land, Polaroid's President and Director of Research. His new kind of positive structure produces an unusually luminous color image and a picture which is stable when it comes out of the camera. Furthermore, the print is finished and does not require coating or washing.[94]

The press release went on to provide many more details on the development effort that went into the new Polacolor film. Interestingly, and for whatever reason, Kodak is never mentioned anywhere in the six-page document, despite the massive contribution it had made over more than a five-year period to bring Polacolor to life.

In response to requests for a comment about Polaroid's announcement, Kodak issued its own press release. Consistent with its long-held philosophy that anything that is good for photography is good for Kodak, the release stated that Kodak "views the introduction of the Polaroid color process as a factor that should help to expand further the market for amateur color photography—a beneficial effect on the total market in which Kodak fully expects to participate."[95] The release noted that Kodak had enjoyed a long supplier-customer relationship with Polaroid, under which it continued to supply a "sizeable amount of materials" for Polaroid's black-and-white film. Kodak confirmed that it would be doing the same for Polaroid's color film. The Kodak announcement, however, made it clear that the company, at least publicly, was not interested in competing directly with Polaroid. Kodak "has no current plans . . . for entering the field that Polaroid has pioneered—the pod-type processing of pictures in still cameras for amateur use."[96]

Demand for Polacolor film far exceeded supply, so its availability to consumers had to be gradually rolled out across the country over the course of 1963. In April, Land admitted that although the film would be available nationwide by the end of June, "fewer than half our camera owners will have had their first roll of color film."[97] While film trickled out, Polaroid's publicity campaign for Polacolor continued in full swing. The story of Polacolor's development appeared everywhere, but nary a mention of Kodak's role appeared anywhere. For example, an article in

Science News Letter described how "fifteen years of intense research by several hundred scientists, engineers and technicians of the Polaroid Corporation" had produced Polacolor.[98] This article introduced a new term into the Polaroid lexicon. Instead of using Land's time-honored term "one-step photography" to describe Polaroid's process, the article was entitled "Instant Color Photography" and referred to Polaroid's "instant color film," a term that would become virtually interchangeable from then on with the "one-step" terminology.

Life, the big glossy American magazine that chronicled the world's events in pictures, ran a major feature on the development of Polacolor the day after Polaroid's announcement, complete with a host of photographs of Land and Rogers and other Polaroid personnel working in the development and then production phases of the program.[99] Nowhere in the entire spread, however, does the word Kodak appear, leaving the impression that Polaroid had done it all by itself. Those who did not regularly read the fine print buried deep in newspaper business pages were left uninformed that Kodak had made the entire venture possible. (See Fig. 5-5.)

In fairness, Polaroid did mention Kodak in a booklet it released to the media around the time of the announcement entitled "Background Information about Polaroid Land Color Film." A section called "The People" described the contributions of Land and Rogers in extensive detail and then provided brief sketches of the individual roles of seven other Polaroid scientists. At the very end of the section are three brief sentences about Eastman Kodak. The booklet notes that Polaroid approached Kodak after the dye developer chemistry had been invented, and after Polaroid "had made full color pictures from multilayer negatives." It went on to explain that a discussion ensued about "a joint program whose objective would be to have Kodak in a position to manufacture color negative for Polaroid." "Kodak was willing," read the report, "and this resulted in a coordinated effort designed to adapt Polaroid's concept to a multilayer negative which could be manufactured commercially in Kodak's . . . facilities."[100]

By any measure, this was a wholly inadequate description of the contribution that Kodak had actually made. While Rogers had clearly invented the dye developer process, and Polaroid had indeed made beautiful color prints in the laboratory before approaching Kodak, there had been a great deal more effort expended in adapting Polaroid's experimental process into a commercial one. Kodak scientists, working in concert with their Polaroid colleagues, had made contributions to virtually every aspect and element of Polaroid's color film. McCune later characterized it as a "really very extraordinary cooperation that took place between the top people in this field at Kodak and our people."[101] For whatever reason, however—vanity, jealousy, fear of Kodak's enormity—Land was not willing to accord Kodak the public credit or appreciation it was surely due.

The slight, whether intentional or not, was not lost on many at Kodak. Many individuals who had worked so hard to help Polaroid believed that they had done a lot more than simply adapt Rogers' imaging process to Kodak's production machinery.[102] Kodak's research and manufacturing people knew just how tough the five-year effort had actually been. In fact, many at Kodak were aware that Bunny Hanson, then the director of Kodak's Color Photographic Division, had to be replaced as the head of the Kodak team working with Polaroid because he had come to the conclusion that they were pursuing an impossible mission, that Polaroid's laboratory film could never be made into a commercial product.[103]

As a result, a sense of resentment toward Land and Polaroid began to grow out of this very public snub that would signal a dramatic change in the relationship between the two companies. It would have profound consequences for decades to come. Some twelve years later, Land revisited the 1957–1963 joint development program in a letter to Polaroid shareholders included in Polaroid's 1974 annual report. He may have been trying belatedly to give Kodak more credit than he originally had done, but, in reality, he only dug the hole deeper.

> Why, then, did we seek out Kodak? The answer is a case history of the value of industrial symbiosis, a happy saga about the way two groups with complementary talents, competences and facilities can cooperate within a framework of mutual respect to bring to the country an extraordinarily useful field, in finished form, five to ten years sooner than it could have become available without the cooperation. The processes which we used at Polaroid to make the negative for the 1957 picture . . . [shown to Kodak] were characterized in some ways by a higher degree of technological elegance than the processes finally adopted, jointly, for the manufacture by Kodak of the Polacolor negative. The task of the cooperation was to translate our early processes into new forms that could be run safely and economically on Kodak's existing production equipment.[104]

To underscore his point that Polaroid had developed a system that made perfectly beautiful photos even before Kodak became involved, Land included on the cover of the annual report a reproduction of the 1957 color photograph he had shown to Kodak's Yutzy. And so, more than a decade later, the embittered feeling that had persisted at Kodak was only further reinforced. Land was still simply unwilling to give his Rochester colleagues the credit that they deserved for the work they had done in turning Polaroid's laboratory process into a commercial product, Polacolor film.[105] This resentment provided an undercurrent to the company's ongoing relationship with Polaroid, and apparently fueled many of the decisions made by Kodak management over the coming years—decisions that ultimately set the two companies on an irreversible collision course that would have historic implications.

CHAPTER 6

ABSOLUTE ONE-STEP PHOTOGRAPHY

Back in 1943, Edwin H. Land had a dream. "I can imagine a camera that is simple and easy to use," he told William McCune. "You simply look through the viewfinder and compose your picture and push the shutter release and out comes the finished dry photograph in full color."[1] As it turned out, this camera and film of Land's dream would be easier to conjure than to create.

Polaroid's initial one-step camera and film system was a revolutionary advance in the amateur photography world, yet it employed a relatively large and heavy camera, and the film had to be manipulated. After snapping the shutter, the film was left in the camera to develop. The photographer timed the processing and then, at the designated moment, the final photograph was peeled away from the negative and other parts of the film unit, which were left in the camera to be discarded. As amazing for its day as it may have been, this system clearly fell short of realizing Land's ideal. Still, by the mid-1960s, the evolution of Polaroid's initial one-step system had progressed. In September 1960, a new black-and-white film was introduced that cut the processing time to ten seconds. Polacolor film had finally brought color to one-step photography. For the first time, instant pictures could be processed out of the camera with the replacement of the roll format with "pack film" that had two opaque film supports. In this new variation of the one-step process, once the shutter was snapped, the entire sandwich of negative and positive was pulled out of the camera. As before, the photographer would time the process before the film sandwich was peeled apart to reveal the image. The used negative and spent pod could be discarded immediately, instead of being stored in the camera until the entire roll was finished. This advance allowed the design of smaller and easier-to-use cameras.

As a result of these advances, business was excellent. Polaroid reported sales in excess of $123 million in 1963 (more than $956 million in 2014 dollars), with net earnings of $24.5 million ($190.5 million in 2014 dollars), both records for the company. At Polaroid's annual shareholders meeting in 1964, Land announced that Polaroid would be going international in the coming year, opening a manufacturing plant in the United Kingdom to serve Europe, Australia, and Canada. More significantly, he disclosed that a low-cost camera aimed at the broad family market was in the Polaroid development pipeline.[2] Although this camera, called the Swinger, would initially be capable of taking only black-and-white photographs, industry analysts believed that its availability would have a "profound effect on the company." "Polaroid's products have essentially been reviewed by the public as curiosity items for affluent customers," reported one prominent analyst. "The advent of the $20 camera should go a long way towards changing this attitude . . . in the minds of consumers from [a process that is] essentially a gimmick to the accepted form of photography."[3]

Despite the company's success, and the optimistic projections, Land never lost sight of his ultimate objective of bringing the one-step system of his dream to reality. He described his continuing pursuit of the "potentialities" of Polaroid's process in terms anyone could understand. "To us it's just like bringing up a child. You don't stop after you've had it."[4] He would not be satisfied until he reached the archetype he had envisioned on that day in Santa Fe. As he told Polaroid's shareholders, he was committed to "revolutionizing our own revolution."[5] On a pragmatic level, Land was also convinced that the market for one-step photography would explode in popularity if the system were easier to use. Accordingly, he set the goals for the ultimate embodiment of the system as "no garbage, no imbibing time and small-size camera."[6]

In pursuit of this objective, starting in the mid-1960s, Land led his research team at Polaroid on a mission to develop this definitive instant system, one he dubbed "absolute one-step photography."[7] In this system, the film unit would be ejected from the camera and remain intact, with no extra paper or plastic remnants left to be discarded. It would develop outside the camera and display the final fixed image in a minute or so with no further physical manipulation. In particular, no peel-apart step would be necessary. Nor would it be necessary to coat the image with a chemical to stabilize it, as had been necessary with the early black-and-white peel-apart film. It would finally bring to fruition Land's original objective. Land and his colleagues believed this innovation could propel Polaroid's process into the mainstream of consumer acceptance, but there was no guarantee of success. For him "the program . . . was like a siren. She never came clean to say whether she meant to succeed or not, but she never let us escape."[8]

Meanwhile, Kodak's attitude toward Polaroid's segment of the photography market was changing. It had not previously considered Polaroid, or its new technology, a threat to its near monopoly in the consumer market. It was, at most, a limited niche, a photographic curiosity. Kodak had made tidy profits from supplying material to Polaroid for more than fifteen years, during which it also gave vital and continuing support to the much smaller company. Many of Kodak's top scientists had played nurturing and mentoring roles to Land personally and to other Polaroid scientists. Publicly, in 1959, and then again in 1963, Kodak had stated it had no intention of competing directly with Polaroid in its field. Privately, a different mind-set, however, had begun evolving during the negotiations leading to the 1957 development and supply agreement. Kodak's interest in competing with Polaroid was rapidly and inexorably becoming a much more compelling objective.

Kodak had been frustrated by Polaroid's refusal in 1957 to grant licenses under its pod and viscous processing reagent patents that would have allowed Kodak at some point to enter the one-step field. At the time, Kodak was unsure what, if anything, would come of the licenses in non-photographic fields that Polaroid had granted. Thus, as early as 1961, while it worked with Polaroid to perfect Polacolor film, Kodak began quietly conducting research on alternative means of creating an in-camera processing system that would not run afoul of Polaroid's patents.[9] These efforts had so far been unsuccessful, and Kodak realized more than ever that licenses from Polaroid under some of its key technology were going to be required for it to develop and market a workable system of its own.[10]

In 1962, Kodak got another opportunity to address the license issue. Under the terms of the 1957 development agreement, Polaroid was obligated to create a secondary source of supply for its negative material. This effort had begun in 1957 when McCune called two of his engineers into his office and told them that they were "going to learn to make color negatives."[11] During 1958, Polaroid looked for property on which to build a plant for this purpose, while its board of directors committed the necessary capital.[12] By 1959, a handpicked group was pursuing this objective in a top-secret Polaroid facility in Waltham, Massachusetts, that McCune dubbed the "mystery building."[13] As work progressed, it became apparent to Polaroid that to make color negatives, it needed some of Kodak's unique chemical coating and other film manufacturing technology. There were provisions in the 1957 agreement to cover this. But Polaroid was not sure those provisions covered all of the Kodak technology it would need. So, on December 20, 1962, Polaroid's Mikulka wrote to Kodak's George Petersen about receiving licenses under three Kodak patents in the negative manufacturing field, licenses arguably not covered under the 1957 agreement.[14]

The Kodak research scientists who were actively working with Polaroid knew that this request would be coming. They saw it as a potential opening, and had already approached Wren Gabel, who had negotiated the 1957 agreement with Polaroid, urging him to use the opportunity to break Polaroid's resistance to licensing its own one-step technology to Kodak. Cyril J. Staud, the Kodak scientist who had supplied Land with the photographic emulsions Polaroid used for its earliest imaging experiments, wrote a letter to Gabel, on behalf of Kodak's Research Laboratories, outlining a possible licensing quid pro quo. In it, he indicated that Polaroid's collection of patents on its technology was formidable. Staud told Gabel that an attorney assigned to the Kodak research labs had already compiled "several hundred Polaroid patents which go back approximately 16 years" and suggested this cache could be used in their strategizing.[15]

After Polaroid introduced its Polacolor film in late January 1963, the discussions at Kodak about one-step photography seemed to take on a new urgency. On February 7, 1963, a meeting on Kodak's "In-Camera Process" was held that included representatives of its research and legal departments.[16] The meeting was based on the premise that "Kodak has definite interests in entering the in-camera color process field as direct suppliers of finished goods—both film packages and cameras—at an early date." A few weeks later, Henry Yutzy, Kodak's head of research and Land's longtime colleague, confirmed to three of his lieutenants that Kodak's primary research objective was to explore creating "a non-infringing Kodak color film assembly for use in present Polaroid roll film cameras." As a "longer-range . . . secondary objective," Kodak was interested in developing a "unique Kodak camera and film combination with no interchangeability with Polaroid." Yutzy also confirmed that, as a parallel measure, Kodak would "begin exploring with . . . [Polaroid] the possibility of Kodak's getting rights under Polaroid patents in exchange for the [film manufacturing] rights they may want."[17]

Shortly thereafter, on February 26, 1963, Petersen responded to Mikulka's December letter. While he reported that the patents in question were available for licensing under Kodak's standing policy, they "cover inventions of unusual value."[18] Accordingly, Petersen told Mikulka that consideration would have to be given to determine a "reasonable *quid pro quo*." He suggested that Polaroid consider licensing Kodak under Polaroid's patents in one-step photography, and Petersen undertook to determine just which patents Kodak had in mind. The discussions continued, and a meeting was set for June 5 in New York. It had been scheduled to take place in Julius Silver's office, but because Petersen's plane was more than two hours late, Mikulka met Petersen at the Sheraton-Tenney Inn opposite La Guardia airport.

At the meeting, Petersen reiterated the critical nature of the patents Polaroid was seeking licenses under and said that Kodak would have to seek "a very substantial royalty on the order of seven and one half percent."[19] In return, if Polaroid would license Kodak to sell products in the one-step photography field, Kodak would agree to a similar royalty on Polaroid's license of its patents. Mikulka told Petersen that he found the proposal "inequitable." Leaving aside Polaroid's concern that it did not want to have to compete with Kodak, Mikulka pointed out that Polaroid was "requesting a license to use a process in connection with the manufacture of a component of a product developed by us and dominated by our patents, and . . . [Kodak was] asking for a license to manufacture our basic line of products."[20] The meeting ended without a resolution.

The next chapter in this tug-of-war occurred at a high-level meeting in Rochester on October 2, 1963. Land, McCune, and Mikulka made the trip for Polaroid. Kodak's president at that time, William Vaughn, was present, together with Gabel. Yutzy and Staud represented the research department and Kodak's top in-house attorneys, Harmer Brereton and Petersen, were present as well. The Polaroid contingent had requested the meeting.[21] Land began by reviewing for the group the cooperative history between Kodak and Polaroid that had evolved into Polaroid's becoming one of Kodak's best customers. He explained that Polaroid was not interested in getting into manufacturing in a "massive" way as its talents lay elsewhere, particularly in pursuing new domains of technology through research. Land assured his Kodak colleagues that his company was committed to having the Kodak-Polaroid customer-supplier relationship continue. However, because of the requirements of the 1957 agreement, as well as Polaroid's interest in keeping abreast of developments in film manufacturing, he intended Polaroid itself to become the second source of color negative.[22] Accordingly, Polaroid was going to need licenses under the Kodak film coating and emulsion technology. Land also sought to address issues relating to the price Kodak was charging for Polacolor negative, as well as an extension, from two years to five, of the cancellation notice period for production under the 1957 agreement.[23]

When it was Kodak's turn to respond, Gabel acknowledged the history of the two companies' relationship. However, he made it clear, as he had on earlier occasions, that Kodak was not forever going to be content with merely being a supplier to Polaroid. He expressed Kodak's disappointment in "not finding a market" for any products permitted under the limited licenses Polaroid had agreed to in the 1957 agreement. Once again, he expressed Kodak's interest in being allowed to enter the "in-camera" field that Polaroid had pioneered and, so far at least, had kept to itself with the help of its extensive patent portfolio. Gabel went so far as to equate the respective desires of the two companies: it was as "important [for Kodak] to participate in the 'in-camera' field as Polaroid was finding it important to participate in the production of negative material."[24] Gabel made it clear that Kodak wanted licenses from Polaroid to enter the very fields from which it had been excluded in the 1957 negotiation—that is, the "in-camera pod processing" that would allow Kodak to manufacture its own film to be used in Polaroid's cameras.[25]

The result of the ensuing discussion was a matter of dispute. According to the Kodak contingent present at the meeting, Land finally agreed that Polaroid would grant to Kodak the licenses it sought, in exchange for the licenses Polaroid needed to manufacture its own negatives. As Kodak put it, Kodak's executives were "surprised" by Land's capitulation. Petersen said that both McCune and Mikulka looked "taken aback" and "very upset" with what Petersen considered to be Land's agreement.[26] Shortly after the meeting, Kodak's Petersen prepared meeting minutes and sent a draft to Mikulka for Polaroid's review, as was the practice after all the technical meetings that had occurred during the five-year development process.

After a few days, Mikulka called Petersen and told him that he and his colleagues had a problem with the minutes because they reported that Land had agreed to grant licenses to Kodak. Mikulka admitted to his counterpart that "we have had a great flurry down here ever since we returned from that meeting." Petersen got the impression that Julius Silver had become involved in the discussions on the Polaroid side. Mikulka told Petersen that Land believed that the Kodak representatives had misunderstood him.[27] According to Mikulka, although Land may have "expressed the opinion that in due course, under appropriate conditions, conceivably a license might be forthcoming to Eastman Kodak," no firm commitment had been made at the meeting.[28] Mikulka told Petersen that he would be revising the meeting minutes to reflect Polaroid's view of what had transpired.

In the end, the final version of the meeting minutes contained no reference to an agreement by Polaroid to issue a license to Kodak. It was merely stated, in the most noncommittal language Mikulka could compose, that there had been "some tentative discussion as to whether consideration might not be given to the possibility ultimately of a nonexclusive license from Polaroid to Kodak containing adequate safeguards to Polaroid."[29] The Kodak side was disappointed and livid. Petersen characterized the redraft as "the grossest kind of recasting of what actually transpired at that meeting."[30] Kodak's executives considered Land's "recanting" of the agreement to be "a complete breach of faith."[31] According to Petersen, they were appalled and surprised that "Dr. Land, under this kind of guise, would go back on his clear word to us under the pressures from others within the Polaroid organization."[32]

Yet, there was little that Kodak could do about it. The supply deal with Polaroid was clearly too valuable to abandon over this incident. Kodak continued its research efforts to find an alternative technology for in-camera processing that would be outside the parameters of Polaroid's patents. The discussions about a license resurfaced occasionally over the next few years, but nothing ever came of them. However, the relationship between the two companies had clearly taken another serious hit, if not immediately in the dollars and cents of their ongoing business dealings but in the hearts and minds of Kodak's executives as they developed increasingly negative views of Land and his company.

From the perspective of Land and top Polaroid executives, another potential disaster, in terms of its invaluable relationship with Kodak, had been averted. With Polacolor launched, and the supply of negative assured from Kodak, they were interested only in moving forward. Tragically, during this period, Land was forced to move ahead without one of his key players. On December 9, 1966, Donald Brown passed away at the age of sixty-six. Although he had struggled with health issues the last few years of his life, Brown had remained an active member of Land's core group until his death. It's hard to overstate what this loss meant to Polaroid's leader. Brown had been there from the beginning, making sure that all of

Land's and, ultimately, Polaroid's work was protected with patent coverage. He and Julius Silver remained Land's closest and most trusted advisors. At the Polaroid shareholders meeting following Brown's death, the jubilant crowd turned stone silent as Land stood before them, having great difficulty in paying tribute to his fallen friend.[33] With Brown's passing, Charles Mikulka, who had worked closely with Brown and Land on patent matters since the early 1940s, stepped up and assumed Brown's role.

In terms of his work, moving forward for Land meant refocusing on the long term, and the fulfillment of his original dream of "absolute one-step photography." Under great secrecy, even within Polaroid, Land had organized the development effort by establishing two separate teams, each working fairly independently of the other. One was aimed at designing the camera, and the other charged with inventing the film unit to be used with it. Land was the single common element to both teams, bouncing back and forth between the two efforts as warranted. In his characteristic work regimen, if one team seemed hung up on a particular obstacle, Land would dive into their efforts and lead his colleagues through marathon sessions until a solution was found. This happened time and time again over the ensuing several years, unnerving some of the researchers. Christopher Ingraham, one of the engineers who worked on the camera side, recalled the choreography of the effort: "When we seemed to be putting all our efforts into camera design, someone would say, 'God damn it, Dr. Land, how about making the film?' And he would reply, 'Oh, that's all taken care of, don't worry about that.' Actually, the film people couldn't believe their ears."[34]

Both efforts started with the elemental requirement that the camera had to be small enough to fit in a man's jacket pocket but capable of simplifying the picture-taking process as Land had envisioned it. The film unit was to be completely "integral"—that is, self-contained and self-processing from start to finish, with no extra components or garbage to throw away. It had to yield flat photographs with images that would remain stable forever. The photograph was to be ejected from the camera automatically, immediately after exposure. This requirement imposed a seemingly impossible system of opacification—a way to prevent light from further exposing the film unit while it was developing outside the camera. Polaroid's original one-step photography system may have been deemed revolutionary when it came out in the late 1940s, but this idealized system, if it could be engineered, would be an achievement well beyond that.

Many aspects of both the camera and film system Land envisioned had never been attempted before, let alone incorporated in a product for mass consumption. Land knew it was going to be a frightfully expensive undertaking. Fortunately, Polaroid had done so well in the early 1960s, and had amassed such a huge pile of cash, that it would be able to finance the entire project without incurring a penny of debt. Still, Land was betting his company on an effort to make his dream come true. As one of Polaroid's chief engineers recalled, Land's basic idea might have seemed straightforward, but making it work was "like trying to run a car without gasoline."[35] But Land was not to be dissuaded. He began to staff his teams with the sharpest and most creative minds at Polaroid, sequester them as best he could, and inspire them with his unique creative vision and can-do-anything attitude. Land's persistence, his personal brand of tunnel vision, was a necessity. As McCune explained, "One thing about Land—when he is doing something wild and risky, he is careful to insulate himself from anyone who's critical. It's very easy in the early stages to have a dream exploded."[36]

The project began with, and was built around, the integral film unit itself. The starting point for that structure was the arrangement of chemical layers in Polacolor film. Since the same Rogers dye developer imaging chemistry was going to be used, the negative's alternating layers of photosensitive emulsions and corresponding dye developers would essentially be the same, save for any improvements within each element of the film that Polaroid's researchers could come up with. For example, beginning in 1965, Rogers conducted an extensive investigation to come up with more stable dyes to use in his dye developer molecules, ones that would provide a more brilliant and saturated image and be resistant to degradation. Rogers came up with the approach of incorporating a metal ion into the dye, and the resulting "metallic dye developers" he eventually perfected provided, according to Sheldon Buckler, a Polaroid vice president of research, "more stability than any others used in photography."[37] Since the new color film would eliminate the peel-apart step required in using Polacolor, curing that old "headache," members of the team working with Land on developing the integral unit began calling Land's structural format for the film "Aspirin."[38]

The integral film unit also had to be thinner and thus use less processing composition; the pod built into the wide border on one end of the film unit could not be too thick. This requirement posed two thorny problems:

spreading sufficient processing composition within the layers of the film to start development and dealing with the alkaline processing composition that was left over and permanently sealed into what would become the photograph. Land explained:

> We had to learn precision in filling the pods far beyond what we had previously regarded as precision. We had to learn to make a thin spread between layers already in place. We had to learn to make a square spread. We had to learn to keep the residue at the end of the spread miniscule, and we had to learn to conceal this miniscule residue.[39]

Land and his team addressed the other components that would be required to make an integral film unit possible. To deal with the excess processing composition, they created a structure called “a trap,” which was used at the opposite end of the film unit from the pod to capture and to neutralize the otherwise potentially harmful excess viscous reagent. A mordant, the chemical layer contained in the positive print to which the dyes attached in forming the image, had to be developed that would work with the new class of metallized dye developers Rogers was working on. To help time the processing of the film, and to stabilize the final image from chemical damage, Land looked to the polymeric acid layer, or L-Coat, he had developed to eliminate the need for a print coater as was used with Polacolor film. He discovered that a modified version of the L-Coat turned out to be effective in the new integral film unit as well.

The work aimed at ensuring that the final photograph would forever remain flat was particularly counterintuitive. Initially, it was believed that the intact film unit would need to dry out. One way or another, it needed to be permeable so that any existing water vapor could evaporate. Early experiments with the forever-sealed film units using permeable film supports proved disastrous. The photographs all ended up horribly curled. It was Land himself who eventually dove into this problem and ultimately discovered that the film unit could remain “wet forever” without affecting the stability of the image, principally because of the action of the L-Coat.[40] As a result, Land determined that if the film supports used for the integral unit were “dimensionally stable,” even though they might be relatively impermeable to water, they would produce a permanently flat photograph as long as both supports on either side of the film unit were symmetrical with respect to permeability and other properties.

Work on designing the basic structure and dimensions of the film unit had to be completed before any serious development work on the camera could begin. On January 13, 1966, Land and an associate, Al Batchelder, conducted the first spreading tests on the basic configuration of an integral film unit Land believed would fit the bill.[41] The entire film unit was only 9/10,000th of an inch thick. It had the now familiar white border around an image area of three inches square. The white border was wider on the bottom edge of the film unit, the end that would exit the camera first, so it could mask the slim pod within it that would contain the viscous processing reagent.

Although considerable work still had to be done on many components of the film unit, the basic format was one Land was willing to commit to. Land assigned Richard Wareham, one of his mechanical engineers, to head up the camera design effort.[42] Land showed Wareham the film unit and explained how it would work, that it would be exposed and viewed from the same side, and that it could be ejected from the camera into the light immediately after exposure. Land explained that the film team would be creating some kind of chemical "dark room" that would make this possible. Even though work on the opacification system to block out ambient light had not yet begun in earnest, Land nonetheless wanted to base the camera design efforts on the assumption that an effective technique would eventually be realized. (See Fig. 6-1.)

Land also reminded Wareham that the camera had to be small, his objective being that it would fit inside a man's jacket pocket or a woman's purse.[43] For years, this led to the inside joke around the camera design team that the most important person in the whole project was Land's tailor.[44] To emphasize this point, Land had a wooden model carved of the size of camera he envisioned. He covered it with tan vinyl and stuck a lens on one end. This camera model and the experimental film unit became the standards Wareham and the other Polaroid engineers would use to design the absolute instant camera.[45]

The other requirements Land imposed did not make that task any easier. He wanted a lens that would focus from a mere ten inches to infinity, as well as automated shutter and exposure electronics and film transport systems. The latter had to handle the moving of a single film unit at a time through a set of rollers that would burst the pod and produce a perfect spread of processing reagent distributed evenly over the entire area of the image. In addition to designing a camera with all of those mechanical and electronic features, Wareham and his colleagues were tasked with designing some kind of container or pack for the ten film units that would be packaged for sale. The film pack had to be easy for the consumer to handle, while remaining completely impervious to light until safely loaded into the camera. It also had to house the battery that operated the SX-70 camera.

By far the most challenging obstacle in the entire program was finding a means to allow the film to be processed in the light after having been ejected from the camera. Further light had to be prevented from reaching the negative and ruining the picture. As opposed to the second generation of Polacolor pack film that had opaque paper supports on both sides, the integral unit was going to have one clear plastic side. This was necessary in order to allow the photosensitive layers to be exposed by the camera and, later, to have the image viewed. To enable this, an opaque "chemical curtain" had to be drawn as the film unit emerged from the camera, keeping the negative part of the unit in total darkness. Once development was complete, however, the curtain would need to disappear completely so that the picture could be viewed.

VP of research Buckler recalled that "the problem was, not just that . . . [something like that] didn't yet exist, but that we didn't even know the phenomenon that would give it the foundation to come into existence." The effort to find a solution was, in Buckler's experience, "the toughest job I've been involved with in my life," admitting that "there were some doubts."[46] Even Land later admitted that this "was probably our most adventurous step. We designed the camera as if we had solved the problem, and carried the camera all the way through engineering while doing the basic research on the chemical task of bringing the picture directly into the light."[47] (See Fig. 6-2.)

One of the first approaches tried by the camera team was a concept that Land had for a "scanning camera."[48] In this device the film would be moved across a narrow slit through which light would be transmitted by a lens. By manipulating the light with mirrors, it was theoretically possible to reduce the front-to-back dimensions of the camera. However, other problems quickly arose in trying to adapt this technique, and it was dropped. In early 1967, another design requirement for the camera was imposed on Wareham and his colleagues. By that time, the general concept of the opacifying system had been conceived, even though work had yet to begin to find suitable materials. Since the opacifying material was to be incorporated in the processing solution, it became apparent that it needed to take effect before the film unit was exposed to light when it emerged from the spread rollers and exited the camera.

On February 15, Rogers, acting as Land's emissary between the two teams, told Wareham and his principal associate, Alfred Bellows, that some additional time was going to be needed for the film unit to remain shielded from light "before the blackening chemical fully works."[49] A few months after that, on June 26, Land himself called Bellows to reiterate that the camera needed to be designed in such a way as to keep the film unit in the dark "for about an inch" to allow the chemical opacifier to become completely opaque.

As the design effort to realize all of Land's requirements within a pocket-size camera continued, Bellows pointed out, some time in 1967, that if the light path of the image being photographed was folded somehow within the camera—bounced from mirror to mirror to elongate it within a confined space—the focusing and other optical requirements Land had imposed could be achieved.[50] Almost immediately, work on the camera design moved in that direction. By early 1968, the very first prototype for a camera was finished. On January 11, it was evaluated in Land's laboratory.[51] Work continued on various aspects of this camera design, and by September, another iteration, called PT-II, had replaced it. This prototype had motors mounted in the shutter housing in the front of the camera, above the processing rollers, and used a "front picking" system to transport the film units.[52] It required small holes to be placed in the edge of each film unit so that a small pick could enter the hole and pull the film unit out of its cassette, through the rollers and out of the camera. The front-picking approach was a natural extension of all previous one-step systems, in which the film was always pulled out of the camera, albeit by hand.

The effort to find the right materials for the film unit's opacification system began in 1967 under the direction of one of Land's best and brightest. Stanley Bloom had earned his bachelor's degree in chemistry from MIT and a doctorate from Harvard, where he was an associate of Robert Woodward, the chemist who had worked at Polaroid on the synthesis of quinine during World War II. After doing some postdoctoral work at Columbia and at Harvard Medical School, Bloom had taught chemistry at Smith College before joining Polaroid's stable of talented scientists in 1960. To conduct the search for a workable opacification system, Bloom assembled a team that, at times, had close to twenty chemists working on the problem. (See Fig. 6-3.)

The genesis of his solution resided in one of chemistry's basic principles. "You never start with nothing. We knew that some materials color in alkali (the environment that processing takes place in)," Bloom later explained.[53]

> But one thing wasn't in the [scientific] literature: we had to find a material that would totally switch itself off. It's easy to find a substance that will decolorize in acid, but we wanted it to have color in a high alkali content, then turn transparent when the alkalinity dropped just a little.[54]

The work went on in a back room in Land's facility at 2 Osborn Street. It was a phenomenal effort to overcome a phenomenal problem. Howard Rogers observed the Herculean effort being undertaken to make Land's Aspirin design for the integral film unit work. As Bloom's effort was really heating up, still with no guarantee of success, Rogers came up with a wholly different approach to the problem. He sketched out the structure of an alternative format for an integral one-step film unit, one that would not require the kind of sophisticated opacification system that Bloom was desperately trying to create.[55] The sketch was dated January 1, 1968.[56]

Rogers' alternative film unit was designed to be exposed from one side, and to have the image appear on the opposite side. Thus, unwanted further exposure of the photosensitive negative could be prevented by simply including a totally opaque substance like carbon black in the processing reagent in the pod. When spread between the clear support and the negative layers of the film unit, the carbon black would create its own opaque curtain. On the opposite side of the film unit would be a preformed white layer that would become the background for the image once the dye developers migrated to it. Rogers' design was quickly dubbed "Excedrin," since it was an alternative to alleviating the headache of creating an integral one-step film unit.[57] However, given that Land had already committed substantial resources to the design and manufacture of a camera based on his Aspirin format, Polaroid never pursued Rogers' idea for the Excedrin approach.[58] (See Fig. 6-4.)

Eventually, after designing and synthesizing and trying hundreds of compounds, Bloom and his team came up with an opacifier that met the requirements Land had set out. In the end, the chemical curtain was created by a combination of pH-sensitive phthalein dyes in combination with a white substance, titanium dioxide, which was included in the processing reagent to serve as the white background against which the image would ultimately be viewed. Land called it a "synergistic optical effect."[59] Bloom called it "a miraculously complex material."[60] The technique was first demonstrated by Bloom on November 4, 1969, to about fifty Polaroid researchers working on the project. Because everyone present knew how crucial the success of Bloom's work was to the whole endeavor, it was "an emotion-charged scene." One report of the event conveyed the relief that must have been experienced by all who were present:

> As the exposed film, under the direct rays of two sun guns (equal [in intensity] to sunlight at the top of Mount Everest) began to clarify, excitement mounted. And when it became obvious that the color negative had developed, a great cheer went up, backs were slapped, and impromptu speeches were delivered.[61]

Although the efficacy of the system had finally been demonstrated, there was still more work to be done. Bloom and his team knew that, beyond proving the concept in a laboratory setting, he was also charged with the task of figuring out how to manufacture "not grams of these materials but tons of these materials . . . at a reasonable price."[62] The work toward this end continued unabated for more than two years. When at last Bloom's team produced the first batch of eighty pounds of the opacifying material, thus overcoming the most critical obstacle to making his dream of an integral film unit a reality, Land presented them with a cake inscribed: "From darkness there shall come light."[63] When asked later how he could have taken such a risk—having committed to a camera design and to building manufacturing facilities before a final solution to the project's greatest challenge had been found—Land reminded a reporter of one of his basic tenets, steeped in his personal brand of indefatigable confidence: "If you can state a problem, then you can solve it. From then on it's just hard work."[64]

CHAPTER 7

THE BIG PUSH—THE BIGGER DIVIDE

With progress seemingly being made on both the camera and the film fronts, Land decided in early 1968 that the time had come to bring Kodak into the picture. The object was to secure a commitment from its business partner of two decades to manufacture the negative for Polaroid's new film, as Kodak had done for every previous Polaroid one-step product. In the preceding months, during the regular meetings held with Kodak to monitor the continuing supply agreement for Polacolor negatives, there had been some hints of a new project on the horizon at Polaroid.[1] But now Land was ready to put Polaroid's new integral film firmly into the forefront of the conversation. Land invited Kodak's Wren Gabel, by then an executive vice president and member of Kodak's board of directors and executive committee, and George Petersen, the in-house attorney who had risen to associate director of Kodak's patent department, to visit on February 9.

Polaroid's objective was clear. Although McCune had started the process of putting Polaroid into the position of being able to manufacture its own negatives at some point, thus getting the company out from its almost total reliance on Kodak, it remained a long-term objective. Land and his colleagues were not confident that a Polaroid plant could be ready to produce sufficient quantities of negative to enable an introduction of the new absolute one-step photography system in the time frame it was looking at—namely, 1971 or 1972 at the very latest. Accordingly, Polaroid needed to secure Kodak's commitment to bridge the transition.

At the meeting, Land remained extremely secretive. He told his Kodak colleagues only that Polaroid had developed a "new process," which for purposes of their discussions was code-named P-110.[2] The two existing

Polaroid color films that Kodak had worked on were known in intercompany parlance as P-108 (roll film for in-camera processing) and P-109 (pack film for out-of-camera processing). The new P-110 process, Land advised, was going to require an "improved color negative," and Polaroid wanted Kodak to embark on another developmental program in order to supply that material to Polaroid. Land disclosed nothing else about the nature of the film, the camera, or the overall system Polaroid was working on.

As Land and the other Polaroid executives might have suspected, it was a very different Kodak, at least in its attitude toward Polaroid, which they were approaching for help. For a variety of reasons, the attitude at Kodak was no longer that of the paternalistic mentor anxious to help ambitious little Polaroid with its curiosity of a photographic system. The incredible success that Polaroid was experiencing in the market clearly had caught Kodak's attention. In 1968, Polaroid was in the midst of a decade that would see its sales grow from $129 million in the year Polacolor was introduced to $571 million. This growth was attributable not just to Polacolor. As predicted by industry analysts, Polaroid's introduction of the low-cost Swinger camera line had spurred sales immensely, even though at first the system took only black-and-white photographs. Polaroid had introduced the first Swinger model in 1965 at a retail price of around fourteen dollars, squarely in the heart of Kodak's mass-market domain. Seven million Swingers were sold in the first three years, driving Polaroid's sales and profits to new heights, and continuing to change Kodak's executives' perception of Polaroid and its market segment growth potential.[3]

During this period, Kodak also enjoyed success with its own innovation in amateur photography. The Instamatic was Kodak's first system in which the film was contained in a cartridge that could simply be dropped into the camera. No more threading, no more loading film in the dark. The basic Instamatic model sold for around twenty dollars, and Kodak sold nearly fifty million units around the world in the first few years after its introduction in 1963, just months after Polaroid had introduced Polacolor film. The Kodak Instamatic was a product so successful that *Fortune* suggested that one had to "look back to Singer's foot-powered sewing machine or Hoover's vacuum cleaner for a comparable example."[4]

Yet, the unparalleled success of the Instamatic and Kodak's apparently unassailable market dominance did not stop it from casting a wary eye on Polaroid's continuing progress. While Kodak may have viewed Polaroid's early instant system as "a toy of limited commercial appeal,"[5] its reassessment of Polaroid was highlighted in a *Fortune* feature detailing the history of the two companies:

Many Kodak executives thought in-camera processing would never amount to much.

> In Kodak's view, the Land camera was . . . everything that a camera for the mass market should not be. It is too big, too expensive, and too difficult to use. A quick sequence of photographs is impossible. A Land camera user leaves behind a trail of refuse. Additional prints and enlargements are inconvenient to obtain. The film seems expensive. And despite elegant innovations that are a delight to the technically minded, it forgives few errors. Not only does conventional film have greater latitude, but many mistakes by Instamatic users can be corrected during processing.[6]

Despite these reservations, Kodak executives had been, up to that point, delighted to be "taking a free ride on Polaroid's coattails," Land's company having become a valuable customer accounting for $50 million a year in sales.[7] As a result of the success of Polaroid's Swinger, however, "Kodak's chiefs were finally toppled from their complacency," and a resolve to get directly into Polaroid's field grew with renewed vigor.[8] Despite the faults they attributed to the Polaroid system, a Kodak executive admitted: "No matter what . . . [we] think, sales are a better barometer than . . . [our] opinion."[9] As a report in *Time* put it, "Land was no longer simply an ingenious inventor and customer; he was an enlarging and possibly troublesome competitor."[10] This would be especially true were Land to eliminate all those aspects of the original one-step photography system that Kodak disdained. Although Kodak was still in the dark as to what exactly Land and his colleagues were up to with their P-110 "new process," they had to suspect that this might be a possibility and thus may have begun to come to the same conclusion as *Fortune*: "If all or most of its liabilities could be eliminated, instant photography might be a devastating competitor to the [Kodak] conventional system."[11]

Adding to the hardening of Kodak's negotiating attitude toward Polaroid was the personal frustration that had built up over the years among some of Kodak's key executives over their dealings with Polaroid. Most of this was directed squarely at Land himself. Some at Kodak had begun to become wary of Land as far back as 1948, when Polaroid introduced its first one-step camera and film system. As the *Wall Street Journal* recalled, "while still a tiny company . . . Land . . . declared that taking pictures in the instant way was 'the only way' to take a photograph, implying that Kodak's huge operation was more or less obsolete."[12] More recently, the resentment over the lack of credit Land had given to Kodak's technical team for the commercialization of Polacolor was a wound that had never fully healed for those involved.

Even years later, Land had never departed from his position that Polaroid's color film had been wholly invented by Polaroid, which, he liked to point out, "took all the risks" involved in marketing it.[13] It is not hard to see how some of the Kodak scientists, who had labored for five years to take Polaroid's process from the laboratory to the marketplace, might resent Land's view. Similarly, the company executives who had authorized huge expenditures of manpower and financial resources, with no guarantee that a successful market would ever emerge for the program's output, assuming it actually succeeded, also felt underappreciated.[14] From Rochester's perspective, it was not enough to say, as Land did, that the supply of negative to Polaroid had actually turned into a good business for Kodak. Kodak had always wanted more. If a meaningful photography market segment existed, it felt entitled to sell its own products into it, packaged in its iconic yellow box. Land's attitude had done much to diminish the warm feelings of camaraderie that had existed between the Polaroid head and his Kodak contacts in the early days of their relationship.

Finally, and perhaps most importantly, there had just been a regime change at Kodak, and the incoming chief executive appeared to be antagonistic to Land from day one. Gabel had lost out in a corporate horse race to succeed William Vaughn. The winner was Louis Eilers, the son of an Illinois grocer who had joined Kodak in 1934 after earning a doctorate in chemistry at Northwestern University. He was appointed Kodak's president as of January 1, 1967, and would add the title of CEO on January 1, 1969. Eilers took the reins during an unprecedented period of success at Kodak. The company had enjoyed twelve consecutive years of higher sales and fourteen straight years of higher profits, its per-share earnings having climbed from $1.32 to $4.37. To Eilers, maintaining that growth was a formidable challenge, yet he was determined to continue the trend.[15]

Eilers was considered a more hard-nosed businessman than Gabel.[16] He believed that there was a "necessity [at Kodak] for an aggressive leadership."[17] Eilers seemed to harbor a particular enmity for Polaroid, and for Land in particular. He was known to refer mockingly to Land as "Mr. Polaroid" and, reportedly, liked "to twit" Land's public pronouncements.[18] He would call Polaroid asking for "Mr. Land," in blatant disparagement of his honorary degree(s).[19] As a direct result, it was no surprise that Kodak's new management team became, according to *Fortune*, "slightly more belligerent" in its attitude toward Polaroid. This was apparent as the discussions on Polaroid's new P-110 system progressed. Kodak simply was not inclined to embark on another "developmental process," as Land had described what Polaroid was after, without, at last, some firm guarantee of a satisfactory quid pro quo.

Not surprisingly, that quid pro quo involved a license that would allow Kodak to enter the one-step field in direct competition with Polaroid. Kodak had sought this, unsuccessfully, from the very beginning of the companies' relationship but had repeatedly been frustrated by the limited rights it had been able to wrest out of a reluctant Polaroid. Again, much of this frustration was directed at Land personally. Kodak's Petersen fumed over having "found out that all of the glowing promises of the usefulness of . . . [the] rights [granted] to us [under the 1957 agreement were] . . . an empty bag. The bait was held out to us, principally by Dr. Land . . . that [these rights were] going to . . . be useful . . . [but they turned out to be] an empty box, that either Dr. Land had believed . . . would, indeed, be useful and had been wrong, or we were being led down the garden path."[20] Six years later, Land again disappointed Petersen and his colleagues when they believed that he recanted an offer to license Kodak made during the 1963 discussions. All of this growing frustration was exacerbated by the lack of success Kodak was having in its own effort to come up with an original one-step process so that it could market a film for existing Polaroid cameras. No less a researcher than Henry Yutzy lamented the lack of success: "There must be some solution—worth many millions of dollars! So, I struggle," he wrote in a March 1965 memorandum entitled "Image Color Transfer a la Polaroid."[21]

As a result of this history, Gabel, who remained on Kodak's negotiating committee, made it clear to Land and the other Polaroid representatives at the February 1968 meeting that an agreement on some kind of license would be a prerequisite to further consideration of Polaroid's request. To his surprise, Gabel came away from the meeting with the understanding that Polaroid would agree to "license the improved P-110 process to others, including Kodak, with suitable waiting periods predicated on the success of the system."[22] Given that the devil of such an agreement was obviously going to be in the details of what would be considered a suitable waiting period and definition of success, the rest of the meeting was spent on discussing possible parameters for those crucial terms. Kodak made no commitments, and the meeting ended with both sides agreeing to review the situation further.

To follow up, Land wrote a letter to Gabel on April 9 to put forward Polaroid's proposal more formally. "We have advised you that we have, in an advanced stage of development, a new color film which uses a . . . [one-step] process to produce positive reflection prints in a new camera." Remaining vague about specifics, Land merely noted that the "negative component in this new film differs significantly in structure and chemistry from the color negative you are presently supplying to us." Land acknowledged that a "major effort" would be "essential" to put Kodak in a position to produce commercial quantities of the new negative. "We are desirous of ascertaining whether you would be prepared to undertake such major effort and to become a supplier to Polaroid of the new negative material."

Land admitted that Polaroid intended (in accord with its earlier agreement) to create a facility that would enable it to manufacture this negative as well but assured Kodak that "you shall continue to act as a substantial supplier of the new color negative until you shall have amortised your development costs and enjoyed a reasonable return for your contributions to the program." With respect to Kodak's request for a license, Land would only offer a characteristically qualified response: "Subject to approval by our Board of Directors of a definitive program for licensing, it is our intention to offer . . . licenses [that] would become effective a predefined 'number of years' after we have 'successfully commercialized' the new color film."[23]

A response to Land was carefully considered in Rochester, amid the change in its executive management that was already in motion. Gabel drafted a letter to Land in concert with Petersen and other Kodak executives. It was essentially a positive and friendly response. Gabel expressed "the confident hope that we will be able to work out mutually satisfactory terms" and invited Land and his contingent to Rochester for further discussions.[24] Before sending the letter, Gabel shared a draft with Eilers. The new Kodak chief executive apparently thought the letter might not be determined enough. Eilers "didn't understand why we should commit our production resources to someone else's business."[25] Although Eilers did not propose any changes in the draft—it was, after all, merely a polite and noncommittal invitation for further discussion—he laid out the strategy he wanted Kodak to pursue, and the rationale for it, in a handwritten note to Gabel:

> In this cooperative effort with Polaroid we will be spending much money to develop a satisfactory product. In doing so we will tap

> the reservoir of knowledge which has taken us years and millions to build up.
>
> The profit on the sale of this new material as a *joint* supplier is not enough. We need a written agreement and contract when we enter into this development work on the time when we can produce a competitive product under a license from them.
>
> I want to keep well informed of the negotiations and to see the final agreement before signing. (emphasis in original)[26]

Gabel's letter went out to Land unaltered on May 8, but that was to be the last of Gabel's involvement in the negotiations.[27] Having lost the power struggle with Eilers, Gabel left Kodak in late August, retiring after a thirty-eight-year career at the company. Gabel stated that his resignation was based on "personal consideration[s] including a wish for early retirement." Although Kodak's directors accepted his resignation "with sincere regret," there was some awkwardness attached to the moment, with the *Wall Street Journal* reporting that "Kodak officials, apparently finding all inquiries on the resignation very sensitive, declined to specify what Mr. Gabel's responsibilities were in recent years."[28] Likely as a result of Gabel's departure, nothing much happened on the Polaroid front over that summer.

In September, Land contacted Eilers directly to see if he could get the negotiation back on track. Land tried to break the ice by discussing generally with Eilers his "ideas for providing accessible photography for individuals." He then cut to the point and asked Eilers whether Kodak was interested in continuing the dialogue that Land had begun with Gabel and his team. In his characteristic straightforward manner, Eilers assured Land that while Kodak was interested in cooperating with Polaroid, he expressed his view that, in a supply deal like the two companies had with respect to Polacolor, "the amount we receive in return, i.e., the profit on the sale of raw materials, does not compensate Eastman [Kodak] for the large inventory of research and development work" the company contributed. Instead, even though he still had no firm idea of what Polaroid's new system was, Eilers once again insisted that Kodak "be permitted to manufacture a comparable product which would work in the Polaroid camera."

Land did not turn Eilers down flat but instead reiterated Polaroid's stance that a license might be available at some point once Polaroid had made "a success" of its new system. On this basis, Eilers agreed to allow his team, now led by Petersen, to continue negotiations to determine the two key factors: a definition of success and the time frame after which a license under Polaroid patents would be issued. Eilers advised Land that he would "remain in the background and try to expedite matters and solve bottlenecks."[29] The next move was up to Land to call a meeting to continue the process.

That meeting was scheduled for October 29, 1968, to be held in Cambridge. In the meantime, however, Kodak decided to ratchet up the pressure on Polaroid. In a letter dated October 9, Petersen wrote to Mikulka canceling the provisions of the 1957 agreement that related to the joint development work and the related patent licensing. Despite the fact that Kodak was far from agreeing to accept Land's invitation to participate in the new development project, Petersen told Mikulka, that in light of Land's proposal,

> it appears to us that many problems can be avoided if we do not have development work going on under the provisions of two concurrent agreements which would raise questions as to what work was done under which agreement. We think a more orderly arrangement would be to have development work performed under only one agreement at any given time so that patent rights under the two agreements do not get confused.[30]

Now, with even greater consequences potentially stemming from the October 29 meeting, Land and his Polaroid colleagues considered their strategy. They were intent on trying to make the project more attractive to Kodak, without agreeing to the kind of license that would jeopardize Polaroid's very existence. Their view remained that it would just not be feasible for Polaroid to compete head to head with Kodak in the same field, the only field that Polaroid had. While Kodak was a diversified photographic and chemical behemoth, Polaroid was, in many respects, a one-trick pony. Instant photography was its only real business. Polaroid executives were not the only ones who recognized how challenging a proposition it would be to compete with Kodak. As *Fortune* characterized it, Kodak "towers so impressively over the core of its business, conventional amateur cameras and film, that it is difficult to understand why anybody else bothers."[31]

Given Kodak's undisputed dominance over its existing markets, the Polaroid contingent couldn't come to grips with the notion of why it would even want to get directly involved in Polaroid's niche of one-step photography. "They don't know what they're getting into," Land warned.[32] In any event, Kodak seemed determined, and McCune recognized that the tone of the negotiations had changed with Gabel's departure. He felt that Eilers' "tactics were to beat us into" a license.[33] Ever the optimist, Land apparently went into the meeting believing that if Polaroid could make the supply prospects attractive enough, Kodak would drop, or at least moderate, its licensing demands.

Land opened the meeting by telling his Kodak guests that he expected Polaroid's new system to "revolutionize" photography, leading perhaps to film demands based on an estimated usage of "a picture a day a camera." As a result, Polaroid was going to need Kodak to supply a "substantial" amount of negative material, the specific estimates of which McCune laid out in great detail for a ten-year period. Land stated that he hoped that the anticipated high volume of these supply requirements "would minimize the pressure for licenses from Polaroid."

In response, the Kodak position, as articulated earlier by Eilers in his note to Gabel, was once again presented to the Polaroid executives. Petersen emphasized Kodak's primary business objective of producing its own products for distribution and sale, and although a supply arrangement as the two companies already had was of some interest, it was simply not enough. Moreover, Kodak did not want Polaroid to be dependent on it "for any longer period than reasonably necessary." Since Polaroid was already committed to taking on a portion of its own negative supply over time, Kodak did not consider a temporary or limited supply arrangement as a satisfactory "*quid pro quo* for engaging in the research and development work that would be required on our part." Accordingly, Kodak considered a license under Polaroid patents "at some reasonable time in the future as an essential part of the arrangement."[34]

Land said he understood but hoped that Kodak could be satisfied by "a general undertaking to make licenses available . . . without a specific commitment as to dates and the nature of such licenses." Kodak persisted, and a discussion of some parameters for a license ensued. Under guidelines Land proposed, it became apparent that he had in mind a time period of twelve to fourteen years before any license would be granted. Petersen made it clear that such a time frame would have only "highly speculative value" and would be insufficient. He insisted on a "fixed date." Clearly, no agreement was going to be reached on that occasion.

Before adjourning, Kodak's Rudolph E. Damschroder, the representative from its technical team who had worked with Polaroid most directly in bringing Polacolor to realization, pointed out that Kodak was still operating in the dark as to the precise nature of the system Polaroid was developing. According to Damschroder, he needed more of a technical disclosure on the nature of the new negative in order "to evaluate the magnitude of the R&D effort required" and thus, ultimately, the dimension of the remuneration that Kodak could be content with receiving in return. Land said he understood, and another meeting in Cambridge was scheduled for November 5 for that purpose.[35]

Under the circumstances, Land decided that he needed to give Kodak its first look at what Polaroid had up its sleeve by showing his visitors some photographs made with the new process. He believed it was critical to do that since it was still "not at all clear that Kodak would want to participate" in the new project.[36] Although he already had one prototype print made using the new integral film unit, Land had a second one made on November 4. Both photographs were shown to the Kodak visitors at the November 5 meeting.[37] The photographs were carefully masked with a cardboard frame so as not to divulge too many details about the physical structure or the operation of the film unit.[38] While Land may have been trying to be secretive about Polaroid's new film, it was "a difficult exercise in self-restraint. He . . . [was] openly delighted with what he and his co-workers [had] wrought."[39] Land recalled that the Kodak representatives were "impressed" and that one asked if he could take one of the photographs back to Rochester to show others.[40] According to Land, the picture was taken back to Kodak headquarters and then returned to him shortly thereafter. A follow-up meeting was held in Rochester on November 13, but no progress was made on the licensing discussions.[41]

In the ensuing weeks, it became clear to Kodak executives that they were seeing something new and important; something that clearly represented a quantum advance in the state of one-step photography—and perhaps in the history of amateur photography in general. Less than two weeks later, in a November 18 status report, Petersen noted that, based on what he had been shown by Land in Cambridge, Polaroid's "laboratory results give picture quality approaching [Kodak conventional color] prints."[42] Although Polaroid still had not revealed all of the details of its new system, Kodak by this time had a pretty good idea of what it was, both from the Cambridge demonstration and from early word concerning one or more Belgian patents that had recently been issued to Polaroid on a new "integral negative-positive system."

Within days, when Kodak secured and then reviewed one of the Polaroid Belgian patents in its entirety, it was able to make an educated guess that the new "system . . . would bring about a complete transfer of the image within the camera and would eliminate the in-camera waiting period for development, the special stripping outside the camera now required and the waste to be disposed of by the customer under the present system."[43] Thus, it appeared that Polaroid was about to rid its one-step process of all of the objectionable elements Kodak had long believed limited its commercial potential. Kodak was rapidly approaching an absolute tipping point. As described years later by David Eisendrath, a prominent consultant on the photography industry for *Time* and *Modern Photography*, "Kodak finally realized what Polaroid knew from the start—that there are people who want to take good pictures, and other people who want to see them as fast as possible. The latter group is much larger than the former."[44] The time was coming for Kodak to go after those people too.

Kodak continued to assess the state of the Polaroid discussions. Its marketing people concluded that the relatively high quantities Polaroid had projected for its negative needs—based upon its view that the new system would "greatly expand the amateur photographic market"—might indeed be correct.[45] Walter Fallon, a Kodak executive who was on the negotiating team as the head of Kodak's film manufacturing operations, did not believe, however, that Kodak could begin producing the negative as quickly as Polaroid wanted. Kodak also began making assessments of the impact of Polaroid's new system on its own sales of conventional color roll film, as well as on the value to Kodak of the ongoing license discussions, with respect to both Polaroid's current peel-apart film and its new P-110 system. As a result of these assessments, Kodak came to a few key conclusions, embodied in Petersen's November 18 status report.

First, Petersen reported that Polaroid's new system "will have a tremendous [negative] impact on the growth of Kodak's roll film business whenever it is introduced."[46] The dimension of that potential impact, as Kodak calculated it, was dramatic. Kodak executives were advised that P-110 could potentially reduce Kodak's sale of roll film by 817 million units over a ten-year period, resulting in a sales loss in excess of $6 billion.[47] Second, it was concluded that the rights to produce a peel-apart film for use in Polaroid cameras "would be of little value to Kodak" in the time frame contemplated, mostly because it was agreed by both companies that the new system would eventually render peel-apart technology obsolete and soon supplant it.[48] Third, the memo acknowledged that, despite all the research that had been done at Kodak for more than a decade to find an alternative to Polaroid's one-step technology, there was "no basis to say that Kodak could design a competing system of its own at all without licenses from Polaroid."[49] Petersen predicted that because of the Polaroid patents that were starting to issue, the integral one-step field might "not be open to Kodak" for two decades.

In light of these conclusions, a major meeting of Kodak's top executives, including Eilers, was held on November 27, 1968, to ascertain "management's views . . . before proceeding with further discussion with Polaroid."[50] The view of the negotiating committee was that Polaroid was "stringing it along" with regard to the discussion of a license within the time frame Kodak wanted and that those discussions with Polaroid would remain "fruitless."[51] Kodak would have to find another way. The minutes of this pivotal meeting of Kodak top management reflect the considerations underlying what was clearly a crucial turning point in the history of the two companies:

> Preliminary market research analysis of the [new Polaroid] system suggests that Polaroid's anticipated sales estimates . . . may not be overly optimistic. Convenience and ease of use may outweigh quality in future customer appeal, and cost is considered less of a deterrent for the future.
>
> *If* the system works, *and* Polaroid's estimates should be realized, the impact of such a Polaroid system on Kodak's sales of conventional color films . . . could be substantial. . . .
>
> As we see it therefore, *if* P-110 is workable, Polaroid will have a tremendously popular, new and perhaps revolutionary system with or without assistance from Kodak . . . and, *if* our preliminary market research analysis is correct, our conventional color films run the risk of suffering severely in the market whether or not we go forward with an arrangement with Polaroid. Up to this point in the negotiations, the only substantial reward to Kodak for helping accomplish this . . . would be the supply business for the negative component during the initial years of the introduction of the new system. Thus far, Polaroid had indicated that it does not contemplate granting licenses to Kodak . . . under patents covering the new system until its own "success" has been assured; and, in terms of Polaroid's current thinking, this would not be less than 10 to 12 years from now. . . .
>
> Those present did not believe that we should contribute a major research and development effort and provide our manufacturing

> capabilities for such a program on this basis. It was felt that in order to protect Kodak stockholders, Kodak . . . must have licenses under Polaroid's P-110 patents at a date substantially ahead of the periods now proposed by Polaroid. . . .
>
> Therefore, unless Polaroid substantially modifies its current extreme views on the dates and nature of licenses, we may not be able to reach an agreement. . . .
>
> Assuming the potential success of P-110 therefore, the only certain answer to avoid serious consequences to Kodak's future in the sale of amateur color films is for Kodak itself to develop its own in-camera processed color film system. The effort to bring this about will be begun immediately.[52]

Because of Kodak's stance, and Polaroid's intransigence on the subject of licensing Kodak, the discussions went nowhere. Two additional meetings were held between Eilers and Land on January 7 and March 28, 1969. The gap between Kodak's determination to get into the one-step field, and Polaroid's absolute fear of that prospect, was one that appeared to be unbridgeable. As McCune later described, the challenge remained the same:

> One of the stumbling blocks always in these discussions was that we couldn't see how giving Kodak a license to make products competitive with ours would let us survive . . . if we're trying to compete directly with . . . [them] in this field, that it's our bread and butter and it's the only thing we've got, and there's no way that the company can survive on the basis of royalties.
>
> And so every discussion we ever had about licensing . . . came down to the same question: You help us understand how Polaroid Corporation can survive against the power of Kodak in the marketplace, and we'll be perfectly willing, under suitable terms, to give you a license. But we never were able to solve that problem.[53]

In view of this stalemate, McCune had already concluded that Kodak was not going to undertake the project to help Polaroid with its new system. Polaroid was going to have to manufacture its own negative. When McCune told Land of his conclusion, Land was angry but turned to McCune and asked, "Can you do it?"[54] McCune, who had already taken steps in anticipation that this day might actually come, said, "Yes, we can do it. I won't guarantee we will make all you said you wanted from them, but we'll make enough so you'll never be short of it."[55] With no real options, Land agreed and McCune began a full-tilt operation to create Polaroid's own negative manufacturing facility, despite the daunting challenges inherent in that undertaking.

The denouement of this tussle between the two companies occurred the following spring. Once news of Polaroid's initiative to build a plant to manufacture its own negatives became public, Eilers took action. First, in an April 9 phone call and then an April 11, 1969, letter, Robert Miller, one of Eilers' lieutenants, informed Polaroid that Kodak was immediately increasing the price for the Polacolor negative ten percent domestically and fifty percent internationally. Land was informed immediately. He was provided with an analysis concluding that "the impact that this price increase has on our costs is of considerable magnitude."[56]

After that salvo, Eilers sent a letter by registered mail to Land on April 22 terminating Kodak's agreement to supply negatives for Polaroid's Polacolor film, the only color film Polaroid had on the market.[57] Eilers told Land that while legally he didn't believe any prior notice was necessary, to avoid any questions, he was providing two years' notice in accordance with the original terms of the 1957 agreement. That would end the supply of negative in April 1971, well before Polaroid could be in a position to replace the supply, either internally or from another company. In words that resonated as an undeclared declaration of war, Kodak's chairman told Land that "this clarification of our position will enable each of us to pursue our respective future courses untrammeled as much as possible by commitments made almost 12 years ago."

Polaroid's executives were astounded. McCune wrote next to the first paragraph on his copy of Eilers' letter: "Unethical statement—'for the record'—they have known all along and have been aware of these plans from other negotiations."[58] It was clear to McCune and others within Polaroid management that "a major change [was] taking place in our relationship with Kodak."[59] John Rumsey, a prominent Rochester journalist, in recalling these events, concluded that "from then on, Kodak and Polaroid were on a collision course."[60]

Land was stunned. The letter, especially Kodak's new hard-nosed tactic, came as a complete surprise to Polaroid management.[61] The issue was no longer the possibility of Kodak participating in Polaroid's new venture. It had become Polaroid's very survival if Kodak cut off its supply of color negative. Land wrote back to Eilers on May 13 assuring him that the announcement of Polaroid's program to manufacture its own negative "was not intended to alter our policy of reliance on Kodak as the

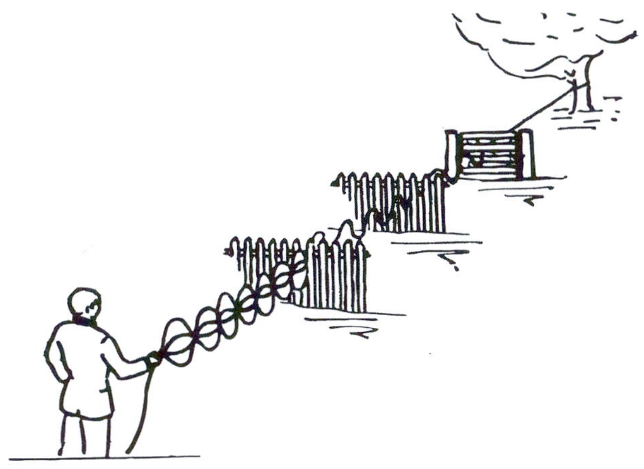

Figure 1-1: The "Rope Analogy" was an early attempt by Land to explain the effect of polarizers. The rope, initially vibrating in all directions, represents a beam of light, and the fences and then the gate represent polarizers that eliminate all vibrations but those in the desired single plane. Years later, Land pointed out that this depiction is inaccurate because the light vibrations that pass through are actually those at 90 degrees to the pickets, but the "combing" effect remains the same.

Reprinted from Journal of the Franklin Institute, *Vol. 224, Issue 3, Land, Edwin H., "Polaroid and the Headlight Problem," September 1937, with permission from Elsevier*

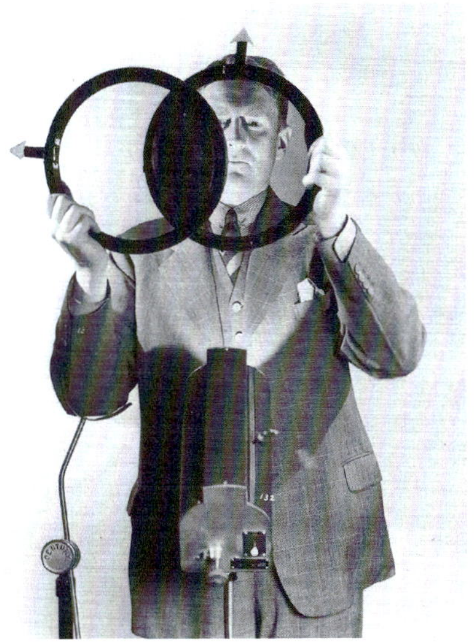

Figure 1-2: George Wheelwright, Land's physics instructor at Harvard and his initial business partner, in 1937 holding two polarizing sheets.

Courtesy of Polaroid Corporation Archives

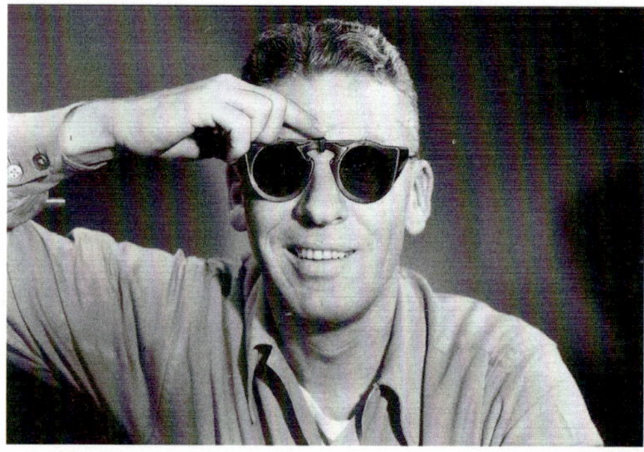

(Top) Figure 2-1: Polaroid advertising photo showing operation of its adjustable sunglasses.

Courtesy of Polaroid Corporation Archives

(Below) Figure 2-2: A prototype pair of Polaroid's adjustable "Polamatic" polarized sunglasses.

Author's collection

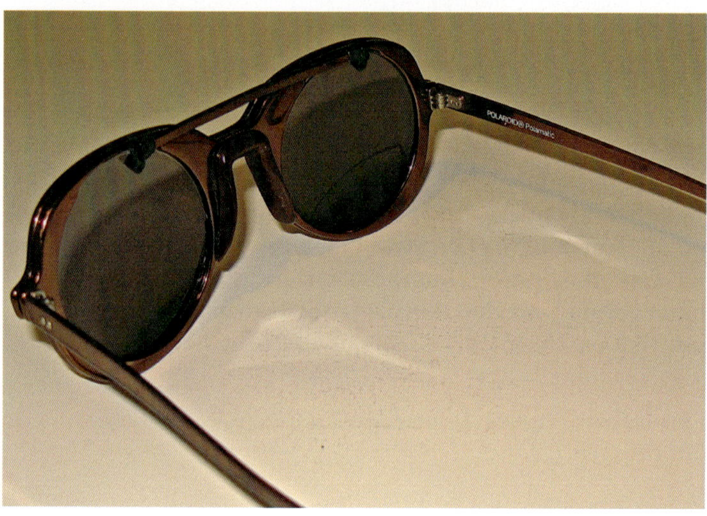

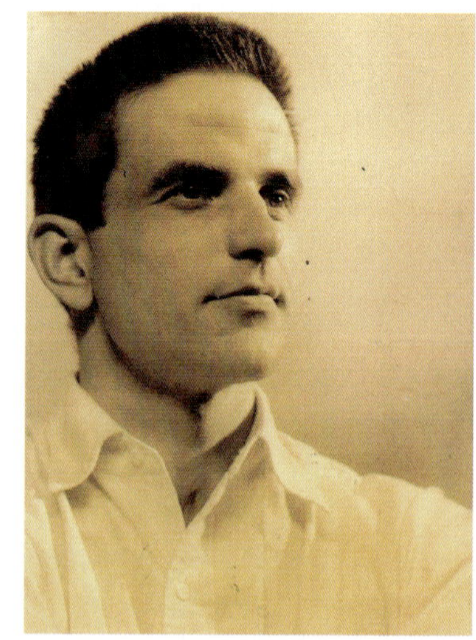

Figure 2-3: Howard Rogers as a young man—an early hire by Land at his new company, Rogers went on to become one of his key collaborators for the rest of his career.

Courtesy of Polaroid Corporation Archives

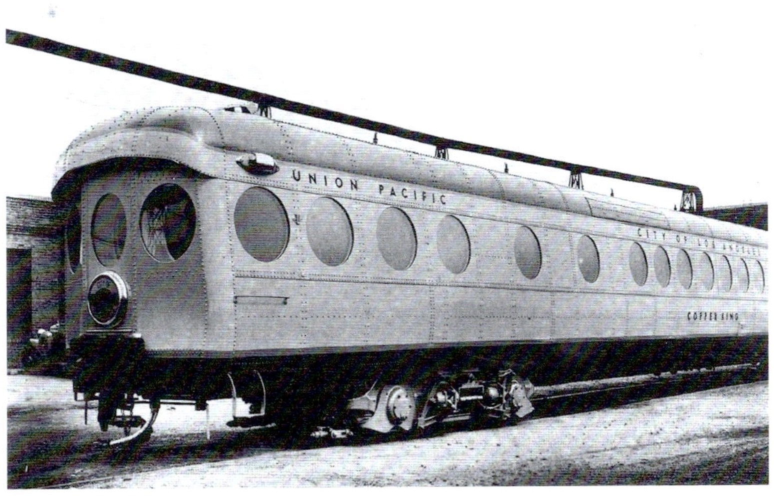

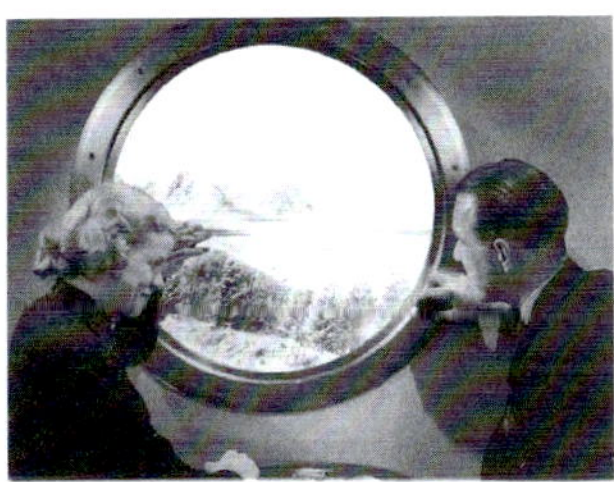

Figure 2-4: Polaroid variable-density windows in the "Copper King" observation rail car, an early and dramatic application of its polarizer technology.

Courtesy of Polaroid Corporation Archives

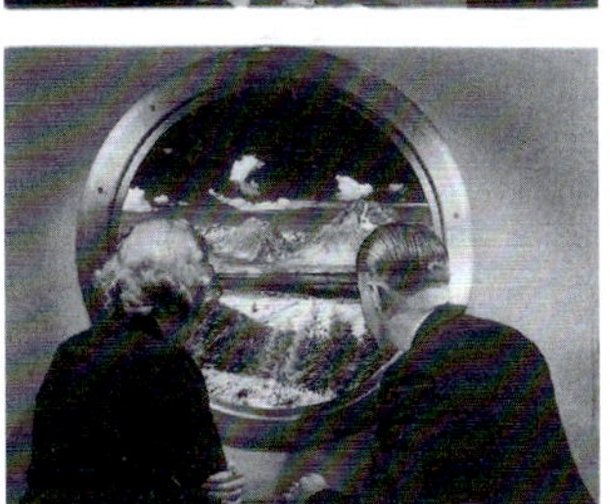

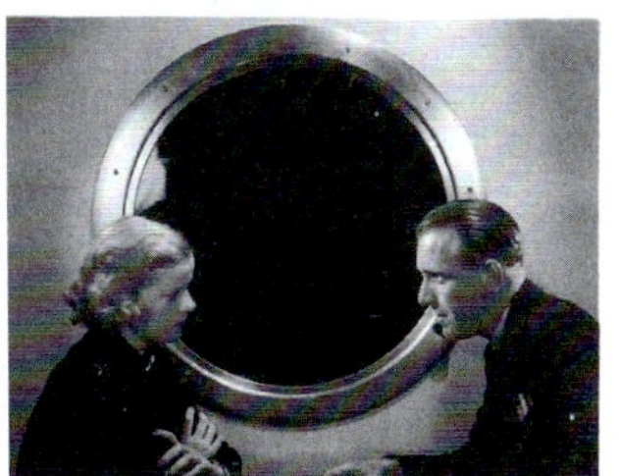

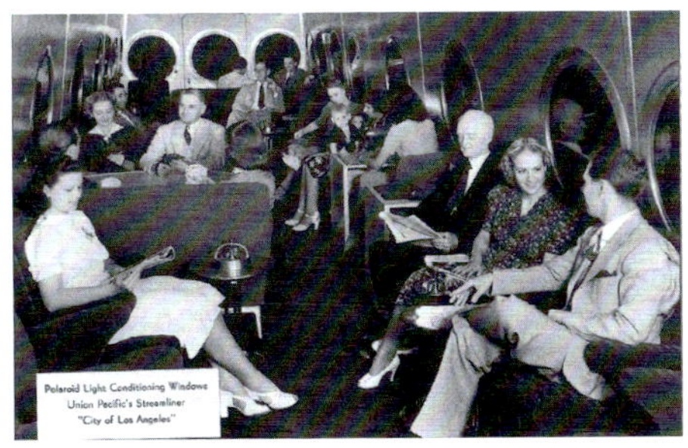

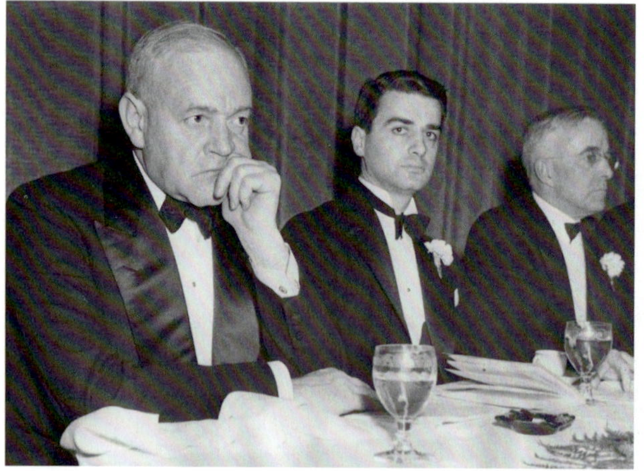

Figure 2-5: Land, just 30 years old in 1940, sitting between Donaldson Brown, vice-chairman of General Motors (left), and Dr. Irving Langmuir of General Electric Research Labs (right) as they were among those honored at the Modern Pioneers banquet.

Courtesy Corbis Images

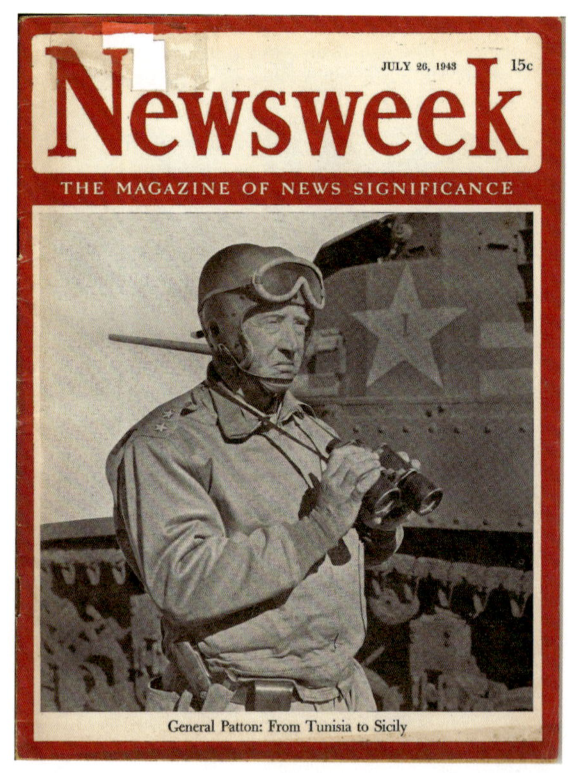

Figure 2-6: General George Patton outfitted with Polaroid goggles as he appeared on the July 26, 1943, cover of *Newsweek*.

Photo by Dr. Otto Bettman, Courtesy Corbis Images

Newsweek cover courtesy of PARS International

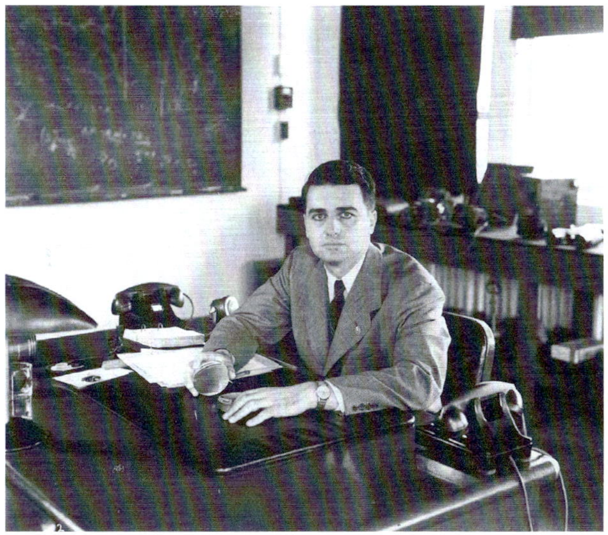

Figure 3-1: Land, still in his early thirties, at his desk in 1943.

Courtesy of Polaroid Corporation

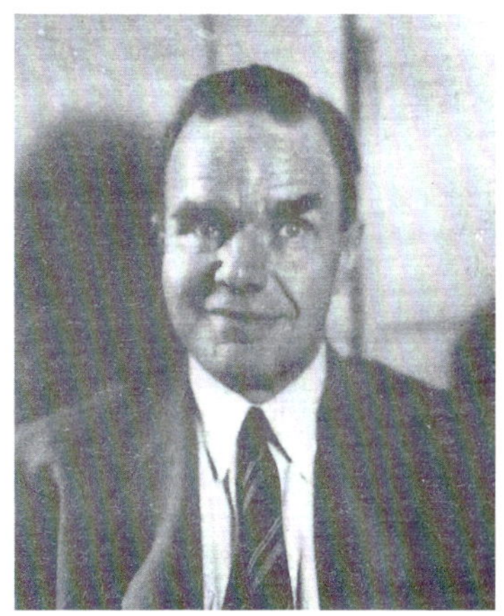

Figure 3-2: Donald Brown, the attorney who built Land's and Polaroid's patent portfolio, depicted in March 1944 in an experimental print made with an early image transfer process.

Courtesy of Polaroid Corporation Archives

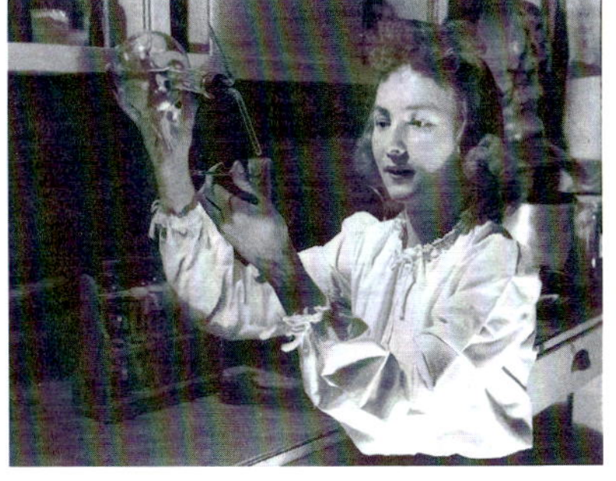

Figure 3-3: Eudoxia Muller, who performed the first experiments in one-step photography under Land's supervision in December 1943.

Courtesy of Polaroid Corporation Archives

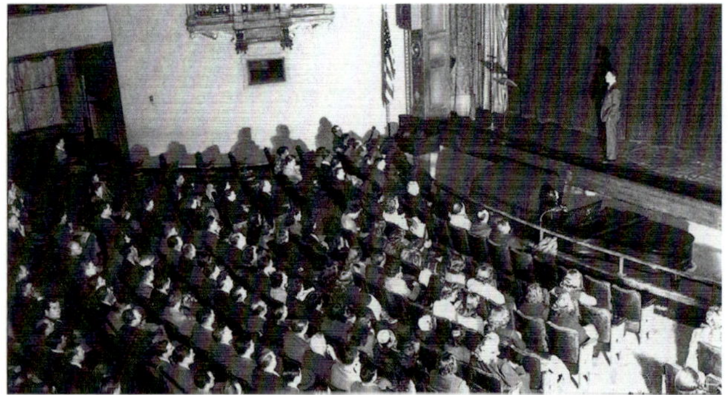

Figure 3-4: Land on stage in a Cambridge movie theater giving his mysterious 1946 Christmas Eve presentation of SX-70 to Polaroid employees.

Courtesy of Polaroid Corporation Archives

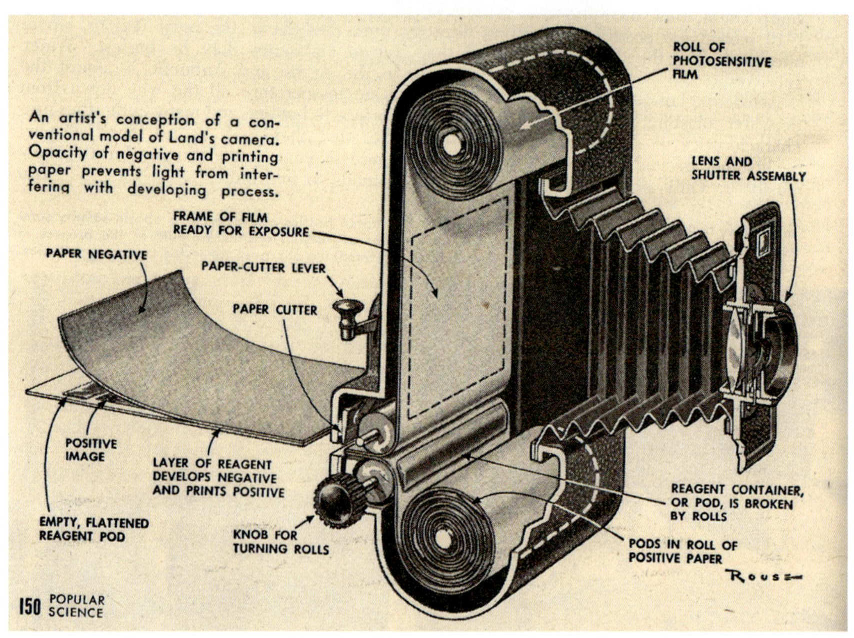

Figure 3-5: Artist's rendering of the one-step camera demonstrated by Land in February 1947 at the Optical Society meeting in New York City.

Reprinted from Popular Science, *May 1947, p. 150, courtesy of Bonnier Corporation*

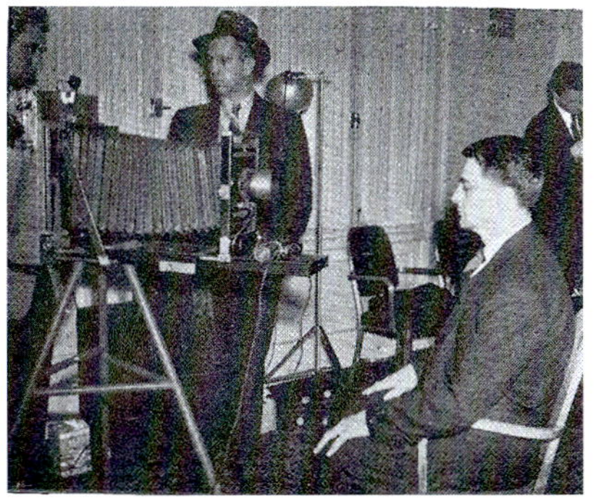

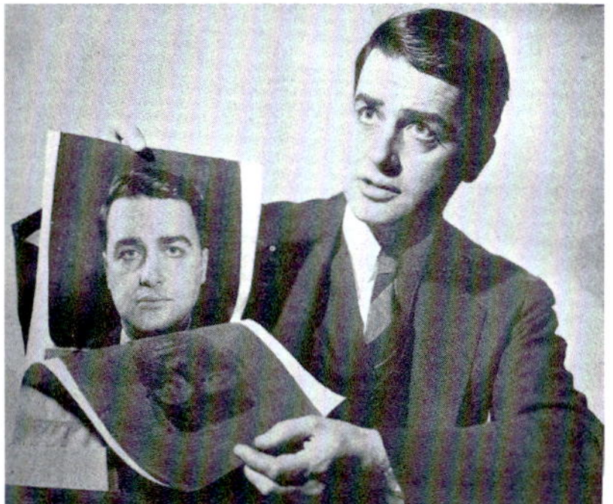

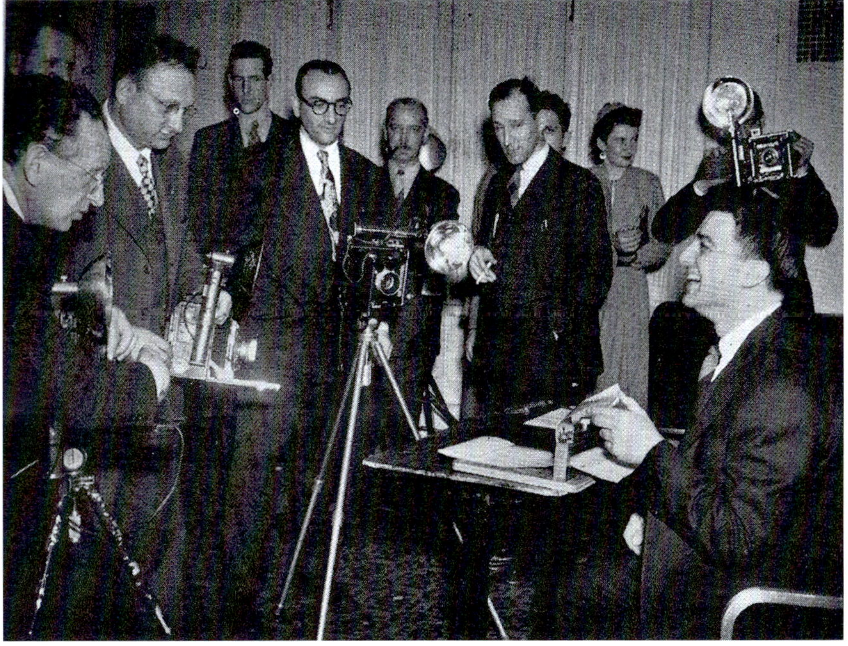

Figure 3-6: Land's February 1947 demonstration of one-step photography to the Optical Society in New York. Land poses for a self-portrait (top left), displays his portrait (top right), and jokes with press photographers asking to see their work "immediately" (bottom).

Reprinted from Minicam Photography, *May 1947, courtesy of Bonnier Corporation*

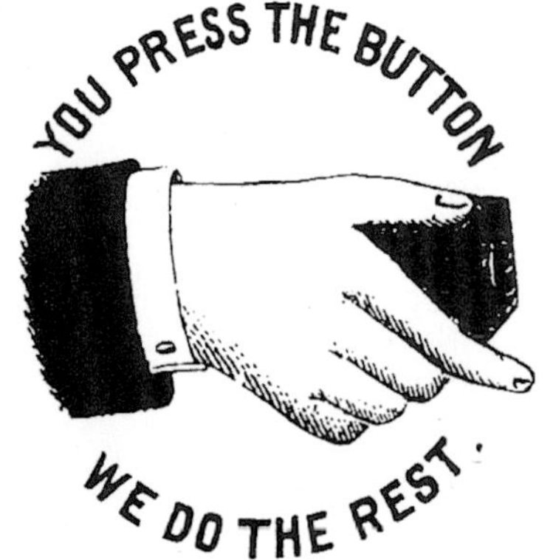

Figure 3-7: Kodak's slogan, conceived by George Eastman, as it appeared in one of the company's first camera advertisements in the September 29, 1888, issue of *Scientific American.*

Reprinted from Collins, Douglas, The Story of Kodak, *Harry N. Abrams Inc. (1990), p. 46, image public domain*

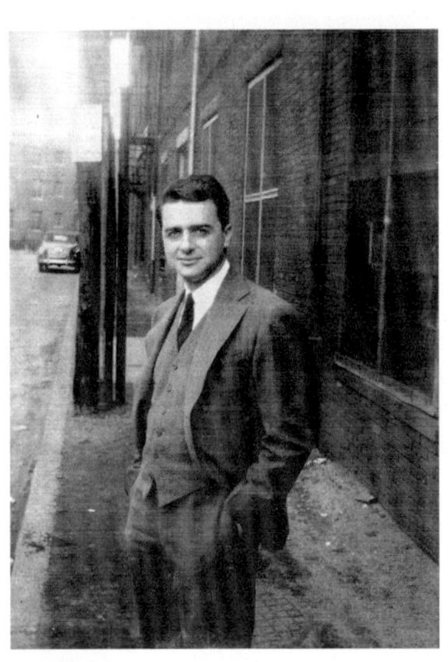

Figure 4.1: Land outside his Osborn Street laboratory in 1947.

Courtesy of Polaroid Corporation Archives

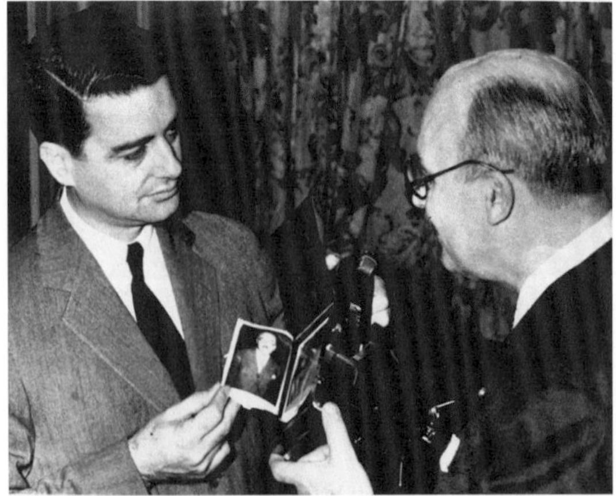

Figure 4-3: Land peels a picture of Charles B. Phelps, president of the Photographic Society of America, from the back of a Model 95 Polaroid Land camera at the PSA convention in Cincinnati on November 5, 1948, just prior to their going on sale to the public.

Courtesy of the Associated Press

prime source for our present type of color negative material. Substantial quantities of this material will be needed for many years to come."[62] Land recounted for Eilers the long history of Kodak's cooperation with Polaroid and the genesis of the mutually agreed idea that Polaroid would someday develop the capacity to manufacture its own negative. He predicted that it would be "many years" before Polaroid would be in a position to assume that task itself. Accordingly, he pointed out to Eilers that Polaroid's "dependence on Kodak for our commercial position . . . is total." Land explained that even the two-year notice was "totally inadequate" and "urged Kodak to recognize its obligations to continue as a source of supply under terms consistent with its dominant position in the field and the absence of practical alternatives available to Polaroid."

Eilers did not respond directly. Instead, Kodak fired another salvo by informing Polaroid in late May that it was unwilling to accept purchase orders submitted by Polaroid to buy increased quantities of Polacolor negative.[63] When pressed by McCune, Robert Miller, a Kodak vice president, admitted that Kodak was refusing to increase its production until the talks between Eilers and Land had come to some definitive conclusion about the future of Kodak's role as a supplier. Obfuscating the real matter at the heart of the dispute, he did not mention anything about the underlying issue of Kodak's demand for a license. Miller merely claimed that he and his colleagues were "confused," despite the detailed explanation provided by Land in his letter to Eilers. "We really don't know what's going on; we read in the papers about your negative facility and then we hear something from you and we're just confused."[64]

At Polaroid's request, the two sides met on June 27 at New York's Plaza Hotel to address the crisis. Land, McCune, and Mikulka attended on behalf of Polaroid. Eilers attended with Petersen and Kodak's assistant general counsel, Ira Werle. It was a dramatic affair. Land began by reiterating the substance of his May letter—that Polaroid was "entirely dependent" on Kodak for the supply of its color negative, that it intended for Kodak to remain a "major supplier" despite Polaroid's emergence as a second source, and that as early as 1963 Polaroid had told Kodak that two years' notice for cancellation of the supply agreement was "inadequate." To back this up, Land offered to enter into an agreement that guaranteed the purchase of a minimum of fifty million square feet of negative per year for three to five years, with the possibility of the order going as high as seventy-five million square feet, quantities beyond any amount that had been contemplated in the past.

When it was Eilers' turn, he coolly responded by noting that Kodak had "many opportunities . . . which needed careful examination in order to be sure that whatever Kodak did was for the best interest of the company and the stockholders." He quickly added that industrial suppliers generally made lower profits than companies selling products directly to the public. Thus, "Kodak wanted primarily to be in the consumer business and to sell products in the yellow box." While the company was willing to continue supplying negative to Polaroid, it only made sense if Kodak could also sell the same product itself. When the subject of Polaroid's increased purchase orders came up, Eilers "expressed the concern" that they were an effort to build up inventories in light of his letter terminating the supply agreement. Land denied this. Eilers pressed his position: "he was not saying that [Kodak] would not supply [Polaroid] with negative, but . . . any further supply commitment . . . required some proposal from [Polaroid] which would include a license to Kodak to make a product for [Polaroid] cameras." In response, the Polaroid contingent "consistently stated that it would be clearly imprudent for us to license Kodak to sell goods competitive with ourselves while we were dependent upon them as a source of color negative."[65] Around and around they went.

As the meeting went on, McCune expressed his displeasure to Eilers. He later recalled, "Kodak was taking advantage of having us over a barrel as a sole source of supply and our complete dependence on them for our existence for the next few years to try to squeeze a license out of us."[66] Eventually, the conversation became more pointed, and one of the Kodak attendees "questioned" Polaroid's "good faith in our earlier discussions of . . . a license."[67] At this point, Land "strongly objected to . . . [Kodak's] tactics in pressing for a license in this way." But Eilers was not about to back down. He reiterated that he "had to make the choice between supplying all of [Polaroid's] requirements and using the same facilities to supply Kodak needs, and that faced with this choice he could not very well neglect Kodak needs for [Polaroid's]."[68] The meeting adjourned with no resolution but with the Polaroid executives under the impression that a further meeting might be held only after Polaroid came forward with a proposal for a license agreement.

This impression was confirmed just days later. On June 30, Eilers wrote to Land reiterating much of what he had said at the meeting.[69] While Eilers acknowledged that Kodak's role as a supplier had been "a welcome one and . . . profitable for Kodak," he stressed the benefit that Polaroid had realized as a result of the relationship between the companies. Kodak's "contribution has helped to establish Polaroid firmly as the only producer of in-camera photography in the world and as a principal, if not the principal, supplier of amateur still color photographic film and apparatus in the United States." As a result, Eilers insisted that it was not "unreasonable [for Kodak] to posit any future relationship between us for the continued supply of your negative materials on a more balanced basis." That basis was "a patent license to Kodak."

The jockeying over this issue continued for the next several months. Letters went back and forth between Land and Eilers. Another meeting was held in Rochester on September 11. Eilers went right to the heart of the matter. "To be blunt, do you intend to give us a license? If not, there is no point in going on."[70] Eilers accused Polaroid of "dragging its feet" for years on the subject of a license, maintaining that Kodak had been pushed so far that Eilers had no choice but to cancel the supply contract. "It is normal business practice," Eilers maintained.

Land replied that it was his intent to offer Kodak a license on terms to be determined that would allow it to manufacture peel-apart film for Polaroid cameras. It was not going to be a license to produce Polaroid's new film, as Kodak had requested, but at least it was something. He advised Eilers that he had received approval from "representative members" of Polaroid's board for him to discuss the terms of such an agreement. Surprisingly, at that point, Eilers pulled out a letter he had drafted. It was a letter to himself for Land's signature that went somewhat further than what Land had just told him. The draft letter stated that although the Polaroid board "had not yet approved a definitive program of licensing," Land assured Eilers that once the "efficacy" of Polaroid's product "has been evaluated in the competitive marketplace . . . appropriate license under our relevant patents will be forthcoming on reasonable terms."[71]

Eilers and his colleagues apparently expected Land to sign the letter on the spot. Land resisted and tried to turn the conversation to other subjects. The Kodak side pressed. Harking back to previous discussions, in particular the misunderstanding over what Land had promised in the 1963 meeting, Werle said, "It is valuable to us to have you say in writing 'we intend to give you a license' because we think we have heard you say it, but we have never had it in writing, all we have had is a lot of gobblegook."[72] Land continued to resist, and it became apparent to all that a signature at that meeting was not going to be forthcoming.

Nonetheless, Werle asked Land if Kodak could assume that for the rest of the discussions about negative supply he had "in mind the principles established in the letter." Land replied, "In a broad sense, yes." But that did not satisfy the Kodak representatives, and the choreography of push and resist continued for the rest of the day. When asked what motivated Kodak to press so hard, Eilers provided perhaps the most insightful explanation so far: "To put it simply," he said, "the field [of one-step photography] is something we contributed to greatly and we feel we should share in it." The meeting ended with the Kodak executives making it clear that further discussions on a supply agreement could only take place if the letter was signed and returned to them.

Land never signed the letter drafted by Eilers. Yet the correspondence continued for the next several weeks. Finally, at a meeting on October 4 in Rochester, Polaroid finally agreed to license Kodak to make Polacolor peel-apart film.[73] In an apparent softening of its position, Kodak accepted. Over the next two months, the terms of an agreement were finalized, with the same kind of drama as had enveloped the almost two years of discussions leading up to it. Eilers' cohort, Gerald Zornow, admitted to *Fortune* that Kodak had been tough in its negotiations. "We're not in business to serve other manufacturers," he said. "You don't hold their hand forever."[74]

In any event, the companies were able to negotiate an agreement that assured Polaroid of a supply of Polacolor negative for an acceptable period of time. The agreement was signed on December 10, 1969, in a New York City ceremony attended by Land and Eilers.[75] In return, Kodak was granted a license to manufacture a peel-apart film for use in Polaroid cameras anytime after 1975, license rights that McCune admitted later he knew weren't "going to be any good to [Kodak]."[76] Although Kodak had initially held out for a license for the new technology Polaroid was developing, it did not get it. But neither did Polaroid get Kodak's help in developing or in manufacturing its new integral film. (See Fig. 7-1.)

Although Kodak was aware that the "license agreement [they received] was not really going to be meaningful" because it covered technology that was "rapidly going to be obsolete," Petersen had convinced Eilers to let him pursue it "as a matter of pride." Petersen was still frustrated because "the bait [of a license] had been held out . . . from time to time" as early as 1957 but had been withdrawn on each occasion.[77] Finally, useful or not, Kodak had what it had long pursued, some kind of license from Polaroid to its one-step technology. It was a symbolic if not a strategic victory. The 1969 agreement provided for payment of a flat fee per unit payment in the early years of any Kodak product manufactured and sold under the agreement. The long-term royalty rate set forth ranged from five to ten percent, depending on the number of Polaroid patents that Kodak actually used.[78]

Publicly, Kodak put a positive spin on the outcome. In a joint press release marking the supply and license agreements, Eilers stated, "The real significance to us of the new agreement with Polaroid is this: it broadens considerably the option of Kodak's participation in picture-in-a-minute photography—not only as a Polaroid supplier—but with a potential market entry under Polaroid patents as early as 1975, or through an earlier entry which could result from extensive work of our own."[79] The press reacted to Eilers' message as being Kodak's first public acknowledgment of its interest in competing with Polaroid. For example, the *British Journal of Photography* "deduced that Eastman Kodak will themselves eventually enter the in-camera processed film market; although, in the past Eastman have repeatedly denied that work was in progress on its own in-camera developing system."[80] After noting that the agreement only gave Kodak rights to make peel-apart film, and not film for the "entirely new" system under development at Polaroid, the *British Journal* cautioned that "introduction of . . . [other] Kodak colour diffusion materials will probably depend on Polaroid patent rights."

From his perspective, Land characterized the agreement in the joint press release as having achieved Polaroid's objectives, without putting the company at jeopardy:

> The new supply contract provides a guaranteed quantity of color negative supply adequate to meet anticipated needs we may have. . . . It also provides backup support and flexibility even after the completion of the negative manufacturing plant we are now building.
>
> The license to Kodak does not cover any rights under Polaroid patents for present or future cameras, nor does it include rights to a fundamentally new type of Polaroid color film presently in development which will utilize a new color negative.[81]

Nearly two years of contentious haggling between the companies was finally at an end. With the immediate crisis seemingly averted, McCune moved full speed ahead with the design and construction of Polaroid's film manufacturing plant for its new integral film while Polaroid's business continued, assured of its supply of Polacolor negative from Kodak. Nonetheless, the episode severed once and for all the cooperative relationship Polaroid and Kodak had enjoyed for two decades and pit them against each other in dramatic fashion over the two decades to come.

During the same period in early 1968 when Land commenced discussions with Kodak, he also sought to move the development process for the new absolute one-step system toward commercialization. He decided to bring on board someone who could assume the role of program manager and help shepherd the still secret—both within and without Polaroid—camera and film programs from the laboratories to the manufacturing plants and, ultimately, the marketplace. For this job, Land personally selected and hired Robert Duncan of NASA. Duncan, a Naval Academy graduate with a doctorate in aeronautical engineering from MIT, had been serving as chief of the Guidance and Control Division at the Manned Spacecraft Center in Houston, Texas. Duncan joined Polaroid in September 1968 and immediately focused on the camera development project with Richard Wareham and his small group of engineers reporting to him.

When Duncan came onboard, he joined the ongoing effort to complete the PT-II prototype camera. Finally, by the spring of 1969, the camera was operational.[82] At Land's direction, Duncan began the process of taking the camera from the design group, and putting it in the hands of the preproduction engineers who would ready it for production. During the course of that work, many incremental changes were made in the design, and so the prototype became known as PT-III.[83] By late summer, plans to manufacture the camera were well advanced. On August 29, Duncan held a meeting with his entire staff to discuss the status of the program. Reports were given on the state of the negotiations with several subcontractors who would be contributing components for the camera, including General Electric, Fairchild Semiconductors, and Ray-O-Vac.[84] Manufacturing drawings were prepared, and an all-day presentation of the product for Polaroid's Executive Management Committee was scheduled for September 30, with a rehearsal set for four days earlier.[85] Everyone was working toward a commercial introduction in the spring of 1971.[86]

Unfortunately, continued testing of the PT-III prototype revealed persistent and troubling problems with its operation.[87] In particular, the front-picking mechanism was proving to be very unreliable. Many times, it missed its mark and the pick failed to engage the small hole in the film unit, leaving the film unmoved in the camera. Occasionally, the pick would grab more than one film unit, causing the mechanism to jam and unexposed film units to be ruined. Duncan quickly came to the conclusion that it would be imprudent to continue with the project as then conceived. Duncan later recalled:

> We had so many problems and the rate of progress that we were making in solving these problems was such that we were not going to meet the scheduled introduction date of the system. I was very concerned about this and I was particularly concerned because we were making very large commitments to many vendors, many large companies . . . and I could see great difficulty in being able to engineer solutions to these problems in the time frame that we were working to.[88]

As a result of his conclusion, on September 12, 1969, Duncan wrote a long and detailed letter to Land, with no other copy recipients, that outlined his concerns. He recommended that the introduction date "be slipped" by as much as a year in order to "engineer properly new concepts." Duncan expressed his view that "new program dates [could be established] without any loss in inertia, excitement, sense of urgency, or discipline on the part of the program team." In sum, he assured Land that they would "end up with a better product if we take the time to do careful engineering at this time."[89]

Land got it. Moving quickly, he brought a small number of engineers, including Wareham, together into a room in his Osborn Street facility and joined with them to tackle the problems with the camera. The bulk of the camera group remained in Polaroid's engineering department across the street. As Wareham later recalled with a grin, Land said, "You're allowed one engineering failure and you've just had it."[90] They went to work. The room was dubbed "U-2" because of work on the reconnaissance airplane that Land had done there earlier. It was close to his office, so Land could be in and out, but as Duncan later observed, "he spent more time in our room than he did in the office."[91] The first camera model that emerged from the effort in the spring of 1970 was likewise given the name U-2.[92]

While the U-2 was larger than the ideal camera Land wanted, it did solve several of the problems that had plagued the PT-III. In particular, during the latter part of 1969, Land devised an alternative film transport system that picked each individual film unit from the rear and pushed, rather than pulled, it forward into the spinning rollers and then out of the camera.[93] Land preferred this technique because of several advantages it offered over the front-picking arrangement.[94] Eliminating the hole in the film unit necessary for a front pick was an aesthetic improvement in the appearance of the film unit and also avoided any chance that the pick might perforate the pod. The rear pick also meant that the slot in the film cassette through which the individual film units were engaged could be moved from the front of the camera to the rear, thus helping reduce the chance that unwanted light could reach a film unit before the opacification system took effect. The rear-pick transport system was incorporated into the U-2 prototype and later the U-3 that succeeded it.[95]

Although originally Land had not wanted the camera to open up or to fold, it became inevitable that, in order to satisfy the optical and size requirements he had imposed, the design had to move in that direction. Land's commitment to a camera that could fit in one's pocket was a sine qua non of the project, perhaps, some speculated, driven by Kodak's 1963 introduction of its tremendously successful pocket-size Instamatic camera line.[96] To accomplish this objective, more engineers were going to be needed to pursue the design for a folding camera, and so Land personally went to the engineering department while its head, McCune, was vacationing in Switzerland. Land told McCune's deputy, Milton Dietz, that he would be transferring some of his engineers to the still secret project. He would not answer any questions about what project they were being moved to. When McCune returned, he convened a meeting of his engineering group and discovered for the first time that several key men were missing. When he learned about what Land had done, he was livid, both about the personnel move that had been made without any consultation and about responsibility for the camera design having been denied to his department.[97] But from that day forward, the entirety of the camera design effort would be under Land's total control.

To deal with the optics of the folding camera design, Land brought in Harvard's James Baker, a colleague from his military intelligence activities. Baker designed a very thin, four-element lens that was focused simply by moving the front piece of glass very minute distances.[98] Then, in February 1970, Land decided that the camera should be a single-lens reflex (SLR) camera—that is, one in which the image is viewed by the photographer through the same lens that takes the picture.[99] Serious photographers vastly prefer this type of arrangement because of the advantages it provides in framing and focusing the picture. Basically, what you see (through the lens) is what you get (on the film). This was a feature suggested by Ansel Adams, and Land wanted to prove to his friend and Polaroid consultant that it could be done.[100]

To accomplish this, and to accommodate the wide range of focusing Land insisted on—ten inches to infinity—a viewfinder arrangement of "unusual accuracy" became necessary. Land told his engineers that he wanted to duplicate the effect obtained from a ground-glass viewfinder in a conventional SLR camera, where the image "comes magically out of the clouds. It doesn't require explanation, and a child can do the adjustment."[101] Wareham and his team tried several approaches but "failed miserably," leading them to believe that the folding body design might have to be abandoned. Finally, Land told Wareham: "Forget it. Keep your heads down on the camera. I'll invent the viewfinder."[102] Wareham thought that sounded like "a good deal," so he "grabbed it."

Land went back to Baker, and with another of Polaroid's optical scientists, William Plummer, they set out to tackle the problem. Eventually, they came up with an ingenious set of mirrors, some of which moved within the camera, to accomplish the task. "It . . . [took] virtually every optical trick we could think of," remarked one of the engineers who participated.[103] Although they now had a suitable design, the other major challenge, as with so many other components of the project, was to find a way to mass-produce this optical system inexpensively. Some of the optical elements had to be manufactured to an accuracy of 20/1,000,000th of an inch. This effort pushed the design and manufacture of optical components to levels never before attempted on a mass-produced product. (See Fig. 7-2.)

The new optical system, together with many of the other improvements Wareham's team had been working on, were finally incorporated into a camera model known as U-4 in early March 1970.[104] Land called a special meeting of his key project people for Saturday, March 7. The primary purpose of the meeting was to enable Land to demonstrate the new prototype, particularly the unique optics that Baker and Plummer had developed. The weekend meeting became an even more memorable occasion when Land paused so that everyone could go outside to watch the total solar eclipse that day.

There was great relief from the entire team that this major hurdle had been surmounted. And yet, further work on several aspects of the camera design still had to be done, in particular with respect to the location of the motor and the mechanism for driving all the moving parts. A prototype based on the U-4, named U-42, was eventually completed in August.[105] It was a folding SLR camera that used the optics Baker and Plummer had developed and included Land's rear-pick film transport system. The motor was located in the rear of the camera, with a gear train running along the length of the camera body to drive the front rollers. This prototype ended up being the basic progenitor of the absolute one-step commercial camera Polaroid would produce, one that would finally make Land's dream a reality.[106]

In the end, the camera was truly a miracle of modern engineering. The complicated electronics system, which used 300 transistors to control the myriad fully mechanized and automated functions of the camera, was as brilliant an achievement as the optical system. Folded, the camera measured seven inches in length, four inches in width, and just slightly more than one inch high. Land proudly declared that the final product represented "the dreams and efforts of hundreds of people in dozens of engineering, manufacturing and research teams" who had designed the revolutionary film and camera, "developed extraordinary manufacturing processes," and then "built the factories for the automatic assembly."[107] It was Land, as the conductor of this orchestra, who had led the group to success. He characterized the process in these terms:

> We have been writing without even realizing it a tremendous technological symphony. You can find early themes, you can find the reiteration, you can find the expansion of the theme, you can find the mystic way in which a group of people become almost an individual as they compose a mighty technological, almost musical undertaking.[108]

The team research experience, under the leadership and inspiration of Land, was perhaps best described by Plummer years later when recounting this special episode of his life:

> The experience of working with Edwin Land . . . will rank right up with the best I ever expect to experience in my professional career. He expressed an uncompromising technical need, bordering on a kind of fantasy, in an enormously expensive and valuable development program. We were able to bring together a superb collection of diversely and incredibly skilled people who were able to support each other's concepts with complementary skills, and we were given the chance to be successful. For most of us involved, that fine line between "impossible" and "very difficult" was moved a little bit. Dr. Land had an uncanny way of making an individual an expert by telling him that it was his job to be that expert, that important decisions would depend on the results of his work, and that the results were needed quickly. Yet the intensity of the work was balanced by good humor, by intervals

> of celebration, by time taken to contemplate technical and social consequences on all sides, and by the sense that the work was a shared responsibility.[109]

But there was more work to be done. Although tremendous progress had been made by the second half of 1970, the program was not quite complete, and work continued on many critical aspects—on both the product development and the manufacturing facility fronts. It was, as Land described it, "a period of high technological drama."[110] The rabbit was not quite ready to come out of the hat, but, with an anxious and increasingly excited magician running the show, it couldn't stay hidden for long.

CHAPTER 8

ALADDIN EMERGES

America was a troubled and distracted country in early 1971. Richard Nixon was starting the third year of his first term as president and continued to pursue the war in Vietnam vigorously. Protests grew against Nixon and his increasingly unpopular policies, from students on college campuses who had endured the Kent State Massacre the previous year, to the mainstream media in which Archie Bunker began his TV reign as a personification of the average guy enduring the "pinko" rants of his "dumb Polack" son-in-law, to the national political arena where Senator George McGovern of South Dakota announced his candidacy for president on an anti-war platform. The tide would not turn against the Vietnam "conflict," as it was called, until June 30, when the *New York Times* published the first of the Pentagon Papers. But Vietnam was the event that, during this period, dominated the attention of the nation, its people, and its media. It was in this environment that Edwin Land continued to work toward realizing his almost thirty-year quest to provide the world with the ultimate instant photography system. And in early 1971, it appeared that the goal was indeed finally within reach.

Polaroid, led by the charismatic and enigmatic Land, had become the Apple Computer of the mid-twentieth century. It was arguably the most admired and glamorous technology company in the world. A loyal fan base of consumers and investors eagerly anticipated its new products. Six years earlier, Land had made the daring gamble of committing more than half a billion dollars of Polaroid laboratory research funds to the pursuit of his ultimate instant photography system. He did so without having done a single dollar's worth of market research to determine whether people would actually buy the product. This leap of faith would later be characterized as "the biggest gamble ever made on a consumer product."[1]

Articulating a philosophy that would later become the mantra of Apple's Steve Jobs, Land proclaimed, "We don't do market surveys. We create the markets with our products."[2]

At this point, work had not yet been concluded on several remaining technical issues for Polaroid's new system. For example, a "light-piping" problem that occurred during the film unit's exit from the camera still persisted.[3] That is, unwanted light would "pipe" through the top transparent support of the film unit as it emerged from the camera exit, and move back through the plastic support into the camera behind the spread rollers where it would reach the yet unprotected photosensitive layers of the film. This light was reaching the film unit before the chemical opacification system incorporated into the processing solution could completely take effect, causing unwanted extra exposure that ruined the picture. It all happened in a fraction of a second, but it was a potentially disastrous defect. This was a problem first identified by Land and Rogers two years earlier, but it had not yet been eliminated by any of the design changes that had been made to the camera prototype as it evolved from model to model.

More importantly, Polaroid's new facilities for manufacturing the camera and film were still taking shape, with no guarantee of an actual completion date.[4] Manufacturing the film was a particularly daunting prospect, with no guarantee of success. Although Polaroid had years of experience making the positive components of its film, it had never made photographic negative, the more challenging part of the operation. And now, Kodak refused to manufacture the negative for the new system. As *Fortune* pointed out, "only a handful of companies in the world have been commercially successful at making even conventional color film. No less a company than DuPont tried and gave up."[5] Kodak had its doubts about Polaroid's prospects as well. "As one in the business," Louis Eilers had told *Business Week*, "it seems a prodigious task to me. They can hardly afford to hire all PhD's to run their machines. I would predict quite a few headaches in startup of their production."[6]

Kodak had never allowed Polaroid's scientists to see its negative-coating facility and had steadfastly kept its negative coating technology secret. While progress had been made in Polaroid's pilot plant in Waltham, Massachusetts, it was still a huge leap of faith to assume that the effort could be scaled up sufficiently to supply the entire negative requirement for the commercial release of Polaroid's new system. For that purpose, Polaroid had committed $60 million to build a new "fantastically complicated" plant in New Bedford, Massachusetts, but that facility was not yet on line. In addition to manufacturing the negative, the New Bedford plant was designed to assemble the film, a "terribly difficult" task, according to McCune, that had to be conducted in the dark. As McCune described the challenge: "The question is not whether we can do it, but whether we can do it consistently and with high yields."[7] In early 1971, that question was yet to be answered.

Despite all this uncertainty surrounding the project, Land could not repress his desire to give the world its first glimpse of what he had come to refer to as "Aladdin." Many of his Polaroid executives argued that any public demonstration of Aladdin was premature. Despite many problems yet to be solved, Land, as was his nature, simply assumed that any remaining issues would be overcome by the hard work and determination of the supremely talented team of researchers he had assembled. His optimism was based on a decades-long track record of success in similar circumstances. And so, frustrating and even angering many around him, Land decided to go ahead and embark on a campaign to whet the photographic world's appetite for the introduction of Aladdin, his absolute one-step photography system that would ultimately be marketed as SX-70 almost two years later. As Land often said, there was "[no] virtue in waiting."[8]

The supreme confidence underlying Land's decision to move ahead had always been his singular attribute, and in many ways the real secret to all of his and Polaroid's success. So now, despite the uncertainty of others at Polaroid about the company's ability to pull off the projected commercial launch of its new integral system in the fall of 1972, Land determined that it was, indeed, "show time." Literally. Land clearly reveled in the dramatic. "There's no scientist I know who wouldn't rather be a charlatan," he once explained, "and when circumstances allow you to be both, why it's great fun."[9] What made this admission so paradoxical is that it came from someone who, by this point in his career, had become renowned for his secrecy and aloofness, a man who worked within a very tight circle of handpicked colleagues.

Land was determined to unveil Aladdin in a fashion similar to his dramatic announcement of the advent of the first one-step photography system, staged in a movie theater with great dramatic flair twenty-five years earlier. His aim was to craft a gradual, but equally theatrical, public introduction for his new masterpiece. It was a script that would ultimately, as described by the *Wall Street Journal*, "create an aura of breathless anticipation" about Polaroid's new and revolutionary photography system.[10] As McCune so aptly described Land, "if he hadn't been an inventor, he would have been a marvelous playwright, because he has a great sense of drama, and a great sense of staging things for gaining the effective command of his audience and taking them where he wants to go."[11] He apparently enjoyed the experience, taking "to the footlights like a regular trouper."[12]

Act One of Land's presentation was a tease. The performance occurred on April 27, 1971, when he surprised the Polaroid faithful and the assembled media at the company's shareholders meeting with an unannounced and unofficial glimpse of the future. The event was staged in a massive, as yet unoccupied, factory Polaroid had built in the Boston suburb of Norwood to manufacture its new and mysterious camera. For the first time since 1947, when he had introduced the original peel-apart system, the sixty-one-year-old Land conducted a press conference prior to the start of the meeting.[13] As reported by an observer, Land took the podium and casually took a "smallish camera out of his pocket, held it aloft and put it back in his pocket."[14] That was it. He did not demonstrate its operation; he didn't even open it.[15] He did comment, however: "The way it opens is unprecedented and is a delight."[16] To a reporter from the *New York Times*, the camera appeared to be "about three inches by six inches and . . . somewhat thicker than a package of cigarettes."[17] Land's tease was done, as described by another eyewitness, "with flare that was part showman, part magician . . . [and] was the show-stopping event of the annual meeting and brought loud applause."[18]

Land declared that Aladdin was "utterly revolutionary" and predicted that it would become "the basic camera of the future," ultimately forging a "complete restatement" of photography.[19] Suggesting that the new camera and film system would be ready for the public in a year's time, Land warned that "it will be like an invincible steamroller coming down the next block."[20] When asked later about the camera he had flashed to the audience, Land disclosed that he called it Aladdin "because it has in it those things which Aladdin in his wildest and most intoxicated moments wouldn't have had the nerve to ask his genie for."[21] What no one knew at the time, other than a few of his closest associates, was that the Aladdin in Land's pocket was made of wood.[22]

Land's stunt significantly heightened the growing curiosity surrounding Polaroid's new system that, until then, had only been the subject of rumors. As reported by the *Wall Street Journal*, just on the basis of that quick glimpse: "The excitement was mounting. Polaroid was taking on the trappings of a glamour stock. Analysts feverishly sought out photographic experts for any information on the dramatic new invention."[23]

Also fueling the excitement was the financial world's recognition that amateur photography was "one of the fastest growing sectors" in the U.S. economy. A study by Chase Manhattan Bank observed that "[r]ising personal incomes, favorable demographic trends, stable film prices and declining camera prices" all contributed to the bullish outlook.[24] Developments in photo technology, such as Polaroid's instant systems that, according to Chase, made taking pictures "foolproof and fun," helped stoke enthusiasm on Wall Street.[25] Land was aware that interested parties were using every possible means to learn more about the secrets of Aladdin. He was determined to maintain the public's excitement while also using every means to protect its secrets.

This commitment to secrecy was well illustrated by one of the patent practices Polaroid followed during this period. It was imperative, of course, that the company continue to file for patents on the various innovations that were embodied in the new system, but anyone reading the patent literature might be able to learn too much about the new camera. Accordingly, Polaroid filed for patents on many key Aladdin components by suggesting they might be incorporated into other, unrelated camera bodies. This method, which was perfectly legal although clearly cryptic and evasive, made it harder to pierce the veil of secrecy that shrouded Aladdin.

The day of the 1971 Polaroid shareholders meeting was, coincidentally, also the day of Eastman Kodak's own shareholders meeting, held in Flemington, New Jersey. By this time, Kodak was well along its path toward entering the instant photography field to compete with Polaroid, and the bonds of cooperation between the two companies, so essential to Polaroid's success, had frayed. Kodak was determined to find a way to sell instant cameras and film in its trademark yellow box. After all, Land had predicted: "There would be as many [of the new cameras sold] . . . as there are telephones."[26]

Fearful that Land might be right, Kodak had conducted a market analysis in late 1968, and concluded that Polaroid's new integral system could have a major negative impact on Kodak's conventional amateur color photography market.[27] Kodak believed it could no longer simply sit by. As one journalist observed, it "was only a matter of time before Goliath decided to compete with David."[28] Acknowledging that Polaroid had "a 3–5 year start," the attitude at Kodak was that the time had come "to get serious . . . [by] 'girding for battle.'"[29] As a result, in early 1969 Kodak ramped up its previous efforts and launched a massive research program to develop its own instant photography products to compete with Polaroid's. When news of this effort began to become public, concern for Polaroid's ability to compete with Kodak immediately started to materialize. A *Newsday* journalist reported that Kodak, "by far the largest [photography company] in the business, has long had a well-deserved reputation as one of the best research and marketing companies in the country, and some analysts fear that when . . . [it] comes out with an instant camera . . . it could be an almost fatal blow to Polaroid."[30] A Wall Street analyst predicted that whenever Kodak markets its instant system, "it can't help but seriously hurt Polaroid."[31]

By that April day in 1971 when the shareholders of both companies met, Kodak's instant photography program comprised a double-barreled effort. Both of these efforts received significant new momentum following Land's stunt. At Kodak's shareholders meeting, Gerald B. Zornow, its president, addressed Kodak's interest in instant photography. Zornow confirmed that indeed the company had under "active development" two separate programs to introduce an instant product at an unspecified future date. First, he described Kodak's ongoing effort toward developing an integral system. In his words, this effort was aimed at introducing "highly portable cameras that produce quality results and reduce or even eliminate bothersome leftover material."[32] This was known as Kodak's P-130 program and was committed to the development of Kodak's own instant integral film unit, based upon its own proprietary technology, and the camera to go with it. Together, these would comprise a Kodak system to compete with Polaroid's Aladdin.

This approach, however, was fraught with legal difficulties, and Kodak's executives knew it. As early as 1970, in considering its options for getting into one-step photography, Kodak management had recognized that this approach "might be blocked" by Polaroid patents.[33] Just a few months after the annual meeting, in August 1971, Kodak executives met with in-house and outside patent counsel to review specific Polaroid patents. On a list of those patents, Walter Fallon, the senior Kodak executive then in charge of its film manufacturing operations, placed the notation "block" next to one patent, and "big block" next to another Polaroid patent on the list.[34] The latter notation was apparently made even though Kodak's in-house counsel had already provided its opinion that the patent in question had "no inventive merit" and thus might not stand up if challenged in court.[35] This led Fallon to advise Kodak chairman Louis Eilers that, because of Polaroid patents, Kodak was in a "difficult position" with respect to its P-130 project.[36] Nevertheless, Kodak's P-130 program continued and was now a matter of public knowledge.

Second, Zornow acknowledged that Kodak was continuing to look at a way to "participate in the market for instant pictures before [an integral] system of our own becomes available" by manufacturing and marketing Kodak films for Polaroid's existing peel-apart cameras.[37] Under the 1969 agreement, Kodak had the option to produce a peel-apart film like Polacolor using Polaroid technology. It was a product Kodak could manufacture with legal impunity. But because of the effective dates of the license, and the cost of royalties thereunder, Kodak elected to try to develop a peel-apart film for Polaroid cameras using its own technology.[38] This was known within Kodak as the P-129 project. It was the latest incarnation of the effort to develop a peel-apart film for use in Polaroid cameras that Kodak had been working at for years. Zornow said that Kodak's decision to market the film for Polaroid cameras "depended on a number of external factors," but he did not elaborate on what they might be.[39]

Word of Kodak's statement apparently spread quickly to some attendees of the Polaroid meeting going on simultaneously. When, during a question-and-answer period, a Polaroid shareholder asked Land about Kodak's statement, Land exuded complete confidence in Polaroid's position and genuine skepticism about Kodak's ability to deliver on its promises any time soon. "I am the last person in the world to undersell or underestimate Kodak," Land acknowledged, but "we are so far out ahead in conceptualization and insight and understanding." Land paused, then added, "[and] in patents," as the audience broke into laughter, "that we can not only hold the lead, but move out well, well ahead of anyone else in the domain . . . [of instant photography]."[40]

While Kodak worked diligently along both of these tracks, it had little concrete idea of what the new Polaroid system would actually look like, or how it would operate, other than that it would be a non-peel-apart system, as suggested by the Polaroid patents they were following, and by the film Land had shown Petersen and others back in 1968. Kodak was anxious to know more. Information was being collected and transmitted to Kodak management from whatever sources, public and otherwise, it could find. For example, some information came from a report prepared at Kodak on what Land had displayed at the Polaroid shareholders meeting. Additional information arrived via "a rundown" received by Fallon from "a source" he had in Massachusetts.[41] Despite these bits and pieces of information, in the spring of 1971, Kodak still had no way of knowing the precise nature of the competition it was going to face if it was able to develop its own system for the instant photography market. Indeed, as it turned out, the proverbial light at the end of Kodak's instant photography tunnel was a massive freight train.

It was not until a full year after the 1971 Polaroid shareholders meeting that the world in general, and the scientists and executives at Eastman Kodak in particular, would get the next glimpse of Aladdin. Then, for the first time, Kodak would finally be able to observe the full dimension of what it was up against in Land's magical creation. By the spring of 1972, all of the lingering development problems had been resolved. Stanley Bloom had perfected the chemical opacification system and, twenty-nine months after first demonstrating it, had finally delivered the first mass-produced batch of his special chemical formulation.

The light-piping problem had been resolved in the fall of 1971 by Richard Wareham and an associate, Richard Paglia, who together had devised a light shield structure that was added to the camera.[42] Instead of allowing the film unit to exit the camera in a straight path, this apparatus deflected it downward toward an exit at the bottom of the camera, thus lengthening the path the film unit traveled before exiting into the light. This reduced the light-piping phenomenon to an acceptable level. The deflection of the film unit also had the synergistic advantage of improving the spreading action of the processing fluid within the film unit.[43] Contrary to Polaroid's long-held belief that the film unit needed to exit directly out of the camera, deflection caused a squaring up of the leading edge of the spread, thus helping the processing fluid to reach the far corners of the film unit.

Ultimately, this advantage allowed the absolute minimum amount of processing fluid to be used, thus allowing for further reductions in film unit thickness and, as a result, in the dimensions of every other component dependent thereon, such as the film cassette and ultimately the camera itself.[44] Perhaps most importantly, by the spring of 1972, the New Bedford film manufacturing plant was close to becoming fully operational. Some problems persisted regarding the synthesis of sufficient quantities of some chemical components—notably magenta dye developers and green photosensitive emulsion.[45] Those would be quickly addressed, however, resulting in a facility deemed the "most advanced negative factory in the world," according to McCune.[46] In light of these developments, everything now appeared to be in place for Land to continue with his unveiling.

For the second act of his two-year drama, Land orchestrated a double-barreled back-to-back scenario. First, he decided to debut his new revelatory product to 3,000 Polaroid enthusiasts at the April 25, 1972, Polaroid shareholders meeting. At this meeting, he would wow the audience with a demonstration of what the new system could do, but not allow anyone close enough to get too detailed a look so as to maintain the mystery and continue to build anticipation. Second, Land planned a detailed technical presentation about the system at a scientific conference to be held just two weeks later. With the approbation of the scientific community at that event, the magnitude of the technological achievement embodied in Aladdin would be fully disclosed and, hopefully, recognized. The cumulative effect of these two events would surely signal the introduction of a truly revolutionary product in the history of photography and in the annals of American technology in general. Land sought the world's attention for his new masterpiece. The plan went off perfectly.

The 1972 shareholders meeting was held in another "cavernous" Polaroid warehouse, this time in the Boston suburb of Needham, Massachusetts. Land and his assistants had rehearsed this performance extensively over the previous weekend, and many of the props, including the film units that were to be used, were only completed at the last moment. After braving a massive traffic jam coming into the facility, and then passing through "elaborate" security measures that required them to leave their cameras at the door, the Polaroid faithful gathered with great expectation. Land was impatient for the normal business portion of the meeting to conclude so that he could start his demonstration. Reflecting his clear preference for the miracles of technological achievement over the mundane world of corporate finance, Land remarked to those around him, "I don't want what you are about to see to be involved in this monetary domain."[47]

The business agenda completed, Land finally took the stage with an air of drama and wonderment that, many years later, surely inspired the legendary presentations conducted annually by Apple Computer's Steve Jobs, an admitted Land disciple, at his company's annual meetings. Land underscored the landmark nature of what they were about to see by telling the audience that he was there to finally fulfill his promise of introducing a new camera and film system that "revolutionizes the revolution" in photography he had begun in 1947 with his original instant photography system.[48] "This day is one that will be a turning point in the history of photography. Photography will never be the same after today," Land announced.[49] "With the gargantuan effort of bringing [absolute one-step photography] into being, the company has come fully of age."[50]

With pride and confidence, Land pulled Aladdin out of his jacket pocket, and joked, "There's the camera . . . a record success for concise presentation of a new product."[51] The demonstration that followed was simple yet miraculous and well described by one journalist:

> Unfolding a leather-covered box to form a vaguely triangular Polaroid camera, Land focused on his oversized meerschaum pipe and pushed the shutter button five quick times in succession. About a second after each touch, a 3-in. by 3-in. blank plastic square shot out. Slowly and almost magically, like invisible ink being activated, they turned into color prints.[52]

The audience was amazed by the "spellbinding display."[53] It was impressed by the quality of the color prints that were noted by *Time's* photography correspondent as being "superior to any previous Polaroid process." "Unlike the damp prints that emerge from present models, the new ones—which are made of plastic, not paper—feel completely dry, even during the remarkable, outside-the-camera developing process," continued the impressed journalist.[54] Land pointed out that the new film unit allowed the user to "spill martinis, drop it in the bathtub, put it in your pocket, stack it, give it away."[55] The absence of "garbage" from the film unit that needed to be discarded after use, as was the case with the current peel-apart system, was another notable major improvement.

Later, in his "cluttered" office during a demonstration of the film unit for a *New York Times* reporter, Land described the process that went on within the film unit as "an elaborate chemical balancing act performed by dozens of ingredients. They migrate, react, dissolve, oxidize, reduce and combine with precision—then all activity ceases when the picture is finished. It occurs within 17 microscopically thin chemical layers sandwiched between a clear plastic cover and a black backing."[56] When asked why the camera was not quite shirt-pocket-size, Land responded in good spirit, "It's like a husband and wife forgiving the last 10 pounds because they love each other so."

Time summed up the reaction to Land's Aladdin best when it reported that "industry leaders are unanimous that it is a stunning technological achievement."[57] *Life* called it "a marvelous toy . . . [and] a daring challenge to Kodak for supremacy in the $4 billion-a-year U.S. photo industry."[58] Land predicted that the public would find the camera irresistible: "You just can't stop using it once you start." In jest he added, "We may plan to ask for government subsidies—this addiction is preferable to heroin."[59] Repeating the prediction he made to Kodak executives in November 1968 when he first showed them an SX-70 film prototype, Land admitted that "my fantasy is that this camera will be as widely used as the telephone."[60]

After Land's demonstration, the attendees were able to wander among twelve octagonal platforms set up in the warehouse to watch Polaroid employees use the camera in different applications. Twelve thousand tulips had been flown in from Holland and arranged around the platforms, each housing its own scene including love ducks splashing in a pond, some children playing at a party, and a painter working at her easel.[61] Admiring the photos being taken, Land described their color as "astonishing, having qualities for which we have no names."[62] (See Fig. 8-1.)

Many attendees, however, were frustrated because Land did not allow the audience to photograph or handle the camera and photographs, or even to get too close.[63] No measurements or specifications about the camera were released.[64] Land was just following his basic equation: more secrecy created more mystery, and that, in turn, created more excitement. There was also the basic worry that a film unit might fall into the wrong hands, with Kodak presenting the biggest threat.[65] Filled with pride and enthusiasm, Land summed up the appeal of Polaroid's new system to the assembly: "Our camera is intellectually complicated and operationally simple. All you have to do to have the picture is to will it."[66]

Land announced that the new system would be available in stores by year's end, in time for the holiday buying season. Apparently, the shareholders and other observers from the investment community were duly impressed, and the demonstration had its intended effect. Within the next few days Polaroid stock, already one of the "nifty-fifty" high-flying issues of the era, shot up more than twelve percent to new highs for the year.[67] Erich Hirschfield, president of Willoughby-Peerless, one of the country's largest photography retailers, hailed Polaroid's new instant system: "Fantastic! Fantastic! It's the greatest thing we've ever seen. Photography will never be the same again."[68]

Two weeks later, on the evening of May 10, 1972, Land conducted a full technical demonstration of Aladdin at a special meeting of the Society of Photographic Engineers and Scientists (SPSE) held in the Continental Ballroom of the San Francisco Hilton Hotel. While the shareholders meeting had been a chance to demonstrate in broad strokes the operation of the system, this event was Land's real moment of glory. It was his chance to review in great scientific detail many of the groundbreaking innovations achieved during the development of his new pride and joy. With this audience, Land was much more comfortably in his element. The group included America's foremost photographic scientists, as well as many of its most prominent photographers. Ansel Adams, Land's friend and Polaroid consultant, personally escorted Mrs. Land into the ballroom.

The venue was set up with a podium facing a large seating area. Behind the podium was a stage on which three octagonal platforms were erected for the various technical demonstrations that were to be conducted. Surrounding the entire area were displays of photographs taken with the new system, perfectly illuminated with specialized lighting.

Land took the podium to thunderous applause, energized by his pride and satisfaction in having accomplished a long-term challenge. "It is an extraordinary pleasure to share with so many old friends the result of a program which has lured us since 1944, captivated us, entranced us. As with all sirens, this program never came clean to say whether she meant it or she didn't mean it, but she gave us no chance to escape. We had to go on."[69]

Harking back to the genesis of his scientific journey, his daughter's question in Santa Fe back in 1943 about why she couldn't see the pictures he had taken right away, Land explained that, based upon work he had done on the 3-D Vectograph technology primarily developed for military applications, he had conceptualized an instant photography system right away. With a nod to the magnitude of the endeavor, and with a dose of his dry humor rarely seen by those outside his laboratory, Land noted: "Frankly, by the end of that walk [in Santa Fe], the solution to the problem had been pretty well formulated. I would say that everything had been done except those few details that took from 1943 to [now]."[70]

Land acknowledged that the initial peel-apart system was but a stepping stone towards his idealized concept, admitting that Polaroid's first system "didn't quite make it as you know, although we did do well enough to keep our company going and growing, well enough so that we had the luxury during the last ten years of going back to the full-grown dream of a system in which all you did was look at the image in your finder, press a button, have something come out through a pair of steel rollers, and end up with a dry, finished, shiny, beautiful picture. And, that's what we're going to show you tonight."[71]

With that, Land reached into a briefcase next to him and removed an Aladdin camera. He joked that regrettably he had erred because he had meant to take it out of his suit pocket but quickly showed that it would indeed fit there. After demonstrating how the camera easily unfolded into its operating position, Land briefly described the complex single-lens reflex (SLR) optics, crediting Ansel Adams' challenge with inspiring it.

Following a description of the flash unit and some of the camera's electronics, Land pulled out the same oversized pipe that he had used as a subject for his demonstration at the shareholders meeting. Immediately realizing that he had forgotten the "necessary" step of loading the camera with film, he paused to relate the old "war story, which most of you are too young to remember, about the English girl who said to the American driver when he turned the heater on in the car, 'don't you Americans ever do anything by hand?' You do have to load this by hand. Loading is fun."[72] Land was clearly enjoying himself.

Returning to aim the camera at his pipe, Land took five photographs in rapid succession just as he had done at the shareholders meeting, with each film unit ejecting automatically from the front of the camera immediately after exposure. He held them up for all to see, noting how they were "completely dry" to the touch, "like a playing card," and that each had an image of the pipe coming into being, something that would occur "whether it's here or outside on Mont Blanc."[73] In just eighty seconds the developing images became clearly visible to the audience, and loud applause erupted. Land then turned and immersed one of the photographs into his water pitcher. Removing it and showing that no damage had occurred, he explained to his audience that the water does not penetrate the film unit, which means that it is no problem to take pictures in the rain. Irrepressibly, he noted that you could even leave the picture in the water. "In the days when many of my best friends lived in Rochester [headquarters of Eastman Kodak], the important thing would bc . . . could you leave it in martinis?"[74]

At this point, it was time for the audience to see the camera and film system work up close. Land described each of the octagonal stages set up behind him. The first stage had an artist painting brilliantly colored images that could be photographed. The second had an array of still-life objects for the same purpose. The third stage was set up with a microscope that allowed for a demonstration of how pictures could be taken with various magnifications of a subject image. While the audience members could observe photographs being taken, they were not allowed to take home samples, and after viewing, all of the film units were punched with a hole and stored on a wooden spike.

The audience milled about for a period of time, visiting each of the demonstrations, and eventually was asked to return to their seats. Land resumed his talk, describing in further detail elements of the development and operation of the optical and electrical systems of the camera. The structure and chemistry of the film unit's photographic process was described in a much less detailed manner, after which Land entertained questions. During this session, he disclosed that the cost of the camera would be in the range of $100–$175 and that the cost per picture of the film would be the same as the color peel-apart system then on the market. He reiterated that it would be available by Christmas.

With the program at an end, Land graciously paid tribute to the Kodak scientists who had inspired and encouraged him during the early years of his pursuit in this unique "field of industry in which science and art and applied science and chemistry and physics all converge." Crediting his team of Polaroid research colleagues, he concluded by noting how the quest for developing Aladdin, the ultimate embodiment of instant photography, was a "wonderful inspiring undertaking and [gave us] a sense of feeling of what you can do in the United States if you pick the right people and put them together."[75] He received a standing ovation. It had been the triumphant presentation Land had hoped for.

Back in Rochester, Kodak engineers working on the P-130 project continued to work on many ideas for Kodak's own integral instant photography system. During the previous year alone, at least five major camera designs had been built and tested.[76] These embodiments had names like the "January" camera, the "Card" camera, the "Roll-up Waste" camera, and the "Frame" camera. On September 8, 1971, a major review of the P-130 project was held for Kodak top management.[77] It was presented by a "Mini Task Force" that had been assigned to evaluate Kodak's progress in developing its own system following the dramatic glimpse of the new Aladdin camera Land had provided at Polaroid's annual meeting that April.[78] While Land's tease had provided little detailed information about Aladdin, other than its basic dimensions as something that could fit in a man's suit pocket, the Mini Task Force had "extrapolated" information as best they could about Polaroid's work-in-progress "from foreign and domestic patents, annual meeting comments, and investment analysts' reports."[79]

By early 1972, as a result of the Mini Task Force's analysis, Kodak had settled on a particular embodiment that was called the Lanyard Camera.[80] This device was operated by pulling a string to move the film unit through a pair of rollers and then out of the camera. This action simultaneously removed functional parts from the film unit, primarily the pod that had held the development chemicals.[81] Those parts had to be discarded from the camera later on. This design was "as close to garbage free" as Kodak was able to design at that point.[82] The remaining film unit then developed outside the camera. This camera and film system was successfully demonstrated to the P-130 committee on February 4, 1972.[83]

In early March, just weeks before Land's unveiling of Aladdin, the Lanyard Camera was demonstrated to Kodak management.[84] Everyone was apparently happy with what they saw, particularly with its size and configuration. In a March 17 memo, Albert Sieg, Kodak's corporate coordinator for its instant photography program, reported that the camera as then configured met the goal of producing acceptably sized prints, acknowledging that "it is desirable to have the smallest possible camera consistent with good design and engineering."[85] He went on to confirm that the Kodak marketing gurus "were of the opinion that the camera size as presently known meets this goal and that its present size is not too large."[86] As the date of Polaroid's shareholder meeting approached, work continued in Rochester on the Lanyard camera and film combination with which Kodak intended to compete with Polaroid. Fine-tuning of the optical, mechanical, and photometric components was under way, and target dates were set for early May for the completion of these improvements.[87]

But all of that work was about to come to a sudden halt. All hell broke loose in labs and executive suites across Kodak's Rochester headquarters when Land finally demonstrated Polaroid's Aladdin camera and film system at that April 1972 shareholders meeting. Kodak had sent several executives and engineers to the meeting, including Sieg and his P-130 program manager, V.H. Jungjohann.[88] The Kodak contingent was "anxious to find out what . . . the consuming public would be seeing."[89] Upon his return, Jungjohann provided a surprisingly detailed report on the Aladdin camera and film, despite the general nature of the demonstration, and the fact that the audience had not been allowed to examine either up close. Among other things he noted were the size, white border, and black backing of the film unit and the timing of the development process.[90] For the camera, he provided details on its size and configuration, its lens, shutter, exposure control, battery system, and, particularly, the motorized and automatic film advance system.[91]

Following the Polaroid annual shareholders meeting, Sieg sent an even larger contingent, headed by Donald Delwiche, the supervisor of Kodak's Systems Laboratories, to observe Land's technical demonstration at the May 10 SPSE meeting in San Francisco.[92] The entire event was tape-recorded and later transcribed as part of a highly detailed twenty-seven-page report, complete with diagrams of the meeting room and various aspects of the film and camera.[93] Nothing escaped the Kodak team's scrutiny, including the holes punched through the film units during the demonstration so they could be collected on wooden spikes. When some of those punches fell to the floor, Kodak's people picked them up and brought them back to Rochester for microscopic analysis.

There is no doubt that these Kodak engineers were amazed and disturbed by what they witnessed Land demonstrate at the two events: a sleek metallic rectangle that unfolded into a triangular-shaped camera that, with the push of a button, snapped the shutter and then ejected the film unit from the camera with the help of a small motor. The film unit was square and thin and intact from start to finish. No components were removed by the camera or by the user. It was an elegant system of camera and film that the Kodak engineers ultimately dubbed "a masterpiece of engineering." It was readily apparent to all that Polaroid's system was far superior to the product Kodak had in development at the time.

Land's demonstrations had an immediate impact in Kodak's Rochester research labs. In a report comparing what Land had displayed to Kodak's Lanyard Camera model then in development, Sieg summarized the change of perception at Kodak that had occurred in just a few short weeks. Noting the compactness of the "Brand X camera," he conceded that "obviously, our projected camera is too large by comparison."[94] He went on to point out that the necessity of stripping components from the film unit as it exited the camera in Kodak's model was "a nuisance." Finally, he admitted that "our camera is now unappealing in appearance" and that designers and engineers needed to "make it somewhat smaller, lighter and more colorful."[95]

Kodak's P-130 committee met on May 5, during the two weeks between Land's Norwood and San Francisco demonstrations. By then, Kodak engineers had fashioned wooden models of the Aladdin camera based upon Jungjohann's observations at the Polaroid shareholders meeting. They were examined and compared with Kodak's Lanyard Camera. The committee quickly came to the same conclusion that Sieg had reported three days earlier: "the P-130 camera as presently conceived appears too large." Recognizing that this camera size was largely a function of the size of the film pack, and ultimately the photo print size, they concluded that "to effect a camera size reduction will most likely require changes in the size of the cartridge and picture unit." Thus, the entire Lanyard Camera system, both camera and film, was no longer viable. Sieg concluded that Kodak's research and engineering teams should take three months "to brainstorm" this and to figure out how Kodak's program could be altered to come up with a more comparable and ultimately competitive product.[96] Kodak engineers were clearly going back to the drawing boards. It was no small undertaking.

It wasn't long before the instant photography dilemma in Kodak's research facilities also reached its executive suites and board of directors. On May 12, just two days after Land's full demonstration of Aladdin at the

SPSE meeting in San Francisco, top Kodak management met again, this time to review the details of Land's presentation in the context of Kodak's ongoing efforts. Their conclusion was as clear as it was disastrous: "The P-130 Program as currently defined has lost its impact from a marketing point of view. It was generally concluded that a new picture format is required that would allow a smaller cartridge and camera to be produced."[97] The very next week, on May 19, Kodak's board of directors shook up the company's top management. It made the surprising decision to name Walter Fallon as its president and chief operating officer. Louis Eilers stepped down. Gerald Zornow was named chairman of the board of directors.

Fallon was a chemist educated at Union College in Schenectady, New York, and Rensselaer Polytechnic Institute in Troy, New York. He had joined the company in 1941 directly out of graduate school and had worked his way up the management ladder through Kodak's research and technical divisions. He was then serving as the general manager of Kodak's Canadian and U.S. photographic divisions, responsible for the manufacture, marketing, and distribution of Kodak film. It was a job he had been assigned to by Eilers in 1970 in a previous surprise promotion, bypassing his boss at the time and signaling his rapid ascension.[98] In this position, Fallon had already become fully familiar with the relationship between Kodak and Polaroid. He had joined the Kodak team that met with Polaroid in late 1968 and had previously worked directly with McCune on many of the details relating to Kodak's development and then supply of negative to Polaroid. In fact, McCune wrote to Fallon upon hearing the news of his promotion, expressing his "heartiest congratulations and very best wishes. Many of us who worked with you during the early phases of our color negative program developed a keen appreciation for your competence and effectiveness, and for your personal warmth. We share vicariously the pleasure and satisfaction that I am sure you feel." The letter was signed "Bill."[99]

What made the appointment of Fallon surprising among Kodak observers is that the board bypassed Gerald Zornow, the marketing executive who had been serving as president under Louis Eilers. Most Kodak watchers had considered Zornow the heir apparent for the top job, having received substantial credit for the tremendous success of Kodak's Instamatic camera and film system. Yet in a move some at Kodak called "the Zornow affair," Fallon's appointment puzzled many Kodak followers.[100] Perhaps the decision was one of personality and style more than substance. Kodak was well known for being an extremely conservative organization, patterned after its "exceedingly straight-laced bachelor" founder, George

Eastman. For example, during this period, Kodak executives were prohibited from including alcoholic beverages on their expense reports, and if a secretary was asked to stay past five p.m., a chaperone had to be called in. Fallon was considered "a typical Kodak product . . . [who] avoids the limelight and lives in a modest colonial house in a Rochester suburb."[101] In contrast, Zornow, although described as "brilliant," was considered "a veritable hippie" by Kodak standards because of his penchant for wearing loud shirts and ties and his use of colorful language.[102]

More likely, Fallon's appointment signaled an apparent belief among board members that the challenges facing the company were centered not so much in marketing, where it continued to dominate its competitors, but in the area of research and development. For the most part, scientists or scholars had always run the company, and a sentiment likely existed that this was not the time to change that tradition. With respect to the field of one-step photography, Kodak's effort to develop a competing system was disappointing, if not embarrassing, especially given its research prowess. The Kodak Board had apparently concluded that the company needed a scientist, not a marketer, to lead it.

What was not surprising about the Kodak management shuffle was that the board kept control of the company tightly within the grasp of insiders. Fallon personified the "Kodak man." Hailing from upstate New York, he had spent his entire professional life working within the insular environment of Kodak's Rochester research facilities and corporate headquarters. Kodak had long been the predominant economic force in Rochester. By the early 1970s, about one out of every six members of the Rochester community was on its payroll. There was little that the company could do wrong in its hometown. Kodak was the king of its domain. It treated its employees well, "dousing . . . [them] with liberal fringe benefits," according to the *Wall Street Journal*.[103] In the process, Kodak was able to keep unions away. On another level, "to the myriad small companies that look to Kodak for a living, it is 'the great yellow father,'" noted the *Journal*.[104] Operating in this kind of environment, isolated to a great extent from the mainstream of the business world, Kodak and its executives gained a certain bravado and feeling of infallibility. *Fortune* once characterized the "proprietary" perspective from which Kodak operated as its "lofty pinnacle above the fray."[105]

A burly man of fifty-three, Fallon had a reputation within the company of being hard-nosed and authoritarian, exuding that confidence and air of Kodak superiority and inevitability of market dominance that had long characterized its management.[106] While his wife had described the Fallons as "just average-living people," Walter, in fact, brought some changes to the corporate lifestyle of Kodak's chief executive that perhaps only further isolated him from reality, and the rest of his company. A private dining room was built so that he didn't have to eat with other employees, and a private terminal was created for Kodak's corporate jets. When he flew, Fallon apparently insisted that a bulkhead be installed to physically separate him from anyone else flying on the plane.[107] It would remain to be seen what impact, if any, Fallon's attitude and management style would have on the key decisions he would confront in the months and years to come.

Clearly, high up on Fallon's agenda was instant photography. Kodak had to respond to Polaroid's breakthrough. Characterized by one of his top lieutenants as "a very decisive and aggressive chap," Fallon, like Eilers before him, gave the impression to others that he was not enamored of Land and that he had made up his mind to find a way to compete with him on his home turf.[108] Having been involved with Kodak's photography divisions, Fallon was already fully familiar with the state of affairs in its P-130 project. Following his appointment, he agreed with others in Kodak management that its program had to be completely redefined. As Fallon later recalled, when he and his colleagues "looked at our camera versus SX-70 . . . [we] decided that it was too large and bulky . . . so . . . we should go back to the drawing boards."[109] In a meeting with his P-130 team managers on May 31, Sieg broke the news of Fallon's decision and advised them: "In view of what is expected to be in the market place . . . the P-130 program as earlier defined is no longer desirable."[110] Fallon had decreed that it was time for Kodak to move its whole program in a new direction, and it set about doing exactly that.

Given these developments, Kodak's efforts to compete with Polaroid in instant photography became a sensitive subject of interest around Rochester. A reporter from the *Wall Street Journal* found that of the 175 buildings at Kodak's manufacturing complex, the building housing its instant photography research effort was "a bit less anonymous than the rest" because of the guards posted at the door and the "sign that warns: 'No Visitors.'"[111] Leonard Sloan, a noted journalist on photography from the *New York Times*, was sent to Rochester that summer of 1972 to investigate the matter. He quickly found that it was "a virtual 'no-no' when corporate officials at the Eastman Kodak Company talk to outsiders [to discuss] . . . Kodak's forthcoming entry into the field of instant photography as a challenger to the company that created the field in 1948 and monopolizes it to this day—the Polaroid Corporation."[112] But when pressed publicly, Kodak did not deny its continuing determination to enter the instant photography market. Fallon acknowledged to Sloan that the company was still pursuing a double strategy. On the one hand, it was "interested in a [peel-apart] product that would fit existing non-Kodak [that is, Polaroid] cameras." With Polaroid's recently demonstrated Aladdin product in mind, Fallon also admitted to being "interested in a Kodak camera for in-camera processing."[113]

Analysts found confirmation of Kodak's resolve in the company's financial statements. It was noted that despite cutbacks in many spending areas, Kodak's research budget over the past five years had increased by fifty-nine percent to $188 million. Maybe not coincidentally, that period dated back to Land's 1968 revelation to Kodak of Polaroid's move toward a new system of one-step photography. No one missed the fact that Kodak was making a gigantic effort to enter the field. Instant photography had evolved, as Sloane described it, "from a concept that perhaps many at Kodak . . . [and others] originally considered a toy, [to] become a key factor in the business with a potential that boggles many imaginations."[114] But frustration mounted as Kodak continued to be unsuccessful despite expending huge amounts of research and development capital.

Meanwhile, everything was not running so smoothly at Polaroid. It became clear over the summer of 1972 that a full national commercial introduction of the Aladdin camera and film system was not going to be possible by Christmas. The newly built Polaroid manufacturing operation in Norwood was unable to turn out enough cameras in time. Land faced the expected barrage of second-guessers from some of his Polaroid executives—those who had previously argued the imprudence of prematurely announcing Aladdin. But, as usual, he plunged ahead nonetheless. Rather than cancel the release of Aladdin completely, Polaroid announced in early September that the national introduction was going to be scrapped in favor of a regional distribution plan, starting in southern Florida.[115]

To set up the Florida introduction, Polaroid dispatched public relations personnel in early October 1972 to select national news publications to provide private demonstrations of what the *Wall Street Journal* dubbed "the much-ballyhooed but still mysterious Polaroid instant-picture color camera."[116] Land personally gave presentations in his office to key journalists, including Robert Reinhold of the *New York Times*.[117] A yearlong television advertising campaign that featured legendary British actor Sir

Laurence Olivier, estimated to cost $20 million, was created and scheduled. At some point Land dropped the name Aladdin in favor of SX-70, which had long ago been his code for the secret program within Polaroid to develop the very earliest one-step photography system. The new SX-70 system represented the fulfillment of the dream that had started Land and Polaroid down this road, and so the name was apt. Some believed that Land had abandoned Aladdin because, despite the explanation involving the genie that he had provided at the 1971 shareholders meeting, he was concerned that people would dissect the name into "a la din."[118] "Din" was the nickname used for Land by his closest friends and colleagues.

With anticipation having built to a high pitch, Polaroid was finally ready to release SX-70 to the public. By this point, it was referred to in the media as the "most highly publicized camera ever made."[119] To trumpet the retail launch of his "dream" system, Land once again orchestrated a big and dramatic event, gathering 1,200 camera dealers at the Fontainebleau Hotel in Miami Beach on October 26, 1972. Searchlights crossed the Miami evening sky while an airplane trailed an SX-70 banner along the beach. Although Land himself did not attend the Miami event, he had written the entire script for it and declared, in a *Time* cover story, that SX-70 could "have the same impact as the telephone on the way people live."[120]

The reaction was everything he could have hoped for. While there was some concern among industry commentators about the system's relatively high retail price, the brilliance of the technology embodied in the camera and the film was undeniable. *Life*'s description of SX-70 as "both a marvelous toy and a stunning technological achievement"[121] was typical of the press plaudits. The SX-70 system was indeed big news and, together with Land, it received the massive media coverage they had sought, including the covers of both *Time* and *Life*.[122] (See Figs. 8-2 and 8-3.)

The public release of SX-70 only intensified the misery at Kodak. With SX-70 finally available in some stores immediately after the Miami launch, Kodak engineers obtained a large quantity of Polaroid products and rushed them back to Rochester for analysis. Though Kodak was dismayed by the formidable challenge posed by the new system, it was even more determined, under Fallon's leadership, to retreat from its current position well down the wrong road toward its goal of competing with Polaroid.

First, it decided to scrap the P-129 project to develop a color film that could be used in Polaroid's existing peel-apart cameras. Up to this point, Kodak had already spent $94 million on this effort.[123] As recently as its April 1972 shareholders meeting, then chief executive Gerald Zornow had proclaimed that Kodak "definitely" planned to sell the peel-apart film.[124] But it was apparent to Fallon and his colleagues that, in light of the impressive integral SX-70 system, peel-apart technology was not going to be the answer.[125] In a November 6 memo to Kodak management, a senior researcher noted that if Kodak continued with its peel-apart film, "We would be introducing an essentially obsolete product, in a dying phase of . . . [the instant photography] market with little or no possibility of recouping the additional capital required to achieve a marketable . . . [peel-apart] product."[126] Kodak would instead "concentrate its efforts" on coming up with products to compete directly with SX-70.

On November 17, 1972, Kodak went public with its decision, announcing "that it had changed its mind and no longer planned to market its own self-developing [peel-apart] film for use in Polaroid Corporation cameras."[127] The *Wall Street Journal* termed the decision "an intriguing and unexpected turn" in the "fast-growing instant photography market."[128] It reported that "Kodak's about-face left . . . analysts generally confused . . . [and] institutional investors . . . bewildered."[129] Initially, Fallon tried to put a positive spin on the development by publicly assigning at least part of the rationale to anticipated demand for Kodak's new and smaller version of the very successful 1963 Instamatic. Other Kodak executives insisted that there were technical issues relating to the P-129 film—the processing composition stained anything it inadvertently touched.[130]

But these ploys did not fool observers, one journalist wondering whether "this decision didn't simply represent Kodak's acknowledgment that it had been outflanked by the competition," namely Polaroid's release of SX-70.[131] Within a couple of days, Fallon conceded that Kodak was "unwilling to divert further effort and funds from the development of our own instant system into a secondary and more limited marketing opportunity."[132] Yet Kodak executives stubbornly denied any connection to Polaroid's introduction: "We don't respond to outside effects like the SX-70," claimed Colby Chandler, Kodak's executive vice president of its U.S. and Canadian photographic divisions. "We watch the competition for information, but we don't respond to it."[133]

In addition to abandoning the P-129 project, Kodak recognized that it had to change course radically for its development program to have any chance of producing a product that would be competitive with the Polaroid system, even if that product were not released until 1975. But this was a formidable undertaking. In yet another review of the situation held at the beginning of 1973, Kodak researchers and executives despaired in awe of Polaroid's SX-70, noting that Polaroid had "set the standard and level of expectation for all future products of a similar type."[134] By this time, Kodak's study of Polaroid's integral instant system had resulted in the acquisition of seventy-one SX-70 cameras and more than 3,000 film units.[135] An internal Kodak report acknowledged that despite whatever efforts Kodak could reasonably expect to muster, it was likely that the best it could expect to produce would be a "me too" system "no more than equal to" and "in some ways . . . less than equal to" Polaroid's products.[136] Kodak's marketing executives were forced to admit the obvious: "We see no unique consumer benefits in the proposed Kodak program at this time."[137]

Nonetheless, Fallon was determined. A new management directive dictated that research efforts had to continue because of "the desirability of participating in the field of instant photography."[138] It was now four years too late for Kodak to participate by accepting Polaroid's invitation for it to serve as the supplier of SX-70 negative. Yet, in its view, it certainly was no longer an option for Kodak to leave Polaroid alone on its increasingly popular and profitable turf. When questioned at the April 1973 Kodak shareholders meeting, Fallon signaled the company's persistence, stating that it was still progressing "toward a viable Kodak approach to instant photography . . . [in a] mechanism which will deliver dry prints of high Kodak quality without waste and at a price the mass market can afford."[139]

Behind the scenes, as work at Kodak continued through 1973, the magnitude of the challenge to out-engineer Land and his Polaroid colleagues in the field they had created and pioneered for decades created a sense of desperation—between 1,300 and 1,400 researchers labored on its P-130 project.[140] In September 1973, Kodak formed yet another new group, the Kodak Park Development Committee (KPDC), to oversee the work being done in ten key areas. At the initial KPDC meeting, researchers were encouraged to try potential solutions even if they had serious doubts about them. "In major questionable areas, it was suggested that a letter stating reasons for not continuing a development concept be issued and decisions reached at a higher council."[141]

But that same committee issued another directive, one that foreshadowed what was to become one of the most important legal battles over technology in the history of the United States. As observed many years later by industry commentators, Kodak, feeling "hemmed in by Polaroid's vast portfolio of patents," had indeed "panicked."[142] In apparent desperation, a

KPDC memo directed Kodak engineers to "not be constrained by what an individual feels is a potential patent infringement."[143] Although the memo did go on to direct the researchers to "consult" the patent department in such instances, the excerpted passage would later become the signal call for Polaroid and its legal team for years to come—all the way up to the U.S. Supreme Court.

CHAPTER 9

THE COMING STORM

Even though Polaroid experienced some bugs in the SX-70 as 1973 began, as Bill McCune later explained, the introduction continued because "we felt it was important to get the product on the market as rapidly as possible so that we would learn from the customer's use of it and from the marketing aspects and customer service."[1] At the Polaroid annual shareholders meeting on April 24, 1973, Land reluctantly admitted that due to the slow pace of manufacture, the full national distribution of SX-70 was still but a goal for the fourth quarter of that year. Yet, with his characteristic indefatigability, Land's attitude was, "what's six months' delay in the course of a revolution?"[2] Despite the popular consensus that its new instant photography system was a technological marvel with unlimited commercial potential, these issues began to have an adverse impact on the perception of Polaroid in the financial community.

Throughout this period, the costs for working out the persisting issues continued to build and unavoidably affected Polaroid's bottom line. At a presentation for analysts held in October 1973 at the Science Museum in Cambridge, Massachusetts, that was meant to inspire confidence in the company, Polaroid executives were forced to admit that the company's earnings for the year would fall below previous estimates.[3] The company's stock fell more than ten percent the next day.[4] By the end of the year, the stock had lost more than fifty percent of its value from the highs achieved during the initial introduction of SX-70.[5] Some of this was attributable to a generally difficult economic climate that saw the stock market in decline across the board. But there was no doubt that the sheen on Polaroid in the Wall Street community, initially created by the dramatic launch of SX-70, was wearing off in light of the logistical problems it had met in bringing its revolutionary product to market.

Land refused to accept any blame for the falling stock price. When a journalist suggested that perhaps he had contributed to an inflated valuation of Polaroid immediately following the SX-70 announcement because of his "ebullience," Land refused to accept any culpability, suggesting instead that the analysts had created their own dilemma. "I agree that their disappointments come from exaggerated expectations. But they're theirs."[6] In his view, the technical problems the company experienced in getting the new system to the marketplace were merely to be expected. "Anything we put out is new enough, different enough, daring enough, that you plan on a family of technical problems out in the market," Land insisted.[7]

Business Week recounted how the SX-70 project represented the full transition of Polaroid's manufacturing philosophy. In the early days, Polaroid had outsourced as much as possible the manufacture of its products, particularly its cameras. Now, as Land described it, Polaroid was engaged in the "high technological drama" of manufacturing almost everything itself.[8] At least publicly, Land asserted that Polaroid was pursuing this approach "because these cameras and film are technologically unique, involving new science at each point of manufacture." A senior Polaroid executive went even further, according to *Business Week*, comparing "the complexity of Polaroid's project with the Apollo moon shot."[9] A report in *Fortune* seemed to concur: "However the venture turns out commercially, the mere production of the SX-70 must already be counted as one of the most remarkable accomplishments in industrial history."[10]

While personally disappointed that Polaroid's manufacturing facility was still turning out only 5,000 cameras a day, only half of what he had hoped for by this time, Land went on to point out that the delays would eventually end and that the company's investment to date was sound, in terms of both the source of the funds and its expected return. The vast majority of the development capital for SX-70 had been taken from Polaroid earnings, and a large portion of the expenditures had created tangible assets in the form of manufacturing plants that were, as Land characterized them, "three times as valuable as when they were created."[11]

Ever the optimist, Land remained confident as his team continued to work through the issues in camera and film manufacturing and brought more and more SX-70 systems to the market every day. Internally, Polaroid predicted sales of six million cameras over the next few years, and while the media chronicled the company's problems, the brilliance and historic nature of SX-70 remained undeniable. Peter Wensberg, a Polaroid senior vice president who ran the marketing operation behind SX-70, predicted that the company would reach the break-even point on the SX-70 camera and film system during the first quarter of 1974.[12] Indeed, there was every indication that Polaroid had a successful product that consumers embraced despite its relatively high price. The 461,000 cameras produced in 1973 had been virtually all snapped up by an eager public by Christmas. It was clear that demand was exceeding supply.

During this problematic period, some Polaroid executives urged Land to focus on the introduction of a less expensive camera that would use its SX-70 film.[13] By expanding its product line to the mass consumer, Polaroid could broaden the instant photography phenomenon and increase its profits through the sale of more cameras, as well as from the film that they would use. It would also beat Kodak to the punch, since it had always been Kodak's stated philosophy and intention to direct its products, including its instant photography system, to the average consumer. But Land had little interest in such a system, preferring the elegance of the leather-encased, fully automated folding SX-70 camera made of metal, not plastic.[14] Although some development work was conducted at Polaroid toward a more economical model, Land did not consider it a priority at this juncture. The effort received neither his attention nor his enthusiasm, and languished accordingly. Analysts and journalists sought answers from Polaroid executives about why the company had not made this seemingly obvious move, but they received no response.[15] The best Polaroid could or would do in terms of the low-end market was to announce a new camera, The Zip, that would take small black-and-white pictures using a peel-apart film.[16]

Despite these internal issues, it was clearly the continued determination by Kodak to perfect an instant photography system of its own that posed the long-term danger to Polaroid's survival. In November 1973, Kodak chairman Gerald Zornow reassured an analyst's meeting of his company's commitment: "We know where we're going and we know how to get there."[17] Zornow also stressed that Kodak's camera would retail "for a price the mass market can afford," a clear knock against SX-70's list price of $180. Another camera company, Berkey Photo, had also announced its intention to introduce an instant camera that would use Polaroid's film. But Berkey, as a company and a competitor, did not strike fear into the heart of the Polaroid community. It was a relatively small company, founded in the early 1920s by its eponymous leader Ben Berkey, and was still run as a family business.[18] Only Kodak had the technological and marketing capability to swamp Polaroid.

The more successful the SX-70 became, the bigger the potential of the instant photography field grew and thus the more certain and imminent the challenge from Kodak became. Some Polaroid executives saw a silver lining in that challenge; they recognized the possibility that the entire market for instant products, and for photography in general, might actually be expanded by Kodak's entry.[19] While such hopeful views were speculative, the reality was that Polaroid had never gone head to head against Kodak, or any other competitor for that matter. The marketplace would be the ultimate field of battle, and the place where Polaroid's products would have to prove their superiority. Kodak had for years trumpeted the revolutionary achievements of the researchers in its instant photography development program. Land and his Polaroid colleagues did not underestimate the ability or the potential of Kodak's scientists, many of whom they had worked with for decades. Kodak was perfectly capable of coming out with an instant system that would leapfrog over Polaroid's technology. This would be the ultimate worst-case scenario.

In other quarters within Polaroid, however, executives believed that the struggle with Kodak would take place in the courts, not in the marketplace, and that the Polaroid patent portfolio, rather than its cameras and film, was in Kodak's crosshairs. The first clues came from abroad. When an international technology company like Polaroid decides to secure patent protection for a development, it will generally seek worldwide protection. This means filing corresponding applications on the same invention in most major foreign countries in which it does or may do business. Kodak had begun a campaign to challenge certain Polaroid patents in foreign countries where either the local patent system allowed for third parties to dispute an application while it was still under consideration or where the patent process worked faster than in the U.S. and thus patents had already issued and could be attacked by seeking revocation. As these Polaroid executives saw it, there was every chance that the particular patents and applications Kodak chose to target abroad provided clues about the instant system it was intending to release. At the very least, Kodak's foreign patent activities suggested it still believed that certain Polaroid patents stood in its way.

Polaroid's portfolio of patents covering the instant photography field was undeniably extensive. When addressing technical challenges along the development trail, Polaroid scientists often came up with several alternative solutions to a given problem or for a given feature of the camera or film unit. If these solutions embodied patentable inventions, Polaroid filed applications for patents on them, whether or not the particular invention found its way into Polaroid's commercial products. Land adopted this strategy from the very earliest days of the company; the lawyers in Polaroid's patent department dutifully carried out the practice under a system that remained a legacy of Donald Brown's early guidance. The breadth and intensity of Polaroid's research over the years thus resulted in a correspondingly comprehensive collection of patents covering both the technology embodied in its commercial SX-70 system and alternative approaches to many of the key camera and film features. Polaroid regarded this as a legitimate intellectual property strategy for securing the fruits of its labors.

Kodak, however, believed that Polaroid was imposing a patent blockade that effectively prevented any competitor from developing an instant photography system, a blockade that it might have to blast its way through. Given the developments outside the United States, it was beginning to appear that Kodak might indeed be initiating a strategy to do just that. The great irony in Kodak's position was that it had essentially followed the same patent strategy with regard to its own research and development efforts. As one student of patent law and history has pointed out, "More than sixty years before Edwin Land applied for his first patent, George Eastman applied in the United States and Europe for patents on his own pathbreaking photographic inventions, ultimately erecting an intellectual property fortress as impregnable in its domain as was Polaroid's."[20] In fact, as of 1975, Kodak had over 6,000 unexpired patents, all of which it considered valid, and all of which it expected others to respect.[21]

The stage was set for an inevitable legal battle. Land and his fellow executives at Polaroid recognized this, and, by the end of 1973, they knew that it was time to retain outside legal counsel. Julius Silver, although based in New York, was at this point general counsel to Polaroid and a member of its board of directors. As his longtime friend, confidant, and legal representative, Silver remained the one individual Land most trusted and listened to.[22] In late 1973, Robert Peck was Polaroid's in-house patent counsel. He had worked at Polaroid since June 1959. While Peck's department continued to handle most of Polaroid's patent application work, Polaroid also used an outside firm, Boston-based Fish & Richardson, to handle some of its more difficult patent application ("prosecution") matters.

William Rymer, professor of patent law at Harvard University, was the Fish & Richardson partner in charge of Polaroid matters. Rymer had first worked with Polaroid in 1963, handling a prosecution litigation involving one of its patents in the polarized lens field. Donald Brown and his protégé, Charles Mikulka, had supervised Rymer's efforts on that case and had praised him "for doing a very good job."[23] Mikulka remained at Polaroid as a senior vice president and a member of its executive committee. He was, in effect, Polaroid's chief in-house legal officer, leaving the day-to-day running of the patent department to Peck. Now, a decade later, it was to Rymer that Silver, Mikulka, and Peck turned for help in finding counsel sufficiently qualified and experienced to handle what could possibly become a life-or-death litigation for the company. Although the Fish & Richardson firm had a long history in the patent world, Rymer knew that the coming legal battle was likely going to be more than it could handle, given its specialty in the prosecution, rather than the litigation, end of the business. As a result, he turned immediately to Fish & Neave, a New York litigation powerhouse that shared a common history with his firm.

The firm of Fish, Richardson, & Neave had been leading lawyers in the protection of intellectual property since the industrial and technological explosion in nineteenth-century America. It had served as patent counsel for the Wright brothers, Alexander Graham Bell, and Thomas Edison. The firm's origin is traceable to 1878, when Frederick P. Fish, the son of a New Bedford, Massachusetts, sea captain, was admitted to the Massachusetts Bar and opened a Boston office. Fish was the patent lawyer who appeared before the U.S. Supreme Court to defend Edison's patent on his first electric light bulb. He was later instrumental in the 1893 merger of Edison's company with another of his clients, the Thomson-Houston Company, to form the General Electric Company. During this period, Fish handled patent matters for Bell and his invention of the telephone, winning over 600 suits for infringement in the decade 1883–1893. He also served as president of AT&T from 1901 until 1907 when he returned to the firm.[24]

Fish was perhaps best known for having prosecuted the famous "wing-warping" patent infringement case on behalf of the Wright brothers against archrival Glenn Curtiss and his company.[25] He was brought in to handle the case after an initial setback and was ultimately able to persuade the courts that the Wright's 1906 patent on their "Flying Machine" was valid and had been infringed. Ironically, it was Fish to whom Wilbur Wright had written from the office above the bicycle shop he shared with his brother, complaining about the pace of the litigation. This letter was written shortly before Wilbur died of typhoid fever in May 1912.[26]

To extend their roots from Boston, in 1895 Fish had sent his partner, Charles Neave, to New York to work with some of the great companies developing there, including AT&T and General Electric. Mirroring Fish's involvement in the genesis of General Electric, Neave played a pivotal role in the formation of another American technology leader. Following the end of World War I in 1918, several U.S. companies were busy litigating patent rights for the fledgling radio industry while European companies took the lead in the developing technology. The government suggested that the American companies join forces in order to better compete. Adopting this idea, Neave was instrumental in getting several key U.S. competitors to pool their talents and patent property into a new company, known as Radio Corporation of America (RCA), under the leadership of David Sarnoff.[27]

Neave went on to become one of the great litigators of his era, working with the giants of America's industrial age, including Allied Chemical, DuPont, U.S. Steel, Bethlehem Steel, General Motors, and B.F. Goodrich. In 1931, *Fortune* characterized Neave as "the recognized head" of the patent bar, and he received plaudits from U.S. Supreme Court Justice Oliver Wendell Holmes and esteemed federal judge Learned Hand.[28] Neave's firm continued as Fish Richardson & Neave with offices in New York and Boston until 1969 but eventually split into two separate firms because of client conflicts.[29] Fish & Neave, the New York branch, continued Neave's legacy as one of the most prestigious litigation firms in the patent area, litigating many notable patent disputes, including cases involving Marconi's radio, Xerox's document copiers, and Gillette's safety razors.

In fact, Neave's reputation had long ago made a significant impression upon the Polaroid legal brain trust. In 1963, Rymer shared with Donald Brown a copy of Learned Hand's tribute to Neave, written in 1938 following Neave's death on September 10, 1937. Hand had written:

> With the courage which only comes of justified self-confidence, he dared to rest his case upon its strongest point, and so avoided that appearance of weakness and uncertainty which comes of a clutter of arguments. Few lawyers are willing to do this; it is the mark of the most distinguished talent.[30]

Brown told Rymer at the time that he was "very interested in Judge Hand's eulogy of Mr. Neave. This is an exceptional statement for any judge to make with respect to the ability of counsel appearing before him, and for Learned Hand to say it makes it even more striking."[31] Although Brown was no longer alive when the Kodak threat materialized, the time had finally come for Polaroid to avail itself of the law firm that had evolved from Neave's legacy. In early December 1973, Rymer contacted Rynn Berry, a senior partner at Fish & Neave, to ascertain whether the firm was free to represent Polaroid in a possible action for infringement of its patents against Kodak and other potential competitors to the company's "picture in a minute or less" system.[32] After confirming internally that Fish & Neave had no client conflicts of interest that would prevent it from representing Polaroid in litigation against Kodak, Berry responded in the affirmative on December 12, and Polaroid had its litigation counsel. It was not a moment too soon.

Within weeks, the potential threat to Polaroid truly began to materialize. In March of 1974, Kodak announced in its annual report that it had finally come up with a "feasible" film of its own that would challenge Polaroid.[33] The company declared: "The basic decisions have been reached concerning the chemical and physical configurations within which Kodak products will be realized."[34] It also announced that "productive work had made feasible a film that will yield dry prints of high quality without waste, to be used in equipment priced for a wide spectrum of consumers."[35] At its annual shareholders meeting in April, Kodak's leader, Walter Fallon, left no doubt about Kodak's commitment to develop its own products for instant photography, declaring: "We intend to commit to this program whatever it requires."[36] Since Kodak had been short of specifics about the camera that would use its film, industry observers speculated that Kodak's entry into the instant photography market was still years away.[37] Rymer and Berry watched the developments with interest.[38]

Polaroid stock was battered by the news nonetheless.[39] The *New York Times* reported that Polaroid's "sharp price plunge" occurred because of "the critical market judgment—what's good for Kodak is bad for Polaroid."[40] At the 1974 Polaroid annual shareholders meeting, Land did his best to reignite excitement over SX-70. He suggested that the system's full potential had not nearly been reached and remained to be discovered as the new technology matured and was used more and more by the public. "What good is a new invention whose potential is not yet realized?" he asked rhetorically. "What good is a new-born baby?"[41] When a special presentation was made to Land of the one millionth SX-70 camera to come off the production line, Land was moved to tears. "The satisfaction in doing a job comes from the great joy we get in doing it," he explained. "Such a joy, that it brings tears to our eyes."[42]

Analysts, however, were concerned about Kodak's forthcoming challenge, as well as the continuing issues with SX-70's introduction. Though

Polaroid characterized its sales in the first quarter of 1974, up from $135.3 million to $146.7 million, as "excellent,"[43] concern remained about the high price of the system. There was still no more-affordable Polaroid camera model in sight. Also, a new technical issue, destined to dampen analyst enthusiasm, crept into the picture. Polaroid had given an outside supplier the task of developing a battery for the film pack that would last a year, but consumers were discovering that the actual shelf life of those batteries was closer to five months. Thus, unless a pack was sold and used within that time frame, it might prove to be useless when loaded into the camera. This had the potential to cause significant returns of useless film packs with the attendant customer dissatisfaction. As soon as this problem surfaced, Polaroid took steps to address what McCune later characterized as "one of our worst stumbling blocks."[44] In the short term Polaroid's distribution team worked to keep fresh film stocks on store shelves,[45] while long term it "reluctantly" took over the manufacturing of batteries itself.[46]

This battery issue increased the persisting downward pressure on Polaroid's profit picture that lingered from the initial SX-70 production problems and was intensified by a weakening U.S. economy. The result was a dramatic decrease in Polaroid's net profits for the second quarter of 1974, despite the rising sales of its SX-70 line.[47] The cumulative effect of these issues was that Polaroid was not benefiting as much as it could have from the successful introduction of its new system. At a time when it should have been reaping tremendous economic benefit so as to buttress itself against Kodak's eventual entry into the market, Polaroid was appearing more vulnerable than insulated. Polaroid executives announced moves to address its bottom line, including "streamlining" its work force through attrition and even layoffs, but analysts were not convinced it would be enough.[48] Even the perks of top management were affected: Polaroid canceled its contract with a private jet aviation company so its executives could save the company money by flying commercial airlines.[49] Land continued to insist that these issues were only the expected birth pains of a truly revolutionary product. But Polaroid found itself in a precarious position. Some observers, former believers in Polaroid as a company and as an investment, were starting to jump ship, fearful that, as expressed by analyst Louis Rusitzky, who had followed the company since 1957, "the impact on Polaroid could be disastrous when Kodak comes out with its instant film system."[50]

Land and his colleagues still feared that Kodak might introduce a "next generation" instant system against which Polaroid might not be able to compete. Developments on the legal front, however, continued to suggest that Kodak's approach would be to go *through* Polaroid's technology instead of around, over, or beyond it. The legal initiatives abroad continued, and even began to shed some light on the possible technical direction of Kodak's system. For example, in early 1974, it was reported that a dormant legal proceeding Kodak had brought in 1972 in the United Kingdom to have one of Polaroid's key instant photography patents revoked had "suddenly . . . been activated."[51] The patent covered Howard Rogers' "Excedrin" film format, the alternative film unit structure to the "Aspirin" format that Land had invented and that Polaroid was using in its SX-70 film. Pressed for an explanation, Kodak merely insisted that the proceeding had "no effect on previously announced plans to make and market films and equipment for instant photography," Kodak intending only to "challenge what we consider invalid patents that could have an undefined effect at some point in time."[52] In reality, this was Kodak's first shot across Polaroid's bow in its attempt to break the patent blockade—and give its own, potentially infringing, products clear sailing into the market.

In terms of legal strategy, Berry advised Rymer that there were important considerations to keep in mind while defending the U.K. patent revocation action. In the long run, if Polaroid were going to bring a lawsuit against Kodak for patent infringement, it would be to its great advantage if it could have control over the timing and locale of the action. Litigating the matter in a Boston courtroom, as opposed to one in Kodak's hometown of Rochester, would be a major benefit to Polaroid, giving it the legal equivalent of a "home court" advantage. The fear Berry expressed to Rymer was that Polaroid might do something in the United Kingdom that would give Kodak grounds to bring its own anticipatory lawsuit against Polaroid in the United States. Such a case would seek to have a court declare Polaroid's relevant patents invalid prior to Kodak's releasing its own instant photography system.[53]

Grounds for Kodak to bring this kind of action, known as a declaratory judgment action, could be created inadvertently if, in the United Kingdom, Polaroid asserted that Kodak's forthcoming products infringed or were likely to infringe Polaroid's corresponding U.S. patents. Such a pronouncement might create the "justiciable controversy" that is the prerequisite for a declaratory judgment action. A Polaroid accusation, even one made abroad, might very well put Kodak in a position to claim that it was under the threat of being sued in the United States and thus entitled to a declaration that a given patent or patents were invalid before it risked the capital necessary to bring its allegedly infringing product to market. Under this scenario, it would be Kodak who could choose the jurisdiction it thought most favorable and, in all likelihood, the entire litigation with Polaroid would end up taking place there.

As the year progressed, with the certainty of a major legal battle against Kodak growing, the time had come to find the right man to lead the litigation team. Berry was the senior partner at Fish & Neave, but he had expressed reluctance, both to other members of the firm and to Polaroid, to handle the litigation.[54] By now, Berry knew that, within Polaroid management, the executive considered to have real power with respect to the company's legal strategy, and a direct channel to Land, was Charles Mikulka, by this time a member of the Management Executive Committee. The MEC, as it was known, operated within Polaroid as an advisory board to the corporation's president and chairman of the board. On July 24, 1974, Berry received a call from Rymer advising him that Mikulka wanted to have a meeting to discuss the litigation situation, and that he wanted to talk to William K. Kerr, another Fish & Neave senior partner.[55] Given that this was Mikulka's first active involvement in the preparations for the litigation, it is likely that he was, at least in part, doing Land's and Silver's bidding.

If what Mikulka and Land wanted was a tiger to head Polaroid's litigation team, Bill Kerr was the logical choice. The contrast between him and Berry could not have been starker. Berry was a mild-mannered academic. Kerr was a tough, no-nonsense scrappy street fighter, born in rural Wilkinsburg, Pennsylvania, in 1916. Unlike the big-city upbringing of many of his partners, he grew up on the rural campus of Colgate University in upstate New York, where his father, Andy Kerr, was the longtime football coach. Colgate's Kerr Stadium is a testament to his father's historic tenure at the school. Kerr attended Colgate as an undergraduate, unlike most of his partners who had Ivy League backgrounds. Tough and determined, he worked his way into Harvard Law School and then served during World War II as an FBI agent and as a Naval Intelligence Officer in the Pacific before joining Fish & Neave in 1947.[56]

With a masterful command of the English language, and a clear and concise writing style, Kerr quickly became one of the rising stars at the firm. Not physically large in stature, his carefully slicked red hair, perfectly tailored English suits, and cuff-linked dress shirts helped him perfect the role of the distinguished Park Avenue lawyer. Ruddy-faced, with a confident, tough, steely demeanor, and a straightforward approach, he was a force on the litigation battlefield as well as within the firm, where he became the managing partner. He was clearly the right man to lead the Polaroid litigation, and it appears that Rymer had so advised Mikulka. In a meeting on August 8 at Polaroid's offices in Cambridge's Technology Square, Berry and Kerr met with Mikulka, Rymer, and Peck.[57] From that day on, Kerr assumed command.

Kerr's first task was to help Mikulka and Peck reexamine all of the strategic options they had been considering with respect to Kodak and, to a much lesser extent, Berkey. While it seemed likely that Berkey would beat Kodak to market with a competing instant product, it was developing a camera that was designed to use Polaroid film. There was no threat of Berkey's introducing its own film. Although Polaroid was determined to keep the field of instant photography to itself, the threat from a low-cost Berkey camera, one that would surely pale in comparison technologically to the SX-70, was not a major one in terms of the marketplace. One could even argue that the economical Berkey camera would help Polaroid. It would fill the low-end market niche that Polaroid had not exploited and would increase sales of SX-70 film and generate profits.

But Polaroid was not ready to license Berkey and thus legitimize its entry into the market. First, despite Land's lack of enthusiasm for it, work continued at Polaroid toward its own mass-market camera. Even more importantly, Polaroid could not play favorites in licensing its technology without creating even more legal complications. If it licensed Berkey, it would surely have to license Kodak and others, and thus open up the entire field, something it definitely did not want to do. Accordingly, Kerr knew that when Berkey actually introduced its camera, some action would have to be taken to protect Polaroid's exclusivity in the market and to enforce its patent rights.

The main focus, though, had to remain on Kodak, the competitor with the capability to crush Polaroid—if it came out with the right system. If, however, Kodak's system was not the technological breakthrough that Land and others at Polaroid feared most and turned out to be something drawn from the kind of technology Polaroid had pioneered, then an infringement suit against Kodak would be unavoidable. Polaroid's counsel prepared accordingly and had to consider several interrelated strategic aspects of the developing legal situation. As Berry had previously advised, it was imperative that Polaroid avoid giving Kodak the critical advantage of time and place with respect to the litigation. But waiting to sue Kodak until its products actually hit the market, a necessity if it was to sue for infringement, also posed a threat to Polaroid's ability to try the case in

Boston. By delaying action as it waited for this perfect legal scenario to develop, Polaroid gave Kodak more time in which it could take some kind of preemptive legal action in a court of its choosing, most likely Rochester. As a result, there were some within the Polaroid in-house team who wondered whether Polaroid should act proactively and file a suit against Kodak for "threatened" infringement, based upon assumptions about the Kodak products gleaned from the overseas legal actions Kodak had commenced.[58]

Apart from those concerned about the patent aspects of the legal situation, there were some members of Polaroid management who wanted to consider using the antitrust laws to defend the company against Kodak's challenge.[59] Could Kodak's determination to enter the instant photography market to Polaroid's detriment be legally sustained as its attempt to maintain a virtual monopoly in the amateur photography market and thus be deemed the kind of anticompetitive behavior that might be considered an antitrust violation? This notion was complicated and fraught with problems. But it seemed to be of particular interest to Richard DeLima, a Polaroid vice president.

Kodak was already the subject of antitrust suits brought by smaller competitors in the photography industry, notably Berkey Photo and Bell & Howell. The thinking among Polaroid executives who subscribed to this view was that Kodak might be sensitive to the prospect of having yet another such action brought by Polaroid. A former Kodak executive, Cameron Foote, who had moved over to work at Polaroid, had advised his new colleagues of Kodak's extreme sensitivity on the subject. "Unless one has worked in marketing at Kodak, it is hard to realize just how serious a threat they consider anti-trust action . . . to be," Foote wrote in a memorandum that was circulated to Land, McCune, DeLima, Mikulka, and others in Polaroid's top management. "The company's rhetoric and actions are loaded with examples which I would be happy to discuss at length."[60]

Even if the antitrust threat did not discourage Kodak from entering the instant photography market, the thinking went, it might give Polaroid significant advantage in a negotiated settlement. Although keeping Kodak out of instant photography altogether was Polaroid's preference, there was always the chance of a compromise that included licensing Kodak to use Polaroid patents. It was something the company might someday elect or even be forced to do. Such an outcome did have some superficial appeal to Polaroid. It would avoid the uncertainties of litigation and the possibility that Polaroid, if it lost, might have its patent portfolio vitiated forever. That would leave the company in the disastrous position of having to watch Kodak and others come into its field without compensating Polaroid either in the form of patent license royalties or damages for infringement of its patents. There was good reason to want to avoid this scenario.

But others within Polaroid's patent department, as well as Kerr, were dubious about the antitrust approach. First, there was doubt about whether any of Kodak's actions or Polaroid's supposed grievances amounted to a legally sustainable antitrust cause of action. Second, it was hard to believe that the threat of an antitrust action would affect Kodak's plans or position, given the huge expenditure of time and money it had already made to get into the instant photography business. Third, given that Polaroid had steadfastly maintained a policy against licensing any other companies with respect to its core technology, there was a concern in some company quarters that antitrust complaints about Kodak's alleged attempts to monopolize the photography field in general could somehow be turned against Polaroid because of its actions to maintain its exclusive domain over integral instant photography in particular.

As the discussion over Polaroid's legal options continued through the fall of 1974, the significance of the exercise was underscored when Kodak unexpectedly asked for a meeting with Polaroid. The senior managements of the two companies met in Land's office on October 30, 1974.[61] Walter Fallon, Kodak's president, led a contingent down from Rochester at the suggestion of Bunny Hanson of Kodak's research department. It was to be what Fallon later characterized as a "courtesy call."[62] Whether he intended it or not, Fallon was there to deliver what was a chilling message to Polaroid. In addition to Land and Mikulka, also present for Polaroid was Bill McCune, who was then serving as executive vice president.

Fallon was apparently very straightforward in his presentation to Land and his associates. He told the Polaroid contingent that Kodak would be entering the instant photography market with a product based on a new chemistry it had developed and gave them a copy of a French patent on that chemistry issued to a Kodak chemist, Lee Fleckenstein.[63] Kodak believed that it had successfully avoided infringing any of Polaroid's patents.[64] Fallon maintained that there was nothing that Polaroid could do to stop this move. Although Fallon did not request a license under any specific Polaroid patents, he said that he was interested in exploring whether or not Polaroid had reconsidered its position on licensing Kodak under its integral instant photography patents.[65] Everyone in the room was fully aware of the two companies' history concerning this contentious issue. One of Fallon's colleagues added, perhaps facetiously, that Kodak would actually prefer a non-assertion agreement with respect to Polaroid's patents instead of a license, clearly conveying the message that Kodak did not attach a significant economic value to whatever license royalty might be negotiated. McCune asked Fallon to identify the patents Kodak had in mind but received no response.[66]

In effect, Kodak wanted a ticket into the instant photography market for something close to free. While Kodak might be willing to pay to avoid litigation, it was also clear that the amount was a nuisance value, as opposed to the significant sums that would be generated by a license under Polaroid patents, should Polaroid be willing to grant one.[67] It was a truly audacious posture that Kodak was taking. As one chronicler of Kodak's history has observed, "the company's success [through the 1960s and early 1970s] fueled an arrogance that, in hindsight, is truly remarkable."[68] There was no formal reply from Polaroid to Kodak's inquiries, either at the meeting or thereafter.[69] But the Kodak executives left with the clear impression that Polaroid was unwilling to make a deal along the lines that Kodak had in mind.[70] One can only wonder whether Fallon and his cohorts actually expected a different reaction or whether they had traveled to Cambridge merely to deliver the message that Kodak was coming, like it or not.

On November 15, 1974, a meeting was convened at Polaroid for Kerr to review the various strategic options under consideration with Mikulka, Peck, and DeLima.[71] At the end of the session, Kerr was introduced to two of Polaroid's key executives whom he had not yet met, Bill McCune and Thomas Wyman, a senior vice president on the marketing side of the business. Conspicuously absent from this critical meeting was Land himself, a curious circumstance given that he was Polaroid's chairman, chief executive officer, president, and director of research. This was not unexpected, at least by his Polaroid colleagues. In a practical sense, Mikulka and McCune were his stand-ins. Land could question them and secure whatever information he wanted about the legal developments. Land's focus, as always, remained in the lab. There was no doubt he wanted Polaroid's patent portfolio protected and expected a vigorous legal campaign to be waged against Kodak, or any other transgressors. But he was unwilling at this point to be distracted from his work to participate personally in the deliberations over the strategy Polaroid should adopt.

In fact, the Polaroid lawyers who knew him best confided that they had some doubt as to whether Land would ever play *any* active role in *any* aspect of the legal battle. Would he ever be willing to surrender his precious privacy to participate in the public arena of a lawsuit? Up to this point, he hadn't even seen fit to meet Polaroid's trial counsel. Could he be relied on to become a witness on behalf of his company and subject himself to examination in pretrial depositions or in a courtroom? Under the circumstances, the Polaroid legal team had to consider its strategic options, fully aware of the real possibility that the case would have to be conducted without the active help of the company's most famous figure. In fact, this was a possibility Kodak may have considered—and a factor that some believed played an active role in the development of its strategy.[72]

At the meeting, Kerr expressed his clear preference for filing an action for patent infringement immediately upon Kodak's release of its products. He asked that a study be commenced internally at Polaroid to select potential candidates for a lawsuit from Polaroid's patent portfolio based on whatever knowledge Polaroid might have about the expected Kodak products. He endorsed the idea of bringing infringement actions concurrently in other countries, including Canada, France, West Germany, Italy, and Japan, and agreed to help find suitable counsel for Polaroid in each of those jurisdictions. Finally, he agreed to at least look into the idea of bringing an action for threatened infringement.[73] Within weeks, Kerr decided not to employ this tactic, at least for the time being. He provided Peck, Mikulka, and DeLima with some legal research on the subject in early December and diplomatically advised "in epitome, the message which I get from the authorities is that an action for threatened infringement is permissible, but that the time has not arrived to commence such an action against Eastman."[74]

With a legal strategy beginning to take shape, problems continued for Polaroid in early 1975. The delays Polaroid had experienced in getting its SX-70 system into the marketplace had already taken a heavy toll on the company, in terms of morale and public perception as well as financial performance. The company's stock price languished in the mid-teens, precipitously down from its 1973 high of $143.50 at the time of the introduction of SX-70. Not all of the adversity was within Polaroid's control. The country was suffering from its worst recession since World War II. Richard Nixon had resigned the presidency in disgrace in July 1974, following Watergate. Gerald Ford, his successor, inherited "the worst inflation in the country's peacetime history, the highest interest rates in a century, the consequent severe slump in housing, sinking and utterly demoralized securities markets, a stagnant economy with large-scale unemployment in prospect and a worsening international trade and payments position," according to the *New York Times*.[75] In his January 1975 address to Congress, Ford admitted that "the state of the union is not good."[76] This economic environment, pessimists pointed out, was not conducive to finding lots of new buyers for what was essentially a luxury-priced camera.[77]

With Polaroid nearing the end of its fourth decade, Land undeniably remained the heart and soul of the company, a seemingly irreplaceable icon leading all of its research efforts. As noted by *Fortune*, "there are no carbon copies of Land. He combines the prophet and the promoter, the egghead and the executive, in a way that is unique."[78] Land himself could hardly contemplate the notion that a worthy successor might even exist. Once asked by a journalist "what qualities his successor should have," Land provided "a description of such a paragon of talent, intelligence, and virtue that even he paused, laughed, and said, 'We're making him down in the laboratory.'"[79] Yet, whispers began to be heard within Polaroid that perhaps the company might benefit from having someone else assume its day-to-day leadership. This view was also expressed in the media, most alarmingly when the *Wall Street Journal* quoted a Michigan State business professor, Eugene C. Jennings, as declaring that "Land is a genius, but the time has come for him to get the hell out. He's going the way of . . . other builders who stayed too long."[80]

A feeling was developing that the company had simply grown too big to be managed by one person with the kind of almost dictatorial control that Land exercised. Although Polaroid had many talented and able people in its executive ranks, Land had always made all key decisions. The technical and production problems with the introduction of SX-70, and Land's bulldog approach to solving them, often without the kind of conventional cost analysis one would expect, had, in the end, strained his relationship with company managers and even his board of directors. It was becoming apparent to all that there was a need for some kind of change in Polaroid's management.

Land, apparently, got the message. On January 22, 1975, it was announced that Land had nominated his longtime comrade, Bill McCune, to the position of president and chief operating officer.[81] (See Fig. 9-1.) The news "sent shock waves through Polaroid's ranks."[82] Although Land did not speak directly to the press, he rationalized the change: "In view of the company's size and complexity, Mr. McCune's assumption of additional operating responsibilities will enable me to devote myself more fully than ever to those areas of technological innovation in which the company has made major contributions."[83] Land's spin on the development was clearly aimed at assuring the financial community and other observers on two fronts. First, despite the fact that Land had just turned sixty-five, he was not going anywhere, and would remain at the company as chairman of the board, chief executive officer, and director of research, focusing on what he did best. Second, the company was addressing any perceived management issues by elevating McCune and allowing him to focus on the more mundane (at least to Land) yet critical tasks of production and marketing. For his part, McCune paid tribute to the company's iconic figure:

> Polaroid Corporation is a very special company. It was founded and built by an extraordinary man with whom it has been my pleasure and good fortune to work for the past 35 years. I am deeply honored . . . [by this nomination] and I am undertaking the assignment with a great deal of enthusiasm.[84]

McCune's selection was a big surprise inside and outside Polaroid. To many, the obvious choice for the top job was Wyman, whom many thought had been in line to assume the Polaroid presidency.[85] But there had been a tug-of-war between the marketing and the technical contingents within Polaroid's executive ranks. McCune told Land that Wyman lacked the technical background to lead the company and suffered from a poor rapport with many key executives.[86] McCune later admitted that Wyman had issued an ultimatum. "Wyman was getting restless and pushing for the presidency, so the question came up of whether I would work for him and I said no because I believed that I was better for that job and I had a better understanding of the company."[87]

In the end, greatly influenced by Land, the scientific division won out when the Polaroid board of directors finally acted. It was not a coincidence that upon McCune's appointment, it was simultaneously announced by the Green Giant Company of Le Sueur, Minnesota, that Wyman was leaving Polaroid to assume the job as its chief executive. Putting a positive spin on the circumstances, Wyman admitted: "It was unbelievably hard for me to pull up stakes at Polaroid. . . . The company has been very special to me . . . [but] there's nothing mysterious about my leaving. . . . I just wanted to take advantage of an opportunity to grow and broaden my horizons."[88] In 1980, Wyman went on to become president and chief executive of CBS and in later years served on the board of directors of General Motors.

Although Land had resisted this move, and had been forced begrudgingly to relinquish some control over running the company he had created, he took comfort from the fact that it was one of his protégés that had succeeded him as only the second president in Polaroid history. "The fact that Mr. McCune and I have been working together almost since the beginning of Polaroid Corp. means that the individual character of this innovative company will be reinforced," Land acknowledged.[89] Others echoed this view. In a warm letter congratulating McCune on his appointment, Ansel Adams expressed his confidence:

> I am aware of the magnitude of this job, but I am sure you can carry it through with distinction. The extraordinary technical, professional and procedural ethics of Polaroid places it at the top of American organizations, and I have considered it a great privilege to have been associated with it in limited and specific ways over many years.
>
> I am sure that under your presidency, the aims and ideals of Polaroid will be carried on to maximum degree. You have my best wishes and warmest personal regard.[90]

It was, nonetheless, a difficult experience for Land. At the annual shareholders meeting held just a few weeks later, Land introduced McCune. According to one observer, "Land was near tears, his head bowed as he concluded his speech . . . introduced his new president . . . and in doing so . . . faced squarely and painfully the problem of his own succession." In turning over the podium to McCune, Land said: "You're at bat now."[91]

Under Polaroid's new management structure, the top operating executives who had previously reported to Land would now report to McCune. Although McCune reported to Land, the new Polaroid president had authority, according to an interview given to the *Boston Globe*, "to make some decisions himself."[92] To McCune, this was going to be a major improvement with job responsibilities clearly defined and measured by concrete goals. McCune had admitted that, previously, Polaroid's power structure below Land had been "a bit vague and mushy."[93] (See Fig. 9-2.) This impression was a function not only of Polaroid's organization chart, or lack thereof, but also a result of the relative leadership styles of the two executives. An unnamed Polaroid executive contrasted the two for the *Wall Street Journal*:

> When Land wants you to do something, he makes you feel like you're the greatest thing since ice cream. You walk away knowing that if you fail you'll let him down personally and the whole

> company as well. Only, half the time, you walk away not quite sure just what it is Land wants you to do.
>
> McCune's orders are normally clear, but if they aren't I can say, "Bill, what are you trying to tell me?" "Exactly what do you want me to do?" I would never dare ask Land that.[94]

Unfortunately, outsiders didn't seem to recognize the succession from Land to McCune as a transition in management style. Nor were they necessarily reassured by it. At least one Wall Street observer opined that it was "not at all obvious to us that the company needs a man with a background similar to that of Dr. Land" under the difficult conditions it found itself in, especially in anticipation of the imminent threat of competition it faced from Kodak.[95] But in reality, McCune was a different sort of man than Land, with a different skill set. An avid outdoorsman, with a ski chalet in Zermatt, Switzerland, McCune did not share the laboratory tunnel vision that had long characterized Land, and pursued other interests in his private life. He had climbed the Matterhorn, studied and collected wine, tinkered with his collection of Porsche sports cars, and had even had taken up silversmithing, learning how to make fine jewelry. "I'm perfectly happy to work seven days a week, all hours of the day and night on emergencies, but that isn't the way I'm going to work all my life," McCune told an interviewer.[96] A legend of his "toughness" sprang from an incident in which he was confronted in a Polaroid parking lot by a would-be kidnapper armed with a shotgun.[97] Instead of getting into a van as ordered, McCune grabbed the gun and overcame the man, who fled.

Educated as an MIT-trained scientist, McCune had joined Polaroid in 1939, just two years after its formation. After initially working on quality control systems for the production of polarizer materials, he had worked for decades on the design and production of Polaroid cameras. McCune had served as the virtual head of all Polaroid engineering since 1950. In recent years, McCune had managed effectively some of the company's most critical operations, including the task of designing and building the facility outside Boston where Polaroid would manufacture the film for its SX-70 cameras. "The boss gave him that job because it had to happen," a Polaroid engineer told the *Wall Street Journal*.[98] He had succeeded in bringing that plant in ahead of schedule and below the budgeted cost.

As a result, McCune had earned his dual reputation as an engineer and as a business executive. He acted decisively upon taking office. He immediately appointed Robert Duncan, the ex-NASA executive who had supervised the final stages of the SX-70 project for Land, to chair a special Overhead Control Committee to eliminate unnecessary expenses in the short term and to study ways of reorganizing the company's operations to achieve "long-term economies."[99] McCune was clearly grasping the reins of the company firmly. It would remain to be seen if he had the ability to solve the problems that continued to prevent the company from realizing the full potential of its revolutionary SX-70 product, while at the same time dealing with the daunting prospect of competing head to head in the years to come with the unchallenged leader of the photographic industry.

McCune did not have much time to settle into his new job before facing some of the long-anticipated challenges to Polaroid's exclusive domain in instant photography. On March 19, 1975, Kodak released its 1974 annual report in which it disclosed that it had finally "completed the design of its own instant cameras, [had] . . . made final the format and characteristics of a litter-free film for instant prints," and had begun a "major building program" to manufacture them.[100] The disclosure made headlines, and although it was acknowledged that Kodak's announcement had long been expected at some point, the *New York Times* noted how the "hard news from Rochester nevertheless weighed heavily on [Polaroid] investors."[101] Although no firm date for the introduction of Kodak's products was mentioned, Walter Fallon claimed, "That moment is within our reach."[102] Polaroid stock dropped almost ten percent after news broke that it would soon no longer be the sole presence in what had become an annual $500 million instant photography market.

The very next day, Berkey Photo invited a small group of industry analysts to its New York offices to announce that it intended to introduce later that year a relatively low-priced camera, to be called the Wizard, that would use Polaroid's SX-70 film.[103] In fact, there was already a history between the two companies. In October 1972, Berkey had released, without prior arrangement with Polaroid, a camera that used three types of Polaroid color and black-and-white peel-apart film already on the market. At that time, no legal action was taken immediately against Berkey, essentially because peel-apart technology was already destined for obsolescence. Instead, senior management for both companies had engaged in a dialogue over the possibility of Berkey taking a license from Polaroid to resolve any patent issues.[104] The talks dragged on for some time, but an agreement was ultimately reached, and a license was announced on December 16, 1974, more than two years later.[105] The Berkey contingent apparently believed that the same scenario would play out if it introduced a camera using Polaroid's integral SX-70 film. Such was not to be the case.

On March 17, 1975, just prior to its announcement, Sam Zausner, Berkey's president, made a surprise courtesy call to McCune.[106] The Polaroid president was unavailable, and so Zausner left him a message advising him of the forthcoming announcement, and inviting him to lunch to discuss the matter. When press reports of Berkey's announcement stated that the company had obtained prior approval for its camera, Zausner tried once again to reach McCune by phone "to personally apologize for the misstatement," claiming that it was the journalist's error and not a representation they had made.[107]

Notwithstanding Berkey's attempt to mitigate the circumstances, Polaroid reacted strongly to the Berkey announcement. Disclosing that Berkey had requested but had been denied approval to manufacture these cameras, a Polaroid spokesman noted, "We haven't offered licenses to anyone under our SX-70 patents."[108] The spokesman admitted that Polaroid had not yet seen the Berkey products, so it could not have possibly formulated an opinion as to whether there were any patent infringement issues. For his part, Ben Berkey told the press, perhaps hopefully, "We don't anticipate any patent problems with Polaroid."[109] But, in fact, patent problems there were. Despite the relatively benign threat the Berkey introduction posed to Polaroid from a business perspective, Polaroid would have to act aggressively if it turned out that the Berkey cameras indeed infringed Polaroid patents. This was especially true if the patents turned out to be the same as those that would have to be asserted later against Kodak when it entered the market. To not sue Berkey, thereby giving it preferential treatment, would subject Polaroid to potential legal complications in the main event that was sure to come, the case against Kodak.

By this time, Kerr had brought on board one of his younger partners to serve as his right-hand man in handling all of the Polaroid litigation. Herbert F. Schwartz had joined the predecessor firm of Fish, Richardson & Neave in 1964 after completing his law degree at the University of Pennsylvania. With an undergraduate degree in engineering from MIT, a master of arts in economics from the University of Pennsylvania, and some real-world experience in the business world at Philco, Schwartz had the perfect credentials for a career in patent litigation. But when he sought a position at Fish, Richardson & Neave, culturally, it was a foreign world he was seeking to enter. Throughout its early years, the Fish firm was the epitome of the "white-shoe" Park Avenue law firm. It had never employed a Jewish attorney. Having demonstrated his intellect and ability during a stint at the firm as a summer associate in 1963, Schwartz was able to overcome whatever remnants of anti-Semitism may have existed at the firm to land the only starting position offered that year. He was told years later that it was Alexander Neave, the son of founding member Charles Neave, who himself had decreed that the firm finally hire a Jewish attorney so as to erase definitively any hint of prejudice that may have lingered from its past.[110]

If the barrier was to be broken, Schwartz was the right man at the right time for this old-line institution. Once there, working with several of the key partners, Schwartz quickly distinguished himself and went on to become, according to the *National Law Journal*, "one of the nation's most prominent and respected intellectual property litigators."[111] He later taught intellectual property and patent law at the University of Pennsylvania and New York University law schools. An avid sailor, Schwartz was more the tall, rumpled professor than the short, taut pugilist that was Bill Kerr, but together they were indeed a formidable pair.

The first task for Schwartz was to figure out which patents from Polaroid's extensive portfolio could be asserted against Berkey. The criteria he applied to the selection process were dictated by the fundamentals of patent law litigation. There are primarily two legal defenses to a charge of patent infringement. First, the accused infringer can argue that the invention disclosed and described in the patent is not used in its product. This is known as "noninfringement." Second, the alleged infringer can argue that the patent being asserted is "invalid" because the U.S. Patent Office should never have issued it in the first place. In that case, the invalid patent would not be enforceable. "Invalidity," as this is known, can be established for one of several reasons. The most common reason is a lack of invention, which is the failure of the patent to meet the quantitative requirements of the patent law. Under the federal patent law statute, a patent can be granted only if the inventor discloses something new in the application—that is, something not previously disclosed in earlier patents or scientific literature, cumulatively known in the patent world as the "prior art." In addition, not only must the alleged invention be considered new, or "novel," but it also has to be deemed "nonobvious" to qualify as a patentable invention. Under this standard, the advance claimed for patent protection must not have been "obvious" to a person reasonably skilled in the technology in question at the time the invention was made.

Even though the U.S. Patent Office may determine at the time of application that a disclosed invention meets these requirements, and thus issues a patent, this determination is not forever conclusive and may be challenged in court. Issued patents begin with a presumption of validity, but this presumption can be overcome through various means, such as demonstrating that the patent examiner did not consider all the pertinent prior art when the application was evaluated.

In addition to identifying which Polaroid patents had been infringed by the Berkey camera, it was imperative to determine whether any of these infringed patents had any potential validity issues. Schwartz set out to work with the Polaroid patent department to review the Polaroid portfolio and to find the best candidates. In doing so, he worked closely with Robert M. Ford, a key member of the Polaroid in-house patent department who was intimately familiar with instant photography technology, the Polaroid patent portfolio, and the workings of the Polaroid cameras and film. Ford was diminutive, brilliant, and a chain-smoker of unfiltered Camel cigarettes, with tar-stained fingers to show for it. He was truly an engineer/lawyer—and wore his pocket protector proudly. Ford knew this area of technology and understood the Polaroid patent labyrinth better than anyone. Not coincidentally, he had personally drafted and prosecuted many of Polaroid's key patent applications covering instant integral photography. Ford was a tremendous asset for Schwartz to have, and he proved his worth many times over in the years of litigation to come. (See Fig. 9-3.)

Beyond the normal considerations that a patent litigator would confront in any case, there was one truly unique criterion that Schwartz and Kerr had to consider in selecting patents for the Berkey lawsuit: should the list include one or more patents on which Land himself was named as an inventor? The answer to this question was not as obvious as one might expect. In reality it was a difficult call indeed. Land had never met Kerr or Schwartz. To the extent he was involved behind the scenes, he was acting through his intermediaries, Silver and Mikulka. While no one doubted that Land wanted to defend vigorously the validity of Polaroid's patents, there were reasons, mentioned previously, that made his colleagues, as well as his adversaries, wonder whether he would ever be willing to emerge from his cocoon of privacy to participate in any legal action.[112] If Polaroid asserted one of his patents, he would have no choice but to come forward and be thrust into the center of the proceedings. As a named inventor, he would likely have his laboratory records and office searched for relevant material and, more importantly, he would surely be called upon to testify at a pretrial deposition, as well as at the trial itself. These were all very intrusive and public impositions, the kind that he avoided at virtually all costs. In fact, Schwartz was informed that Polaroid had previously conceded a foreign patent application challenge known as an "interference" when it appeared that it would be necessary for Land to participate actively if Polaroid wanted to prevail.[113] Rather than go through the trouble of asking Land to appear, Polaroid actually gave up the case.

So what were they to do in the Berkey case? Kerr and Schwartz had no idea whether or not Land would agree to take part in the lawsuit. Land apparently did nothing to clarify this. It had become clear to Schwartz that Polaroid's in-house lawyers were mortally afraid of having to raise this matter with the company's chief executive.[114] Eventually, Schwartz and Kerr concluded that if the Berkey situation could be handled without resorting to asserting a Land patent, it might be the easiest and most desirable way to go.[115] Keenly aware that Berkey was a necessary but clearly peripheral matter, all they really cared about was the coming Kodak battle.

Schwartz's review of the Polaroid patents that could possibly be asserted against Berkey was completed in the first two weeks of May 1975. A representative of Berkey had delivered to Polaroid in mid-April a sample of the forthcoming Wizard camera, together with a printed instruction manual.[116] This was apparently done in the hope that a licensing arrangement legitimizing the camera might come once Polaroid saw for itself that the Wizard was no match for its SX-70. Instead, having the potentially infringing product to examine was a major help in the patent selection process. Initially, Schwartz issued to Kerr a list of eleven potential candidates on May 9.[117] This list did not include a Land patent. They decided, however, that their client would have to make the final decision on whether to include a Land patent in the Berkey complaint. Accordingly, when Kerr presented a list on May 12 of potential patents for Polaroid to consider, it numbered twenty in all and included four on which Land was named as an inventor.[118]

After further consultations about the list, Peck asked for a formal proposal that could be presented to the Polaroid board of directors at its upcoming meeting on June 10.[119] Schwartz completed a detailed review of the application file histories for each of the final contenders to ensure as best he could that no invalidity issues were lurking. Afterward, a meeting was held on June 3 with the full Polaroid in-house team, including Peck, Mikulka, and DeLima, to make some final decisions on what should be presented to the Polaroid board.[120] The list of patents and the strategic implications of proceeding with or without Land were discussed. Not unexpectedly, it was decided to go forward against Berkey Photo without including a Land patent. Following the meeting, Kerr provided Peck with a formal letter outlining his firm's conclusion that Berkey indeed "flagrantly copies a number of [SX-70] features in its proposed camera and, in doing so, has infringed a number of Polaroid's patents."[121] He recommended that Polaroid sue Berkey on eleven patents he enumerated in an attachment to the letter.

The Polaroid board endorsed Kerr's recommendation. However, an additional study was conducted of the patents to be asserted, and a couple of changes to the list were made.[122] Yet there was still no Land patent among them. On July 3, 1975, even though Berkey had yet to release its Wizard camera to the public, Polaroid filed suit in the U.S. District Court for the District of Delaware, alleging infringement and threatened infringement of ten patents and asking the court to enjoin Berkey from manufacturing or selling its cameras within the United States.[123] Berkey made the expected public pronouncements of innocence, maintaining that its camera did not infringe any valid claims of Polaroid patents and was "the product of independent development by Berkey engineers and had many of its own patentable features."[124] But the first shot in the overt patent war over instant photography had been fired, if not at the real enemy, then at a stand-in. This would warn Kodak, and the rest of the photography industry, that Polaroid was ready, willing, and determined to defend its patent positions.

CHAPTER 10

A FIGHT FOR SURVIVAL

With the Berkey Photo lawsuit filed, and the campaign to defend its turf begun, Polaroid had other news to feel good about as 1975 progressed. Second-quarter earnings looked as if they were going to be almost double what analysts had predicted.[1] This meant that some rehiring would be possible, reinstating about one-third of the employees who had been laid off during the difficult economic days of 1974. By the end of the year, Polaroid was finally able to announce the news that many had been waiting for: in early 1976 it would introduce a non-folding instant camera with a molded plastic body designed to use SX-70 film.[2] The camera, dubbed Pronto, had a suggested retail price of sixty-six dollars but was widely discounted by retailers to forty-nine dollars, just a few dollars over its wholesale cost.[3] It was a much more affordable model than the original SX-70, which bore a list price of $180.

Although Polaroid had released two somewhat lower-priced versions of the SX-70 in 1974, neither was as economical as the Pronto, and so neither had its potential to increase sales of instant film.[4] As *Time* pointed out: "Almost from the time George Eastman fathered the snapshot, the biggest profits in the photography business have come from selling not cameras but film."[5] Polaroid had been slow to take advantage of this truism, a fact not lost on analysts, some of whom "chided Polaroid for wasting time in descending from the unnecessarily complex SX-70 original to the simpler Pronto."[6] Now, at last, the company was finally poised to release the mass-market camera for which many both inside and outside Polaroid had long been clamoring.

Polaroid staged an "upbeat, lavish affair" on January 13, 1976, to introduce Pronto.[7] Never enthusiastic about the concept of a low-price camera, Land was conspicuously absent from the champagne gala held at New York's Plaza Hotel. Instead, Bill McCune made his first high-profile public appearance since ascending to the presidency the previous year.[8] A $4 million advertising campaign was announced featuring Alan Alda, star of the then popular TV series *M*A*S*H*, and the actress Candace Bergen. The campaign was planned to debut on Oscar night in a few weeks' time.[9] McCune was able to announce very favorable financial news about the company to the 450 industry analysts and reporters who attended the event. SX-70 film sales were up forty percent in 1975, a year that had seen another one million cameras sold. Earnings for the first nine months of 1975 would be reported at $1.15 per share, compared with fifty-eight cents in 1974. Eventually, it turned out that Polaroid's profits for 1975 would jump 166 percent from the previous year, aided by fourth-quarter sales nearly triple the level achieved in the same quarter a year before.[10]

Despite this good news, the imminent threat of Kodak's entry into the instant photography field hovered like a dark storm forming over the distant horizon. In October 1975, Walter Fallon had announced at a meeting of industry analysts in Atlanta that Kodak's program to develop and market its own instant film and camera was "moving ahead on schedule and [it was] encouraged by the prospects."[11] He had noted that a specific target date had been selected for the introduction of the products, but he declined to disclose it. And he had made it clear that unlike the Berkey camera, Kodak's cameras were not going to use Polaroid film. The two systems were going to be completely incompatible. In January 1976, at the same time that Polaroid was heralding its new camera model, reports were circulating that a crew from the major advertising firm of J. Walter Thompson was already in California shooting commercials in great secrecy for Kodak's entry into instant photography.[12]

The concerns at Polaroid over what was coming through the Kodak pipeline persisted given the lack of hard information available. The possibility remained that Kodak might surpass Polaroid's technology, given the massive research and development efforts that had been under way for seven years, just a few months short of the time it had taken America to land a man on the moon. Some analysts had already concluded that, based on what they had seen and heard, Kodak's "instant photo system may well be technically superior to Polaroid's."[13] The photography giant had already made the claim that its film "would be [of] characteristic Kodak quality."[14] An analyst for L.F. Rothschild offered his "best guess" that while the Kodak camera might be large in size, "we are betting that its [film] quality will be substantially better than that of Polaroid."[15]

It was also clear that Kodak was going to have a substantial marketing advantage over Polaroid, particularly with respect to mass-market consumers and the photographic sales industry that served them. Because of its historic reluctance to discount its products and to offer the kinds of incentives and bonuses that were the norm in the photographic industry, Polaroid had never been particularly popular with retailers. "I carry Polaroid only because I have to," snapped one dealer in what was termed by the *Wall Street Journal* "a typical comment."[16] Even more fundamentally, instant photography started with a basic disadvantage with shop owners: its self-developing system eliminated the photo-processing business that in many cases was a substantial portion of retailers' income. As the industry stood, Polaroid occupied a unique niche within the amateur photography world, and so stores carried its products to meet customer demand. But Kodak's entry would surely be met with preferential treatment from retailers from the outset.

In short, there was an increasingly vocal consensus, at least in the retail world, and to a growing extent in the financial community, that "Polaroid is in deep trouble."[17] The direst prognosticators analogized the coming battle to those previously fought by CBS and RCA over technologies such as the long-playing record and color television, in which only one system ultimately prevailed. Even Polaroid's biggest supporters had to acknowledge that despite the possibility that Kodak's entry into the field might expand the instant photography market overall, the larger company would likely capture fifty to sixty percent of that market, severely impeding Polaroid's ability to grow and to reap the rewards of the technology it had pioneered.[18]

In the face of this uncertainty, nervous anticipation continued to build within the Polaroid legal department over Kodak's coming challenge. On February 20, 1976, Kerr and Schwartz were summoned to meet with Bob Peck in Cambridge to review the situation.[19] At Peck's urging, the agenda included another discussion of whether Polaroid should sue Kodak prior to its actual commercial introduction on the basis of threatened infringement, as had been done in the Berkey case. This possibility seemed to have some merit in that Kodak remained cagey with respect to its release date. Speculation had it that the introduction would occur within a matter of weeks, but there was no definitive confirmation from anyone at

Kodak.[20] Walter Fallon still refused as late as March 17 to disclose the product's release date, even under direct questioning from analysts.[21]

However, the essential factor that made the strategy of attacking threatened infringement problematic was that, unlike the Berkey situation in which a sample of the camera had actually been delivered in advance to Polaroid, no one on the Polaroid side had actually seen the Kodak camera or film. Kodak had trumpeted the fact that its products were going to be the result of a massive research program that resulted in the development of, among other things, "new chemistry and emulsion technology."[22] It had assured analysts that "there will not be a patent infringement."[23] Even though this view was not shared by Polaroid or its counsel, who knew just how well Polaroid's patent portfolio covered the instant photography terrain, there was still insufficient hard evidence on which Polaroid could base a claim of threatened infringement.

The tension mounted quickly. On March 19, just two days after Fallon had evaded questions from analysts about a release date, Kodak released its 1975 annual report in anticipation of its upcoming April 1976 shareholders meeting. The full-color report, mailed to over 237,000 Kodak shareholders, confirmed once and for all that Kodak was finally about to deliver on its 1969 commitment to enter the instant photography market. While it included a reproduction of a photograph taken with its new system, the image was carefully positioned so as to not reveal its exact dimensions or physical structure. The annual report also included a rudimentary description of the imaging chemistry used in the Kodak film units. These tidbits only further fed the rampant curiosity. As the *Wall Street Journal* reported, "because so little is known about Kodak's newest product, analysts were putting their rulers to that reproduction with all the zeal of private eyes."[24]

Despite the lack of detail, the story spread fast and hit hard. Everyone at Polaroid's headquarters in Technology Square took note of the headline that appeared ominously that Friday in the *Boston Herald American*: "Kodak Says Its Instant Color Photography Ready for Market."[25] Surely, the article delivered the long-dreaded news that Kodak's program was "moving quickly from advanced development toward marketing reality" and that the introduction date had been fixed. The newspaper coverage galvanized legal activity at Polaroid, like the blare of an alarm at a firehouse. Faced with what appeared to be an immediate and devastating threat, and based on the assumption that some Polaroid patents were sure to have been violated by whatever Kodak released, Peck told Schwartz that Polaroid would like to be prepared to sue Kodak on the very day its products were released to the public, an event he thought would occur by May 1.[26] At a meeting of Polaroid's board of directors held later that same day, the board endorsed Peck's recommendation that the company be prepared to file suit as soon as the Kodak instant products hit the market.[27]

Based on the disclosures in the Kodak annual report, Polaroid counsel finally had some public materials that described at least some aspects of Kodak's instant products. They could now focus on the process of identifying those Polaroid patents that deserved careful study as potential candidates on which to sue. The disclosures were not significantly detailed, but it was at least a start. Peck asked Schwartz to begin that process. It was decided to convene a meeting the following week in New York with Kerr, Schwartz, Peck, Mikulka, and Julius Silver to discuss Polaroid's strategy. The Polaroid board was still considering the possibility of bringing an antitrust action against Kodak in addition to the patent infringement suit. And so, Hammond Chaffetz of the Chicago firm of Kirkland & Ellis, who had done some work for Polaroid in that area, was to be present as well.[28] Kerr and Schwartz had always believed that there were no legitimate grounds for any kind of antitrust action against Kodak, but Richard DeLima forced that item to remain on the strategic agenda. Kerr never veered from his view that DeLima was "tilting at windmills" as regards an antitrust claim.[29]

Even though the release of its annual report had created a media ballyhoo, Kodak remained taciturn. *Fortune* noted how the "atmosphere around Eastman Kodak . . . [was] so cautious . . . you'd never guess the company has embarked on a bold challenge to . . . [Polaroid]."[30] The next bit of public information did not come until March 30, when an article in the local Rochester newspaper leaked word that Kodak's instant products would be unveiled at the annual shareholders meeting in Flemington, New Jersey, on April 27. The article, based on unnamed sources within the company, disclosed that the system would first be marketed in Canada in May, with a U.S. release to follow in June. It reported that the company had been assembling cameras for at least two months and that Kodak employees admitted that "security is unusually tight."[31] Still, no formal announcement from Kodak was forthcoming.

With only sketchy bits of tangible information on Kodak's still mysterious instant products available, the only thing Polaroid could do was focus on the best Polaroid patents covering features of instant integral film units that Kodak's product *might* utilize. Bob Ford had previously made some preliminary educated guesses as to these elements. By the end of March, with the new information available from the annual report, he was able to update his review. Thereafter, Peck conveyed Ford's selection of key Polaroid film patents to Kerr for Schwartz to review.[32]

By this time, there was no longer any doubt in the minds of Kerr and Schwartz that any lawsuit against Kodak was going to have to include patents attributable to Land. While they could keep Land out of it in the relatively minor Berkey action, it would be impossible to engage in a life-or-death struggle with Kodak without including in a prominent way the man who was, in essence, the father of the technology.[33] Kerr and Schwartz had to hope that Land would not want to be anywhere else but in the midst of the battle. Unfortunately, none of the attorneys inside the company actually knew what Land's attitude was going to be. There were, in fact, some discouraging signs. Up to that point, Land had spurned requests from Polaroid executives who wanted to "use [him] as a weapon" in the public relations "war with Kodak."[34] Land preferred, instead, to maintain his privacy. Nonetheless, Polaroid's legal team hoped that when the time came, Land would step forward into the center of one of the most important battles over intellectual property of all time.

Based on everything he had seen, Bob Ford believed that the technology of the Kodak system was going to be highly vulnerable to an infringement suit. But Ford also knew that one of his main objectives was the identification of which of Land's patents to include in the case. As it turned out, this was not going to be an easy task. As exemplified in the elegant SX-70 camera and film system, Land's work tended toward the more ambitious, and often more complicated, approaches to a problem. The evidence available suggested that Kodak had followed its traditional inclination to make its products as simple and user-friendly as possible. The question was, did the Kodak camera and film employ solutions that had previously been discovered by Land, or did they adopt alternative features and techniques discovered and patented by Land colleagues working in Polaroid's laboratories?

Among the patents Ford gave Peck to transmit to Fish & Neave were several Land patents relating to various means of stabilizing the photographic image. Land's work in this area had resulted in a variety of techniques for stopping the development process within the closed film unit at a predetermined time and then stabilizing the image from fading or otherwise deteriorating. Most of these techniques involved placing a neutralizing layer of polymeric acid somewhere in the film unit. Since the processing fluid was alkaline, this acid would bring the pH in the film unit back to neutral in a timed fashion, thus gradually bringing the processing reaction to a halt. As noted earlier, this type of layer had become known as the L-Coat, short for Land coat. While Ford couldn't be totally sure that Kodak used such a technique or, if it did, where in the film unit the layer might be located, he suspected that they had, and guessed that one of Land's patents would cover whatever it actually did.

The other primary batch of patents selected by Ford for transmittal to Fish & Neave covered the actual structure of the film unit. These described the ordering and arrangement of the various layers of chemicals and physical components within what amounted to a self-contained photo-processing lab. During Polaroid's development of SX-70, Land had worked on and patented a film format that became known as "Aspirin." This film unit structure was adopted for use in Polaroid's SX-70 film. Aspirin was an elegant but technologically challenging solution that required years of research headed by Land and his colleague Stanley Bloom to find a chemical opacification system.

Unfortunately, if anyone had hoped to be able to assert one of Land's Aspirin film format patents against Kodak, those hopes were dashed early on. From what he had seen, it appeared to Bob Ford that Kodak had likely taken an approach for its film unit that was less technologically challenging and didn't require the kind of sophisticated opacification system employed in Aspirin formats like SX-70. Instead, he concluded that Kodak had probably adopted the alternative "Excedrin" format invented by Howard Rogers. Rogers' Excedrin film unit utilized clear plastic sheets on both outside surfaces of the film unit. It was thus arranged so that the unit was exposed on one side and viewed from the other side. This allowed the negative or photosensitive side of the film unit to be protected from unwanted extra exposure by including carbon black—a totally opaque but chemically inert material—in the processing fluid that was spread under the transparent outside sheet as the film unit exited the camera. Since the image was viewed from the other side of the unit, this opaque curtain did not have to disappear as in the Aspirin configuration, avoiding the necessity of having a sophisticated temporary opacification system as was employed in SX-70. The basic Excedrin patent was issued to Rogers as U.S. Patent 3,594,165 on July 20, 1971. A second patent, which described the same kind of film unit, but with the addition of an L-Coat, was issued a year later, on September 5, 1972, as U.S. Patent 3,689,262.

The first clue that Kodak's film unit might have a structure similar to Rogers' Excedrin film unit had come from the United Kingdom. In

August 1972, Kodak had filed suit to have the British counterpart to Rogers' U.S. Excedrin patent declared invalid. That effort was still pending. Years later, an article in *Fortune* reporting on Kodak's introduction of its new products concluded, much as Bob Ford had done as he watched these events unfold: "There is some evidence that Kodak may have felt hemmed in by Polaroid's vast portfolio of patents . . . [in that] it sued to have Polaroid's British . . . [Excedrin patent] declared invalid, perhaps because it describes a film that is exposed from the back, has opaque black and white layers in the front, and bears other uncanny resemblances to what finally emerged from Rochester."[35] While Ford and Peck were confident that one or both of Rogers' Excedrin patents would cover the Kodak film unit, there was certainly some disappointment that this key invention was registered in the name of someone other than Land himself.

The same would turn out to be true of the chemical process utilized to form the image. Instant photography is a color diffusion transfer process. That is, color dyes diffuse away from their original location in a layer within the negative part of the film unit and transfer to a receiving layer where they become attached and form the image. This diffusion takes place on a microscopic level. It occurs in a pattern corresponding to the frequencies and shape of the light hitting the photosensitive layer of the film unit as reflected by the object or scene being photographed. In this way, the film reproduces the image of the photographic subject.

Two basic photographic chemistries for conducting this process were invented and patented at Polaroid during its research on color instant photography. Both approaches use chemical compounds known as "dye developers." These compounds are composed of a dye attached to a "developer" molecule that can control the dye's movement. Rogers had pioneered both approaches when he was given the task by Land, back in the late 1940s, of bringing color to Polaroid's one-step photography. A "positive" dye developer process employs dye developers that are initially soluble in the processing composition. In unexposed regions of the negative, they are free to move and diffuse to the image-receiving layer to take part in forming the image. On the other hand, dye developer molecules exposed to light are developed and thereby rendered immobile. They remain in their original locations and do not become part of the image. This control of dye transfer was at the very heart of the image-making system. It was the kind of imaging chemistry employed in every Polaroid color film.

During the same period of investigation, Howard Rogers came up with an alternative, mirror-image process—a "negative" dye developer system.

In this system, dye developers are initially insoluble and thus immobile compounds. In exposed regions, they are developed and, as a result split off a mobile dye that can diffuse to form the image. All of the negative dye developers that are not exposed remain in place and thus do not become part of the final picture. The basic Negative Dye Developer process was covered in U.S. Patent 3,245,789, issued to Rogers on April 12, 1966. Polaroid never used it in a commercial product. But these two dye developer diffusion transfer methods, and the patents obtained by Rogers on them, were the basic processes used in the technology of making instant photographic images, the progenitors of all adaptations to come. One might change the specific nature of the dye molecules or the developer molecules, but the fundamental technique of using combined dye developer compounds in this way was a breakthrough accomplished by Rogers at Polaroid during his pioneering work in the field. The question became, which of these two basic color diffusion transfer methods, if either, did Kodak employ?

Ford had a pretty good idea. At the October 1974 meeting during which Kodak's Walter Fallon informed Land and other Polaroid executives of his company's intention to enter the instant photography market, Fallon had handed his Polaroid hosts a copy of a patent issued to Kodak scientist Leo Fleckenstein. Fallon had claimed that this patent embodied the film chemistry that Kodak was going to use in its instant film units.[36] Kodak made other similar assertions in its annual report that were picked up in the press.[37] Even though Kodak's patent reflected its belief that the chemistry used in its film units was something new, it appeared to Ford that the Fleckenstein chemistry was merely a specific embodiment of Rogers' basic negative dye developer process. As a result, Peck included the Rogers negative dye developer patent in the group sent to Kerr. All of the attorneys met on April 1 to review the situation. As Schwartz, working with Ford, continued his detailed study of all the potential candidate patents to assert against Kodak, one disappointing conclusion was becoming more and more apparent: of the patents that appeared to be infringed, Rogers' patents might actually prove to be more significant than Land's. It was not the result the two men had hoped for, but one they would have to make the most of.

Official word on the unveiling of the Kodak instant photography system finally came from its Rochester headquarters on April 12, 1976.[38] Kodak's announcement set off a frantic effort at Polaroid to secure a sample of the Kodak camera and film so that they could be evaluated. For Polaroid scientists, like Land, there was a desperate need to see what

Kodak's formidable research team had created. For Polaroid marketers, there was an urgent desire to measure the competition. For Kerr and Schwartz, together with Peck and Ford, there was the anxiety of finding out, once and for all, whether the basis for an infringement lawsuit would actually exist. Or, was Kodak bringing out a system that would be so utterly different that there was no case for infringement?

Initially, Polaroid tried a straightforward approach. Jon Holmes from Polaroid's publicity department called one of his counterparts in Rochester and asked that Polaroid be provided with an invitation to the New York City event. The simple request was turned down without explanation. Kodak was unwilling officially to invite anyone from Polaroid to attend its press conference. Undeterred, Don Dery, then director of publicity and communications for Polaroid, enlisted the help of an employee who was also a contributing editor at *Family Circle* magazine. A normal press invitation was arranged for this person using the *Family Circle* credentials, and she attended the press conference, calling Dery to report on the event from inside the hotel. A second Polaroid employee was able to gain entrance to the event by using the invitation sent to a friend, a Chicago-based broker who apparently had no interest in going himself.[39]

Kodak's April 20 event was "a lavish presentation" in the grand ballroom of the posh Hotel Pierre.[40] Two camera models were unveiled: a lower-priced EK-4 version listed at $53.50 that operated via a hand crank and an automated EK-6 model listed at $69.50 that included a motor to eject the film unit from the camera after exposure. Both Kodak cameras were obviously aimed at the same market as that of Polaroid's Pronto ($66). Kodak did announce that a more sophisticated folding model, apparently aimed at the SX-70 camera market, would be introduced later in the year. That model, known as EK-8, was to be produced in Germany and was planned for a list price of $140.[41] The Kodak PR-10 instant film, like the Polaroid's SX-70, came packaged in a cartridge of ten units, but had a list price of $7.45 as opposed to SX-70's $6.99.[42] The new cameras and film, it was announced, would be available in stores by the end of June.[43]

A demonstration of the camera and film was conducted, and then Albert Sieg, the man who had led Kodak's P-130 program to develop the system, sat for interviews about the operation of the products and their development.[44] In what was called a "carefully structured" event, Walter Fallon, Kodak's president, read a statement but "disdained to answer many questions."[45] He stressed the company's concern for the consumer by pointing out that the "new instant cameras and film could very well be the most thoroughly tested products Kodak has ever introduced."[46] And he also reiterated the company line that Kodak had developed its own system. (See Fig. 10-1.)

Kodak did what it could to stress this position. As first claimed in the annual report, and then echoed in a press release issued at the launch, Kodak trumpeted the fact that the quality of its film prints was the result of "fundamental breakthroughs in imaging chemistry."[47] For Polaroid, this same press release also included a single sentence that provided the first confirmation that Bob Ford had been accurate in his prediction that the Kodak film unit would utilize the basic chemistry process and film structure invented by Howard Rogers. Providing significant details for the first time, the Kodak release disclosed: "Kodak instant print film is exposed through the back of the film, and during development imaging dyes are released for direct migration to the front or viewing surface."[48] There, in one simple statement, was a strong indication that Kodak indeed used a version of Rogers' basic negative dye developer chemistry in a film unit that sounded a lot like his Excedrin configuration.

By 12:45 p.m., reports about the details of the Kodak event began arriving at Polaroid by telephone.[49] It would be only a matter of hours before Land and his Polaroid colleagues would get their hands on the Kodak products and see for themselves what the Rochester giant had wrought. At the end of the Kodak event, each of the attendees received an "inspection kit" comprising an EK-6 camera and five film packs. Both Polaroid employees who had infiltrated the event promptly brought the cameras and film back to Cambridge on the one p.m. shuttle flight from New York to Boston.[50] The key players were all assembled in a boardroom at Polaroid's headquarters in Technology Square in Cambridge.[51] Present were Land and McCune, together with key members of the Polaroid legal team, including Kerr and Schwartz, who had earlier lunched with the company's Management Executive Committee.[52] In fact, this was the first time that either of the Fish & Neave lawyers had ever met Land, despite the fact that they had been working toward this day for more than two years and had already filed a patent infringement action on behalf of Polaroid against Berkey Photo.

When the Kodak camera and film arrived, Land quickly took some pictures, and breathed a huge sigh of relief. It was clear immediately that the Kodak system was no big deal, considering the buildup and what one might have expected a company like Kodak to come up with.[53] "Well, we expected more of Kodak," Land said immediately.[54] This was not the innovative system that many had predicted—and many at Polaroid had feared. While it was yet to be determined what legal action could be taken to protect the company's turf, at least everyone in the room knew at that moment that they would be competing in the marketplace with a Kodak system that was roughly comparable to Polaroid's, and likely inferior.

By the end of that afternoon, Polaroid public relations executive Donald Dery issued a statement that was picked up verbatim by the media, expressing the company's relief and confidence: "We have had a chance to make a brief comparison between the Polaroid instant-picture system and the new Kodak system. The comparison renews our confidence that our leadership in the field of instant photography remains unchallenged."[55] The headline in the *Boston Globe* about what it had termed the arrival of "doomsday" said it all: "Kodak enters the instant photography market . . . Polaroid officials breathe easier."[56] "Our first reaction was surprise," recalled a senior Polaroid executive. "With the length of time they were working on it, their capability, and the number of people on the project—it might have been something superior. Then our people started smiling. I'd rather be selling our product than theirs."[57] Surprise quickly turned into derision, as the Kodak EK-6 camera soon became known as "the parking meter" around Polaroid's Osborn Street labs because of its bulky size.[58] (See Fig. 10-2.)

Polaroid's confident public reaction to the unveiling was initially shared by the media, and by the financial community as well. Polaroid's attendees at Kodak's unveiling collected a host of immediate reactions and reported them back to Cambridge. Tom Owen of *Playboy* commented that the Kodak "camera looks old-fashioned and klutzy."[59] An official from the New York Stock Exchange declared: "Polaroid has a superior product. I was not impressed with Kodak's picture quality."[60] Other journalists recognized at least some kind of similarity to Polaroid's products. "Kodak's cameras are probably a rip off," said one. "They are so close to . . . [Polaroid's] it is scary," said another.[61]

Noting that Kodak had not introduced "any startling technological innovations," the *Wall Street Journal* reported: "the lack of any technical breakthroughs seemed to disappoint stock market traders."[62] While it was apparent that the Kodak camera and film did not operate exactly like the Polaroid SX-70 in some specific aspects, it was also obvious that in its grand strokes, and in the eyes of the average consumer, the Kodak system was merely a "copy cat."[63] More than one commentator termed it a "me too" system, while a New York television newscaster reportedly referred to the new "Kodak Polaroid" before correcting himself.[64] The reaction on Wall Street was immediate. Kodak's stock dropped while Polaroid's jumped more than twelve percent. However, this sense of optimism about Polaroid's prospects was short-lived indeed.

Within twenty-four hours the same commentators realized that, although Kodak had not delivered on its potential or its promise to introduce a radically improved instant picture system, its entry into the market as a direct competitor was still likely to be a serious problem for Polaroid. Several camera retailers expressed the view that although "the design and the mechanics are a big disappointment," and "any other company except Kodak would have bombed-out with cameras like that," these will sell because "Kodak's doing it."[65] Suddenly, several longtime Polaroid boosters changed their positions and starting recommending the sale of Polaroid stock, citing the inevitable decline in market share that Polaroid would suffer, whatever the comparison with the Kodak system. Other analysts pointed out Kodak's announced willingness to license other companies to manufacture cameras that would use Kodak's film, something that Polaroid had never done.[66] "What's really going to hurt Polaroid is the decision by Kodak to let other manufacturers make . . . camera[s] [that can use Kodak film]," noted Harvey Zucker, editor of *Penthouse Photo World*.[67] In short, even though Kodak had failed to introduce a technologically superior product, many of the negative consequences for Polaroid that had been speculated about during the months leading up to Kodak's entry had nonetheless materialized.

Land's initial sigh of relief about Kodak's system also quickly turned into something else. After studying Kodak's products, he became surprised and outraged by what he saw. A senior Polaroid executive, who was present while Land examined the newly introduced Kodak cameras and film, noted that he was "astonished" at the similarities to Polaroid's system, having "expected to see something new and unusual" from Kodak.[68] This took seven years to develop? On a technical level Land was underwhelmed by what Kodak had produced, but on an emotional level he quickly became enraged and agitated as he realized that the company that had once been his mentor and his biggest supporter had essentially ripped off his company's technology to compete with Polaroid in the field it had created. "Everything that was done used techniques extremely familiar to us," Land observed. "It has turned out that it took all their genius to make it possible to play the game with us at all."[69] It was clear to Land that Kodak's system was based on the foundation he and his Polaroid colleagues had developed: "It seems to me that with an uncharacteristic lack of grace [Kodak has] . . . combined some good but not quite finished chemical ideas with a number of techniques and devices that derive quite directly from our activities."[70]

Land was annoyed, and concerned. In what was for him an unusual step, he went public with his complaints. Given Kodak's enormous marketing power, and its virtual stranglehold over the conventional photography industry, Land knew that the very survival of Polaroid, a company almost completely dependent on the "ardent cultivation" of its instant photography business, was in jeopardy.[71] Given the companies' previous relationship, Land felt personally aggrieved by his former mentor's actions and clearly regretted the metamorphosis of their relationship. "The extraordinary, arms-length relationship that we had with Kodak from 1937 . . . is an historic model of the benign and proper behavior of the great corporation in intelligent perceptiveness and spiritual support of significant creativity outside of its own corporate structure," he lamented to the *Boston Globe.*[72] Land believed that Kodak should have been satisfied with a continuation of that relationship, supporting Polaroid's continuing innovation instead of deciding to become its direct competitor in the unique area of photography Polaroid had created. "Kodak just walked deliberately into our field with a callous pretension that we didn't exist," he seethed to *Business Week.*[73]

With Kodak's products now unveiled, it was clear that Polaroid had to take legal action. Apparently, no one knew that better than Land. As during decades of discussions over Kodak's attempts to secure a license to compete with Polaroid, he still feared that it would be extremely difficult for Polaroid to survive long term in a head-to-head marketing battle with the Rochester powerhouse. Kodak was a diversified company that not only controlled the worldwide amateur photography business but also had major operations in other industries like chemicals and textiles. Instant photography was all Polaroid really had. If Kodak had entered the field with original and superior technology, there would have been little that Land or his company could say or do. But that had not happened. Polaroid had availed itself of the protection of the patent system for just this kind of eventuality. It had meticulously patented the results of all of its developmental work, whether that work was actually utilized in its commercial products or not. Now that Kodak had become a direct threat to Polaroid's survival by misappropriating Polaroid's intellectual property, it was something neither Land nor his colleagues could or would abide.

Land summed up the necessity for suing Kodak in a rare but wonderfully articulate interview published in the *Boston Globe*:

> This would be our obligation even if one-step photography were but one component of our business. Where it is our whole field and where we have dedicated our whole scientific and industrial career to bringing this previously non-existent field to full technological and commercial fruition, our manifest duty to our shareholders is vigorously to assert our patents.[74]

Thus, with this manifesto, the stage was set for what was to become one of the most historic patent battles in American legal history.

Kerr and Schwartz had been directed to have a patent infringement suit filed against Kodak before the Polaroid annual shareholders meeting scheduled for April 27. That gave them exactly six days to complete their study of the Polaroid patent portfolio and to choose the patents to be asserted. While they had a good head start on the process, it was a monumental undertaking. They had to figure out which patents were actually infringed, and then ensure that the application file histories did not contain any flaws that could be raised by Kodak to invalidate them. Most importantly, they had to get it all right. The rush to the Boston federal courthouse was clearly partly emotional, but there were strategic reasons to move quickly as well. Kerr and Schwartz knew that Polaroid would have a substantial advantage if the legal battle could be fought on its home turf. They wanted to avoid at all cost having to try a lawsuit in a Rochester venue. Given the circumstances, if Kodak suspected that Polaroid was going to sue for infringement, it could take the preemptive move of filing an action in its home court seeking a declaratory judgment that its products were free of violating any of Polaroid's patents.

Why Kodak did not make this proactive move is open to conjecture. It's possible that its executives and its attorneys were simply unsure about how Polaroid might react. There were several factors that may have led to this uncertainty. For one, Kodak may have believed that Polaroid would be reluctant to risk asserting its older patents covering the fundamental processes of integral instant photography against the arguably newer variations of those techniques that Kodak had utilized in its products. Kodak had promoted a public position that its instant system was based on its own breakthrough technology. Its executives may have believed that it would be difficult for Polaroid to prove that those cornerstone Polaroid patents covered the specific and allegedly new Kodak features.

Second, Kodak was confident that it could defeat any patents Polaroid might assert against it. It had legal opinion letters in hand to that effect. Word that Kodak was relying on legal advice that it was not infringing any valid patents had even become public by this point. According to a report published in *Time*, Kodak's patent attorneys had so advised Kodak's management after having studied Polaroid's patents for six or seven years.[75] In fact, Kodak's litigation counsel was convinced that Polaroid would not sue and ultimately lost a bet to its director of research on that score.[76] Finally, Kodak might have still believed that the publicity-shy Land would never participate in a lawsuit and never agree to appear at a trial. For whatever reasons, in the end, Kodak did nothing and left the first move in the legal battle to Polaroid.

By five p.m. on the day of Kodak's April 20 demonstration in New York, at least one of the cameras obtained there by Polaroid employees was already undergoing testing and examination in Polaroid's laboratories. Pictures were taken with Kodak film units until two a.m. to provide adequate samples for investigation. Over the next few days, another camera was disassembled and mounted on a display board by a Polaroid technician.[77] Having the actual Kodak products to examine finally gave Ford the information he needed to focus on which of the inventions in the patents under consideration were actually used by Kodak.

In patent law, the Doctrine of Equivalents allows a patentee to sue even if the accused product does not copy exactly every detail of the invention as defined in the patent but is deemed the equivalent and thus, basically, close enough. In establishing this principle, the U.S. Supreme Court long ago determined that "one device is an infringement of another . . . if the two devices do the same work in substantially the same way, and accomplish substantially the same result . . . even though they differ in name, form or shape."[78] As a tactical decision, Kerr and Schwartz had rejected that approach for this case. They were determined to limit the patents asserted against Kodak to those that were literally infringed.[79] This had eliminated ninety-five percent of Polaroid's patent collection and reduced the real number of contenders to between twenty and twenty-five.

This decision also made it even more challenging to find patents developed by Land that could be included in the lawsuit, an ongoing major objective so that he could rightfully take his place as the central figure in the case. The examination of Kodak's film units quickly confirmed Ford's early hypothesis that Kodak had utilized alternative approaches to the film unit structure and the imaging chemistry from those utilized in the

SX-70 film unit. It became clear that, rather than Land's basic patents on integral instant photography, Rogers' Excedrin film unit and negative dye developer patents covered the Kodak film and so would play a prominent role in the lawsuit. However, Polaroid's analysis of Kodak's PR-10 film unit resulted in some good news with respect to the search for Land patents to include. Apparently, the Kodak film unit did employ a polymeric acid layer to halt processing and to stabilize the image in a location taught and claimed in one of Land's patents. But the Land L-Coat patent that best described Kodak's particular technique had evolved out of work done in 1962 to eliminate the print-coater initially planned for Polacolor film and was issued on January 9, 1968, as U.S. Patent 3,362,821. One of his earliest and most basic patents in that technology, it was not the ideal candidate, but Ford and Schwartz knew that they had to include it nonetheless.[80]

Even after its decision to assert the L-Coat patent, the Polaroid team wanted to have more of Land's work included, if possible. Yet they continued to have trouble finding Land patents that were literally infringed by the Kodak film unit or cameras. One other possibility remained in the pile. Land had a patent addressing the problem of keeping the film unit forever flat despite all of the liquid and chemical reactions within its closed environment. This patent was known internally at Polaroid as the "Symmetrical Supports" patent. Basically, it taught that both outside sheets of the film unit needed to be dimensionally stable and exhibit substantially the same physical response to moisture. It had been issued to Land as U.S. Patent 3,578,540 on May 11, 1971.

Despite their desire to find another Land patent, Ford and Schwartz decided not to assert this one against Kodak in the lawsuit. Their gut feeling was that the invention disclosed in this Land patent, at least on first glance, did not embody as much technology as the other patents they were selecting.[81] The lawyers were being hypercautious; they had to be sure that any patent asserted would be substantial enough to withstand the all-out attack Kodak would be sure to bring against it. Under the rush and pressure of the moment, the Symmetrical Supports patent did not appear to be among Land's most sophisticated work.

Fortunately, the inspection of the Kodak cameras gave them another option. Because of the extensive studies done of the Polaroid portfolio leading up to the Berkey suit, Schwartz had a solid list of camera patents to work with. Once these patents were compared with the Kodak cameras, they quickly determined that three of the patents asserted against Berkey would also be good candidates for the Kodak case. But none of these was attributable to Land. However, the Kodak EK-6 camera seemingly infringed one of Land's basic camera/film system patents that had been originally investigated for the Berkey matter. The so-called Rear Pick patent taught a motor-driven method for processing a film unit after exposure. The film unit is engaged at its rear edge and ejected from the top of the pack of film units within the camera and out through a pair of rollers that bursts the pod of processing fluid contained in the film unit, spreading the fluid evenly between the two layers of the unit so that development can take place. U.S. Patent 3,753,392 on this basic architecture and mechanism for an automated instant camera and film system was issued on August 21, 1973, to Edwin H. Land.

The Rear Pick patent had ultimately been omitted from the Berkey suit because of the decision to leave Land out of that case, but it also had a problem with respect to the scope of the invention described and claimed. Because it broadly taught and claimed as its invention the entire process basic to the operation of an instant camera with a pack of film units, it was susceptible to attack as being too broadly defined. In Germany, a corresponding Polaroid patent had in fact been declared invalid on that basis. There was a concern that the same thing could happen in the United States if the Rear Pick patent was asserted against Kodak. But Schwartz knew that there was a remedy and advised the Polaroid patent attorneys that they could "disclaim" the portions of the patent that were objectionable for being too broad and still proceed on the balance of the patent.[82] Although the decision had been made not to assert this patent in the Berkey case, steps had been initiated to cure the U.S. version of Land's Rear Pick patent.[83] Thus, it was determined that the Land patent could be added to the Kodak list as a fourth camera patent. While Kerr and Schwartz wished that the Land patents they had to work with were stronger, they at least had two they could use.

At 4:59 p.m. on April 26, 1976, six days after the commercial introduction of Kodak's instant camera and film products, and the day before both Polaroid's and Kodak's annual shareholders meetings, a complaint detailing Polaroid's action for patent infringement against Eastman Kodak was hand-delivered to Boston's U.S. District Court for the District of Massachusetts.[84] The lawsuit was officially commenced when the court clerk stamped Polaroid's papers at 5:07 p.m.[85] Under the court's rotation system, the case was immediately assigned to Judge Frank J. Murray, one of the most senior judges on the court. Polaroid's complaint charged infringement of ten patents, six covering various specific facets of the film and four relating to the camera technology. Asserted in the lawsuit were Land's L-Coat and Rear Pick patents, the two Rogers patents on the Excedrin film unit format, and a third on Rogers' Negative Dye Developer process. The other five patents covered basic features of instant camera and film technology developed by other members of the Polaroid research team:

The Trap patent—integral film units contain more processing fluid in the pod than the absolute minimum needed. The extra fluid ensures that processing fluid is spread over the entire surface of the photosensitive layer. Excess alkaline processing fluid is contained, neutralized, and immobilized, and the film unit is "spaced apart" by an acid-impregnated trap spacer element included in the trailing end of the film unit. The Trap Patent asserted against Kodak, U.S. Patent 3,761,269, was issued on September 25, 1973, to Polaroid scientist John E. Campbell.

The Mordant patent—the image-receiving layer within the film unit includes a chemical known as a "mordant." The mordant holds the dyes in the image-receiving layer after processing, thereby preventing the dyes from migrating back into the rest of the film unit. From its analysis of the Kodak film unit, Polaroid believed that Kodak's PR-10 used a mordant covered by U.S. Patent 3,770,439 issued to Polaroid's Lloyd D. Taylor on November 6, 1973.

The Light-Shield/Deflector patent—in the Polaroid SX-70 and the Kodak instant cameras, as the film unit passes through the rollers on its way out of the camera, it is deflected to control spreading of the processing fluid while being shielded temporarily from light that might fog the picture before the opacification system became effective. Polaroid engineers Richard Wareham and Richard Paglia developed the structure that performed both of those functions and received U.S. Patent 3,810,211 on the apparatus on May 7, 1974.

The Gear Train patent—the Polaroid SX-70 camera and the Kodak EK-6 camera, its automated version, both utilize a motor-driven mechanism that advances the film unit through the rollers and out of the camera following exposure. A patent on this mechanical apparatus was issued as U.S. Patent 3,709,122 on January 9, 1973, to Polaroid engineers Igor Blinow and Robert Leduc.

The Detachable Spread Housing patent—in the Polaroid SX-70 and both Kodak instant cameras, the rollers are included in a specially mounted housing section of the camera that can be detached from the camera to enable service and cleaning of the rollers and other camera parts. Polaroid's Richard Paglia received U.S. Patent 3,810,220 on May 7, 1974, for this invention.

In its court papers, Polaroid sought preliminary and permanent injunctions against the infringement of its patents that would prohibit the further sale of Kodak's products. It also sought money damages for infringement and reimbursement for the costs of bringing the legal action, including attorneys' fees. If it could be proven that Kodak's infringement was "willful and deliberate," Polaroid asked the court to treble the damages assessed.

Kerr and Schwartz had delivered on their commission to have the lawsuit filed within six days. This gave Land the opportunity the next morning to use the public platform of the previously scheduled Polaroid annual shareholders meeting on April 27 to rail against his former mentor and business partner from Rochester. In what was dubbed a "virtuoso performance," Land passionately described to the 3,000 shareholders assembled at Polaroid's Needham, Massachusetts, facility the desperate position in which the company found itself and the true nature of the battle it faced.[86] Reprising his role as the champion of patents, this time in a real-life instead of an academic context, Land proclaimed, "The only thing that is keeping us alive is our brilliance. The only way to protect our brilliance is patents. . . . This is our very soul we're involved in . . . our whole life. For them it's just another field."[87] Paraphrasing Lord Byron's observation that for a man, love is 'a thing apart' but for a woman, love is 'her whole existence,' Land exclaimed: "To the rest of the photographic industry, instant photography is a thing apart. To Polaroid, it is the whole of life."[88] Angrily, he declared to a cheering audience: "We intend to stay in our own lot and protect that lot."[89]

Land's address to the Polaroid faithful "was an emotional session . . . a strange blend of ridicule and outrage."[90] Land could not resist the opportunity to point out, somewhat sarcastically, the technical disappointment inherent in Kodak's new instant products. Making fun of the relatively large, bulky and non-folding Kodak cameras, he drew chuckles and applause from his audience, observing, "The new guys would like to confine its use to cocktail parties."[91] He noted that since the Kodak demonstration in New York the previous week, he and his Cambridge colleagues had been "in a considerable state of euphoria" because of the relief they felt when they finally tried out the Kodak system.[92] He acknowledged that there had been some fear about what Kodak might have released after such a long development program, but now confided that he and his colleagues "could all relax."[93]

Acknowledging that several of the inventions Kodak used were not employed in SX-70, Land noted that the Kodak products "might incorporate some of the really brilliant ideas we've had but have never incorporated ourselves."[94] Nonetheless, he pointed out how there was, overall, great systemic similarity between the Polaroid and the Kodak systems, or as he characterized it, "an overlap of their way onto our way. How serious this is remains for the courts to decide."[95]

Polaroid's court complaint against Kodak was the first salvo in, as the press had already characterized it, David's attempt to battle back against Goliath's invasion of its domain, a Goliath with 1975 annual sales that dwarfed Polaroid's by a factor of eight-to-one. It was estimated that Kodak controlled ninety percent of the conventional amateur film market and eighty-five percent of the consumer camera market.[96] In terms of corporate resources, this was clearly a mismatch. While Polaroid was surely an underdog in this battle, it was a fight it could neither avoid nor back away from. Perhaps more significantly, this was becoming Land's personal crusade, with his entire life's work and the very survival of the company he had founded and built at stake. "This was his final effort," a colleague remarked, "the defense of everything he believed in: the patent system, his patents, all the inventions he had made."[97]

CHAPTER 11

KODAK'S VIEW

On the very same day in April 1976 as Land was rallying the troops at Polaroid's shareholders meeting, Kodak's shareholders were gathered in Flemington, New Jersey, for their annual meeting. There was no way to avoid the news of the Polaroid lawsuit that had been filed the day before. From its Rochester headquarters, Kodak had immediately issued a straightforward denial: "Kodak's instant cameras and film are based on our own distinctive technology and don't depend on the patents of others. . . . Kodak has made an extensive study of patent literature in the instant photography field . . . and we are prepared to defend [our] position."[1]

At the annual meeting, Walter Fallon tried to ignore the legal threat and focused instead on delivering a very bullish forecast for the business prospects of Kodak's new instant photography system, reporting that interest from dealers in the trade was "at a high and very positive pitch."[2] However, later in the day, at a lunch held for security analysts, he was unable to avoid the subject of Polaroid's attack. Here he was more defensive, issued a "frosty denial of the charges," and perhaps even provided an advance glimpse of Kodak's legal strategy. "We still believe our patent position is sound. . . . We don't knowingly infringe anybody else's *valid* patents," Fallon declared (emphasis added).[3]

As previously mentioned, when faced with a charge of patent infringement—that is, the unauthorized use of an invention covered by a patent—the accused has two basic defenses, noninfringement and invalidity. It was likely that Kodak had already concluded that it might face great difficulty in prevailing on a noninfringement defense for most of the patents asserted against it. That defense would be based on an argument that its cameras or film did not use the invention of any particular Polaroid patent. Polaroid's patents had been studied thoroughly, and Kodak knew the scope of the inventions that they covered. Since Kodak also knew exactly what was included in its products, it was in an excellent position to make a very accurate assessment about whether its products fell within the scope of Polaroid's patents. Given that Kerr and Schwartz had been adamant about asserting only patents that were literally infringed by Kodak's products, the collection in the lawsuit would be tough to defeat on that score. Kodak had to know that. Although Kodak might assert noninfringement as an alternative argument on one or more specific patents, it was not likely to be its key defense on most. Instead, its central strategy in the case seemed sure to be an attack on the validity of each patent asserted against it. One can only assume that Fallon had this clearly in mind.

In fact, Fallon made his pronouncement armed with the knowledge that his patent counsel had assured Kodak that Polaroid's patents were, in fact, vulnerable on this basis. For eight years leading up to its product launch, Kodak had conducted a comprehensive study of Polaroid's patents and had collected a series of legal opinions stating that the key patents standing in the way of Kodak's entry to the field were flawed.[4] The study included a review of between 200 and 250 Polaroid patents and had resulted in written opinions that sixty-seven key Polaroid patents were either invalid or, if valid, would not be infringed by Kodak's proposed instant photography system.[5] Kodak's longtime outside patent counsel, the New York City law firm of Kenyon & Kenyon Reilly Carr & Chapin, issued these opinions.

The Kenyon firm, an historic Wall Street intellectual property institution with its roots in the late nineteenth century development of the electric power and light industries, was truly the downtown counterpart to Fish & Neave. It had started its practice in 1879 as Browne & Witter. The first Kenyon joined shortly thereafter, and there was at least one Kenyon in the firm thereafter until December 31, 1978, when Houston Kenyon retired.[6] During the years surrounding the turn of the twentieth century, the firm litigated the validity of many pioneer patents covering inventions that made possible an America powered by electricity and gasoline. These included work done by Charles Brush on the arc light and the rechargeable battery, Charles Van DePoele on the trailing trolley pole that brought electric power down from overhead wires, Nicola Tesla on a motor that ran on alternating current, and others.[7] Kenyon & Kenyon, which adopted this simplified name in 1979, had been a force in the law of technology ever since.

Francis T. "Frank" Carr, the epitome of the old-school, pipe-smoking gentleman lawyer, headed the Kodak legal team and, since late 1968, had personally supervised the study of the Polaroid patent portfolio. Carr hailed from a small coal-mining town in Pennsylvania, attending Lehigh University before serving in the Navy during World War II. After the war, Carr attended law school at the University of Virginia and joined the Kenyon firm as its fourteenth attorney immediately thereafter, in 1948. In the early 1960s, he took over the primary responsibility for handling Kodak's work from Theodore Kenyon, who was approaching retirement. Carr had assisted Kenyon since Kodak became a client in 1954.[8] During his representation of Kodak, he settled favorably several important litigations in the chemical and polymer fields and advised the company on patent matters involving many of Kodak's most significant products, including the Carousel slide projector, and its Instamatic camera and film system. His other clients included Bristol-Myers, Coca-Cola, U.S. Steel, Uniroyal, and Xerox. Balding and bespectacled, Carr had one of those pale complexions that seemed to redden like a pH indicator whenever he was aroused. That trait may have been a handicap for him at the poker table, but it did nothing to detract from his fierce advocacy for his client. (See Fig. 11-1.)

Even before the release of its instant photography system, Kodak was already pursuing an attempt to weaken or to destroy elements of Polaroid's patent portfolio around the world. It did not keep this initiative a secret. Kodak advised interested journalists that the company was challenging the validity of Polaroid patents in at least five other countries. This news received broad coverage, including a report in *Newsweek*.[9] There were several fronts to this counterassault, all under the direction of Carr.[10] The proceeding to dispute the British equivalent of Rogers' "Excedrin" patent had been pending since 1972 but had only recently been reactivated.[11] Similar actions were also pending in Japan, Germany, and Australia.[12] Just two weeks before the shareholders meeting, Kodak had filed a suit in Canada to have the Canadian counterparts to several key Polaroid patents revoked for invalidity.[13] A prior action filed in Canada by Kodak in December 1974 to have two other essential Polaroid patents revoked was still pending.[14]

Kodak's foreign strategy was certainly aimed at trying to undermine Polaroid's confidence in the patents it was suing on in the United States. If Kodak could knock out a fundamental patent abroad, or at least raise serious issues with regard to its validity, that might take the wind out of Polaroid's sails. Surely Polaroid did not want to see its patent portfolio vitiated so that every camera and film manufacturer in the world would be free to waltz right in and compete with it. Perhaps the fear of this possibility might convince Polaroid to reconsider the risks of a vigorous legal campaign, one that subjected its patents to attack? Perhaps Kodak could move Polaroid to grant the kind of license at a nominal fee that Kodak had sought when Fallon had visited Land in Cambridge in the fall of 1974?

But Land and Polaroid remained on the offensive, following up the filing of the lawsuit in Boston with similar filings in the courts of Canada and Great Britain, accusing Kodak of infringing its patents. For the most part, these lawsuits were based on counterparts to those in the U.S. suit. The consensus among industry analysts and observers was that this was destined to be a major and protracted legal battle. "We'll be old men before this one is settled," concluded the *Wall Street Journal*.[15] In the United States, it was fairly clear to Polaroid and its counsel that no court would grant Polaroid the immediate remedy it sought—an injunction preventing Kodak from marketing its products even before a trial could be held. Preliminary injunctions of that type are very difficult to obtain in any litigation. It would be the longest of long shots to expect a court to grant one in this case. One analyst was quoted as pointing out that "the case is far too technical and complex to issue a quick, temporary injunction."[16]

Such was not the case in Great Britain, where the standards for securing preliminary relief were apparently different and more lenient than in the United States, at least in the opinion of Polaroid's British counsel. All that was required in Britain to secure the preliminary injunction was a finding that Polaroid had an "arguable case" and that the "balance of convenience" leaned in favor of the injunction. In contrast, a U.S. court would require Polaroid to prove a "substantial probability of success on the merits" before it would issue a preliminary injunction. This is virtually impossible to do in complex cases. So all in all, Kerr and Schwartz knew that whatever occurred in the British lawsuit with regard to preliminary relief, it would have no direct impact on their case. Nonetheless, Polaroid's British counsel persuaded its Boston client, as well as Kerr and Schwartz, that it could secure a preliminary injunction and thus stop Kodak from releasing its cameras and film there.[17] If that occurred, perhaps it might be a major blow to Kodak's confidence in its legal defense worldwide. Accordingly, Polaroid moved for an injunction in Britain to see if it could win an early skirmish in what had become a world war.

In fact, it was beginning to appear as if the U.S. lawsuit was going to have all the earmarks of a real grudge match. Over the previous decade, it had become apparent that a new era in the long joint history of the two companies had clearly begun. There was no denying that their relationship, on a personal as well as a corporate level, had changed forever. Polaroid's longtime mentor had become its archenemy. Although few of Kodak's present executives had warm memories of the many happy years of cooperation, some older, now retired colleagues privately expressed a nostalgic pride over what had once been.[18] Wren Gabel, the former Kodak executive with whom Land and others had worked so closely in the 1950s and 1960s, and who was pushed out of Kodak management in 1969 when he lost a power struggle with Louis Eilers, expressed his disappointment in a reminiscence to Bill McCune. "I have been greatly distressed by the deterioration in our corporate relationship, but my belief in what . . . Din and I and others did to further the photographic industry and each of our companies never has wavered," he wrote.[19] McCune shared the letter with Land, telling Gabel in his response that his letter had given "both Din and me a much needed lift."[20]

McCune knew, however, that the deterioration Gabel referenced stemmed from a growing sense of resentment of Land, and thus Polaroid, by Kodak executives on several fronts. Some believed that the company had never received sufficient credit for its contribution toward the substantial challenge of taking Land's laboratory-scale technology to a mass-production level. Land had always characterized Kodak's contribution as having "adapted" Polaroid's process to Kodak's manufacturing equipment. Kodak personnel who worked on the project knew they had contributed substantially more than that in making Polaroid's system practical.

Others at Kodak were still frustrated and indignant over Polaroid's refusal over the decades to let Kodak into its field through the licenses it had requested repeatedly. Last, but not insignificantly, it seemed as if some Kodak executives simply did not like Land, considering him egotistical and imperious. They certainly could not have appreciated his occasional barbs, for example, when he told Polaroid shareholders that "our systems take such gorgeously intimate close-ups that we may call it the Intimatic," a jab at Kodak's popular Instamatic product line.[21] All of these attitudes, in varying amalgams of toxicity, could now be found in segments of Kodak management. To them, an ungrateful Land was belittling Kodak's development efforts and attacking it by claiming that Kodak had wrongfully appropriated Polaroid technology. Resolute, Kodak continued to claim innocence of any wrongdoing and pushed ahead, publicly announcing that the lawsuit "[has] not changed our marketing plans."[22]

A curious incident occurred just days after Polaroid sued Kodak. It was perhaps innocent, perhaps not. In a letter to Land dated April 30, a relatively low-ranking Kodak executive named R.D. Lorbach informed Polaroid of Kodak's intent to license to others on "reasonable, nondiscriminatory terms" the technology necessary to enable the manufacture of cameras that could utilize its PR-10 film.[23] In return for a requested payment of $10,000, Lorbach, who was serving as director of Kodak's Photographic Program Development, offered to Polaroid all of the necessary drawings and specifications. It's impossible to know precisely how Land and others at Polaroid reacted to this letter, but they must have been perplexed, if not outraged. Was it merely a form letter sent to the chief executives of all companies in the photographic industry? As such, did a junior executive, not aware of the just-filed lawsuit, send it to Polaroid in error? Or was it knowingly sent to irritate Land and his Polaroid colleagues? Land turned the letter over to Bill McCune, who forwarded it to Bill Kerr.[24] While Polaroid made no response, the company and its lawyers were left shaking their heads at what may have been either Kodak's ignorance or arrogance or, perhaps, a combination of both.

Beyond the personal rancor was a clash over fundamental patent issues. Whereas Kodak had initially professed admiration publicly for the Polaroid research and development that had led to one-step photography, resentment had gradually grown in the Kodak labs as its scientists tried to tiptoe through the minefield of Polaroid patents. There was no debate that virtually all of the pioneering work in the field had started at Polaroid in the 1940s. It would be hard to complain about Polaroid's diligent efforts to protect the intellectual property it had developed, including the alternative solutions developed at Polaroid but not actually used in its own commercial products. But the American patent system grants exclusivity on inventions for only a limited period, at that time only seventeen years.

Accordingly, the early patents on much of Polaroid's groundbreaking work had already expired. As the research progressed, a second generation of patents was secured on new advances in the technology. By the mid-1970s, even these patents had begun to expire. Over the years, Polaroid patent lawyers continued to secure patent protection for new improvements and refinements to these basic inventions as they were discovered, in effect perpetuating its web of patent protection. In Polaroid's view, this was a legitimate use of the patent system as its work in the field evolved. As McCune explained, the progression kept the company viable in the field it had pioneered:

> The company was so creative and Land was so creative and he kept driving these ideas into systems so that by the time one set of

> patents had run out and [it] might have been possible for someone to compete . . . we had new patents and new systems that they just couldn't compete with. That just went on and on.[25]

Kodak, however, viewed Polaroid's patent practices as an improper perversion of the law, which requires that each invention satisfy a quantitative level of technical progress to be entitled to patent protection. Kodak believed that most of what was covered by the succeeding Polaroid patents were only incremental advances so minimal that they were not entitled to patent protection in their own right and insufficient to support Polaroid's now challenged monopoly in its technology. Even though the U.S. Patent Office had apparently acquiesced in Polaroid's practice, and had granted patents on these second- and third-generation inventions, the validity of those patents would now be challenged in court.

This strategy was clear in Kodak's Answer and Counterclaims to Polaroid's complaint filed with the court on May 17, 1976. With respect to the ten patents asserted by Polaroid, Kodak's pleading contained the expected defenses of patent invalidity and noninfringement. But it went much further than that. In its response, Kodak sought from the court a ruling that Polaroid should be prevented from asserting its patents because, Kodak claimed, it had misused the patent system in a variety of ways. Surprisingly, this assertion was not limited to the ten patents Polaroid had asserted against it. In its counterclaim, Kodak also made the same charge with respect to an additional nine specific Polaroid patents Kodak described as "related to the same subject matter." Kodak maintained that this relief was necessary because it "is in reasonable apprehension and fearful that plaintiff [Polaroid] will, at a later date, also assert" those patents against it.[26] Then, Kodak took its allegations of patent misuse even further.

Kodak alleged that Polaroid sought to protect its monopoly by abusing the patent system to create a wall of patents impenetrable to anyone interested in entering the instant photography field. Kodak acknowledged that Polaroid had pioneered instant photography and had lawfully patented its work in the field. But it pointed out that these first-generation patents would have expired by this date, and therefore would have become part of the public domain, free for use by any interested party. According to Kodak, Polaroid had attempted to create an "illusion in the public mind" that the subject of the expired patents was covered by the new generation of patents.[27] In view of this wrongful conduct, Kodak asked the court to declare Polaroid's *entire* patent portfolio unenforceable. Kodak even asked for an assessment of damages it had allegedly suffered as a result of that conduct.[28] In fact, Carr had wanted to take this argument even further and allege that the conduct amounted to an antitrust violation by Polaroid. But, cognizant of Kodak's active defense of antitrust claims brought against it, he was overruled by Kodak in-house counsel, who wanted to stay away from that area of the law in the Polaroid suit.[29]

In its court papers, Kodak laid out the equity of its position and the philosophy of its case:

> A United States patent is the grant of an exclusive right intended to benefit the public. An illegally or improperly obtained patent is of no benefit to the public and, in fact, is a burden to the public. Even where such a right is legally and properly obtained, its subsequent abuse or misuse is a detriment to the public. Both the illegal or improper acquisition of a patent and the abuse or misuse of a patent renders it unenforceable against the public. For example, if a patent is properly obtained for one invention, the patentee is entitled to an exclusive right for 17 years for *that* invention. He may not, in the guise of a slightly different environment, gain a second and continuing right for another 17 years, or another, or another, for the same concept. When the first patent expired, the public was entitled to free, unrestricted use of that invention. . . . Such is the flavor of this conflict.[30]

There it was. All of the frustration that had built up over the years at Kodak from having to navigate around Polaroid's patents had become public, aired out in Kodak's official response to Polaroid's charges of infringement. Kodak intended to take on Polaroid's entire patent portfolio in an effort to open up the field of instant photography once and for all. This tactic was widely reported by the press in great detail. The *Wall Street Journal* printed a complete description of the strategies Polaroid had allegedly engaged in, including passages detailing the claimed misconduct quoted straight from Kodak's court papers.[31]

The potential for irony or even hypocrisy in some of Kodak's claims was not lost on Polaroid's lead litigator, Kerr. On a newspaper clipping quoting Kodak's allegation that Polaroid patents were "intentionally drafted to be exceedingly lengthy and obscure," Kerr wrote by hand the comment "they will be crystal clear when relied on as prior art," before passing it on for Schwartz to see.[32] Kerr knew that when Kodak would later want to use those supposedly "obscure" patents in court as part of its invalidity defense, its lawyers might not have the same problem deciphering their meaning. Moreover, the charges of Polaroid's excessive patent activity were being leveled by a company that, by 1972, reportedly had secured 10,000 patents of its own.[33]

On June 28, Polaroid responded officially to Kodak's filing. In so doing, Kerr's and Schwartz's main objective was to keep the lawsuit focused on the real issues as they saw them, the infringement of the specific patents asserted. Accordingly, Polaroid moved to dismiss, or to strike as insufficient in law, Kodak's patent misuse claims.[34] It also moved the court to dismiss the portion of Kodak's counterclaim that added the extra Polaroid patents to the controversy, with one ironic exception. Among the nine additional patents Kodak tried to attack was Land's Symmetrical Supports patent, the very one that Schwartz and Ford had originally considered for the suit but had left out because of doubts concerning the strength of the invention disclosed. Given what had transpired in Kodak's response, the Polaroid team changed its collective mind and decided to add this additional Land patent to the list of patents asserted against Kodak.

There was a clear rationale for making this move. Kerr and Schwartz didn't want to miss the opportunity to have one more Land patent in their portfolio for trial. They were beginning to believe that Land would not only play a major role in the proceedings but would also be a highly effective advocate, particularly when his personal patents were under attack so it was worth taking a chance on the Symmetrical Supports patent. If anyone could explain and defend the invention covered in that patent, it would be Land. At the same time, Polaroid decided to add yet another patent to the lawsuit, one covering the hand-driven mechanism used in the EK-4 camera to eject the film unit after exposure.[35] This work had been done by two Polaroid engineers, Vaito Eloranta and Benjamin Ruggles, and had been issued on September 11, 1973, as U.S. Patent 3,757,657. It was known as the "Crank Patent." So now, there were twelve patents in suit.

Getting rid of Kodak's patent misuse claims was critical for Polaroid. Obviously, on its face, the thrust of the attack was insulting and outrageous to Polaroid. On a more pragmatic tactical level, it was clearly an attempt by Kodak to widen the scope of the matter so as to confuse and complicate the proceedings, as well as to allow it to dig deeper and more broadly into Polaroid's internal records prior to trial. As pointed out in Polaroid's court filings, Kodak was urging "in effect, that it should be given a hunting license to probe in discovery every facet of Polaroid's business [patent or otherwise] simply because Kodak has the economic muscle to devote to harassing Polaroid into submission."[36] As an example,

Polaroid noted that as soon as the case was commenced, Kodak served requests for documents that were "calculated to cause Polaroid to search for millions of documents concerning almost every aspect of Polaroid's business and turn them over to Kodak."[37] This case was going to be massive enough if it remained focused on just the patents Polaroid had sued on. Kodak already had a distinct advantage because of its size relative to Polaroid. It was critical that the legal dispute not be enlarged so that this advantage was not unjustly magnified.

Some within Polaroid, particularly Bob Peck, were sensitive to and somewhat concerned about Kodak's misuse allegations. Polaroid had surely pursued an aggressive policy of patent protection for decades. But Schwartz knew from the outset that Kodak's misuse claims were a diversion in terms of the lawsuit, and he strongly believed that they had no merit.[38] If Kodak could prove specific instances of misconduct by Polaroid during the application process for any of the patents in suit, so be it, they would defend those cases one by one. But the allegations of systematic misuse of the patent system were a different story and seemed to touch a nerve at Polaroid. Schwartz and Kerr had more than a little corporate handholding to do.

Polaroid's concern about the issue really was twofold. First, Polaroid had never licensed any of its key patents in the field of instant photography, despite many requests for it to do so. Indeed, it had licensed some arguably obsolete technology, for example, when it enabled Berkey Photo to manufacture a camera using Polaroid peel-apart film years after that format was introduced. But Polaroid had never licensed anyone to compete with it directly with respect to its state-of-the-art technology. The antitrust laws exist to combat anticompetitive behavior in the business world, and Polaroid management was keenly aware that it had successfully avoided any competition principally because of its patent policy. Polaroid had enjoyed a near monopoly in instant photography for nearly thirty years, and at this point, there was some sensitivity within the company that its licensing policy was beginning to look like stonewalling and would perhaps be vulnerable to attack on some kind of antitrust basis. Even though no one wanted to open the field up to others by licensing Polaroid patents if it could be avoided, there were clearly those at Polaroid who were becoming more and more uncomfortable with the notion that the company could continue to maintain this policy.

Accordingly, whenever the issue of Polaroid's license policy came up, it was handled with great diplomatic care and a defensive posture, though not always with the intended results. For example, the week before Kodak filed its answer in court, McCune and one of Polaroid's other top executives, Richard Young, visited Japan to meet with representatives of the Japanese photo industry. Apparently, McCune and his advisors thought they should not reject absolutely the possibility of licensing its SX-70 technology to others, in particular the Japanese manufacturers who were sure to be interested. Anticipating questions about Polaroid's licensing policy, they went armed with a statement prepared by Polaroid's in-house lawyers but communicated to Fish & Neave only after the fact. Instead of reiterating the company's policy of not licensing the proprietary property that was at the heart of its business—a position it had every right to take—McCune went armed with a far more equivocal response to any questions that might arise: "Polaroid is *reconsidering* its present policy of not granting licenses under its SX-70 patents and is studying various possible arrangements which would allow camera manufacturers to produce cameras which would accept Polaroid SX-70 film" (emphasis added).[39]

There was no intention on the part of Polaroid for this to be an announcement or a formal press release. It was prepared merely as an answer McCune could give in response to direct questions on the subject. Of course, those questions did inevitably come, and the answer was duly given. As might be expected, this became big news, and it did not stay within the borders of Japan. Within a week, on May 17, inquiries in the United States had forced Polaroid to make the same announcement in Cambridge, resulting in a *Wall Street Journal* headline: "Polaroid May License SX-70 Film-Type Camera."[40] Ironically, that headline appeared on the very same day that Kodak filed its response in court to Polaroid's lawsuit. Ultimately, the suggested licensing policy was never adopted, nor was there anyone at Polaroid who wanted to see it adopted. But the whole exercise is an indication of either just how little confidence Polaroid seemed to have in its ability to sustain its policy of not licensing its patents or, at the very least, its sensitivity to the subject.

Polaroid's in-house lawyers were also sensitive to Kodak's misuse allegations because of their familiarity with an ongoing patent litigation between Xerox and IBM over the plain-paper copying field. In some ways, that case paralleled the one Polaroid had brought against Kodak because it involved a much larger corporation, IBM, trying to force its way into what had been the exclusive domain of a much smaller company, Xerox, which had been the first company to produce and market products in the xerography plain-paper field. Like Kodak, IBM had decided that it would accomplish this by putting out its own products, knowing it would get sued. Its plan was to then beat Xerox's patent portfolio through infringement litigation and thus open the field for its products. So, IBM had developed its own machines and put them on the market in April 1970. Xerox, like Polaroid, had immediately sued for patent infringement to protect its turf. Top analyst William G. Prime of Stuyvesant Asset Management Corporation noted at the time, "IBM's threats against Xerox's very strong patent position is the most ominous part . . . from Xerox's point of view."[41]

Like Kodak, IBM had included a claim of patent misuse as a defense against Xerox in its July 1970 response to Xerox's infringement claim. IBM alleged conduct by Xerox that, IBM asserted, made its patent portfolio unenforceable. IBM's claims probably sounded eerily familiar to Peck and to other Polaroid executives: "For many years, Xerox has pursued a planned course of conduct designed to perpetuate, after the expiration of its lawful monopoly, its monopoly of trade and commerce in xerographic office copiers capable of making copies on plain uncoated paper."[42] Again, in language that echoed remarkably that of Kodak's, the *New York Times* described the essence of IBM's claim as being "that Xerox's basic patents have expired and can be freely used by anyone. . . . However, Xerox has tried to bar competition by patenting a large number of minor variations . . . in an attempt to re-establish its expired patent position."[43]

The *Xerox v. IBM* lawsuit was still pending in 1976 when Polaroid sued Kodak, so it was still unclear what impact IBM's patent misuse strategy would actually have in that litigation. Nonetheless, IBM's mere allegations had already hurt Xerox, in that they had apparently contributed to the commencement of an investigation by the U.S. Federal Trade Commission (FTC) of Xerox's patent practices. As a result of that investigation, the FTC had filed a complaint in January 1973 against Xerox, charging it with monopolizing the office copier business. The FTC had eventually settled that complaint in November 1974 with an order forcing Xerox to make its 1,700 patents available to any company that wanted them.[44] Polaroid could not even imagine suffering a similar result—being forced to license its patents broadly.

A certain irony may not have been immediately apparent to the Polaroid contingent concerned with this issue. Fish & Neave was representing IBM in its patent battle against Xerox, and it was none other than Herb Schwartz who, as part of the team on the case, had devised and run IBM's successful patent misuse strategy. He had even been responsible for the "white paper" detailing Xerox's patent misuse that was handed over to

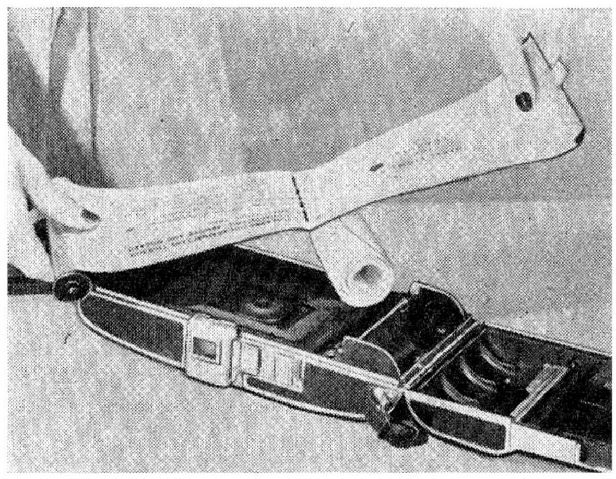

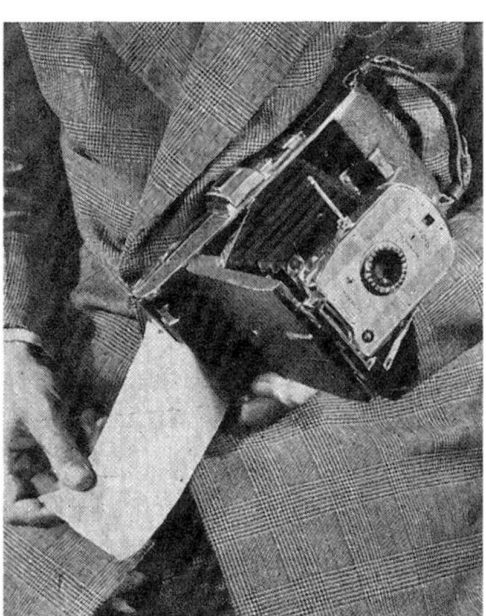

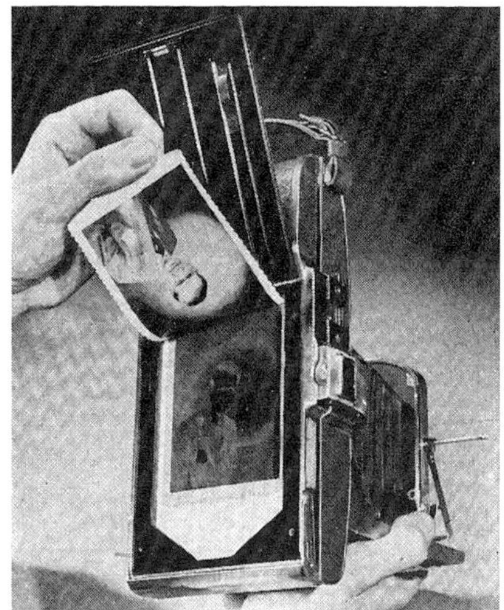

Figure 4-2: Operation of the first commercial Polaroid Land Camera, the Model 95:

(Top) Double-roll film is dropped into the camera back;

(Bottom left) Paper leader is pulled out to position the film for exposure;

(Bottom right) Finished print is peeled off the negative after development inside the camera.

Reprinted from Popular Science, *February 1949, courtesy of Bonnier Corporation*

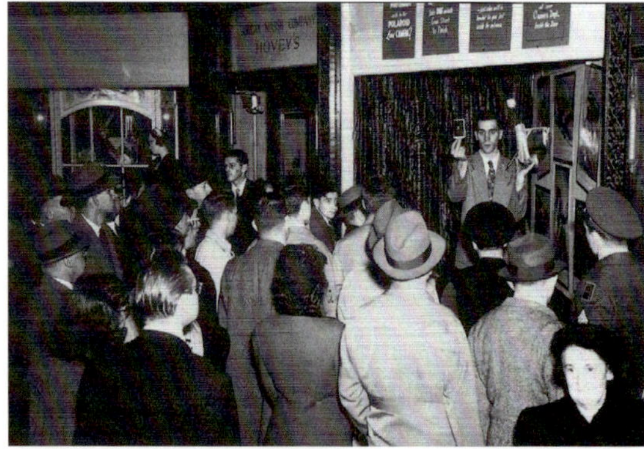

Figure 4-4: A demonstration of the new Polaroid Land camera and film system as it went on sale at Jordan Marsh, November 26, 1948.

Courtesy of Polaroid Corporation Archives

(Right) Figure 4-5: The Model 95 Polaroid Land camera.

Author's collection

(Below) Figure 4-6: A carton containing a roll of the original Polaroid Land Film—Type 40—that produced eight "finished pictures in a minute."

Author's collection

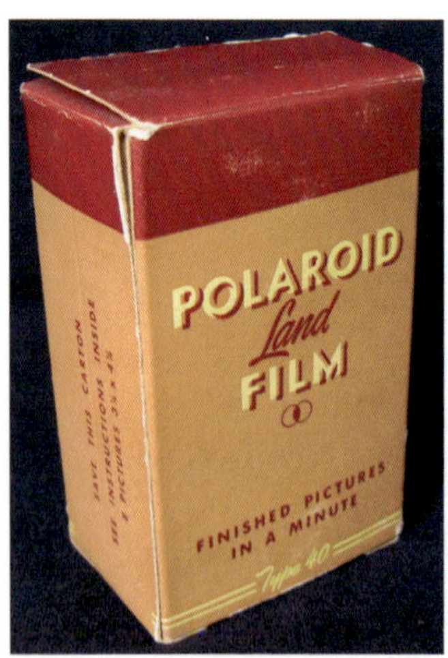

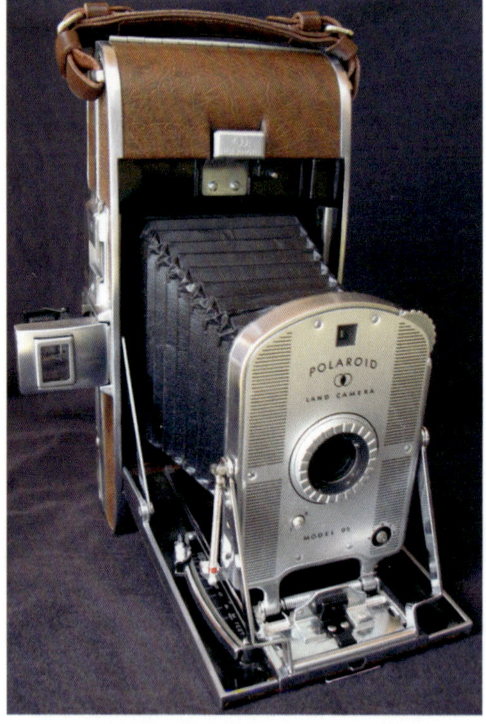

Figure 4-7: Famed Canadian artist Sid Barron weighed in with his take on the new one-step photography system in a cartoon published in April 1947.

Courtesy of Jesi Barron. Polaroid Corporation Administrative Records, Box I.89, f. 29 Press clippings—Cartoons, 1947, Courtesy of Baker Library Historical Collections, Harvard Business School

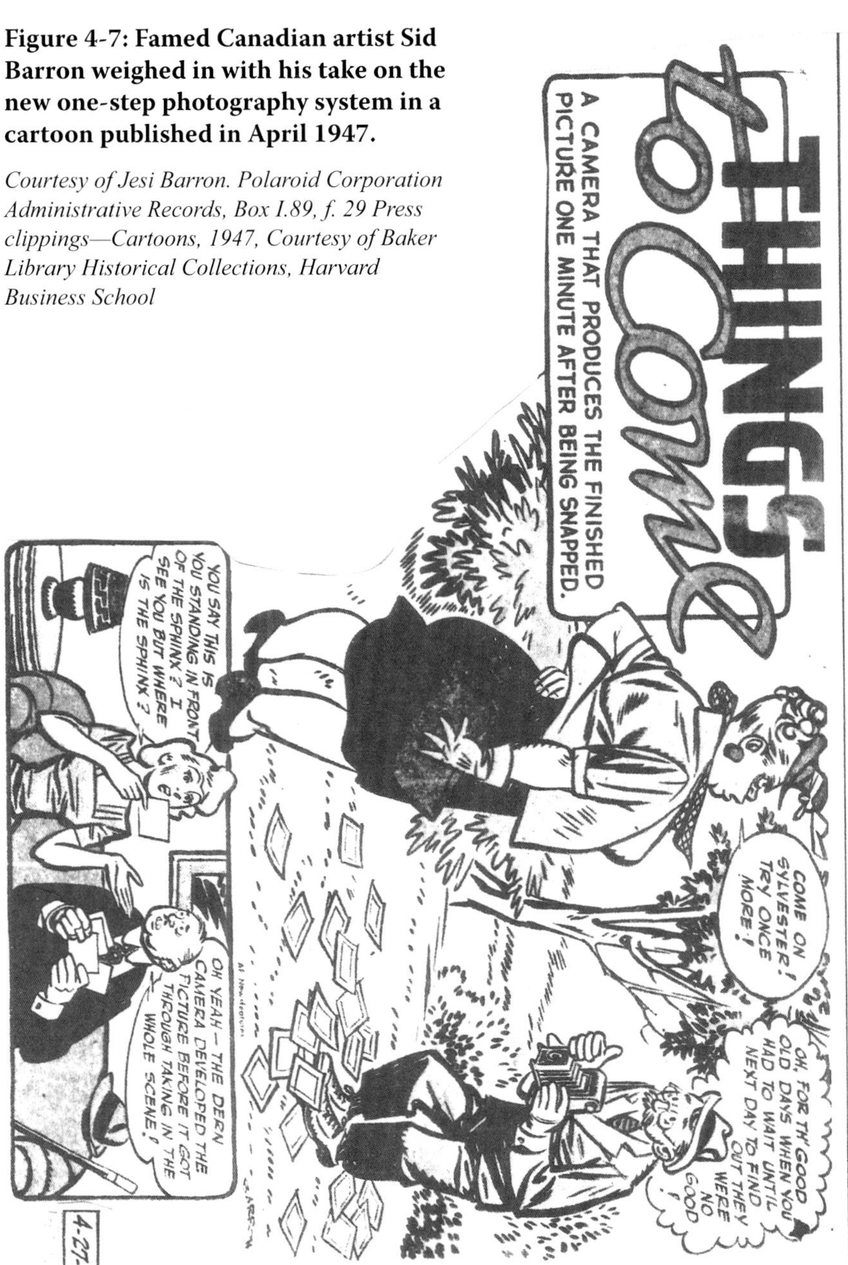

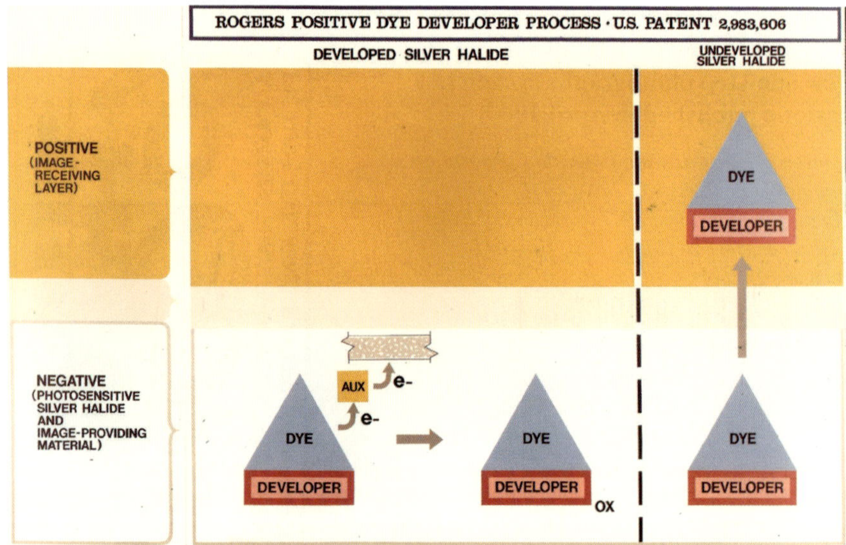

Figure 5-1: Charts depicting the action of a dye developer molecule in Rogers' Positive (top) and Negative (bottom) Dye Developer Processes.

Polaroid Corporation v. Eastman Kodak Company, United States District Court, District of Massachusetts, Civil Action No. 76-1634-Z, Exhibits PT-219 and PT-220

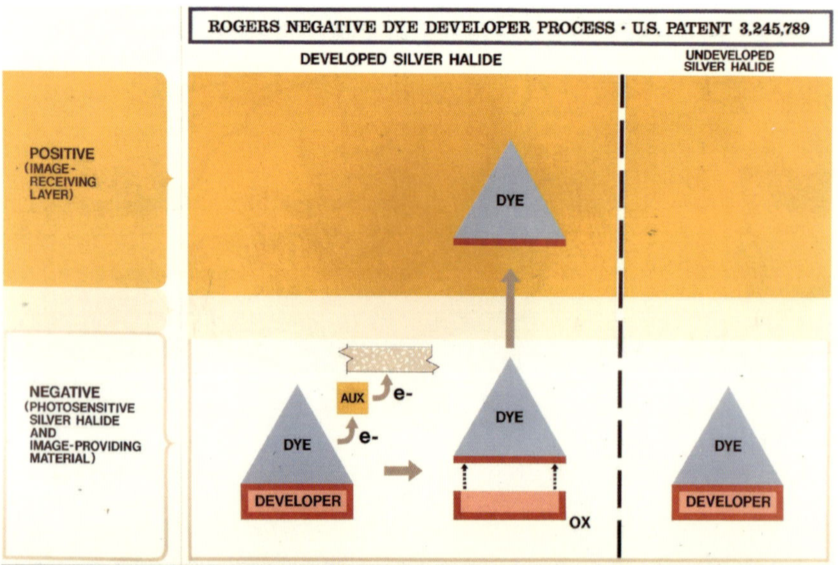

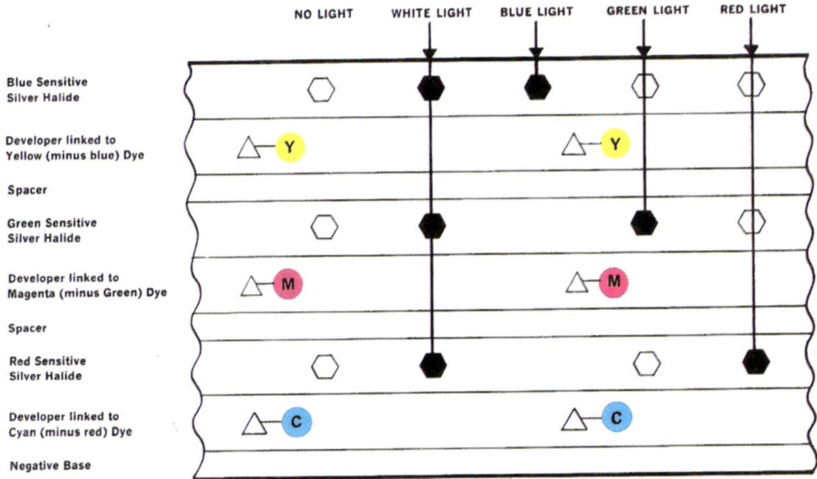

Figure 5-2: These charts depict the structure and operation of the negative for the first Polaroid Polacolor one-step film, which used Rogers' positive dye developer process. Above is the negative during exposure but prior to processing. The diagram below depicts the dye developer process during processing to form the image.

Booklet, Background Information about Polaroid Land Color Film, Polaroid Corporation, January 1963, Author's collection

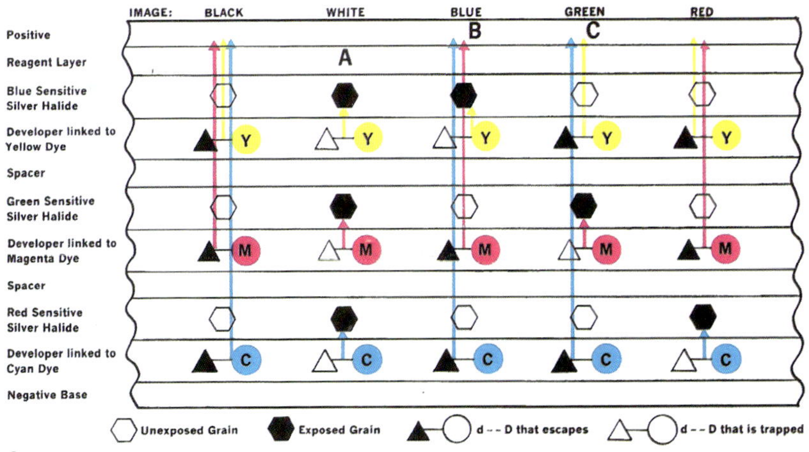

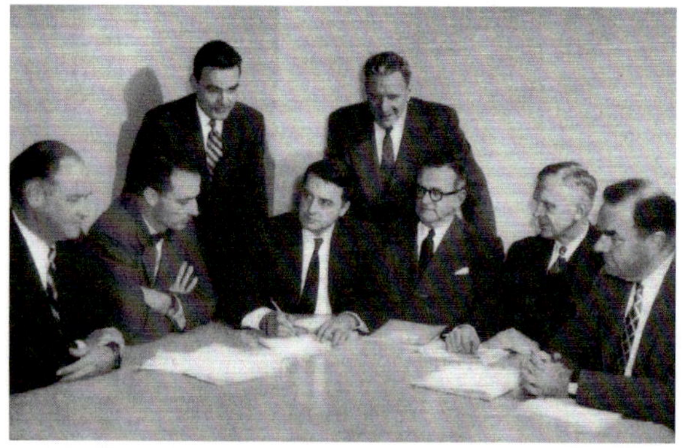

Figure 5-3: Polaroid's board of directors in the mid-1950s. Land is seated at the center, with Julius Silver (wearing glasses) seated to his left. William McCune is second from the left, and Donald Brown is on the far right. The others are Harold Booth (seated far left), Robert Casselman and David Skinner (standing, left-right) and Carlton Fuller (second from right).

Courtesy of Polaroid Corporation Archives

Figure 5-4: Land receiving the National Science Medal from President Lyndon Johnson, February 13, 1968.

Boston Globe, *February 14, 1968, courtesy of Associated Press*

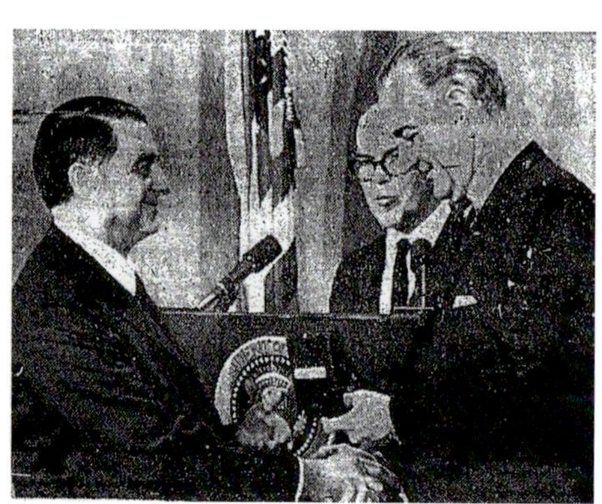

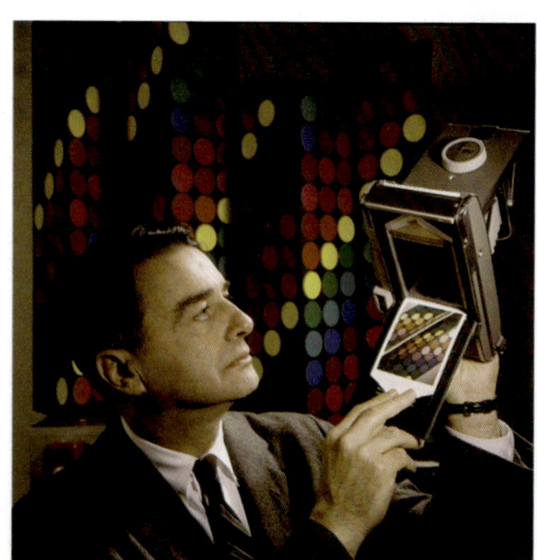

Figure 5-5: Land peeling a color image made with Polacolor film from the back of a Polaroid camera.

Reprinted from Life, *January 25, 1963, courtesy of Getty Images*

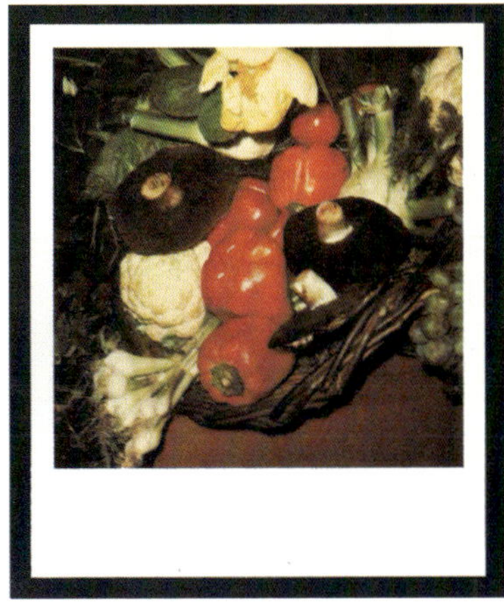

(Left) Figure 6-1: An early prototype of the integral film unit format, showing the now-familiar white border with a wider section at the bottom to cover the pod. The SX-70 camera had to be designed to handle this film unit.

Reprinted from Land, Edwin H., "Absolute One-Step Photography," The Photographic Journal, *July 1974, courtesy of The Royal Photographic Society, UK/www.rps.org*

(Below) Figure 6-3: Land pictured in his Osborn Street office with key members of his SX-70 research team. Back row (left-right): Nicholas Hadzekyriakides, Gordon Kinsman, Thomas McCole, and Dr. Stanley Bloom, who led the search for an opacification system that would allow SX-70 film to develop outside the camera in full light. Front row (left-right): Howard Rogers, Florence Rubin, and Inge Reethof. Nan Chequer is obscured behind Land's left arm.

Photograph by Alfred Eisenstaedt, Courtesy Time Life Pictures/Getty Images

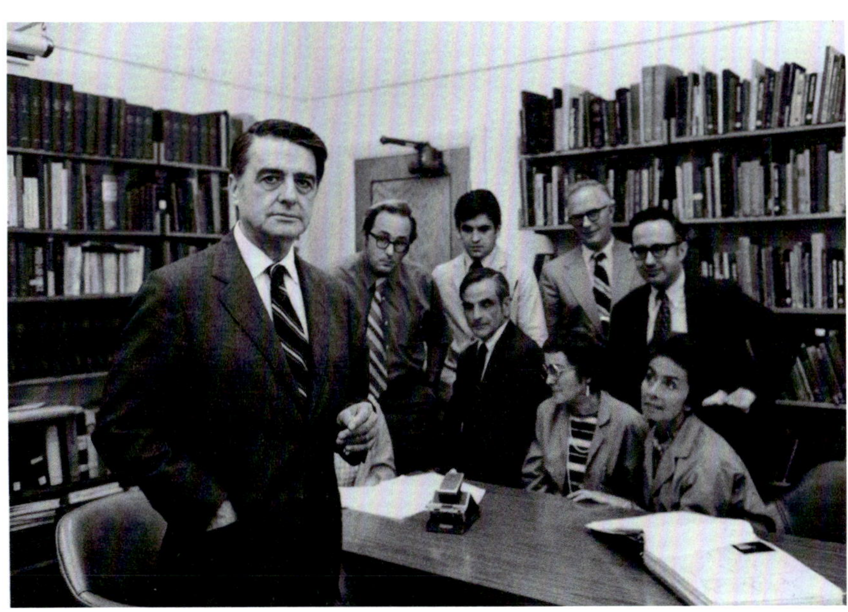

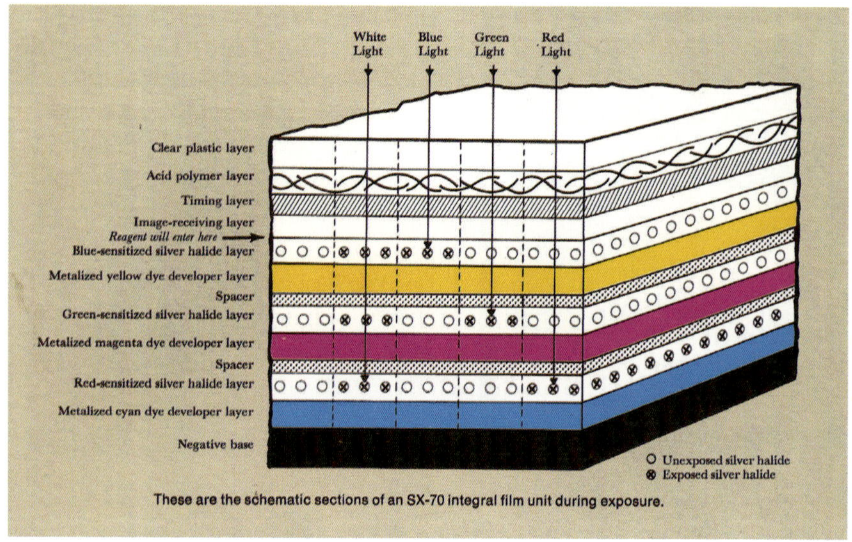

These are the schematic sections of an SX-70 integral film unit during exposure.

Figure 6-2: The integral one-step film unit in operation:

(Top) During exposure;

(Bottom) During processing as image forms after spreading of the temporarily opaque processing composition.

Reprinted from Land, Edwin H., "Absolute One-Step Photography," The Photographic Journal, *July 1974, courtesy of The Royal Photographic Society, UK/www.rps.org*

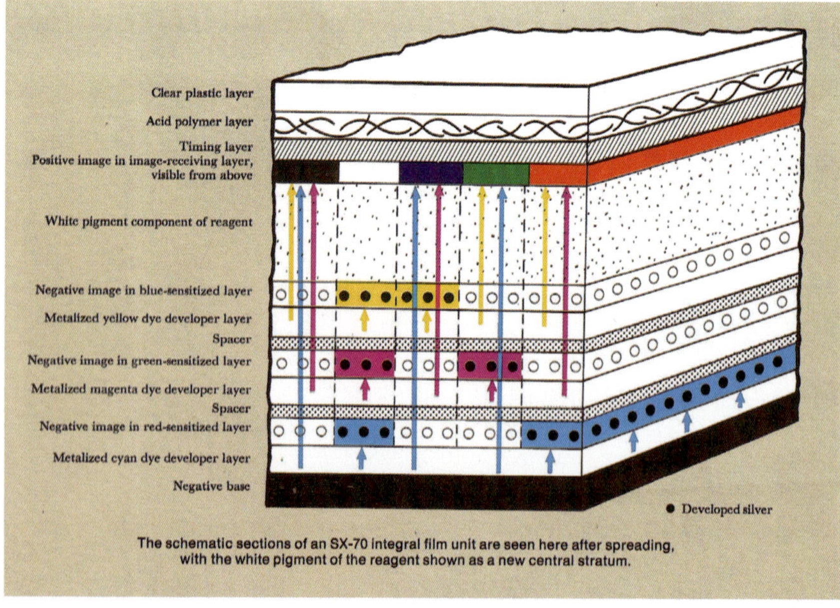

The schematic sections of an SX-70 integral film unit are seen here after spreading, with the white pigment of the reagent shown as a new central stratum.

the U.S. Justice Department—and that had led to the FTC investigation.[45] Thus, in effect having been on the other side of this argument, Schwartz was fully aware of the ins and outs of what Kodak was attempting to do. More importantly, he fully appreciated the distinctions between Polaroid's patent practices and the patent conduct of Xerox that had made it vulnerable to misuse charges.

The fundamental difference between the two cases was the manner in which the two companies had acquired the patent portfolios that enabled them to monopolize their respective fields. Xerox had been actively licensing its technology in copying technologies other than the plain-paper field for years. But in return for those licenses, it had a policy of requiring its licensees to grant licenses back to Xerox for any technology they had in the plain-paper area. Xerox had also simply bought patents relating to this technology from other companies that had actually pioneered the field. This gradually resulted in a situation where Xerox had accumulated essentially all patent rights to plain-paper copying technology. Significantly, those rights had not been acquired as a result of research and development work done at Xerox but, rather, through this aggressive barter patent-licensing policy and outright purchase. Schwartz himself had developed these facts through an ambitious program of deposing executives from scores of companies that Xerox had abused in this fashion. He had also deposed Sol Linowitz, the CEO of Xerox, who had been the architect of its strategy.[46]

It was a very different situation at Polaroid, where the entire field of instant photography had been invented and developed over decades through its own hard work and scientific initiative. Polaroid had not acquired any of its patents from others. This was a major distinction in Schwartz's view. Because Polaroid had secured its patent monopoly based on work done in its own labs, by its own scientists, Schwartz believed that, under the law, it was under no obligation to license that technology to others and had nothing to fear from its practice of refusing to do so. Instead, it was perfectly entitled to enjoy the seventeen-year exclusivity that the patent laws afforded on each invention. On that basis, Polaroid counsel asked the court to dismiss Kodak's patent misuse claims against Polaroid. They were confident that Kodak's patent misuse strategy would fail and that this part of Kodak's defense would go nowhere. They advised Polaroid accordingly, although concern persisted within Polaroid's legal department and executive ranks. Peck and the others seemingly couldn't believe that it would be that easy.[47]

The reaction in the Polaroid camp to Kodak's misuse defense may have dampened Polaroid's initial exhilaration over the initiation of its case against Kodak, but events occurred abroad in those first few weeks that would further undermine some of Polaroid's early internal confidence about the prognosis for its case. In early August 1976, the U.K. court held hearings on Polaroid's motion to secure a preliminary injunction against Kodak's release of its products, pending final resolution of the pending patent infringement suit. Toward the end of those hearings, on August 6, Kodak's stock fell about three percent in one day. The *Wall Street Journal* attributed this precipitous drop to the fact that London observers expected the hearings to end in what was known as a "quick decision" handed down by the court without lengthy legal explanations.[48] Such a decision was apparently imminent.

Sure enough, the next day, Justice John Patrick Graham in the High Court of London found for Polaroid and issued a temporary injunction preventing Kodak from making and selling in the United Kingdom its "instant picture" cameras and film. Under the peculiarities of U.K. law, this finding was not based on the substance of the lawsuit. While the court admitted that Polaroid had presented a "fair, arguable case," there was no final determination on the merits of Polaroid's infringement claims or Kodak's defenses of patent invalidity or unenforceability.[49] Instead, Justice Graham relied on the uniquely U.K. concept known as the "balance of convenience" in concluding that since Kodak's products had not yet been released to the public, it would be less disruptive to prevent them from moving forward than it would be to have to recall the products should they ultimately be found to have infringed Polaroid's patents. He also concluded that Kodak might have done more to avoid this dilemma by either applying to Polaroid for licenses or by using provisions of U.K. law for either a compulsory license or a declaration of noninfringement.

One might have expected this early apparent victory to provide great comfort and encouragement to Polaroid. But such was not the case. First of all, as the *New York Times* reported, the decision "merely imposed a . . . technical holding action" on the release of the products and did not "foreshadow a final resolution in Polaroid's favor."[50] Second, the hearings on the motion in the U.K. case had given the parties some opportunity to outline their respective theories on the merits of the case. In particular, Kodak had described its theories of why the patents were invalid and/or unenforceable because of various defects and why Kodak's products did not infringe them. This preliminary elucidation of Kodak's case provided the first hard information that could be used by observers to evaluate the merits of the U.S. case, and it did nothing to dissuade those who believed that Polaroid was going to have a hard time ultimately prevailing in court. As one Wall Street observer reported to its subscribers, "we doubt that Kodak would have entered the instant market, investing hundreds of millions of dollars to do so, if it had not been reasonably confident that it could win any patent infringement cases or at least settle them at reasonable cost."[51]

One of the leading Wall Street commentators on business litigation, Calvin Crary of Bache Halsey Stuart, issued a detailed report on the U.K. proceeding. His piece was called "*Polaroid v. Kodak*, The British Injunction Proceeding: Window on the U.S. Case."[52] Crary noted that four of the patents in the United Kingdom had direct counterparts in the U.S. litigation. Two of those were Land's patents, the Rear Pick and the L-Coat. After analyzing the detailed arguments asserted in the British court, he concluded that Kodak's position appeared to him "to have merit" with respect to the former and that "Polaroid will have to argue a difficult position" with regard to the latter. Crary also detailed Kodak's case against the counterpart to the critical Rogers' Excedrin film unit patent. All in all, he reported, while "it was fairly evident . . . that the Kodak system isn't exactly revolutionary," he believed that it had a "good chance" of squeaking by Polaroid's claims of infringement in the U.K. lawsuit. He then went on to "make the same assessment" of Kodak's case in the United States. Crary didn't even believe that the preliminary injunction had been properly awarded to Polaroid. He concluded that he was "unpersuaded" by the justice's reasoning in his decision and that he "wouldn't be surprised" to see it reversed on appeal.

Within a few months, Crary was proven right about the preliminary injunction. On November 10, a British appeals court overturned the ruling, allowing Kodak to proceed with the launch of its new instant cameras and film in the United Kingdom. Lord Justice Denys Burton Buckley of the appeals court reasoned that there was no serious danger that Kodak's entry into the market would irreparably injure Polaroid. He stated that Kodak would neither drive Polaroid out of the market, nor cause injury to Polaroid that could not later be compensated for in money damages should Polaroid ultimately prevail on its infringement claims.[53] This latter test was apparently a necessary step in the legal analysis that the lower court justice had skipped.

In retrospect, the U.K. preliminary injunction episode was a chapter in the overall litigation that could have and should have been avoided. Years later, Schwartz admitted that he and Kerr "had really been sold a bill of goods" by Polaroid's British counsel, who had convinced them that U.K. law gave them a good chance of succeeding in securing the preliminary injunction.[54] Had they fully understood the law in the United Kingdom, Schwartz was sure "he would have fought bitterly against bringing" the U.K. motion.[55] In the end, everything this U.K. initiative did was harmful. It allowed Kodak an unnecessary preview of some of Polaroid's arguments on key patents, especially the Excedrin film format, and it unnecessarily sapped Polaroid of momentum and confidence while giving some comfort to the opposition. To Wall Street and business observers, it raised the possibility that someone like Crary might be right about his pessimistic assessment of Polaroid's overall prospects for victory. If Crary had been right about the injunction being reversed, could he also be correct in his prediction that Kodak had a "good chance" of winning the main case?

CHAPTER 12

THE BATTLE IS JOINED

With the legal issues now drawn, the lawsuit entered the pretrial discovery phase, during which both parties try to secure information from each other to bolster their respective cases. Almost immediately, interrogatories were served. These are written questions, usually in many multiple parts, that require written answers. Simultaneously, requests were made for the other side to produce documents from within its files that pertained to the various factual issues in the suit. Later, depositions would be held at which various company engineers and executives would sit, sometimes for days at a time, responding to questions posed by the other side. While each of these procedures is supposed to be a means of discovering the facts of the case, they may also be wielded as strategic weapons in that they impose varying degrees and forms of burden on the opponent.

Everyone knew that it would be a protracted and acrimonious struggle. Even so, the litigation had an incidental benefit to Polaroid—an intimidating effect on the plans of other companies to enter the instant photography market, particularly those other camera manufacturing companies that had intended to take advantage of Kodak's previously announced plans to license camera manufacturers to make products that would use its instant film. By the early summer of 1976, *Modern Photography* noted that the "rush to 'sign up' among many Japanese camera manufacturers . . . was cut to a tiny trickle."[1] Other potential entrants into the instant photography market—Germany's Agfa-Gevaert and Fuji Photo Film Co.—continued to show interest in the field but also took note of the lawsuit in delaying their plans.[2]

Meanwhile, Kodak was having other problems with respect to its launch. In early September, two of Kodak's top executives, Walter Fallon and Executive Vice President Colby Chandler, admitted to a group of analysts that production problems with the EK-4 and EK-6 cameras were being addressed, assuring them that there would be an "adequate supply" for the holiday buying season.[3] On September 10, Kodak announced that it would not be introducing the high-end model in its camera series, the EK-8, in 1976 as previously planned. Kodak explained it was "deferring those plans," ostensibly to concentrate on the lower end of the market where its EK-4 and EK-6 competed with Polaroid's Pronto model.[4]

More alarmingly, hints about potential defects in Kodak's PR-10 film started to surface. Consumer testing seemed to indicate that there was a question concerning the stability of the prints when exposed to light.[5] In fact, *Consumer Reports* suggested that Kodak's instant film might be useful only for photo albums or storage in a drawer, and could not be displayed.[6] This problem received widespread publicity, in the general media as well as in the photographic industry press. Stories spread about Polaroid executives who "get a kick out of taping Kodak prints to their office windows, then exhibiting the results to visitors [because] in a few days, the prints are all but obliterated."[7] Publicly, Kodak was defiant in its defense and asserted that "the stability of Kodak instant prints is entirely satisfactory when such prints are handled, displayed or stored in the usual variety of home and office situations."[8] Yet when Kodak conducted its own testing, its instant film rated "poor" in its stability to light, whereas Polaroid's rated "good," a result that engendered front-page headlines in the Rochester newspaper.[9]

Despite Fallon's claim that Kodak's new instant products "could very well be the most thoroughly tested products Kodak has ever produced,"[10] there was actually some talk about them failing altogether. One "major" New York camera dealer was reported as saying "lots of people are buying Kodak instants because they think they might flop. They want to own one as a curiosity . . . as a kind of Edsel, just in case Kodak doesn't make it."[11] An item in *Newsweek* noted: "Eastman Kodak may dominate the $6 billion-a-year photography business in Olympian fashion, but when it comes to innovation the company has had an embarrassing tendency to lose focus [allowing] Polaroid [to] grab . . . a 30-year lead in the instant photography field."[12] The reaction on Wall Street was a steady drop in Kodak's stock price.

In contrast, the financial news for Polaroid was nothing but positive. Its 1976 third-quarter earnings were projected to be up fifty percent over the previous year, and more analysts began recommending its stock, which was trading near its twelve-month high.[13] Julius Silver made sure that Land saw a report from Drexel Burnham Lambert Research reporting on "a most salutary development—the greater and even faster growth in the market for instant cameras than even the most optimistic observers had assumed."[14] Land was especially anxious to address the financial community, which he thought had treated Polaroid so harshly during the inevitable glitches in the SX-70 introduction and as it weathered the building anticipation of Kodak's entry into the market. Of course, it was easy for him to gloat now that Kodak's products had proven not to be the significant technological leap forward that Polaroid had feared. In an interview published in the *Boston Globe*, Land expressed his relief over this turn of events as well as his anger at those who had made the journey to this point more difficult than it might have been. He said:

> We're really off and running. It's a very exciting time. What I call the analyst albatross has finally gone. You don't know what a pain in the neck it was to pick up the newspaper every morning for four years and to read our stock is down half a point in "view of the fact that Kodak is coming into instant photography this year." None of those who misled the public, however innocently, has had the grace to apologize for the 4-year barrage of that kind of carefree misrepresentation.[15]

The cumulative effect of these developments led some industry observers to the conclusion that things might actually go Polaroid's way, perhaps more quickly than initially anticipated. A lengthy review of the situation in *Barron's* declared that Kodak "has never won a major case in court" and reported a "consensus" that it "will wind up paying Polaroid something in tribute to stay in the market."[16] Louis Rusitzky, the longtime photographic analyst for Boston's financial firm Adams, Harkness & Hill, postulated that Kodak would actually be relieved to settle the Polaroid lawsuit and pay Polaroid a royalty so as to prove to the world that it was not "destructively dominant" in photography, as had been alleged in several antitrust suits pending against it.[17] But these observers badly underestimated Kodak's fierce determination. After investing so much time and money over two decades to get into the instant market, Kodak was not about to give in.

On October 19, 1976, Walter Fallon conducted another presentation for security analysts in San Francisco meant to change perceptions about Kodak's prospects for instant photography. Colby Chandler, a quickly ascending star in the Kodak executive constellation, was also there. Like

Fallon, Chandler was known as a "tough, tight-lipped businessman of the Kodak mold."[18] A veteran of twenty-six years at the company, Chandler was a physicist by training who had risen within the ranks to become Fallon's right-hand man, heading Kodak's U.S. and Canadian photographic divisions. Polaroid, not invited to the event, was eager to know what transpired. It secured a report on the presentation from Ted James, an analyst with the Boston firm of Robertson, Colman, Siebel & Weisel. James also provided information based on a direct conversation he had with Chandler about Polaroid and instant photography.

In his formal presentation, Fallon unequivocally stated, "we see no end to Kodak's interest in instant photography."[19] "No new product in our history has received the attention given to the family of instant products introduced by Kodak in April of this year," Fallon crowed.[20] He disclosed that the company expected to ship more than one million cameras before the end of 1976 and that, in 1977, Kodak would increase its production of instant cameras and film by "a factor of several times."[21] "Regarding Kodak's promotional efforts for their instant products," reported James, "Mr. Chandler emphasized that Polaroid hasn't seen anything yet."[22] James' report on Kodak's bullish and resolute position was circulated to Polaroid's Management Committee and then shared by Bob Peck with Bill Kerr and Herb Schwartz.[23] The message received in Cambridge was loud and clear. Kodak was fully engaged and would not be backing down without a fight.

Fallon's comments at the San Francisco meeting encouraged Julius Silver and Richard DeLima to raise once again the possibility of bringing some kind of antitrust action against Kodak for what they perceived as its "predatory behavior."[24] Key points in such an action would be the tough negotiating position Kodak had taken in the 1969 discussions, when it refused to continue manufacturing negative for Polaroid unless it was granted licenses enabling it to compete in instant photography, and the injury done to Polaroid by Kodak's "repeated premature announcements" of its entry into the instant market over the eight years before it released its products.[25] Kerr and Schwartz made clear once again their conviction that this was a waste of time—an argument without merit from a legal point of view.[26] The idea was finally dropped.

As 1976 drew to a close, both parties were prepared to start producing to each other internal corporate and laboratory documents, as well as written answers to interrogatories in response to their respective discovery requests. Both sides acknowledged that a good portion of this material would contain confidential proprietary information. The general practice in such a situation is for the parties to agree to the terms of a protective order that spells out procedures for handling this material, including limitations on personnel who are allowed access to it. But Kodak would not agree to the stringent restrictions that Polaroid sought.[27] On December 23, 1976, Schwartz made a motion requesting the court to resolve the issue.[28] It was Judge Frank J. Murray's first opportunity to become involved in the litigation and to assess the respective counsel. He ruled quickly, basically adopting the limitations suggested by Polaroid counsel but leaving room in the protective order to expand them should the exigencies of the case demand it. Effective January 5, 1977, a protective order was in place, and both parties could now begin exchanging discovery materials.[29]

Bickering over discovery had little to do with the substance of the lawsuit; it spoke to the styles and the strategies of the lead litigators. As plaintiff's counsel, Bill Kerr's primary interest was getting discovery taken care of as quickly and as efficiently as possible and then trying the case. As defendant's counsel, Carr's strategy was, by haggling over everything, to delay the proceedings as much as possible, putting off the actual trial for as long as he could, maybe forever. Kodak counsel intended to make the discovery process as excruciatingly slow, aggravating, and ultimately expensive for their opponent as possible. In this, they were indeed successful.

After an interchange with Kerr in early January, Carr committed to produce documents in two-week intervals but could offer no prediction as to when the production would be complete.[30] When Kodak finally delivered an installment of documents at the end of the month, they made it clear in the transmittal letter that nothing additional would be forthcoming until Polaroid made a production. Kerr was livid. Before circulating the letter to the other Fish & Neave lawyers now working on the case, he scribbled a note on it: "Chicken shit—but we'll have to go along."[31] The next day he fired off a letter to opposing counsel.

> Let's not play games. There was never any meeting of the minds on the proposition that both sides would produce at two-week intervals. We expect to produce documents to you as they are collected and processed, and we assume that you will do the same, irrespective of whether such productions are made simultaneously or at some regular intervals. This is, at least in my experience, the conventional *modus operandi* for production of documents.[32]

Kerr went on to ask once again for a date on which Kodak expected to complete its production.

It took two weeks for Carr finally to respond to Kerr's letter. Acknowledging that he had "planned to drop this note to you but couldn't seem to get to it," Carr couldn't resist tweaking Kerr: "I'm sure you realize that we don't play games."[33] He reiterated that Kodak could not provide any estimate for completing its production and then advised that he had "a number of problems that we would like to raise with you . . . relating to the nature of the production thus far as well as the responses to the interrogatories and document demands." Translation: we're in no hurry to do or to address any of this, and we will be dragging our feet as well as contesting every piece of paper and interrogatory answer as best we can for as long as we can. This was indeed "chicken shit," but it reflected the temper of the time and the stage of the lawsuit.

This skirmish was an indication that the legal battle would initially be fought in the trenches, in an absolutely grueling discovery process, using armies of lawyers, paralegals, secretaries, and clerical personnel. In these days before fax machines or e-mail, letters between counsel arguing over this or that would fly back and forth, up and down the avenues of New York, from Fish & Neave's Park Avenue offices, to and from the Wall Street location of Kenyon & Kenyon, carried by bicycle messengers, and delivered by hand. While Kodak's strategy of delay was clear, and while there was little Polaroid could do to avoid this, they could certainly attempt to give back as good as they got.

Meanwhile, Kerr and Schwartz worked to move along the pending Berkey Photo case. As part of its defense, Berkey had filed a counterclaim seeking to have the court cancel Polaroid's trademark on the name "Polaroid," based upon allegations that its use was basically anticompetitive.[34] In late December, Kerr and Schwartz succeeded in having the court dismiss this counterclaim from the lawsuit, eliminating an irrelevant but nonetheless potentially troubling issue from the matter.[35] Sidney Neuman, counsel for Berkey, had long been seeking a way to settle this case on behalf of his client. Since at least August 1976, he had been pressing Kerr to inform Polaroid of his client's willingness to accept a license "upon reasonable terms" under the patents in suit.[36]

But if Polaroid were to license Berkey, it would be forced to license others; favoritism in its licensing policy would not withstand legal scrutiny. Ultimately, in no small measure because of Kerr and Schwartz's advice that it was not obligated to license anyone, any decision to move ahead with licensing was shelved in favor of pressing ahead with the infringement litigation on all fronts, including Berkey's.[37] In early

February 1977, Kerr informed Neuman "that Polaroid does not wish to license Berkey under the patents in suit at this time."[38] Then Polaroid went on the offensive and decided to ask the court to sever the issues of patent validity and infringement for trial and to stay any other stray issues that Berkey had dragged into the proceeding. Kerr and Schwartz knew that having separate trials on these issues would significantly raise the expense of litigating this matter through trial. Berkey was already immersed in other expensive litigation, including a major antitrust action it had brought against Kodak. Unlike its relative position in its litigation with Kodak, Polaroid had vastly greater resources than Berkey. Kerr and Schwartz knew that a significant strategic advantage would be gained if the court granted this request. They moved quickly and filed that motion on February 22, 1977.[39]

At the same time, Kerr and Schwartz maneuvered to have the Berkey case brought to trial at the earliest possible date. The action had been brought in Delaware because of the prospects for a "reasonably early trial."[40] The idea was to keep pressure on Berkey in every conceivable way. Independently, during the spring of 1977, Polaroid's engineers helped by making some changes to its SX-70 film pack that prevented the film container from working properly in Berkey's cameras. This modification was made without any notice, and Berkey's counsel complained directly to Kerr, alleging that the action was "designed to improperly exclude Berkey from the market place" and, as such, had caused his client "substantial harm."[41] On this point, Kerr remained silent.

In early 1977, Walter Fallon and Colby Chandler received promotions, thereby cementing their leadership of Kodak. Fallon continued as chief executive officer but added the title chairman of the board of directors. Chandler moved up to president. Despite the optimistic reports given by the two executives to financial analysts the previous fall, the first half of 1977 continued to be a difficult time for the company. The first quarter proved to be a poor one, with Kodak reporting a twenty percent drop in earnings.[42] The company's stock price continued an "alarming" decline to a level about one-half of what it had been a year earlier in the spring of 1976 when it had first introduced its instant products. Kodak's early disappointments in instant photography contributed heavily to the drop. The *New York Times* reported that "the shares have been in a more or less steady decline ever since Kodak introduced another 'me too' product, its instant developer cameras decades after pioneering Polaroid."[43] Susan Black, photography analyst for Donaldson, Lufkin & Jenrette, noted that she had taken Kodak off her buy list the previous fall even though she wasn't sure how significant instant photography and its attendant issues were going to be for Kodak. She now admitted that "as time has passed I am even more convinced that it is important and not the icing on the cake for the company," thus cementing her pessimistic view.[44]

Behind the negativity was the reality of complications for Kodak in the rollout of its instant products. In March, it was announced that the already delayed top-of-the-line EK-8 camera would finally be released in Europe but not in the United States. Kodak was not willing to say exactly why not. The *Wall Street Journal* reported that the U.S. introduction would depend on European sales.[45] According to the *New York Times*, all that Douglass Harvey, a Kodak executive vice president and general manager of its photographic division, would say was that the release would be put off "until later."[46] In April, Kodak announced that it was immediately cutting the work force on its instant photography products by 150 jobs through a program of rotating furloughs.[47] In May, it announced a rebate program for consumers and dealers in an attempt to spur sales of its instant cameras and film.[48] Spokesmen denied that this was a price war with Polaroid, claiming it was merely an attempt to "build up our instant camera base."[49] However, to others it appeared, along with the layoffs, to be another sign of weakness as Kodak struggled with the launch of its instant-photography line.

During this period, another bit of worrisome news for Kodak, and perhaps the most ominous, was the announcement that it had received a civil investigative demand for documents from the U.S. Department of Justice in connection with an investigation into the company's photographic business. Kodak was already on the defensive with regard to its dominance of the consumer photography industry. Three separate companies had lawsuits pending against it for alleged antitrust violations. Berkey Photo, GAF, and the Pavelle Corporation alleged, in essence, that Kodak had monopolized or had attempted to monopolize the markets for film, photofinishing, photofinishing equipment, photofinishing chemicals, and the amateur photography industry as a whole. At least one professional observer believed that these cases meant trouble for Kodak. He commented that the "cases constitute serious challenges to Kodak's business . . . from which significant consequences should be anticipated."[50]

Kodak was well aware that its 1950s run-in with the Justice Department in similar circumstances had led to its being forced to sever its photo-processing operations from the rest of its business. An additional investigation by the government was thus certainly something Kodak did not welcome, especially as a similar inquiry had proven to be problematic for Xerox in its patent battle with IBM. It was yet unclear whether or to what extent this investigation would reach into Kodak's instant photography activities, but its pendency certainly was another source of concern in Rochester.

In stark contrast, the news was better in Cambridge, Massachusetts. As sales continued to be strong for Polaroid in early 1977, the company enjoyed a major public relations boost as well. The year began with the induction of Edwin Land into the National Inventors Hall of Fame. The U.S. Patent Office had established this institution in 1973 and had made Thomas Edison its first inductee. Land was joining an august and highly selective group that included the most important inventors in history, including Alexander Graham Bell (telephone), Eli Whitney (cotton gin), the Wright Brothers (airplane), Guglielmo Marconi (radio telegraphy), Samuel Morse (telegraph), Cyrus McCormack (reaper), Charles Goodyear (vulcanized rubber), and Rudolf Diesel (internal combustion engine).[51] He was the first inductee ever admitted during his lifetime.[52] Land was elected for the sum of his contributions to the fields of sheet polarization as well as one-step photography.[53] In particular, he was cited at the event for his basic patent on the rupturable pod—the container within the film unit for thc processing fluid—basic to every one-step photography system starting with the original, peel-apart process.

Ironically, that 1977 class of five inductees also included George Eastman, the founder of Eastman Kodak, who was honored for his contributions to photography. In 1884, Eastman invented the first photographic film and then went on to build and eventually market his first camera in June 1888. The research, manufacturing, and marketing institution he fashioned based upon those landmark discoveries went on to become one of America's great companies. Known as a quiet man and a bachelor for life, the founder of Kodak lived with his mother in a Rochester mansion he had built. In 1932, at the age of seventy-seven, Eastman committed suicide, leaving behind a note: "To my friends. My work is done. Why wait? G.E."[54]

Wesley Hanson, then Kodak's director of research, represented Eastman at the induction ceremony. He pointed out that Eastman's significance lay "in his sense of a universal need and his efforts to provide the means to meet it."[55] Without referring to the lawsuit, Hanson acknowledged the historic mentoring that his company had provided to Eastman's co-honoree. "I am certain that [Eastman] would take special pride and delight on being here with Dr. Edwin Land. . . . The technology developed by 'Din' Land and the company he founded are clear indications of the impact a brilliant mind and a tireless effort can have in the space of one lifetime."[56] (See Fig. 12-1.)

In his acceptance speech, Land chose to pay tribute to the process of invention by analogy to the basic American sense of adventure and exploration:

> We are becoming a country of scientists, but however much we become a country of scientists, we will always remain first of all that same group of adventurous transcontinental explorers pushing our way from wherever it is comfortable into some more inviting, unknown and dangerous region. Now those regions today are not geographic, they are not the gold mines of the west; they are the gold mines of the intellect. And when the great scientists, and the innumerable scientists of today, respond to that ancient American urge for adventure, then the form that adventure takes is the form of invention; and when an invention is made by this new tribe of highly literate, highly scientific people, new things open up. . . . Always those scientific adventurers have the characteristic, no matter how much you know, no matter how educated you are in science, no matter how imaginative you are, of leading you to say, "I'll be darned, who ever thought that such a domain existed?"[57]

The day ended with another special presentation to Land. At the induction ceremony, Commissioner of Patents C. Marshall Dann awarded Land his 500th patent, granted in connection with his work on flat batteries.[58] Even though he was sixty-seven years old, there would be more than thirty additional patents issued to Land in the coming years before his research career concluded, leaving him third on the list of America's most prolific inventors. Fittingly, his first and the last patents were in the field of light polarization.

Other news for Polaroid was also encouraging. In the spring of 1977, it was announced that Polaroid's earnings for the first quarter of the year had risen by thirty-three percent from the previous year, with per-share net profits setting a company record for the period. This was all part of a longer positive trend in that it represented the ninth consecutive quarter in which Polaroid's profit sharply exceeded the previous year's level. At the fortieth annual shareholders meeting in late April, Land once again made one of his dramatic appearances to introduce a new product. As *Newsweek* observed, "for most of the year, Edwin Land is the reticent, reclusive genius of instant photography—but when he has a dazzling new product for his shareholders, he puts on a show to rival P.T. Barnum."[59]

That year's dazzling new product was Polavision, an instant home-movie camera system. Land's introduction was, as one might have expected, decidedly not understated: "The era of immediately visible living images is now at hand," he declared.[60] Land orchestrated a "carnival-like" environment in which shareholders could try out the new system.[61] The moment was described as a "triumph" for Land, who explained the big "to-do" by telling the 4,000 assembled about "the rule they don't teach you at the Harvard Business School. . . . If anything's worth doing, it's worth doing to excess."[62] The technology writer for the *New York Times* dubbed Polavision a "technological skyrocket," predicting that "it could be the most spectacular [apparatus] that the 67-year old Dr. Land has developed in his decades-long career in instant photography, the billion-dollar industry he invented."[63] (See Fig. 12-2.)

Polavision was not the only product unveiled at the meeting. In a move characterized by the *Wall Street Journal* as perhaps "at least as significant as [Polavision] for Polaroid's financial prospects," Polaroid also used the occasion to introduce a new instant camera called OneStep.[64] This new model was motorized like the SX-70 but was non-folding and was expected to retail for forty dollars, a price point that would further help Polaroid compete against Kodak in the low-cost end of the instant market.

Amidst these events, for the first time Kodak showed something other than a total resolve to fight Polaroid to the finish. In all, it was a puzzling episode. At Kodak's April 26, 1977, annual shareholders meeting, Colby Chandler began with a bullish report on Kodak's instant photography business, noting that the company was "positioning [itself] for the profit turn in instant photography" as the number of Kodak instant cameras in use increased.[65] Part of this optimism stemmed from Kodak's introduction a month earlier of a new camera model—the Handle—that was touted as "easy to carry, easy to hold . . . with no settings to make."[66] But unlike Polaroid's Pronto, the Handle required the user to turn a crank to eject a photo. The Handle retailed for about ten dollars less than Pronto, but the conclusion of a report in *Photo Weekly* was that it could not "compare with the superbly engineered, compact, full focusing . . . [Polaroid] Pronto."[67] The Handle was declared a "brilliant" marketing move, but "no contest" technically.[68]

Then came the news that no one expected. In response to a question from a securities analyst, Walter Fallon surprisingly disclosed that Kodak had "talked informally" with Polaroid to explore the "possible settlement of [its] patent disputes."[69] He cautioned that it was "too early to predict what might emerge from the talks."[70] The *New York Times* characterized Fallon's statement as "an offer . . . to seek an out-of-court settlement."[71] In fact, Fallon had enlisted one of his lieutenants, David Greenlaw, to approach Polaroid early in the year to assess whether there was any possibility of settling the case. Greenlaw was a scientist by training who had worked closely with Fallon, who had "a high regard for Greenlaw's intelligence and ability."[72] Greenlaw had been brought in from Kodak's research and development area by Fallon to advise him on technology licensing matters.[73] He offered a fresh perspective on the litigation with Polaroid that was then over a year old.

When asked about the discussions in view of Fallon's disclosure, a Polaroid spokesman confirmed that they had occurred but described them as "informal, seeking a climate to have formal talks."[74] Kerr and Schwartz never took these discussions seriously.[75] Nevertheless, the national press picked up on them and did treat them as meaningful. Even the U.S. Justice Department was apparently following the news in the context of its investigation of Kodak. The day after Fallon's April 26 statement, Polaroid received a letter from Gerald Rubin, an attorney in the Antitrust Division, noting that the discussions concerning settlement had come to his attention "through public sources." Rubin conveyed his department's "desire" that all documents and information generated by the parties to the lawsuit should "be preserved pending any possible review of the material we may wish to make."[76]

Calvert Crary, the Wall Street analyst most closely following the case, was skeptical, reasoning that "Polaroid's visible motives for settlement are not particularly persuasive." In contrast, although he acknowledged that he "happened to favor the Kodak argument on each of the various patents involved in the . . . infringement litigation," he nonetheless thought that Kodak "would be happy with a world-wide non-exclusive license at a reasonable royalty under any and all patents . . . [under] dispute."[77] He reasoned that despite its stronger position on the merits, in his view, Kodak was in

> a high risk position in the litigation because of the probability of an adverse determination on one or more of the twelve patents which Polaroid has asserted. Moreover, pendency of the litigation itself is expensive, a drain on resources, and a cause of uncertainty both inside Kodak and among the people with whom it deals.[78]

In the end, Crary's overall prognostication was correct, but his analysis of Polaroid's situation was far more accurate than his reading of Kodak's position.

Cecil Quillen, an assistant chief counsel of Kodak's chemicals division, who had been brought up from Kingsport, Tennessee, shortly after Polaroid filed suit to become Kodak's litigation director, was now supervising the case for Kodak. (See Fig. 12-3.) He instructed Greenlaw to make it clear that Kodak was only willing to consider a payment representing a small multiple of what it might cost to conduct the litigation but nothing more.[79] Given this position, it's curious that Kodak had bothered to commence discussions at all. A similar proposal had been totally unacceptable to Polaroid years earlier, but now Polaroid would be even less likely to agree.

Kodak had come under tremendous criticism for its "me too" system. Now that Polaroid had overcome Land's reluctance to produce cameras aimed at the low-price end of the market considered Kodak's marketing sweet spot, Polaroid was doing better than expected in the head-to-head competition. Its bottom line had also improved without the drain on its resources incurred during the problems encountered in the introduction of the SX-70 system. There was also some confidence building with regard to the merits of the lawsuit. Some documents produced by Kodak early on clearly demonstrated pre-commercialization problems at Kodak with aspects of its cameras covered by the patents Polaroid had asserted in the case. Kerr had advised Polaroid that while "these are not block-busters . . . [they] do tend to add stature to patents on what may appear superficially to be simple mechanical devices and arrangements."[80] All in all, Polaroid had some reason to be more confident than it had been earlier and thus was unlikely to be receptive to a settlement on less than very favorable terms. To an observer like Crary, it seemed that Kodak had more to gain from exploring a settlement seriously. Yet, curiously, Kodak was not prepared to have any kind of meaningful negotiation. Accordingly, the talks initiated by Kodak went nowhere, and nothing was heard again about them for quite some time.

Over the summer of 1977, the divergent financial trends of the two companies continued. Contrasting their most recent quarterly reports, the *New York Times* reported that Kodak had announced "a second straight decline in net earnings, while its archrival, the Polaroid Corporation, piled on another quarterly earnings record."[81] In large measure, Polaroid's good performance was attributed to the immediate success of the low-cost OneStep camera it had introduced that spring. Demand for the camera was far outstripping supply. A Polaroid executive told *Fortune* that its factory had been for months "turning out OneSteps three shifts a day, six days a week. . . . The factory is making more than it was originally designed to, but it's still not enough."[82] Also cited were the improving gross profit margins on SX-70 film and the fact that the decline of its peel-apart business was not proving as harmful as had been predicted.

All of this good news, however, was not universally convincing to analysts. While Polaroid clearly had its supporters in the financial community, there were bears in that world as well. Not everyone was convinced that Kodak would remain behind in the head-to-head competition. For example, Charles Elia of the *Wall Street Journal* reported that reduced forecasts by some Wall Street observers "raise questions over the ability of Polaroid to keep expanding sales and profit in the face of dogged competition by Eastman Kodak in the growing instant camera and film market."[83] Clearly, the final result was far from clear in the marketing battle over instant photography.

Equally uncertain was the question of when progress would be made in Polaroid's lawsuit against Kodak. Since the early flurry of charges and counterclaims more than a year earlier, there had been virtually no word or input on the case from the courthouse. None of the pending motions had been decided or even addressed preliminarily by Judge Murray. Kodak had sought to add several patents to the case and had also lodged its broad patent misuse claims covering the entire Polaroid portfolio. Polaroid had moved to dismiss or to stay most of these, but there was no word from the Boston federal courthouse on any of it. The court was simply sitting on a huge pile of paper. In early July, with no prior warning, the parties learned that Judge Murray had resigned from the case. According to Polaroid's local counsel, William Cheeseman, Judge Murray had elected to take senior status.[84] The judge indicated that the federal court backlog of cases in Boston was "hopeless" and that he personally had a docket of 1,000 matters. This move would allow him to elect the cases he actually wanted to deal with. Clearly, he had elected not to deal with *Polaroid v. Kodak*, which would now have to be reassigned to a new judge.

Over the summer of 1978, discovery in the lawsuit continued in the same contentious manner in which it had begun. Neither party was satisfying the other's thirst for material. Kerr knew the litigation was getting nowhere with this paper tug-of-war. Despite the larger volume of material produced by Kodak, the vast majority was useless marketing and advertising materials of no consequence to the case. Polaroid needed some specific information on the kinds of documents prepared at Kodak to memorialize its research activities, documents it would thus be expected to produce. This kind of information could lead to targeted discovery requests from Polaroid that, if unanswered, would serve as the bases for bringing Kodak's recalcitrant delaying maneuvers to the attention of the court. Schwartz recommended that Polaroid notice the deposition of some key Kodak personnel, starting with Albert Sieg, who had run the P-130 program responsible for the development of Kodak's instant products.[85] On September 20, 1977, Polaroid filed with the court a notice of taking deposition naming Sieg, and the next interminable phase of discovery officially began.[86]

Meanwhile, Berkey Photo had spent the summer continuing its effort to get out from under its patent infringement battle with Polaroid. Its antitrust action against Eastman Kodak went to trial in mid-July 1977 in a federal courtroom in Foley Square in Manhattan. This was obviously a major drain of resources for the company. The burden of simultaneously litigating with Polaroid was only getting worse. In late July, Berkey's counsel, Sidney Neuman, used the occasion of a discovery conference to raise with Kerr once more the idea of settling the case by having Berkey pay a royalty to Polaroid under a license to use the patents in question.[87] Kerr and his client remained uninterested in settling the Berkey case this way, a move that would open the licensing can of worms.[88]

Berkey could see, finally, that Polaroid was not backing down from its defense of its patent property, and it was not going to license its technology. Just a few weeks later, the outlook for Berkey in the case dimmed even further. Polaroid won an important procedural skirmish that only increased the litigation's prospective financial burden on Berkey. On October 4, 1977, the Delaware court granted Polaroid's motion for separate trials on the issues of patent validity and infringement, and Berkey's unclean hands defense and antitrust counterclaim.[89] Thus, any economies Berkey may have hoped to achieve from trying all issues at once were now dashed. If Berkey couldn't win the first trial by establishing that Polaroid's patents were invalid, it would then have to go through the expense of an entire second trial on the issues that might exculpate it. This was not the ideal scenario. The trial of Berkey's antitrust case against Kodak plodded on in New York federal court, running up an enormous legal tab daily. That case was nowhere near an end and was expected to last at least through the end of the year. This was a daunting economic commitment for a company of Berkey's size.

Polaroid was on two different tracks in its litigations with Kodak and Berkey. In the former, it was pushing to get the case to trial as soon as possible. In the latter, it was stringing out the pretrial process in a war of attrition against Berkey's resolve and diminishing financial resources. Kerr and Schwartz seemed to believe that they could crush Berkey's resolve to continue and force it to capitulate by exerting against it the relative financial strength of Polaroid. Ironically, this was pretty much the same ploy Kodak was utilizing against Polaroid. If Kodak's war of attrition of Polaroid worked, perhaps Polaroid would give in and decide that an ungratifying negotiated settlement in the form of a nominal license was preferable to a devastating loss of its patent portfolio at trial. The jury, as it were, was still out on which of these strategies would prove effective in the end.

CHAPTER 13

EARLY SKIRMISHES IN DISCOVERY

As the discovery phase of the lawsuit got into full tilt, the substantive strategies of both sides gradually began to emerge. The Polaroid legal team largely directed its efforts to highlight the big picture, what Schwartz referred to as the "Kodak Story": Kodak's early relationship as a mentor and supplier to Polaroid; Kodak's admiration of Polaroid's technological advances, but its perception that the field of instant photography was only a specialty market and no threat to Kodak's dominance of the amateur photography industry; Kodak's growing interest in instant photography as the market expanded and as Polaroid's innovations continued to emerge; and eventually, Kodak's realization that Polaroid might pose a genuine threat to its core business when its integral SX-70 system came out. Polaroid counsel sought to demonstrate that Kodak had made significant changes in its research program, particularly upon the release of the SX-70 system—and that it made those changes without regard to Polaroid's patent property. Kodak had run into some of the same obstacles that Polaroid encountered and ended up having to adopt Polaroid's patented solutions, an argument clearly supporting the substance of the inventions (and patents) in question. Schwartz's approach was to put all of the obligatory technical evidence and arguments of the complex case into a common-sense narrative that a non-technical judge would understand—and ultimately be persuaded by.

The Kodak legal team, on the other hand, pursued a strategy grounded in the technical aspects of each of the inventions at issue, which they claimed were nominal advances that did not rise to the level of patentable inventions. This would require conducting discovery on decades of Polaroid laboratory records and experimental results, versions of engineering models and mock-ups, decaying pieces of film from all kinds of imaging experiments—both successful and unsuccessful—and every note, chart, and graph they could lay their hands on. Depositions of the Polaroid scientists would follow, with Kodak lawyers attempting to have the laboratory work, as well as the prior art, characterized in a way to support their arguments.

To implement its strategy, the Polaroid legal team needed to get Kodak to produce meaningful documents detailing its research efforts. The first step would be to identify the existence and nature of the kind of documents it sought. Polaroid's first deposition of Kodak personnel was aimed toward that end. It took place on November 16 and 17, 1977, at Fish & Neave's Park Avenue offices in New York. The witness was Albert Sieg, the man who had directed Kodak's development program, the logical candidate for this kind of exercise. The significance of this occasion was clearly demonstrated by the array of counsel seated around the table. Conducting the examination for Polaroid was Schwartz himself, with Kerr seated by his side. Also observing was Bob Ford, the Polaroid in-house attorney Schwartz had worked with in selecting the patents to assert in both the Berkey and Kodak cases. For years to come, Ford would closely monitor every significant event in the litigation and would provide valuable technical expertise to Polaroid's trial counsel. Carr was present on behalf of Kodak but allowed one of his partners, Robert Fier, to sit first chair in defense of the witness. If the proceedings became contentious, Fier would do the dirty work. Also present were two additional Kenyon partners who were working on the case.

The proceedings began promptly at ten a.m., with a court reporter present to provide a written transcript. After reviewing Sieg's professional résumé and education, Schwartz quickly established that Sieg had corporate responsibility for Kodak's instant products—film, film pack, and camera—starting from late 1971. As Schwartz moved ahead, Fier occasionally interjected, mostly seeking to clarify questions for his witness. Carr, too, spoke up, also asking Schwartz to rephrase a question "just for clarity of record."[1] This was not the normal routine—generally, double-teaming by attorneys on either side is not considered the appropriate way to conduct a deposition. Carr probably wanted to make it clear—likely to his partner, as well as to Polaroid—that these examinations would be vigorously contested, question by question if necessary, so as to make the going as difficult as possible for Polaroid.

Undeterred, Schwartz pressed ahead, eliciting testimony from Sieg that throughout Kodak's research program, regular status meetings of various committees were held on a weekly basis. Moreover, Sieg admitted that he was responsible for having written reports prepared for the vast majority of those meetings. Schwartz asked Sieg, "Do you know whether or not any of those reports still exist?" "I believe they do," Sieg responded without hesitation or comment from his counsel.[2] This was exactly the sort of material that Schwartz was looking for. A short time later, Schwartz established a fact that, strategically, would be very helpful down the road. He asked Sieg whether anyone within Kodak management received copies of the written status reports. When Sieg answered yes, Schwartz quickly followed up, "Did any of Messrs. Chandler, Fallon or Hanson receive such copies?" "Yes," responded Sieg. "All three?" asked Schwartz. "Yes," replied Sieg, confirming that Kodak's current three most senior executives had been privy to these reports during the crucial time periods when key events in the development process occurred.[3]

The second day went much as the first, with Schwartz establishing details concerning the organizational structure of Kodak's efforts and its practices with respect to record keeping. During the questioning, it became apparent that Kodak had conducted research on a broad array of alternative approaches for an instant camera and film system. Fier repeatedly objected to these probing questions, seeking to limit the examination to work done "in the line leading to a commercial product." Fier's objections were necessary, he maintained, to prevent Polaroid from using the lawsuit as a "vehicle . . . to obtain all of the information that Eastman Kodak has in instant photography."[4] Schwartz insisted that this line of inquiry was legitimate, but Fier persisted in blocking him.

After objections from Fier that cut off a line of questioning on an abandoned alternative Kodak program as being "premature in this lawsuit," Schwartz suddenly switched gears and asked Sieg whether anyone had conducted a search of his files in connection with discovery in this lawsuit. "Yes," answered the witness. "Do you know approximately how long ago such a search was conducted?" asked Schwartz. "Approximately six to nine months ago," was the response.[5] That meant early 1977, almost immediately after Polaroid had lodged its first set of requests asking for documents relevant to Kodak's research activities. Schwartz went on to get Sieg to admit that four to six file drawers' worth of material had been removed from his office at that time. None of this material had yet surfaced in the litigation. When Sieg was asked if he had "any information as to what has happened to those documents since that time," Fier's objection to the question was a beat too late; Sieg had quickly admitted, "No, I don't."[6]

With that, Schwartz adjourned the deposition sine die, Latin for "without day," meaning that it could be resumed at a later date if necessary. As he thanked the witness for his cooperation, Carr spoke up, apparently believing it necessary or advisable to put on the record a statement that all of the documents collected from Sieg were being "methodically" reviewed and would be produced "in due course."[7] At this point, Kerr, who had said little throughout two days of deposition, could no longer resist responding to Carr's blatant self-serving disingenuousness: "Let me say that, as I understand it, we have not gotten production of any copies of any of these status reports or progress reports, Frank, and I would have thought that that would be the kind of category of documents which would be produced to us early in discovery in this case."[8] This was a swipe at Kodak meant clearly for anyone, particularly the judge, who might have occasion to read the transcript. "Here is the witness who is right at the heart of this project . . . at the Eastman Kodak Company and my understanding up until now is that we have gotten maybe a half dozen pieces of paper that this witness wrote, and I am surprised that we haven't gotten more."[9] With this on record, the proceedings were concluded.

There were by now many ongoing disputes concerning the substantive scope of discovery. Each side tried to interpret the other's demands as narrowly as possible so as to limit what it had to produce, while simultaneously urging as expansive an interpretation and underlying justification as possible on its own requests. Broad requests were objected to because they called for the production of information or documents the other side was not entitled to. If a far-reaching request could somehow be justified in terms of relevancy, the other side would argue that the burden of collecting, reviewing, and producing the information outweighed any probative value it might have. All in all, it was a remarkable but wearying exercise in strategic nitpicking by an elite group of lawyers, but it underlined the crucial nature of this phase of the case. At this early stage of the game, when the ground rules were being established, the discovery process required the full attention and involvement of the top lawyers.

One discovery issue was particularly salient: Kodak counsel complained that Polaroid's reading of Kodak's interrogatories and document requests was unnecessarily limited when it came to responding. But Kerr and Schwartz reminded them that they had objected to discovery on "alternative" approaches in research at Kodak during the Sieg deposition.[10] Despite the ensuing debate, both sets of attorneys were fully aware that discovery in the area of alternative research was perfectly legitimate, for both sides. Polaroid was entitled to know what things were tried by Kodak and had failed, while Kodak was entitled to look broadly within Polaroid's research, including alternatives to the solutions embodied in the patents in suit. Despite its relevancy, this was an extremely sensitive area for both sides, fraught with burden for the clients and pitfalls for the lawyers. It was precisely the kind of discovery that might develop information that could influence the outcome of the case. It also posed danger in that it allowed the opponent access to the client's most highly secretive areas. It had to be handled carefully.

Both sets of attorneys recognized a deal had to be reached, and they started jockeying for rules favorable to their side. Carr blinked first. At a November 29 meeting, just as one of his partners was declaring how the burden of supplying such information far outweighed its relevance, Carr spoke up. He offered to recommend to his client broad discovery with respect to "alternative approaches" if Polaroid agreed to provide discovery to Kodak on a corresponding basis. He seemed to believe that Kodak had more to gain from extensive discovery than Polaroid. Kerr responded that he could agree as to the camera patents, but with respect to the film patents, he would have to consult with his client, knowing full well that this would be a touchy issue for Polaroid and especially for Land.

Following that meeting, Schwartz sought to keep up the pressure on Kodak. He sent to Carr on December 8 a detailed list of document requests based upon Sieg's testimony during his deposition and asked Kodak to specify when it would respond to these and other requests.[11] Next came a December 19 letter enclosing a legal memorandum backing up Polaroid's contentions with respect to various discovery disputes under discussion, including the discovery into Kodak's unsuccessful attempts that had been objected to at the Sieg deposition.[12] No meaningful documents materialized. The only response he got to these initiatives during the rest of the month was a notice of taking deposition from Kodak, calling upon Polaroid to produce Richard Young, a senior vice president and a member of Land's inner circle. The ping-pong game of dueling depositions was under way.

═══

Early 1978 saw some dramatic developments on the Berkey front. On January 13, Berkey's antitrust case against Kodak finally went to the jury for deliberation, after 113 days of a trial that had begun the previous July. Berkey's basic argument was that Kodak had wrongfully monopolized aspects of the amateur photography market by "using its technological lead and market dominance to keep the competition in disarray."[13] One instance involved photo processing, one of Berkey's most profitable areas. Berkey alleged that after it had built eight automated plants around the country to process film, Kodak changed the physical film format as well as the film chemistry to render Berkey's equipment unusable for Kodak's newest and, presumably, best products, particularly the new and highly successful 110 Pocket Instamatic cameras and film, introduced in 1972. It took the jury a week to deliberate, but on Saturday, January 21, Judge Marvin Frankel announced that it had found for Berkey. Among other things, the jury ruled that by not giving advance notice of the new system to its competitors, Kodak had unlawfully used its influence and control to monopolize the market.[14] While Berkey was seeking $300 million in damages, which would then be trebled under the antitrust laws, the jury did not decide on damages. It was scheduled to return to the courtroom in a month's time to tackle that aspect of the matter.[15]

The jury's decision looked like an epic win for Berkey. Whatever was paid in damages would be a great boost to Berkey's finances as well as an opportunity to refocus its business model and concentrate on areas that showed the most promise. Berkey's management determined that this meant getting out of the camera manufacturing business and putting its emphasis on film processing and camera wholesaling. It decided, after this victory, that it made no sense to stay in the instant photography field.

Within weeks of the jury verdict in Berkey's antitrust case against Kodak, serious discussions were under way to settle Berkey's Polaroid case on terms essentially dictated by Polaroid. By late February, draft settlement agreements were being circulated among Polaroid counsel.[16] On March 22, the Kodak jury returned with an award of $112.8 million in damages for Berkey, an amount that was later adjusted by the judge to $87 million.[17] One week later, Berkey signed off on a settlement agreement in the Polaroid case that was a clear victory for Polaroid. In a consent judgment, Berkey conceded that each of the patents asserted against it was "in all respects valid and enforceable."[18] It admitted infringing each of the patents and agreed to cease "manufacture, use and sale" of any of its infringing products once its current inventory was exhausted. Finally, its counterclaims against Polaroid relating to alleged misuse of the patent system were dismissed with prejudice. A parallel case in the United Kingdom was settled on the same terms.[19] Berkey paid no damages for its admitted infringement, but Polaroid's victory was manifest. It had vanquished the first company that had dared to challenge its legal patent monopoly in instant photography, with its patent portfolio intact. This would continue to hold at bay the other potential entrants into the field. Polaroid, and Kerr and Schwartz, could now focus full attention on the main attraction, the case against Kodak.

As they worked on the settlement papers for the Berkey case, Kerr and Schwartz continued to work out with Carr the ground rules for discovery that would move Polaroid's case forward. A major obstacle continued to be discovery on alternative research. Each company had lots of secret proprietary work that it was not interested in sharing with its competitor. The lawyers met again on January 12, 1978, to discuss these and other unresolved discovery issues.[20] It was now clear that both parties would have to open up this sensitive area to *some* form of discovery. Kerr and Schwartz worked with Peck and Mikulka to devise parameters for discovery into Polaroid's work that could be offered to Kodak in their negotiations.

On January 23, Schwartz sent Peck a memorandum outlining on a patent-by-patent basis the "potential scope of discovery with respect to 'alternatives' to Polaroid's film patents in suit."[21] He reminded Peck that Kodak was obliged to reciprocate by producing documents covering "its alternative approaches to the commercial solutions which have been accused of infringement." Peck called Schwartz the next day, telling him that he would need him and Kerr to come up to Cambridge to meet with Mikulka and other members of Polaroid management in order to work this out.[22] Clearly, for Kerr and Schwartz, there was going to be a lot of reassuring to do, and a lot of explaining why this kind of intrusion behind Polaroid's traditional wall of secrecy was strategically necessary for the lawsuit. Although Land never participated in these meetings, he was certainly the primary person who had to be reassured. Most of that effort fell to Mikulka. The meeting at Polaroid was held on February 1, 1978.[23] Kerr and Schwartz made their case and left it to their Polaroid colleagues to work out the internal politics that would make the discovery possible.

In the meantime, the Polaroid legal team continued to wrestle with their Kodak counterparts over seemingly every aspect of the discovery process. Through February and March, the lawyers generated a constant flow of correspondence complaining about the other side's rate and quality of production. Kodak counsel was happy to grind it out. But, in fact, Polaroid had also been dragging its feet. Kerr and Schwartz had agreed with their client that discovery with Kodak was going to be on a quid pro quo basis, and until Kodak provided meaningful discovery, significant material was not going to be forthcoming from Polaroid. In mid-April, Kerr wrote to Carr in an attempt to break the logjam. It was a very conciliatory letter, outlining compromises on most of Polaroid's objections to various Kodak requests. The intention was to remove the obstacles to a resumption of document production from Polaroid to Kodak. Kerr undertook to do so, and he expected this gesture to move Kodak off its intransigent posture. "We assume that you are prepared to reciprocate," wrote Kerr, "so that we can proceed to have discovery progress promptly on both sides."[24]

It was April 1978, and the New York Yankees were beginning a new baseball season in defense of their twenty-first World Series victory the previous fall, this time over the Los Angeles Dodgers. At this point, the *Polaroid v. Kodak* lawsuit was basically two years old, but both sides were still awaiting the "first pitch" of meaningful discovery in their own world series of litigation.

Now that Kerr seemed to be making some progress in getting the document production aspect of discovery moving, Carr moved to a different front of the discovery effort to confound and to bog down the process, at least from Polaroid's point of view. On April 27, he wrote to Kerr providing a list of twelve Polaroid researchers whom Kodak proposed to depose on a weekly basis starting in June and running through the summer.[25] Land was not on that list, but Carr made it clear that Land would need to be deposed at a future date. This was no surprise to Polaroid; they had been planning for this for some time.

In July 1977, Kerr and Schwartz had prepared a detailed and brutally realistic assessment for Polaroid of the demands the litigation would have on Land's time.[26] When this information was shared with him, Land knew he needed to prepare. So, on his own, he commissioned Stanley Mervis, his most senior in-house patent attorney, to put together a history of the development of SX-70 from a technical point of view. This project was to be done in coordination with an attorney who had no connection to the litigation, Isaac Barnett, from Julius Silver's New York law firm. When Kerr learned of this initiative, he took steps to make sure that the process was conducted so as to prevent Kodak from getting access to the materials generated.[27]

Even though Land was not on Carr's initial list, the April 27 letter riled Kerr once again. He wrote back immediately advising that he was "troubled" by Carr's deposition proposal and pointed out that both parties would soon be engaged in document production, given the progress he hoped had recently been made in overcoming obstacles to it. Kerr also objected to the large number of individuals Kodak intended to examine in such a concentrated fashion. "In my view, this deposition activity will serve only to impede orderly progress of discovery in this litigation."[28] He suggested that both parties commence substantial document production and start a more manageable number of depositions on both sides. If Kerr was aggravated at this point, it didn't take long before his short-lived satisfaction over breaking the document logjam dissipated altogether.

Two days later, on May 3, Kodak finally resumed producing documents. But the Polaroid lawyers who reviewed the material were surprised and appalled to find there were none of the documents that had been brought to light in Sieg's deposition. That same day, Kerr wrote to Carr about the "disturbing" absence of the periodic progress reports he "had believed, perhaps naively, that Kodak was committed to produce."[29] Carr's response arrived the next day in the form of—not one, but a series of—lengthy and detailed letters.[30] They were expositions on the "whys," the "why nots," and the "what are you talking abouts?" Kerr now knew it was going to be impossible to make meaningful progress on discovery and to get this case to trial without the active involvement of the court to keep Kodak on track. Kerr called Bob Peck to report these developments to his client. His follow-up letter told the whole story. "Herewith copy of May 4, 1978 mail shipment from Frank Carr," wrote Kerr.[31] "The time has come to terminate trial-by-correspondence and to submit the current discovery disputes to the Court." On May 15, Polaroid filed motion papers with the court to compel discovery by Kodak. It also asked the court to issue a protective order holding off the depositions of Polaroid personnel until Kodak satisfied its discovery obligations. Kerr had copies hand-delivered to Carr.

In view of the resignation of Judge Murray, the Boston federal court had recently reassigned the case to a new judge, A. David Mazzone. Polaroid's local Boston counsel advised Kerr and Schwartz that Judge Mazzone "was a young, hard working, intelligent, vigorous jurist who should move things along."[32] This was encouraging, but complicated cases like this one often present too many procedural and internecine disputes for a sitting member of the federal bench to handle. Thus, a court can, in its discretion, appoint another person to serve as its designated deputy to handle such matters. Known as a "Special Master," this person is often a retired judge who has had experience adjudicating discovery disputes.

Frank Carr had been urging Kerr and Schwartz since November 1977 to join him in requesting the court to appoint a Special Master, but their client had some reservations. Peck thought that providing a Special Master would only further enable Kodak's attempts to complicate and obfuscate discovery at every turn. Carr had also proposed that the Special Master be appointed in New York, implying that the parties would be spending a lot of time before him. This idea was particularly troubling. "After all," wrote Peck to Schwartz, "one of the advantages to Polaroid in selecting . . . [Boston federal court to bring the lawsuit] was 'plaintiff's convenience' and it seems to me that Carr is trying to take this away."[33]

Because of Peck's position, Kerr and Schwartz never agreed initially to Carr's proposal and nothing was done to follow up. Kodak responded to Polaroid's May 15 motions two days later with a motion of its own, asking the court to appoint a Special Master. Just two days after that, on May 19, Kodak filed yet another motion, this time asking the court to compel the attendance of the Polaroid witnesses at the depositions it had noticed, and whose attendance Polaroid had sought to delay. Judge Mazzone promptly scheduled a preliminary pretrial conference for June 2 to hear oral argument from both parties on the several matters submitted to him. When counsel appeared in Judge Mazzone's courtroom, the judge announced it was going to be necessary to grant Kodak's request to appoint a Special Master. But to Polaroid's relief, he agreed that the Special Master should be located in Boston, presumably for the court's convenience, and he suggested two retired judges that he thought would do a good job. At his request, both parties approved his suggestions by the end of the day.[34]

When it turned out that neither of the first two choices was available (or willing) to take on the case, Judge Mazzone appointed Walter McLaughlin as the Special Master. Judge McLaughlin was a former Chief Judge of the Massachusetts Superior Court, and was considered to be a very competent choice by William Cheeseman, a partner in Polaroid's local Boston counsel, Foley Hoag & Eliot (Cheeseman was a litigator himself and would later gain notoriety for handling the defense of W.R. Grace as defendant in the Woburn, Massachusetts, water contamination case made famous in the book and motion picture *A Civil Action*). As Judge Mazzone wrote in his June 12 order: "Given the size and complexity of the case it is impossible for this Court to devote the time necessary for adequate supervision of discovery."[35]

That same day, Polaroid issued a press release announcing an end to the "exploratory talks" that had been going on with Kodak periodically since April 1977, when Walter Fallon had made them public.[36] The subject of settlement was always an intriguing one, and shareholders and journalists both inquired about it when they had a chance, which was primarily at the two companies' annual shareholders meetings. Polaroid's recent meeting in April 1978 had been no different. In response to a question from the floor at the annual meeting, McCune had read a prepared statement in which he noted there had been "no substantive progress" toward reaching an agreement with Kodak. Land had also been asked about the attempts to settle. His inscrutable response was that "it is a question of goodwill versus a difficult situation."[37] He expressed some concern over the daunting task of presenting complicated technical problems to a judge or a jury, but nonetheless he managed to convey his determination to tackle the challenge and his commitment to redress Kodak's infringement of his company's work.

In its June announcement, Polaroid noted that conversations with Kodak had failed to produce a "satisfactory" settlement of the case, a matter the company would "continue to prosecute vigorously."[38] Kerr and Schwartz had orchestrated the timing of this statement for several apparent reasons.[39] First, it sent a public message to Kodak once and for all that its long-running attempt to settle the case by paying a nominal sum or nothing at all was not going to work. More importantly, it indirectly informed the recently appointed judge and Special Master that Polaroid had been open to settlement talks, but that since an accommodation could not be reached, it was serious about moving forward diligently with its lawsuit. The idea was to create a perception with Mazzone and McLaughlin that Polaroid was aggrieved yet reasonable and deserved to have their support in going to trial. All concerned believed that announcing an end to what were really pointless settlement negotiations would help Mazzone and McLaughlin regard Polaroid favorably.

The announcement did not surprise the media. Calvert Crary reiterated his already public view that settlement was unlikely. He knew any negotiated resolution would necessarily entail a license that would open "the field up to everybody and his brother, which would be just as bad for Kodak as it would be for Polaroid. . . . Better to keep the patents out there as a credible deterrent even if it means litigation," Crary thought.[40]

In any event, it was starting to look as though the epic patent battle was not negatively affecting sales by either company. Vigorous advertising helped; it created the perception that the instant photography market had expanded, to the benefit of both. Kodak now had captured about one-third of the market.[41] But Polaroid was doing more business than it had before Kodak had "elbowed in."[42] Despite aggressive marketing campaigns, no one expected Kodak to improve on its share because of the technological issues with its products. The fading problem plaguing its PR-10 prints was seen as a true disadvantage when compared with Polaroid's stable prints. The only response from Kodak was the claim that its prints held up perfectly well when stored in wallets and albums.

Polaroid's low-cost OneStep was outselling Kodak's Handle model by a ratio of two-to-one, mostly because Polaroid's camera was motorized while Kodak's required using a hand crank.[43] Polaroid was selling more cameras now than it had been before Kodak introduced its competing products. Now that the period of substantial production ramp-up and launch costs was over, both companies enjoyed improved profit margins. This was particularly true for film, as heavy competition was keeping camera prices at severe discounts. Kodak announced that its first-quarter sales in 1978 increased by fifteen percent and that its earnings for the period were up fifty percent, a record amount.[44] For its part, Polaroid's McCune announced early in the year that the company expected record earnings for all of 1978.[45]

Judge McLaughlin wrote to the parties on June 20 setting June 26 as the date for his first preliminary conference to review the outstanding discovery matters.[46] Kodak responded on June 22 by filing its own motion to compel discovery from Polaroid.[47] This ensured that both parties could come to the first hearing with the Special Master complaining about the lack of production and the unreasonable behavior of the other party. It was truly the first act in a circus that would go on for the next three years, as each party maneuvered to provide as little ammunition to its opponent as possible, while simultaneously trying to extract as much useful information as it could get its hands on.The newly appointed Special Master soon proved more than able to deal firmly with the litigants. At the session on June 26, he patiently listened to the complaints of both parties about the lagging discovery process.[48] On July 6, 1978, he issued his first decision (of many to come) and ordered both parties to complete their respective outstanding discovery commitments within a few weeks.[49] He ordered that Kodak could proceed with its deposition schedule, thus denying Polaroid's attempt to delay the start of that phase of discovery. But McLaughlin was equally tough on both parties, making it plain that he wanted to see progress on all fronts. This was a position he maintained throughout the protracted litigation.

Following McLaughlin's ruling, the Kodak lawyers moved quickly to get their deposition program back on track. They sought to establish a schedule of regular weekly depositions starting that fall and continuing through March 1979.[50] While Kodak counsel had a lengthy list of proposed deponents, scheduling Howard Rogers was a top priority because his work was, arguably, the heart of the lawsuit. Rogers' two patents on the "Excedrin" film format were arguably the most critical in the case and had been the subject of the most attention from Kodak in litigation around the world. In fact, in a development not helpful to Kodak, the Japanese opposition proceeding it had initiated in 1974 had been dismissed in September 1977, paving the way for the issuance in Japan of a counterpart to the Excedrin patent that November.[51] Rogers' patent on the Negative Dye Developer diffusion transfer process was a broad, basic patent in instant photography technology that, if declared valid, Polaroid contended, would cover the film chemistry Kodak employed in its PR-10 film units. Thus, Carr was going to take his most vigorous run at Polaroid's "number two" inventor in an effort to develop evidence that could undermine these patents. Within days of the Special Master's ruling, Carr engaged Kerr in correspondence seeking to schedule Rogers.

For Polaroid, the Special Master's initial order meant that discovery in the area of alternative research was indeed going to proceed, ratcheting up Polaroid's anxiety, not only because of its sensitivity about its secretive research work but also because it raised the possibility that this might turn up damaging evidence. Shortly after the new ruling, Peck wrote a lengthy letter to Kerr characterizing this as their "most serious problem" with the new discovery initiative, thus making it clear that everyone within Polaroid had still not come to grips with this awkward issue.[52] Kerr and Schwartz, well aware how seriously their client took this problem, realized that a great deal of reassurance and gentle but firm persuasion would be required to expedite the process of searching Polaroid's records for material that Kodak was entitled to. But Polaroid could take some comfort from the Special Master's determination to move discovery forward in a meaningful way, which was at last progress toward its ultimate goal of moving the case to trial.

CHAPTER 14

SEARCHING FOR THE FACTS

As the summer of 1978 ran its course, pretrial discovery continued along in fits and starts. At least there was finally a referee onboard who could try to keep the proceedings on track. Judge McLaughlin very quickly had his hands full with a continuous stream of disputes large and small, and often completely ridiculous. The judge once again quickly showed he was up to the task and would not allow any nonsense to go on interminably. For example, after yet another exchange of recriminations between attorneys for Kodak and Polaroid about who was responsible for delaying document production, Judge McLaughlin quickly ordered adherence to the schedule despite the protestations, noting sharply that he was "unimpressed by [the parties'] contention[s]."[1]

The main event that summer, though, was the run-up to the upcoming deposition by Kodak of Howard Rogers. The start of that deposition had been set for early September. For Kodak, it was a chance to develop some evidence it could use to strike deeply at the heart of the key inventions at issue in the case. For Polaroid, it was a matter of protecting one of its most important scientists. Rogers was diminutive in stature, but large in intellect. He had been a close colleague of Land's since joining Land-Wheelwright in 1936 and was beloved by everyone who knew him. "Howie," as almost everyone called him, had an invitingly warm personality, with an almost complete absence of ego and a nonconfrontational style. This made him exactly the kind of person who might be abused, confused, or misled by skilled attorneys. No one knew for sure how he would handle this kind of stress. Rogers later confessed, "It wasn't the most comfortable thing I've ever done."[2] Kerr and Schwartz were determined to protect and to prepare Rogers to the fullest, ensuring he understood the issues in the lawsuit in order to avoid his being misled inadvertently into giving testimony that might be damaging to Polaroid's case.

Actually, Kerr and Schwartz had already decided to make Kodak's slog through the deposition process as difficult as possible. Polaroid did not want to provide Kodak with any more ammunition to use against its patents through testimony elicited in Kodak's examinations. Accordingly, Schwartz decided that he would "defend" these depositions as aggressively as he could, "short of getting [himself] disbarred."[3] Schwartz had made that clear at Kodak's initial questioning of Richard Young, Polaroid's first deponent. Young, a PhD chemist and Polaroid's assistant director of research during the SX-70 program, had worked closely with Land. In an unrelated trial in Canada, Young served as a stand-in for Land, having been proffered by Polaroid as someone who was as competent to testify on the instant photography patents in that suit as Land himself.

From the outset of Young's deposition, and then later with other deponents, Schwartz repeatedly objected to certain of Kodak's questions as inappropriate, and instructed his witness not to answer them. Schwartz's position was that the witnesses were there to provide facts within their knowledge, as opposed to opinions that would more properly be the domain of expert witnesses. Such improper questions, Schwartz maintained, included asking a Polaroid witness to interpret and explain a relevant patent or an internal Polaroid laboratory document that the witness had written or read during the research project, or a request for the witness to draw a diagram to illustrate something he was describing verbally. Schwartz also objected to questions he considered too broad or ambiguous. On September 5, 1978, a frustrated Kodak asked the court for an order compelling witnesses to answer a wide variety of questions Schwartz had objected to and for a ruling "regulating the conduct of counsel for Polaroid to permit the depositions to proceed in an orderly manner without purposeful obstruction."[4]

It was no coincidence that the first session of the Rogers deposition was to be held the very next day, September 6. Schwartz clearly was going to make that examination as difficult as humanly possible, and Carr had to know it. He knew that he would be asking the same kinds of questions of Rogers as Schwartz had previously objected to with the other witnesses. Given the timing, one can only assume that Kodak's motion was intended as a notice from Carr to Schwartz that if he continued with his pertinacious conduct at Rogers' deposition, Carr was prepared to bring in the court or the Special Master to help, if needed. As they gathered in the Kenyon conference room for the deposition, Schwartz sat in the defense counsel's chair next to Rogers; Kerr and Polaroid's Bob Ford were also present. Carr was the examiner on behalf of Kodak.

In fact, if Kodak's motion was filed in anticipation of problems at the Rogers deposition, its apprehension turned out to be prescient. Most of the first day of Rogers' examination was occupied with routine background information on his biography and career at Polaroid. Then Carr began to ask probing questions about the chemical processes used to create some published instant-photographic images with which Rogers had testified he was familiar. Schwartz objected and went on a tirade to block further questioning on the subject. The examination bogged down at that point and the proceeding was adjourned to the following morning.

The next day, Carr first asked Rogers about the operational relationship between Polaroid's black-and-white, sepia, and color commercial films. Rogers told him it "might take quite a bit" of thought to answer accurately. Schwartz objected to Carr's follow-up on the ground that Rogers was being asked for "an opinion rather than a recitation of the fact" and advised his witness that he need not answer.[5] Carr attempted to move on by showing Rogers a copy of a seminal 1947 article written by Land about photographic diffusion transfer processes, in particular Polaroid's first commercial black-and-white one-step system.

Though Rogers and Schwartz had prior notice from Carr that the article would be brought up, when Carr asked Rogers to explain some aspect of the article, Schwartz interceded. "I would object," Schwartz bellowed. "This is an article which he hasn't written. . . . It seems to me you are asking him to express an opinion . . . about a process found in this article. . . . I don't think that's a subject of discovery and I advise him he need not do it."[6] In Schwartz's view, Kodak could question Rogers about relevant events related to his inventions and what else was going on at Polaroid at the time, but only to elicit facts, not opinions. Rogers, Schwartz maintained, was not there to testify as an "expert" witness in the case, someone who is brought in to explain the technology in general, to interpret documents, or to render an opinion on their content.

Carr was furious. Frustrated and red-faced, he proclaimed: "Your not permitting this witness to answer questions on this subject matter is completely barring us on discovery in this litigation."[7] Carr then adjourned the deposition and arranged to bring the dispute to the Special Master's attention immediately. Both legal teams got on the Eastern Air Lines shuttle at LaGuardia and headed for Boston, where at 4:30 p.m. they met with Judge McLaughlin in a conference room in his law offices.

Once again, Judge McLaughlin acted swiftly. After hearing both arguments, McLaughlin ruled that the questions Schwartz had objected to did not seek opinions but rather the witness's "understanding of the operations of films with which he is familiar."[8] Rogers had to answer Carr's questions. McLaughlin stated that Schwartz "was obstructing the testimony of the witness" and that "all" of his instructions to Rogers "were without merit." This was a direct rebuke to Schwartz and an indication that McLaughlin was determined to allow latitude for broad discovery. With the Special Master's ruling in place, Rogers' deposition resumed for six additional days, during which he was asked approximately 400 times to interpret patents or technical publications, more than half of which he had not invented or authored. Schwartz couldn't block this examination, but Rogers was now better prepared and proved to be an adept witness, thoughtfully answering what questions he could and speaking up when he didn't understand the question or was unable to answer, thus avoiding the inadvertent disclosures Carr was hoping to elicit. Meanwhile, Schwartz did his best to make things difficult, but not to the point of triggering additional trips to the Special Master.

Carr, believing that depositions offered the best chance to develop helpful evidence, continued to notice the deposition of Polaroid scientists at a grueling rate. Rogers' testimony alone was scheduled to occupy seven additional weeks, and ten other Polaroid witnesses were called to testify for a total of twenty-two weeks.[9] And this was just the first wave of potential witnesses, which did not yet include Land himself. Polaroid's lawyers fought to limit the number of depositions, the extent of preparation its witnesses would be required to endure, and the scope of testimony Kodak's lawyers would be allowed to pursue but were constrained by the Special Master's indication that he would allow broad interpretations of discovery issues and disputes.

As the discovery slog continued, each law firm had ten or more attorneys committed to the case. Each side had two teams to handle the subject matters, with one focused on the film patents in suit and the other on the camera patents. Everything possible—scheduling of depositions, rules for partial masking of documents, document delivery schedules, interpretations of interrogatories propounded by each side—was contested. Some compromises were made, but often the disputes had to be referred to McLaughlin, who, in most cases, continued to rule in favor of relatively unfettered discovery. Still, he made clear his impatience with the kind of silliness these disputes sometimes devolved into. Early in the fall of 1978, he characterized one flurry of briefs as "much to do about nothing."[10]

Polaroid's management grew concerned that the heavier burden was falling on Polaroid witnesses, particularly Rogers, so on October 3, 1978, Bob Peck asked Kerr and Schwartz to come to a meeting in Boston with McCune and Mikulka. The lawyers were asked to provide an appraisal of the Rogers deposition to date, especially whether anything could be done to curtail its length. Polaroid was also interested in limiting the amount of preparation Rogers, as well as other witnesses, were required to do in advance.[11] The unspoken concern was the impact these issues would have on the inevitable deposition of Land.

The meeting at Polaroid's headquarters took place on October 18. Basically, Kerr and Schwartz were there to assure their Polaroid clients that such was the nature of litigation of this type and that they would do everything possible to counter Kodak's strategy. Polaroid wanted its lawyers to go on the attack and undertake an equally arduous deposition program of Kodak personnel. Schwartz was already prepared to do just that; while this counteroffensive might not yield that much useful evidence, it would blunt Kodak's assault and keep its lawyers busy.[12] A list of potential Kodak candidates was discussed at the meeting.[13] The very next day, Polaroid noticed the depositions of twelve current and retired Kodak employees, to be examined three days per week on a schedule to be mutually discussed and agreed upon.

With the litigation now more than two years old, it was surprising that the two opposing senior partners remained involved in the minutiae of the battle. They communicated directly in exchanges of beautifully written correspondence that testified to their appreciation of the gravity of the case, as well as an understanding that there had to be some limit to the squabbling. A great example of this occurred that November, when Polaroid was scheduled to resume its examination of Kodak's Albert Sieg, the former head of its instant photography program. With the deposition date approaching, Kerr and Carr argued over Kodak's continued failure to comply with requests for documents Sieg had previously testified existed. Letters continued on this subject, back and forth every couple of days, parsing this request, questioning that one, and re-characterizing again and again the correspondence that discussed it all. Finally, Kerr had enough. On December 14, he wrote to Carr in response to his latest missive: "I'm tired of letters of this ilk, and I am confident that you are also. . . . Can we do something to turn off these faucets?"[14]

The fact that senior partners in two of America's most prominent law firms were involved in writing this kind of exchange speaks volumes about the landmark importance of the case and shows how vigorously it was contested at every level. It was clearly not something Kerr enjoyed, especially at that stage of his career. Kerr had once told Schwartz that a lawyer's enthusiasm for this kind of discovery is inversely proportional to his or her age.[15] He was losing his enthusiasm rapidly, so from this point on Schwartz increasingly took on more of the discovery skirmishing. Deep down, Schwartz suspected that Carr acted as he did because he did not expect to ever actually try the lawsuit, with Kodak perhaps believing that, eventually, Polaroid would back down. But Schwartz believed this case would be fought to the bitter end, so he pressed on, doing his best to maneuver Kodak through discovery and into the courtroom.

As 1978 drew to a close, depositions were scheduled to continue in early 1979. One witness from each side was scheduled weekly starting with the first week of January and continuing through the end of April. Polaroid management's pressing concern, however, was Land's eventual deposition. While Land apparently followed the case closely, he had remained distant from the fray, with no direct contact with Peck or Ford, let alone with Kerr or Schwartz. Instead, he would send one of his assistants over to the patent department to inquire about developments, or use Julius Silver and his partner, Isaac Barnett, as his source of information.[16]

Astoundingly, the four lawyers charged with conducting the lawsuit for Polaroid had neither seen nor heard from Land in more than two years—since before the lawsuit had even been brought—and really had no definitive word and little feeling for how Land might react when his number was finally called. Would he resist, unwilling to surrender his privacy, and make the process difficult, or would he join the effort and take his seemingly rightful position at the head of Polaroid's effort? No one really knew for sure. While they were aware that Land had initiated some kind of historical review of his work in the field, no one knew whether that process would encourage or discourage him. Preparing for every contingency, Bob Peck even asked Schwartz to look at the legal consequences should Land refuse to appear at his deposition. Schwartz had one of his associates research the issue, and in late November he supplied Peck with a memorandum and set of legal authorities on the subject.[17]

Early in 1979, Kerr and Schwartz decided to take a run at getting Frank Carr to cut down on the material Polaroid deposition witnesses would have to review, since it imposed an onerous burden on them. This would present a real problem when Land's time came, because he would probably be asked to prepare on more material than any other witness. Kodak lawyers had already succeeded in getting the Special Master to allow questions Polaroid regarded as impermissibly seeking opinion testimony, instead of factual information concerning either patents or publications a witness had authored, or materials that had been cited as relevant prior art in the application process for his patent. But Schwartz and Kerr saw another possible distinction to urge in an effort to limit the breadth of Kodak's examinations. Prior to the next set of deposition sessions, when Kodak counsel asked that Rogers study materials he had neither written nor seen before, including patents on which he was not the inventor, Kerr did not provide those materials to Rogers. As a result, Rogers showed up at the deposition unprepared.[18] As before, Kerr's position was that asking a witness to testify about an unfamiliar document would necessarily require the witness to function as an expert and not as a fact witness. Kodak would not accept this limitation and, at its request, the parties appeared before the Special Master on January 12 to argue this issue.

Polaroid lost again. Judge McLaughlin endorsed Kodak's position that as long as the references were relevant to the technology of the invention covered by the patent asserted against it, Kodak was entitled to examine the inventor about those references so as to elicit his understanding as to the differences between the reference and his invention in suit.[19] McLaughlin considered this factual evidence and not expert opinion. He acknowledged that "an answer to a highly technical, factual question in the field of prior art calls for expertise and learning on the part of the witness to answer it [but] that is not to be construed as an opinion."[20] Perhaps most distressing to the Polaroid team, the Special Master stated that his ruling applied not only to the inventors of the patents in suit, but also to any "officer of . . . [Polaroid] who is expert in the field of technology involved in this litigation." This broad application would clearly impact Land's situation, extending the scope of his deposition beyond his patents in suit to perhaps any area of instant photography under consideration in the lawsuit. The only relief McLaughlin was willing to give Polaroid was a requirement that a copy of any such publication be provided by Kodak counsel to the witness in advance of the deposition "in sufficient time to comfortably familiarize himself or herself with it." He also ordered that the deposing attorney must provide a schedule of the order in which the documents would be addressed and that such order "be adhered to during the deposition."[21]

Carr then forged ahead and prepared a list of twelve documents for review by the next witness, Lloyd Taylor, the inventor of the Mordant patent in suit, prior to his being deposed in early February.[22] Some of these were references that Taylor had neither authored nor seen before. Schwartz was not prepared to give up on this issue. On February 1, Schwartz filed a motion with Judge Mazzone, asking him to overturn the Special Master's ruling and stating that the issue was "whether witnesses should be compelled to study patents with which they are not familiar in order to form an opinion about the meaning of those patents and about the difference between those patents and the patents in suit, and then to so testify."[23]

Polaroid, with Land very much in mind, argued that this was against precedent and placed an "unjustified" burden on Polaroid's employees by "unconscionably expand[ing] the scope of these depositions by requiring witnesses to study and then testify about unfamiliar prior art." Given that Kodak had already cited 186 pieces of prior art as being potentially relevant, "the time involved in study and testimony would be staggering" if Kodak were allowed to ask Polaroid witnesses about materials they had not previously seen.[24]

While waiting for a ruling, Polaroid attorneys continued make things difficult for Kodak, to the point that an exasperated Carr decided to seek the Special Master's direct intervention by asking him to attend the Taylor deposition "to enable me to obtain rulings with respect to the propriety of questions and instructions by counsel as we proceed."[25] Kerr knew this would not be a helpful development. We do not want the "nose of this camel in this tent" he noted on Carr's letter before sending it on to Schwartz.[26] After McLaughlin wrote back indicating that he would be amenable to sitting in at the deposition per Carr's request,[27] Kerr responded that he didn't believe "such a procedure is either necessary or desirable."[28]

However, in a definitive setback for Polaroid, the issue was rendered moot the very next day. On March 6, 1979, Judge Mazzone ruled on the dispute concerning the scope of deposition preparation and examination. Citing "the policy favoring liberal discovery," Mazzone approved and adopted the Special Master's order.[29] The practice initiated by Kodak counsel would be allowed to continue, and there was little more the Polaroid team could do to resist it. Kerr and Schwartz reluctantly accepted that they and their Polaroid deponents would have to suffer broad, burdensome, and, in some ways, clearly unfair examinations from Kodak.

At this point in early 1979, nearly three years into litigation, the teams handling the case had grown into small armies. This massive effort was generating huge legal bills for Polaroid. Unused to litigation and costs on this scale, Polaroid's Peck and Mikulka were growing a bit uncomfortable with the financial hemorrhaging. It was greater than ever expected, even though they knew Polaroid had to stay the course. When, fortuitously, an unrelated story about large-scale litigation appeared in the *New York Times* on March 16, 1979, the day of a status conference with Peck and Mikulka, Kerr handed a copy to Bob Peck.[30] Kerr and Schwartz cited its highly relevant point: it was during discovery, and in particular the deposition stage, "that most trials are won or lost." The piece went on to note: "it is also at this stage that a corporation incurs some of the heaviest legal expenses."[31] To their litigator's relief, this article gave Polaroid executives some comfort in knowing that what they were experiencing was not unusual, and it eased whatever strains may have been developing between client and counsel.

It wasn't just the clients, however, who were showing the strain. The mounting workload, pressure, and real, if unspoken, fear that a tactical mistake could hurt their client's case made all the lawyers involved both apprehensive and testy. In early 1979, one of Carr's young partners, John Fogarty, wrote a long letter to Schwartz complaining that Polaroid was using the pretext of "masking" (blacking out irrelevant or privileged material the other side was not entitled to receive) to unacceptably hamper Kodak's discovery process.[32] In response, Kerr wrote to Carr. "It seems to me that argumentative letters of this character serve no useful purpose. . . . I do not propose to respond in kind on behalf of Polaroid," he concluded, noting that all of the documents Carr's colleague complained of had been inspected by Kodak lawyers twice and that no specific request for unmasking had been made.[33]

The following week, a young Polaroid lawyer complained when, at the last moment, and for the second time, Kodak canceled an inspection of a Polaroid camera model that was to be conducted in Cambridge.[34] One of Kenyon's senior partners, Edward J. Handler, responded not to the young lawyer but to Schwartz, stating that such letters "serve no useful purpose, nor shall I respond in kind on behalf of Kodak."[35]

CHAPTER 15

KODAK'S BIG ASSAULT

Almost three years into it, perhaps the most astonishing thing about the *Polaroid v. Kodak* case was that Edwin Land, the unquestioned father of instant photography and the key figure in one of the most momentous lawsuits of the century, was missing in action from the frontlines of the battle. After publicly condemning Kodak's conduct and asserting Polaroid's obligation to defend its patent rights, the press heard nothing more from Land. The media (and Polaroid's executives and lawyers) could only wonder, given his well-known penchant for privacy, his reclusive nature, and lack of active involvement so far, to what extent, or in what frame of mind, he would be willing to participate personally. Schwartz had never before been involved in a litigation in which the primary inventor of the technology at issue, and the CEO of the corporation that had arisen from it, was not at the forefront of the action, and in the midst of every discussion and decision that was made.[1]

In the far less significant Berkey Photo case, Kerr and Schwartz had been forced to take on and ultimately defeat the first infringer of Polaroid's instant photography technology without Polaroid's iconic head. So far, Land had been a no-show in the Kodak lawsuit as well. But as an inventor of three of the asserted patents, eventually he would be expected to be deposed and to give testimony in the trial. At this advanced juncture, however, no one really knew if he would.

Finally, shortly after the 1979 New Year holiday, Land called for a meeting in Cambridge with his legal team. Present were Bob Peck, as head of the Polaroid patent department, and Bill Kerr.[2] Also present was Joan Clark, an attorney from the firm that was handling the parallel infringement case against Kodak in the Canadian courts. At this meeting, Land made it clear for the first time that he was willing and, in fact, anxious to participate fully in the case. To him this meant that, consistent with his nature of tackling problems head-on and with complete immersion, he was going to prepare himself for the endeavor in a most comprehensive way. He told Kerr and Peck that he was assigning two of his closest associates, Holly Perry and Nan Chequer, to work with him on his preparation. They were going to prepare for him histories of Polaroid and the development of all of the technology at issue in the lawsuit. He also asked Kerr to have position papers prepared for each of the patents in the lawsuit so he could fully understand the issues.

Kerr immediately pointed out that under the rules of evidence in the federal courts, an adverse party is entitled to see any document that a deposition or trial witness uses to refresh his or her recollection. That document can then be used to cross-examine the witness and any portion of it may be introduced into evidence. Thus, any documents Land might create in preparation for testifying would be fair game for Kodak. This was also true of any preparatory material created by Fish & Neave. Kerr was determined to make sure that any such material remained "immune from discovery."[3] He advised Peck that "no copies of . . . [the assistant's] notes should be made, and the original of their notes should remain in their possession or in your possession at all times." Most importantly, he concluded, "the notes themselves should not be shown to Dr. Land."[4] (See Fig. 15-1.)

Unfortunately, the meeting also witnessed the first signs of a clash of personalities between Land and Kerr. These were two men of enormous egos. One would not characterize Land as humble; he could come off as arrogant and he certainly did not suffer fools gladly. Kerr, unlike those who worked with and for Land, was not intimidated by Land's brilliance or his often imperious ways. As a former Marine and FBI agent, and someone who had seen his share of the rough and tumble, no one intimidated Kerr; he considered himself equal to Land and was unwilling to show the deference that Land was accustomed to.[5] For his part, Land didn't like people who challenged his control over situations, and he saw Kerr as definitely in that category.[6] Despite Kerr's standing as a prominent attorney, Land saw him as being stiff and full of his own importance. Both were "alpha males" and so, yoked together in a life-or-death lawsuit, perhaps it was inevitable that these two men would develop, at best, an uneasy attorney-client relationship.

This became clear when Bob Peck called Herb Schwartz to tell him that after the meeting, Land had requested an additional day of meetings with Joan Clark. Land clearly liked Clark. She was bright but unthreatening.[7] Kerr had already left Cambridge on a long-scheduled trip to Hawaii to attend a meeting of the American College of Trial Lawyers. It's unclear whether Kerr was invited to this follow-up meeting, but in any event, he would undoubtedly have refused, unwilling to give up his tee time on what was essentially a glorified golfing expedition. Such was his personal brand of arrogance.

Kerr and Schwartz were troubled by the extra meeting with Clark, wondering what Land expected to get out of it. Was this just an attempt by Land to thumb his nose, as it were, at Kerr and cut his honorable counsel down to size? The Canadian lawsuit, one of several suits defending Polaroid patents in foreign jurisdictions, was ancillary to the primary U.S. action. Kerr and Schwartz made all significant strategy decisions, even for these foreign actions. Nothing important could be decided without them. In addition, one concern about Clark among the Polaroid in-house counsel was that she tended to be more pessimistic about the chance of success than they liked to see, especially around Land. Peck made it clear to Clark that in meeting with Land, she was to remain upbeat.[8] The whole affair remained just another curiosity regarding the often inscrutable Land.

Despite his enormous intellectual gifts and great wealth and success, Land could, however, be very down-to-earth when he wanted to be. The private man had a rather reserved and respectful personality. He loved dogs—he had two bullmastiffs named Per and Se—and generally took a real personal interest in the lives of his closest associates, even though his demand that they show dedication equal to his could make their lives difficult. His affection for those who worked closely with him and whom he respected was manifest—and he repaid their dedication and loyalty in his own way. Still, Land lived in his own world, one in which science demanded total immersion and obliviousness to everything else going on outside his laboratory. One of Land's colleagues loved to tell the story of the time Land called him one morning to complain that no one had shown up for work in his laboratory to continue the experiments they were engaged in.[9] The colleague had to remind Land that it was Thanksgiving Day. What his family made of this is not known.

While he was preparing to enter the legal fray, rare public controversy erupted around Land. Financial analysts were disappointed and disturbed when Polaroid's 1978 earnings were reported on February 22, 1979. Worldwide sales of instant cameras had grown from 5.5 million in 1975 to 14.7 million in 1978, meaning that even more instant film, the real profit center in the business, would be sold. Though Kodak had captured about one-third of the instant market, this was not necessarily at Polaroid's expense. Eugene Glazer, an analyst with Dean Witter Reynolds, was quoted as saying that "Kodak did not take away market share from Polaroid. . . . They created additional market."[10] But these expectations were not borne out in the numbers. Although Polaroid's 1978 revenues rose twenty-eight percent, the company's profits came in on the low end of analysts' projections.

The cause was the burgeoning disaster called Polavision. The instant movie system just wasn't selling. Problems concerning the viability of the system had arisen in the marketplace very quickly following its introduction, among them its lack of sound and a widely criticized, cumbersome processing procedure for the film. Another was the emergence of videotape technology virtually simultaneously with the release of Polavision. Videocassette camera recorders garnered immediate consumer interest, resulting in sales of 420,000 units in 1978 alone. They avoided the limitations of Polaroid's new system—they were easy to use, they had sound, and tapes offered longer recording time. Polavision's losses were dragging down Polaroid's overall fiscal performance. Estimates were that only 100,000 Polavision units were ever built, and only one-half of those were sold.

The company's stock price reacted accordingly. It had been on a roller coaster from its euphoric high of $143 per share following the introduction of SX-70 to its nadir just over fourteen dollars a share in 1974, when the mammoth introduction costs of Polaroid's new system hammered the company's earnings. But more recently, with its sales of instant products holding up despite Kodak's competition, Polaroid stock had settled at a more reasonable value of between fifty and sixty dollars a share. Within days, the latest financial results drove the price to below forty dollars.[11]

The stir was caused not just by the fact that the struggling Polavision system was seen as Land's baby. The company's policy of relative secrecy regarding its financial results, as well as its research and development programs, had led investors to be surprised by the lackluster earnings report. When it came to light that the Rowland Foundation, a not-for-profit entity controlled by Land, had sold 300,000 shares of Polaroid stock in early January at $52.50, just weeks before the unsatisfactory numbers were announced, the financial community, and some Polaroid shareholders, cried foul. In early March, a New York investment firm, Seiden & De Cuevas Inc., filed suit in federal court in New York against Polaroid and Land individually, alleging that they had improperly withheld information on the financial pressures caused by Polavision and the effect they would have on Polaroid's upcoming fiscal report.

The suit alleged that Land had improperly sold the shares with knowledge of the bad news that the public was not privy to.[12] Not all observers saw the actions of Polaroid or Land as improper. Given the extraordinary sales year Polaroid had enjoyed in 1978, some analysts recognized that it seemed like a perfect opportunity for the company to offset losses it had accrued on the Polavision system.[13] The timing for the sale of stock by Land's foundation was mere coincidence; it was justified as necessary to raise capital to acquire property and to begin construction of a research facility. Nonetheless, despite the fact that the legal action went nowhere, the accusations of impropriety had to be a source of embarrassment to Land, only exacerbating the humiliation he was experiencing over the commercial struggles of his latest technological contrivance.

Meanwhile, the press and the public were speculating on when the Kodak lawsuit would go to trial. A *New York Times* article in early April 1979 announced, erroneously, "the case may go to trial later this year."[14] An annoyed Bill Kerr wrote Bob Peck: "As usual, the news reporters seem to know more than trial counsel concerning prospective trial dates."[15] There had been no discussion with the court about a potential timeline, nor had there been any action on the pile of motions pending from the outset of the case. It began to seem as if none of the judges on the Boston federal bench was anxious to preside over a trial of this magnitude and complexity. On April 10, 1979, the case was suddenly, without explanation, reassigned to its third judge.

David S. Nelson, the first African-American ever to be appointed to the federal bench in the state of Massachusetts, was the one who ended up with the hot potato. He fell victim to the custom whereby unpleasant or burdensome cases were assigned to the newest judge as a way to fill up his or her initial docket. Nelson, forty-six years old, had risen from the office of the Massachusetts Attorney General to the Massachusetts Superior Court in 1973 and was confirmed to the federal bench early in 1979. It appeared, almost from the outset, that Nelson wanted nothing to do with this case, and it would not take long before he too would find a way out. But, for now, it was his.

Discovery, meanwhile, continued with full intensity, with arguments raging over almost every conceivable aspect. To this point, Kodak had produced to Polaroid in excess of 180,000 pages of documents in answer to its requests, while Polaroid had produced to Kodak approximately 29,000 pages. Kodak had examined in deposition ten Polaroid witnesses for a total of seventy-three days, while Polaroid had conducted thirty-eight days of depositions of thirteen different Kodak witnesses.[16]

Both sides had responded to hundreds of written interrogatories seeking detailed information on every conceivable, and many inconceivable, aspects of the case. On April 12, Kodak filed formal deposition notices with the court for an additional nine Polaroid witnesses.[17] On that list, for the first time, was Land, director of research. Carr proposed to Kerr that the first two sessions of Land's deposition be set for August 7–10 and September 25–28 at Kenyon's New York office.[18] Although simultaneous depositions had been and would continue to be conducted by both sides, virtually every week, for months on end, the Land deposition was clearly going to be a pivotal event for both parties.

Peck wrote a memo to Land, copying Mikulka and Kerr, informing him of the notice and sending him copies of the formal papers. The combination of timidity and downright fear apparent in this memo is absolutely striking, and illustrates the dread Peck and his colleagues experienced in having to deal with their chief executive. For example, Peck felt the need to write a long paragraph apologizing and trying to explain why Kodak had used only Land's title as director of research in its notice. "Why your other titles were not used for the Deposition Notice is a mystery to us," Peck wrote.[19] He suggested that since Polaroid had noticed the depositions of Kodak's top research executives, Kodak's noticing Land under that title was understandable, while curiously failing to point out that Land would be deposed as an inventor. Peck concluded his memo by letting Land know that the deposition would ultimately be held at his convenience. Kerr "would not expect to have any difficulty" in shifting the actual time and place of the examination.[20]

The 1979 Polaroid shareholders' meeting was held on April 24. The highlight was Land's announcement of a new SX-70 film, later marketed as Time-Zero Supercolor, that would begin to show an image in ten seconds and fully develop in one minute.[21] The current version took about four minutes. However, the upbeat mood at Polaroid soon faded, particularly for those involved in the Kodak litigation. The very next day, with news of the Land deposition still settling on the Polaroid team, Kodak counsel dropped another—but this time an unexpected—bombshell. In what one could characterize as its D-day invasion, Kenyon & Kenyon filed a motion on April 25 for partial summary judgment seeking a declaration that Rogers' "Excedrin" film unit patents were invalid, and thus unenforceable, and that they should never have been issued by the U.S. Patent Office in the first place.[22] In a summary judgment motion, the moving party essentially urges the position that there are no facts in dispute that need to be determined at trial, and thus the judge can make a decision immediately. While more common today, at the time it was not a tactic generally considered appropriate for complex factual cases.

Rogers' film format patents were in many respects the heart of Polaroid's case, arguably the most important patents in the lawsuit.[23] Kodak executives had been aware of their significance for years. In 1971, when the Rogers patents were first issued, they were brought to the attention of Kodak management, including Walter Fallon, who at the time was head of Kodak's film manufacturing operation. That summer, Fallon received a report that Rogers' Excedrin format "covered a construction identical to our planned . . . [film unit],"[24] and thus if Rogers' "was a valid patent, it would be a serious matter."[25] A few weeks later, at an August 1971 meeting that included then Kodak chairman Louis Eilers, Kodak management reviewed a list of Polaroid patents that might affect its instant photography plans. Fallon described Rogers' film unit as "patent property which reads on our . . . [film unit] structure."[26] On his copy of the meeting agenda, next to the Rogers patents, Fallon had written the notation, "Big Block."[27]

This concern persisted. As late as May 1975, less than a year before Kodak introduced its instant system, Kodak management met yet again to review Polaroid patents. Just prior to this management meeting, Kodak in-house counsel met with Carr and with some members of Kodak's research team to "review U.S. and foreign [patent] problems." At that meeting, the Rogers patents were identified as "patents which, if asserted, may present a relatively difficult situation due to technical and legal complexities."[28] A memorandum memorializing the meeting with Kodak management stated on its first page: "As a general comment, it should be noted that the critical issue, worldwide, will be the validity of Polaroid's format patents. . . . It is these [Rogers] patents which must be overcome in order to insure freedom of operation."[29]

Kodak realized that if it could possibly get these patents kicked out of the litigation without a trial, it might severely weaken Polaroid's case and its resolve. That, Kodak could assume, would likely lead to a settlement on very favorable terms. It was a gutsy maneuver, but one that Kodak was prepared for. It had been trying to have those patents declared invalid in foreign jurisdictions including Britain, Germany, and Japan since the early 1970s. While not successful abroad, its arguments against the Rogers patents had been well researched and honed in the process. Together with what had come out from the Rogers depositions, these arguments gave Kodak the confidence to launch the offensive.

Strategically, it also knew that Polaroid and its legal team would be knocked back a bit and put on the defensive, having to craft a response to this critical motion, while at the same time dealing with the ongoing deposition proceedings, including the preparation for Land's deposition. The parties agreed that Polaroid would have until July 16 to file its briefs in opposition to Kodak's motion for summary judgment and that Kodak would thereafter have until September 17 to file its reply to Polaroid's submission. Meanwhile, at Schwartz's request, the dates for the initial session of the Land deposition were set for the week of September 24 at a location in the Boston area, rather than in New York.[30] Polaroid was now fighting a two-front war.

Kodak's argument was essentially that the Excedrin film unit was not entitled to patent protection because its geometry did not rise to the level of invention necessary to qualify for patent protection. To be entitled to a patent, an invention must be new; that is, it must never have been described before in the "prior art." However, even if new under this standard, an invention must also be "nonobvious" to merit a patent. This means that the invention would *not* have been obvious to one of ordinary skill in the art of the technology at the time the invention was made.[31] Kodak contended that the Excedrin film unit merely substituted one element in a combination of elements for another element previously described in other Polaroid patents to serve the same function. Thus, according to Kodak, the Excedrin structure was not really a patentable invention because it would have been obvious to one skilled in the art of making instant film units. It further argued that the combination of all the *old* elements in the Excedrin film unit lacked any surprising or unusual result, and functioned no differently in this particular combination than they did in other earlier Polaroid patents.[32]

Kodak tried to simplify the argument as much as it could. The basic premise was that an integral instant film unit needed a darkroom in order to be developed outside the camera. Earlier Polaroid patents had taught the basic geometry of an instant film unit, and all the Rogers patents did, according to Kodak, was to teach a different arrangement of the layers that could create a different *kind* of darkroom. Kodak urged that the method used in Rogers' patents to create this darkroom—incorporating carbon black into the processing solution to create an opaque layer under the transparent film support after exposure—was also disclosed in prior Polaroid patents.

Ironically, the earlier Polaroid patents cited by Kodak as rendering the Excedrin invention obvious were all patents granted to Land. In essence, Kodak argued that Rogers had just taken information disclosed by Land as early as 1963 and put it together in a way that anyone of ordinary skill in the art of instant photography could have done. In its brief, Kodak cautioned the judge that, "although the . . . [Rogers] patents appear to be very complex documents emanating from a sophisticated and advanced laboratory of high reputation, their actual contribution to the advancement of science and to instant photography is non-existent."[33] Some members of Polaroid's legal team acknowledged that this argument had a certain superficial appeal and that this motion was something to be worried about.[34]

At this point, news around Kodak was good on a variety of fronts, and one can sense a certain air of bullishness underlying several moves it made during this time period, including the summary judgment motion in the Polaroid case. Its chairman, Walter Fallon, had just announced a forty percent increase in its quarterly profits, with sales continuing to rise in all of its photographic divisions, including instant photography.[35] Kodak decided to ramp up its efforts against Polaroid in the marketplace. Within weeks, it "fired another salvo in its battle with Polaroid," as the *Wall Street Journal* described it, by announcing aggressive rebates on its instant cameras and film, a strategy Polaroid was not prepared to follow.[36]

On the legal front, Kodak was clearly on an offensive. Aside from the summary judgment initiative, it was vigorously and optimistically appealing the verdict against it in the antitrust suit brought by Berkey, which had ended in complete disaster for Kodak. First, Berkey was awarded damages of $81.4 million for Kodak's violation of the antitrust laws. Second, and perhaps even more embarrassing, one of its attorneys, Mahlon Perkins Jr., a partner at Kodak's counsel Donovan, Leisure, Newton & Irvine, had been found to have lied and to have hidden pertinent documents from Berkey during the litigation. Perkins had been sentenced to a month in prison for his transgressions.

Kodak's new counsel for the Berkey appeal was William Piel Jr., the senior litigating attorney at the respected Wall Street firm of Sullivan & Cromwell, and a man described by the *New York Times* as a "pillar of the legal establishment."[37] Piel had urged the U.S. Court of Appeals for the Second Circuit to reverse the earlier verdict as a matter of both law and fact. The oral argument on the appeal, delivered to a three-judge panel before a courtroom packed with more than fifty standees, had lasted for just under three hours, one of the longest on record.[38] A decision was pending and expected shortly.

With the Excedrin summary judgment motion filed in court, Kodak wanted to publicize the fact that it had struck this blow to the heart of the Polaroid case. But Kodak had a problem. The litigation was being conducted under a protective order, sought by Polaroid from the outset to protect "trade secrets and other confidential research, development and commercial information" that might be transmitted to Kodak during the lawsuit.[39] Judge Frank J. Murray, who was presiding over the case at the time, had granted Polaroid's request for stringent controls. Kodak had relied upon some confidential documents in its motion papers. As a result, the motion brief fell under the aegis of the protective order and could not be disclosed to the public, or even to Kodak executives who were not qualified under the terms of the protective order to see Polaroid confidential material. To get out from under this restriction, Kodak filed a separate motion to have the court declare its summary judgment motion *not* confidential.[40]

With Polaroid blocking Kodak's desire to let the public see the brief and get its side of the story out before Polaroid could have a chance to respond, an elaborate legal minuet went on through May and June.[41] The music stopped, however, when the Special Master considered the briefs and ruled that Kodak would be allowed to unseal its summary judgment motion on the Excedrin patents immediately.[42] Kerr had worked this issue hard, but he must have known that Polaroid's chance of avoiding publicity on Kodak's attack on Excedrin was doomed to fail. Yet he had an apprehensive client, so he was forced to show he had done all he could. In the end, the ruling had little practical effect. The Special Master's decision was handed down on July 20, just a few days before Polaroid was due to file its brief in opposition to Kodak's summary judgment motion on July 26. So, even though Polaroid was formally rebuffed in its effort, it achieved its overall aim because it was able to release its side of the story within days of Kodak's making the material public.

A high level of anxiety prevailed at Polaroid as the time for Land's deposition drew near. Apart from the legal case, all was not well with Polaroid's business performance. Following close on the heels of the disappointing earnings announced in February, and the resulting shareholder unease over the drop in stock price, Polaroid announced in mid-May the layoff of 800 workers, most of them from its camera production facility in Norwood, Massachusetts. This represented the most significant layoffs at the company since the late 1940s when the war effort had come to a close.[43] "As is usually the case . . . Dr. Land wasn't talking, but his actions . . . said enough," remarked the *New York Times*.[44] Polaroid did its best to downplay the news, claiming that the move was not "any indication that the company is in trouble." In fact, inventories had built up, and so production needed to be slowed. It was open to debate whether the problem was due to Polaroid's competition with Kodak, to the then slowing economy, to waning consumer excitement about instant photography, to the failure of Polavision, or to some combination of all of the above.[45] In any event, the move was a blow to morale around the company and only increased the pressure on those in the patent department internally manning the frontlines of the legal battle with Kodak.

From their perspective as litigation counsel, Kerr and Schwartz decided that Polaroid had been too much on the receiving end lately and knew that significant action was needed in response, both strategically in the court case and to boost their client's frame of mind. On May 11, 1979, they met Mikulka and Peck and decided on a plan.[46] First of all, it was agreed that Polaroid should turn up the pressure on Kodak with respect to the fulfillment of its discovery obligations. On May 17, Kerr advised Carr, "It is becoming increasingly apparent to us that the pace of Kodak's document production is not keeping up with Kodak's deposition program."[47] In his response Carr admitted, "You have raised a number of questions, not all of which I am presently prepared to answer. However," Carr promised, "I will promptly look into thcsc, and gct back to you very shortly."[48] Kerr was apparently surprised with the speed and tenor of Carr's reply. He scribbled on the letter "a temperate response" and sent it along to Schwartz.[49]

But Kerr heard nothing more for weeks and then received on June 7 a long, detailed, and exceedingly argumentative rejoinder disputing almost every aspect of Kerr's initial inquiry and rejecting the basic notion that Kodak was getting more discovery from Polaroid than it was providing.[50] The exchange of letters that followed laid the basis for a series of motions that Polaroid made to the court over the ensuing months for orders compelling Kodak to provide what it believed it was entitled to. Kodak responded by making its own series of motions seeking court-ordered compliance with various discovery requests. By the end of discovery, Kodak had made thirty such motions, and Polaroid had filed fourteen. It's unclear whether either party ultimately uncovered any useful information as a result of all this, but it kept the pressure on for both sides and, if nothing else, made each side's client feel as if its trial counsel was aggressively harassing the opponent.

The May 11 meeting with Mikulka and Peck also included the discussion of a plan for addressing Land about his upcoming deposition.[51] It was decided that Kerr should write to Land and initiate a formal program of preparation sessions. Surprisingly, Polaroid wanted to consider adding another Land patent to the lawsuit, even this late in the game. The leading candidate was a patent that had been considered for inclusion at the outset of the case but ultimately was not one of those asserted. It covered an instant film unit in which the positive, image-bearing sheet is extended so as to include a pod for containing the processing chemicals.[52] Fish & Neave was asked to prepare a memorandum on the patent, and the work leading up to it, to serve as the basis for a discussion about it with Land before a decision was made.

Given how far the case had already proceeded, the idea of adding another patent at this time was hard to figure. Tactically, there was little to gain. Possibly, a pod patent might superficially buttress Polaroid's argument that Kodak, having failed to generate its own unique technology, had merely copied Polaroid's.[53] After all, even Kodak's venerable Kenneth Mees had long ago called the pod the "transcendent invention." But Polaroid was ever anxious to move the case toward trial, and adding another patent at this juncture, more than three years into the litigation, would only be counterproductive in that regard. Further, the real landmine was the possibility this later iteration of a pod patent would be vulnerable to attack consistent with Kodak's broad charge that Polaroid improperly engaged in the practice of re-patenting the same inventions over and over so as to extend its patent monopoly on instant photography technology.

Peck, however, now wanted the original decision reconsidered. His motivation appeared to be mollifying Land when he fully came to the understanding that his patents asserted against Kodak were relatively insignificant compared with the Rogers patents.[54] Land's deposition preparation was likely going to be his first opportunity to focus on this issue and to make this assessment. He had not been consulted on the patent selections before the lawsuit, having decided to stay removed from the process. Although many of Land's significant contributions to integral instant photography were used in Polaroid's SX-70 system, they simply were *not* employed in Kodak's products. Still, Peck seemed apprehensive that Land might consider his role in the case too minor. This apprehension again reflected the paranoia and uncertainty Peck and colleagues felt from having to deal with Land.

As agreed, a letter from Kerr to Land was drafted and submitted to Peck for review. The letter informed Land that his deposition had been noticed and was scheduled to commence September 25. In it, Kerr admitted that he couldn't give a reliable estimate for how long the process would last. He let Land know that "preparation for your testimony must be carried out in a thorough and meticulous fashion," which would require that he and Schwartz spend "several consecutive days with you during each of the months of July, August and September, if this is feasible from your standpoint."[55] The sensitivity they all felt about dealing with Land is underscored by Kerr's deeming it necessary to run a draft of the letter by his in-house counsel, a most unusual circumstance.

Meanwhile, consideration continued with regard to adding the Land pod patent to the case. Richard Barnes, a senior Fish & Neave associate, prepared a draft memorandum on the subject, but it became obvious to all that it was too risky to insert a problematic patent into the litigation just to assuage the possible bruising of Land's ego. The subject was dropped.[56] Also dropped was the idea of sending Kerr's draft letter to Land. The news about Polaroid's recent layoffs was breaking that week, and the timing probably wasn't right for trying to get Land to focus on his deposition. Instead, it was decided that Kerr should have a face-to-face meeting with Land, and that was set for June 27. In the meantime, Polaroid senior management requested a meeting with Kerr and Schwartz to review the status of the whole litigation.[57] Everyone was, apparently, becoming antsy.

The third piece of the Polaroid counteroffensive was the effort to oppose Kodak's summary judgment motion. With the lawyers presently working on the case consumed with discovery activities, Kerr and Schwartz brought in a young partner, John Nathan, to draft Polaroid's reply brief. Richard Barnes was assigned to assist Nathan, and on May 25 they met with Peck and Ford in Cambridge to discuss their ideas on how to approach the opposition.[58] Since Peck and Ford had successfully defended the Excedrin patents in several foreign jurisdictions, they could provide valuable advice. A key element in the effort was to select an expert who would submit an affidavit supporting Polaroid's technical arguments. Kerr believed that neither Land nor Rogers should serve this purpose, since they were destined to be Polaroid's primary experts at trial, and he did not want to expose their positions prematurely on any of the patents in suit, or risk having their expertise sullied in the battle of experts that the summary judgment motion would inevitably engender.[59]

Peck and Ford proposed Wolfgang Berg, a professor emeritus at the Swiss Federal Institute of Technology in Zurich. He had been the head of its photography department from 1961 to 1978, following a career in photographic research in Germany and England, including nine years from

1936 to 1945 as a senior physicist for Kodak at its research labs in Harrow, England. In 1977, Berg had authored an article in a German technical magazine describing Polaroid's SX-70 technology in great detail. In his introduction, Berg noted how instant photography had made Polaroid into one of the top photographic companies in the world "through the foresight, inventiveness, and persistency [*sic*] of the originator, Dr. E.H. Land."[60] Berg was perfect. Polaroid invited him to Cambridge, and Barnes and Nathan set to work with him crafting Polaroid's response.

With work proceeding on all fronts of Polaroid's counterattack, Kerr and Schwartz traveled to Polaroid on June 18 for the comprehensive review of the status of the litigation with senior management.[61] Absent, as usual, was Land. Kerr began the meeting with a detailed review of the litigation from its inception. Kerr told the executives about the scheduled Land deposition and outlined the preparation plans for that important event. He conveyed Schwartz's assessment characterizing the results of discovery to date as being what both sides might ordinarily expect: "Kodak has brought to light certain warts and blemishes in the Polaroid patent position, and Polaroid has made progress in developing the kinds of facts a plaintiff likes to find in the files and minds of a defendant in a patent infringement suit."[62] Without being unrealistically optimistic, Kerr was able to assure his client that "Polaroid still has reason to believe that it can be bullish concerning the prospects of sustaining at least some of its patents."[63] He told them that, in the meantime, Polaroid would be pressing Kodak, through motions to the Special Master, to improve both the quantity and the quality of its document production. Kerr predicted that the deposition program would run for another six to nine months, into mid-1980, after which Polaroid would be able to start seeking a discovery cutoff in the case that would move matters "further along towards trial."[64]

CHAPTER 16

LAND JOINS THE FRAY

On June 25, 1979, Kodak got some very good news. The previous year, Berkey had won an antitrust lawsuit against Kodak, having accused Kodak of using its monopoly wrongfully by introducing a new camera and film system without giving competitors advance notice. Chief Judge Irving Kaufman of the Court of Appeals for the Second Circuit, the legendary Truman appointee who presided over the Rosenberg spy trial during the early 1950s, issued a 121-page opinion upending that decision, reversing almost the entire $81.4 million damages award and ordering a new trial.

Citing numerous errors by the trial judge in his interpretation of the law, the Second Circuit ruled, in what became known as a seminal antitrust opinion, that a company does not violate the Sherman Antitrust Act by "simply . . . reaping the competitive rewards attributable to its size."[1] In language that must have resonated at Polaroid, as it burst a bubble at Berkey, Kaufman held that the antitrust laws had been misapplied in the case and that "the mere possession of monopoly power" was not necessarily illegal.[2] The appeals court rejected Berkey's contention that the antitrust laws required advance notice of the change in film system. "It is the possibility of success in the marketplace, attributable to superior performance, that provides the incentives on which the proper functioning of our competitive economy rests," Kaufman wrote, and a "pre-disclosure" requirement would take away a benefit derived from developing superior products.[3] While Berkey pledged to seek review of this decision in the U.S. Supreme Court, it was clear that Kodak had won a massive victory.[4] After the Supreme Court declined to consider Berkey's petition for review of the Second Circuit's decision, the case was eventually settled for $6.8 million and was never retried.[5]

While Kodak celebrated its victory, Polaroid's lawyers prepared for the long-planned meeting on depositions with Land, set for that same week. They would brief him on the status of the lawsuit and explain what he had to do to prepare himself for deposition. Up to that point, Kerr had briefly met with Land just twice. None of the lawyers involved could, even now, predict how Land was going to react to the disciplines of the witness process, from the preparation to the actual examination. He would have to exercise considerable self-control and demonstrate thorough comprehension of the immensely technical matters involved. There was no doubt that Land could handle the latter. But Land's tendency to stubbornly go his own way was cause for considerable anxiety over the former. How he would comport himself was still an open question over which many hours of sleep were likely lost by Peck and his crew.

The meeting took place on June 27. Joining Kerr and Land were Mikulka, Peck, Ford, Joan Clark, and Nan Chequer, who took notes.[6] Chequer had been working for months with Holly Perry to prepare Land for his coming ordeal, but the legal team knew nothing about the details of this preparatory work. (See Fig. 16-1.)

Kerr spent the first hour providing Land with an overview of what had transpired during the years of discovery and litigation. Land asked about the outlook for a trial. Kerr explained that, while a jury trial had been an option, both parties had adopted the conventional thinking of patent litigators at the time that a bench trial would be preferable. He stated that he expected it to last about eight weeks, although Kodak predicted it would take as long as twelve. If the parties could "clean up" discovery in early 1980, Kerr hoped to get Judge Nelson to try the case before the end of that year. That meant a decision would come at the end of 1981. An appeal would take at least another year, so the earliest possible date for a final decision from the appellate court would be late 1982. Land worried about the appeals process dragging on since only Kodak had the "time and the money to do it all over." When Land asked about the merits of the suit, Kerr told him that "Polaroid has a better than even chance to have some of . . . [its] patents declared valid [and] infringed." Kerr also advised Land, however, that the "prior art is close" on some patents, for example, with respect to Rogers' Excedrin patents where some of Land's "earlier patents hint at [an] integral film." Notwithstanding this situation, "no fatal infirmity [exists] in any of our patents," Kerr concluded.[7]

After a break for lunch, Kerr explained that Land would be asked to explain in his own words, as opposed to the technical language of the patent, what he understood the invention to be in each of his asserted patents and to describe the circumstances surrounding his discoveries. Using Land's Rear Pick patent as an example, Kerr asked Land to describe the invention covered. As Kerr recounted it, "after some minutes of silence while he consulted the claims [the detailed legal terminology at the end of each patent precisely defining the invention covered and thus the exclusivity granted], I ultimately suggested that I would try to help him reach a definition of the invention."[8]

Kerr described the Rear Pick invention in terms the finder of facts, a judge having no technical background, would be able to understand. This explanation was one of the many translations of abstruse technical data into plain English prepared by the Polaroid legal team. In many ways, this is the art of the patent litigator, and it was essential for Kerr to give Land a glimpse of the style and level of discussion and analysis he would be facing. There was no time that day to explore fully even that one patent. But Land now had a better idea of what would be expected of him as a witness, first in deposition and later at trial, if he was going to be an asset in making Polaroid's case. Accordingly, ten more days were scheduled for preparing Land before the first day of his deposition in September.

That meeting was a pivotal moment for Polaroid's litigation counsel. Finally, the long-debated question had been answered much to their relief: Land would participate fully and effectively. Unlike his previous reluctance to getting involved, which had forced Polaroid to rely on substitute witnesses in foreign cases and to omit Land's patents in the Berkey case, Land seemed to be engaged in this process. Peck could not contain his relief when he reported to Kerr and Schwartz about Land's subsequent appearance at Polaroid's board of directors meeting. "Perhaps our sessions with Dr. Land have gone better than even we would allow ourselves to believe if his reported 'bullish' or 'upbeat' mood at the recent Board Meeting, when asked about the lawsuit, aptly characterized his feelings," Peck related.[9]

At this point in July 1979, both sides were conducting depositions virtually every week. Carr decided to fire another shot across Polaroid's bow by seeking to extend the length of Land's deposition, even before the first session had occurred. He asked for Schwartz to schedule two additional weeks. Schwartz and Kerr could not refuse to consider granting the additional time, but they also wanted to send a signal that two could play the game of "abuse by deposition." Schwartz fired back: "We will look into [Land's schedule]," he wrote Carr and added, "I enclose a Notice of Taking Deposition of Mr. Fallon."[10]

As Kodak's chief executive, Walter Fallon embodied his company's corporate bravado, in part fueled by its sheer size, which dwarfed Polaroid in resources and consumer power, as well as by its isolation in Rochester, where Kodak was clearly the master of its own universe. With the Berkey antitrust case reversed and Polaroid on the defensive, Fallon personified what some saw as Kodak's arrogance. Fallon was riding high on his company's success. During his tenure, Kodak's sales increased from $3.48 billion to $10.8 billion, and its worldwide work force grew from 114,800 to 136,500. It had posted its highest levels of sales and profits in the company's ninety-nine-year history, a performance, according to *Forbes*, that made it "likely that Kodak will become the tenth industrial company in American business history to earn $1 billion after taxes this year."[11] Dates for the Fallon deposition were set for late October 1979, with Schwartz acceding to Carr's request that the proceeding take place in Rochester. But even as those dates were set, Schwartz reported only that he was "still looking into" the additional dates requested for Land.[12]

Meanwhile, Polaroid was fighting back against Kodak's "Excedrin" summary judgment motion. The essence of Polaroid's rejoinder was that summary judgment was inappropriate because "numerous genuine material issues of fact exist which are vigorously contested by the parties."[13] Those issues of fact needed to be resolved at trial with the help of "live expert testimony subject to thorough cross-examination." The facts in contention, according to Polaroid, included the specific scope of the Rogers invention as described in the asserted patents, the interpretation of the prior art patents that Kodak alleged taught those inventions, and the differences between those prior patents and the Rogers film unit.

Buttressed by an affidavit from its expert witness, Wolfgang Berg, the former Kodak scientist, Polaroid maintained that neither of the two Land patents cited by Kodak disclosed even the most basic attribute of the Rogers film unit—that is, a film unit that is physically unaltered before, during, and after exposure. One of the patents related to early peel-apart film technology, while the other required a movable opaque sheet to protect the photosensitive part of the film unit after exposure. Based on his interpretation of this technical literature, Berg stated his expert opinion that the Rogers film unit "would not have been obvious to one of ordinary skill [in the art of instant photography] at the time those inventions were made."[14]

Schwartz believed that the summary judgment motion would devolve into a battle of experts, resulting in a stalemate. The real-life story surrounding the patents would ultimately be at least as important as the technical distinctions between the invention and the prior art. Accordingly, Polaroid's brief also disclosed a startling fact: Kodak had itself attempted to patent the Rogers film unit "after receiving the opinion of [its legal] counsel that the subject matter of those inventions was patentable."[15] Apparently, a Kodak patent attorney, Harold Cole, had written an "invention conception" in late 1968, more than six months after Polaroid had filed the application for the first of the two Rogers Excedrin film unit patents. That conception described a film unit identical to the one Rogers had previously conceived on New Year's Day of 1968. Over the next ten months, this conception had been reviewed within Kodak's legal department and compared to the same prior art that Kodak was now asserting invalidated Rogers' patent. In September 1969, Cole had sought permission to have a patent application filed on the film unit, advising his superiors that "this subject matter is believed patentable over the prior art and a patent application is therefore recommended."[16] Permission secured, Kodak had filed a U.S. patent application on October 24, 1969, followed by twenty-one counterpart applications in foreign countries.

It wasn't until several months later, on or about January 6, 1970, that "a disquieting event" had occurred at Kodak: it learned for the first time of the prior Rogers invention when a Polaroid application based on his work had been issued in Belgium as a patent.[17] Quickly, Kodak discovered that a corresponding U.S. application was then still pending in the U.S. Patent Office—and it also found out about all of the other foreign counterparts Polaroid had filed abroad. In light of these discoveries, Kodak was eventually forced to abandon all of its attempts to patent the film unit for itself.[18] But it was caught in the inconsistency of having taken the position that the structure was patentable over the very same prior art it now contended invalidated the invention.

In one final blow, Polaroid went on to allege that once Kodak learned of the pending Rogers applications, it had made two ex parte contacts with the patent examiners handling them in the U.S. Patent Office, urging that they be rejected. Leaving aside questions concerning the propriety of these contacts, the Kodak representative had then offered essentially the same arguments that Kodak was making in its current motion. The patent examiners were not persuaded then, and thus, argued Polaroid, the court should not be persuaded now. Polaroid acknowledged that it had not yet been able to discern all of the facts surrounding these activities, but, in its view, the lack of complete facts on this point only gave more weight to its contention that summary judgment was inappropriate because material facts concerning the matter were still to be discovered and then determined at trial.[19]

Would Polaroid's defense to Kodak's motion work? Granting a motion to dismiss just one or two patents in a twelve-patent lawsuit might seem to a judge like a convenient expedient to move the parties towards settlement. But that judge might not be aware of the special significance these particular patents and inventions had to Polaroid's case. Polaroid, like Kodak, knew that it would have a difficult time proceeding if the Rogers film unit patents were knocked out. In short, everything was at stake on this motion, and having made their best written pitch to avoid disaster, Polaroid counsel could now only sit back and wait for the judge to allow oral argument and then rule in his own good time.

As the date for the Land deposition approached, Kerr and Schwartz made several visits to do preparatory work with a now engaged and enthusiastic Land and his assistant, Nan Chequer. They came armed with binders of materials color-coded for each of Land's patents in suit. With the first session just six weeks away, the Polaroid team, particularly Peck and Ford, were getting nervous that Carr had not yet provided notice of the patents and other scientific literature about which he intended to question Land. Given the battles that Polaroid had fought and lost in an attempt to limit the scope of these materials, they expected a voluminous list. On August 13, Kerr wrote to Carr in a casual tone to inquire. "I recognize that Dr. Land's deposition is not now on the horizon," he began, perhaps with tongue firmly in cheek, but "[I] would, however, appreciate your giving to me as much advance notice as is feasible from your standpoint."[20]

A belated response from Carr was finally received by Kerr, in which Carr assured that he would indeed receive "advance notice of the materials I would like Dr. Land to review." Carr also added, "There may not be a great number of patents and publications required for the first session."[21] Nothing else was heard from Carr for three weeks. Then, on September 12, less than two weeks before the first session with Land was scheduled, Kerr received a letter from Carr listing seventeen prior art references with which "we desire Dr. Land to familiarize himself prior to his deposition."[22] Kerr was not amused, nor would his clients be. He wrote back immediately in a letter delivered by hand addressing the "formidable list" that Carr had proffered.[23] Kerr proposed a compromise. Carr resisted. The struggle to reach an agreement dragged on until the first day of Land's deposition.

On September 24, 1979, the two adversaries met and the real battle began in the Emerson Room of the Parker House hotel in downtown

Boston. Present were Land, Kerr, Schwartz, Peck, and Ford. The opposition sitting across the conference table from Land included Frank Carr and two members of Kodak's in-house legal department, Cecil Quillen and Gerald Battiste. The examination started with the usual inquiries into Land's education and other biographical information, the history of Polaroid, and his career as an inventor. Very quickly Carr began inquiring into the existence of his files and other records at Polaroid, but Land was well prepared for this and answered without revealing any information about their locations or content that the lawyers had not already been aware of. Carr took Land through the genesis of one-step photography and the history of the early developmental work he and his colleagues had done at Polaroid through the 1940s and 1950s and into the 1960s. Land described how he built his research team, adding key members like Meroe Morse, Elkan Blout, and Howard Rogers along the way. But Land made it clear that he was at all times the director of this research on both the camera and film components and that it was he who had handed out most, if not all, research assignments.[24]

This brought the first day of testimony to a close. The second day began with Carr inquiring into the day-to-day working relationship between Land and Rogers. The witness acknowledged that they regularly discussed significant insights as they occurred but that Rogers was primarily focused on the area of color imaging, while Land was involved in all aspects of the film, black-and-white and color, and camera development. Then Carr delved into the first technical part of his examination, a series of questions relating to Land's work with polymeric and other acid materials in wrestling with the problem of image stability. This line of examination related to Land's L-Coat patent, but Carr abruptly and inexplicably dropped the subject and went back to relatively harmless inquiries about various individuals in and around Land's research lab. At five p.m., Carr suddenly switched to a series of questions about a "scanning camera" that had been worked on at Polaroid. Presumably, these surprise questions were relevant to Land's Rear Pick patent invention. Perhaps Carr was trying to catch a fatigued Land in some kind of harmful admission? If so, it did not happen. But there was clearly some parrying and prodding going on as the second day ended.

On the third day of Land's deposition, Thursday, September 27, Carr dug deeply into the details of the technology of instant photography, particularly as it related to Land's L-Coat and Rear Pick inventions.[25] He delved into the reasons that stabilization was necessary and the various means, experimental as well as commercial, that Land and his colleagues had used over the years to address this problem. Carr was attempting to equate the materials and methods used for one type of film in a given instant photography system to those used in others. But the more Carr tried to draw and establish parallels, the more Land explained how the vagaries of each system required different approaches and how, ultimately, it was simply pointless to try to equate them. This assessment came with great authority from someone who had spent a working lifetime grappling with these problems, none of which had been solved easily or rapidly, or by "obvious" means.

By this point, Carr must have finally realized that he was up against a fully engaged and formidable adversary. Carr had long believed that Land's later patents, like those in suit, were merely restatements of his earlier work.[26] Now, for the first time, Carr and his client were having the distinctions between the generations of Polaroid patents explained to them in a way they had perhaps not been able to appreciate. With each successive answer it was becoming more apparent that Land was going to be a force throughout the rest of this legal battle. Schwartz and Kerr knew, once and for all, that he was going to be their star witness if the case ever reached trial.[27] In their view, Carr was no match for Land on his home turf of instant photography technology, although that is a conclusion Kodak's lead counsel may not have allowed himself to consider.

Carr's underestimation of Land might also have been influenced by the contingent at Kodak, including Fallon, who frankly had come to dislike Polaroid's founder and had taken to minimizing his achievements.[28] This attitude was reflected in Carr's questions, early in the deposition process, about things that went beyond the usual biographical details. He made reference—for no apparent reason relevant to the litigation—to the fact that Land had dropped out of school and had never received a degree. "I have always been accustomed to calling you Dr. Land," Carr told the witness. "Is it all right if I do that?" inquired Carr, as Quillen and his colleagues restrained their smirks. "Yes, Mr. Carr," Land replied. "That is a courtesy my friends show me."[29] Land, the Kodak contingent thought, didn't fully appreciate the deprecation and sarcasm behind the question.[30] They were probably right.

On Friday, the first series of deposition dates with Land ended without incident. Carr continued to probe on some technical issues relating to Land's patents but found an increasingly argumentative and cagey adversary parrying back at every turn. The Polaroid in-house legal team was relieved and elated with the result. Peck wrote to Kerr that the deposition had gone "extremely well" and expressed his view that "Dr. Land clearly came out the winner."[31] But Peck knew that the next round "could be quite crucial to the lawsuit," and so he encouraged Kerr to continue the "careful and thoughtful preparation" he and Schwartz had conducted. Clearly, since Polaroid had been on the defensive throughout much of the lawsuit, and at times overwhelmed by an adversary with vastly greater resources, the possibility that the Land deposition might perhaps turn the tide was the most encouraging development for Peck and his team since the outset.

While all this was unfolding, however, Land and his company continued to be embarrassed and distracted by Polavision's negative reception. William McCune later recalled that he and Land "had a long series of discussions about Polavision."[32] While Land "was usually so perceptive . . . in this case he simply didn't want to face it," said McCune. "He was at a stage when he was much more difficult to talk to and he didn't want to listen." Polavision "became a personal thing," McCune observed. Land was devoting great time and energy, spending hours in research, in an effort to rescue Polaroid's motion picture technology. But this effort was characterized as "disconcerting," by the *Wall Street Journal*, whose report lamented that the great man was "spending any part of his twilight years tinkering with what some critics call a gadget." Whether a fair assessment or not, the *Journal* probably got his motivation right. "Land has good cause to tinker," it noted, "[because] he desperately wants to prove his critics wrong again."[33]

The verdict in the media and the photography industry was that Land was desperate and losing perspective. He was a man possessed. Land had long focused on the brilliance of the technology, rather than the commercial potential of Polavision. Repeatedly he had rebuffed inquiries into the cost of the project and his expectations for its profitability. When one analyst insisted on asking Land about "the bottom line" for his new product, Land issued one of his most famous retorts, and one that has found its way into *Bartlett's Familiar Quotations*: "The bottom line?" Land replied. "The bottom line is in heaven."[34]

The outside perception was that all of this was having an effect on Land. While "he is still considered the wealthiest scientist alive," reported the *Wall Street Journal*, "he has become a little more cantankerous . . . and a little more reclusive."[35] An anonymous aide was cited as admitting that, "except for his annual appearances at the stockholder meetings, Land doesn't like dealing directly with the outside world."[36] The fact is that Land's retreat into his increasingly smaller circle of colleagues had already been under way for quite some time. Fortunately, however, the issues with Polavision certainly were not affecting his ability to prepare for his deposition and to perform admirably at that task. But when Polaroid took steps that fall to cut its losses on Polavision by taking a huge $68 million write-down on its unsold inventory, the speculation about the company's future, and Land's future within it, intensified to another order of magnitude.

Announced on October 24, 1979, by McCune, the accounting concession caused a third-quarter loss of more than $23 million, even though the company's sales had actually increased that quarter by two percent. Not everyone on the outside saw this development as a negative. Obviously, the short-term financial hit hurt. But although the company pledged to carry on its effort to market Polavision, it also allowed Polaroid to take its economic punishment on this disappointing product all at once and to move on. Others pointed out that Polaroid remained fundamentally sound financially. "For all his restless inventiveness, Land has always been financially conservative," it was noted in a major feature piece in *Financial World*. "Polaroid has never had a penny of long term debt," it pointed out, and predicted that its current problems would not cause it to "deviate a bit from that very conservative course."[37]

No matter how one viewed Polavision or Land's evolving role, Polaroid's survival depended on its success in instant photography, a market that continued to expand. This, as much as Land's personal vendetta against what he perceived as the intellectual thievery Kodak had perpetrated in barging into the instant photography field, kept the ongoing infringement litigation of paramount importance to the company. With the disappointment of Polavision, maintaining its dominance in instant photography, the field and technology it had developed, was more than ever an indispensable centerpiece to Polaroid's survival strategy.

For the first time though, these events in the fall of 1979 seemed to create the apparition of a future for Polaroid without Land. Having surrendered the presidency of the company more than four years earlier so that he could focus on research, some ascribed his stubborn refusal to give up on Polavision as tainting his performance in that role. Was "Polaroid, under the now 70-year-old Land, losing its magic touch?" asked some observers on Wall Street.[38] An anonymous analyst was quoted in the *New York Times* as saying that "the [Polavision] write-down suggested that the influence of Edwin H. Land . . . might be lessening and that a change in the balance of power at Polaroid might be under way."[39] The press noted that by this point Land's ownership of the company had been reduced to fourteen percent, and so speculation began as to whether Polaroid was in danger of being taken over.[40] The takeover, of course, never occurred. The financial hit from Polavision proved, as analysts predicted, to be a onetime glitch in an otherwise consistently positive performance for the company throughout this era. But the rumblings within Polaroid's management about just what role, if any, was still appropriate for the company's aging patriarch had clearly begun and would continue over the next year.

Preparations continued that fall for the next four-day session of Land's deposition. Shortly after the first session, Carr advised Kerr that the deposition would be held at the law offices of Kodak's local Boston counsel, Choate, Hall & Stewart.[41] Moving the event to the law firm certainly gave Carr more control over what was going on and who was doing what behind the scenes. In a hotel, Land and his team were free to rent as much additional space as they wanted—and have as many support personnel present as they wanted. Kerr couldn't reasonably complain about the relocation, but he did immediately ask Carr to provide a second conference room at the law firm for use by Land and his counsel.

Carr surprised him by sending a letter reporting that "a female representative of Polaroid had already been to the [Choate] office, had inspected the conference room in which Dr. Land's deposition is to be taken and requested [among other things] that we provide an additional room wherein his techni cal assistants could reside during the deposition."[42] This made Kerr look, at the very least, as if he was not in complete control of his troops. The Polaroid "representative" was Chequer, Land's chosen "preparer," responsible to him rather than Kerr. His embarrassment was evident in his response. "Your letter leads me to conclude that a comedy of errors has occurred," Kerr wrote. "The Polaroid representative . . . was not authorized to seek a second conference room for any purpose other than . . . for Dr. Land and his counsel during deposition recesses," Kerr continued. He assured Carr that "all of the other arrangements which may have been requested by Polaroid are cancelled."[43]

This was vintage Land. When he heard about the change of venue, he sent his assistant to survey the battlefield, without advising counsel running the case. But Kerr needed to rein him in, and so he undid all of the arrangements that Land had sought to make. This was an exercise of authority that Land was not accustomed to. It was yet another skirmish in the growing battle of egos between these two men. In fact, the tension was growing in any number of ways. For example, during the deposition sessions in Boston, Land had insisted that his entire entourage travel in a Polaroid company van. Kerr was apparently appalled that they were not using a limousine or at least a town car, but that was just not Land's style.[44] Kerr suffered silently the indignity of riding in a van, at least in public, but privately his resentment in having to bow to Land's preference on this mode of transportation was apparent to his colleagues.

For the December round of deposition sessions, Carr had developed another formidable list of materials on which he intended to examine Land. This list included internal Polaroid correspondence and memoranda dating as far back as 1957 and fifteen patents.[45] Ford transmitted all of these to Land and Chequer in late November, and they reviewed them in detail well beyond the formal sessions conducted by Kerr and Schwartz.[46] Land proved to be a voracious study, perhaps not unexpectedly. The upshot of this was that he was prepared for every trap that Carr attempted to lay for him and was able to turn the remainder of his deposition into an intellectual, highly technical debate and exchange that frustrated Carr time and time again. In the end, no damage was done to the Polaroid case during any of the remaining Land deposition sessions, including a final four days of examination conducted the following spring. His stellar performance made the Polaroid team wonder if it had given Kodak second thoughts about taking the case to trial.

While the Land deposition went on during late 1979, other discovery inched along, slowed by niggling obstructionism on both sides, enabled by a Special Master who believed that an important case like this deserved a full exposition of the facts and that these parties were big boys who could afford as much discovery as they could endure. During the Land deposition session in December, at a time when it was clear that Schwartz and Kerr would be out of town and otherwise engaged, one of Carr's partners, Ken Madsen, called Terry Barrett, a senior associate working on the case at Fish & Neave, asking for an extension of time in which to file a reply brief to Polaroid's objection to one of Kodak's discovery motions. After consulting with Schwartz, Barrett reminded Madsen that Kodak was not entitled to file this additional brief without permission of the court and, further, that "it would be difficult" for Polaroid to grant the request given that Kodak had opposed Polaroid's request to file an additional brief on the summary judgment motion.[47] Madsen told Barrett he understood and undertook to discuss the situation with Carr.

With no further communication, Kodak went ahead and filed its reply brief on the discovery motion without permission from the court and without prior notice to Polaroid's counsel. When Polaroid complained, the Special Master acknowledged the violation of the rules by Kodak but concluded that he would accept the filing because "all briefings and memoranda which I can get on this motion would be of assistance to me."[48] In light of this decision, Polaroid was granted permission to file yet another brief, this time a rejoinder brief to Kodak's reply. But when Polaroid filed its rejoinder, Kodak went back to the Special Master on December 11 asking for permission to file yet another reply and for the Special Master "to withhold decision" on Kodak's motion until then. An exasperated Kerr noted on his copy of Kodak's letter: "It's never ending."[49]

A junior Polaroid lawyer, Robert J. Goldman, sent out a holiday card to friends and family that December reporting, "The Polaroid-Kodak suit is three and a half years old. No one seems to know whether it will go to trial or to kindergarten next year."[50]

CHAPTER 17

THE LONG, SLOW ROAD TO THE COURTHOUSE

With the turn of the calendar to 1980, discovery accelerated to a furious and tenacious pace. Depositions went on virtually every week on both sides, with Kodak continuing to take advantage of the Special Master's liberal rulings to force Polaroid witnesses to review massive amounts of material. At this point, Kerr, who was approaching retirement, began to distance himself from the process, leaving Schwartz, who had largely set the original strategy, to fight this battle of attrition.

By this time, both parties had responded to hundreds of interrogatories, most with numerous subparts, and had produced hundreds of thousands of pages of documents to each other, with the aid of a small army of supporting lawyers and paralegals. This mass of material had been copied multiple times, cataloged chronologically, and organized into sets correlated to the patents in suit, to discovery requests, and to depositions in which they had been used as exhibits. The deposition testimony itself was similarly broken down, digested, and cataloged for easy access, if necessary. In these days before computers were in mainstream use, all of this work was done by hand, producing voluminous amounts of paper in a tremendously time-consuming and costly operation.

Kodak continued to drag things out to step up the economic pressure on Polaroid, while Polaroid met every Kodak demand with one of its own. Ultimately, motions filed by both contestants kept the Special Master busy reading submissions and writing rulings to keep the process moving. Yet, he was the one who had opened the floodgates—and kept them open. In his defense, the Special Master asserted that this was a litigation involving "twelve critical Polaroid patents" and thus "extensive discovery within the bounds of reason and judicial discretion is necessary for both parties to adequately prepare for trial."[1] Unfortunately, he did nothing to put on the brakes and exercise any judicial discretion.

In one significant incident, Schwartz called Kodak's bluff when it attempted to hinder the production of detailed laboratory notebooks relating to its research and development activities by making the task too onerous. After some foot-dragging, Kodak allowed one of Schwartz's associates to come to Rochester in early February 1980 to review the material in situ.[2] When he appeared at Kodak Park, the Polaroid lawyer was told Kodak was producing a collection of 2,300 notebooks but that he was only being shown two rolling bookcases containing a small portion of them. Kodak obviously intended to bog down the review so that Polaroid's lawyer would give up the search. When the young lawyer suggested that he survey the entire collection first so as to select a relatively small number that he would review in detail thereafter, Kodak abruptly terminated the search and sent the lawyer packing.[3]

Schwartz concluded that Kodak shut the process down in order to buy time to rethink its position. But Kodak's counsel had agreed to produce this material. Accordingly, after the usual niggling back and forth, the inspection was recommenced two weeks later and, ultimately, resulted in a demand that Kodak produce selected pages from 225 of the 2,300 notebooks.[4] Among those pages were some of the most important documents that Polaroid was to secure from Kodak throughout the discovery phase leading to trial.

As always, Polaroid's urgent and overriding imperative was to get this case to trial. It needed the judge to move expeditiously—but had no power to compel him to do so even as the case neared its fourth anniversary. There were motions still pending from the very beginning of the action relating to, among other things, Kodak's patent misuse defense covering Polaroid's entire patent portfolio and Polaroid's attempt to dismiss it. In addition, Kodak's crucial summary judgment motion on the Rogers "Excedrin" film unit patents, a motion that had the potential to blow Polaroid's case apart, was also fully briefed and suffering judicial inaction. These matters required the judge's attention, but only the judge could take the initiative to deal with them. None of the three judges who had presided over the case since its inception had shown the slightest inclination to take control of the litigation and to deal with these matters. Neither had any of them shown any inclination to put the case on some reasonable schedule aimed at moving it towards trial and a final resolution.

The judge assigned to the case at that time, David Nelson, had held a conference with the attorneys for both sides a few months earlier, in November 1979, for the sole purpose of educating himself about the matter. He specifically told both parties beforehand that he was not willing to hear arguments on any pending motions at that time.[5] Kerr and Carr each presented their view of the status of things, including a summary of the various motions pending decision by the court. For his part, Kerr stressed that the suit was more than three and a half years old, that discovery had been pursued "vigorously" by both parties throughout that time period, and that "Polaroid is anxious to get the case tried as promptly as possible."[6] In that connection he pointed out that "with each passing year, Kodak's invasion of the instant photography field—a field created and pioneered by Polaroid—has increased," and that "against the marketing muscle of Kodak, Polaroid must depend upon its patents and upon the patent system."[7] Kerr suggested a timetable that would have included a cutoff of discovery on June 30, 1980, and a trial commencing on November 17, 1980. Instead, to everyone's surprise, in early December, all counsel were notified by Judge Nelson's law clerk about "certain facts that raise a question concerning the advisability or necessity of [Judge Nelson's] recusal" in the case.[8]

Land had made a contribution of $2,500 some ten years earlier to Judge Nelson's campaign for Congress. The judge disavowed "any personal bias or prejudice in favor of Dr. Land or Polaroid as a result of this political contribution," and he believed that recusal was not mandatory.[9] He just wanted both sides to know the facts. Kerr and Schwartz determined that they should respond to the judge as soon as possible, and certainly before Kodak had a chance to respond, advising him that the facts he had disclosed did not present a problem for Polaroid.[10] Land apparently concurred. On December 18, Polaroid advised the clerk that it believed "there exists no reasonable basis for questioning Judge Nelson's impartiality in this case."[11]

Kodak waited for two days after Polaroid had delivered its response, and only then asked for a meeting with the judge and counsel for Polaroid to discuss the matter.[12] Because of the holiday season, that conference was not scheduled until January 11, 1980. At that hearing, Kodak contended that recusal was "required" under the applicable law that directed a judge to step aside in any case in which "his impartiality might reasonably be questioned." Kodak cited the size of the contribution and Land's central role in the case, contending these factors "could cause a reasonable person to question the court's impartiality."[13] Judge Nelson conceded that it was "a close and difficult question" but noted that there was a "lack of consistency" in the applicable case law.[14] Nonetheless, whether his ultimate motivation was judicial caution or a strong desire to get out from under the weight of so important and difficult a case, he concluded shortly after the conference that the sum of the "considerations suffice to tip the balance in favor of the court's disqualification."[15]

Curiously enough, he did not announce his decision until March 17, later explaining that the delay was due to his attempt to get the chief judge of the federal court to assign a senior judge to the case, "having in mind the urgent need to have this matter reached for trial this year."[16] Senior judges are retired judges who remain active by accepting a lesser workload at the court. But the chief judge was unsuccessful in finding a senior judge willing to take the case, and so the assignment would have to be made by the random selection of another active judge on Boston's federal court. Once again, the judiciary was tossing this hot potato from one hand to another—while Polaroid desperately wanted to get to trial.

On the very day that Nelson rendered his decision, the parties learned that the case had been reassigned to Judge Rya Zobel.[17] Zobel was, like Nelson, another new judge, having just been appointed by President Carter the previous year. She had been born in Germany and had survived the Nazi era with papers that hid her Jewish identity. When East Germany fell into Russian hands, Zobel's mother was sent to Siberia for ten years but later was freed and joined her daughter, who had made her way to the United States. Zobel built a new life in Massachusetts, attending Radcliffe College and then Harvard Law School. Early in her career, she faced the kind of pervasive sexism so many women encountered in the practice of law at the time, and she initially practiced matrimonial law. To escape, she got a position clerking for a state judge for five years, and from there went on to practice trusts and estates law, eventually landing a position at the prestigious Boston law firm of Goodwin, Proctor & Hoar. Her appointment by President Carter was reputed to be at least in part because he was anxious to have a woman on the bench, and there were few qualified candidates in the major law firms.[18] She had a reputation for being "industrious and active."[19] She also proved to be decisive when, almost immediately, questions were raised about whether she too should be disqualified from presiding over this case. (See Fig. 17-1.)

This time, unlike the situation with Judge Nelson in which he had raised the issue himself, Polaroid initiated the inquiry. The Goodwin firm at which Zobel had practiced had represented Polaroid and Land in various matters. Within days of her assignment, Polaroid's senior local counsel, Lawrence Fordham, wrote to Judge Zobel, noting that these facts raised "the prospect that . . . Kodak might urge your recusal in this case." Noting that even Judge Nelson had recognized in his disqualification opinion "the urgent need to have this matter reached for trial this year," he asked her to give "early consideration to the issue of whether recusal should be addressed" to avoid further delay.[20]

Kodak took its time in responding, treading carefully so as not to impugn Judge Zobel's integrity and thus alienate her in the event she remained on the case. A week later Kodak finally delivered its "initial response" to the inquiry. It took the position that what was currently known was not a cause for recusal. Only Judge Zobel or Polaroid would be aware of any additional facts that might change the determination, such as whether Zobel had had a significant attorney-client relationship with Polaroid or Land or if she retained a financial interest in the law firm. Absent "any such facts we believe that Kodak has no ground for requesting that Judge Zobel recuse herself."[21]

Deftly played by Kodak—but Polaroid received the assurance of Samuel Hoar, the senior partner at the Goodwin firm, that while at the firm "Rya Zobel did not have any connection whatsoever with any work being done for Polaroid Corporation, Dr. Land" or any related entity, nor does she have any "continuing interest of any nature in this office."[22] On April 9, the judge reached the same conclusion and determined that her recusal was "neither required nor appropriate."[23] This swift and firm response gave Polaroid hope that they at last had a judge who would halt the seemingly endless discovery process and get on with the trial.

Accordingly, Kerr and Schwartz promptly filed an application with the court on April 11, asking it to set "an early pre-trial conference" to lay out the pending issues, set an end to discovery and schedule a trial date.[24] A date of September 30 was proposed by Polaroid for a cutoff of discovery, with the trial to commence in early December. Predictably, the April 16 response from the Kodak camp was completely at odds with the view of Polaroid's counsel. While Kodak acknowledged that a pretrial conference would be useful to familiarize the court with the case, it went on to assert that the main order of business would be for the court to come to grips with what had to be done to dispose of all the pending motions, as well as all of the additional motions that Kodak intended to bring. The nature of those decisions would greatly affect the future path of the case, noted the lawyers for

Kodak. They might expand the scope of discovery, thus requiring additional time to probe new subject-matter areas. On the other hand, a favorable decision for Kodak on its pending partial summary judgment motion, as well as on other summary judgment motions that it was "seriously considering," might also change the course of the case.[25]

While Kodak counsel resisted the "consideration of any specific schedule, such as proposed by Polaroid," they made a very surprising concession. Perhaps wanting to convey some modicum of reasonableness, they admitted that, "barring unforeseen events," Kodak counsel believed that discovery could be completed by the end of the year. Again, no one involved could seriously believe that even this date would ultimately hold, but the mere discussion of possible dates represented a large step forward for Polaroid.

When no response to this correspondence had been received from the court for three weeks, however, Polaroid renewed its push for a conference by writing directly to Judge Zobel in early May. The letter asked for an "early conference" with the court, "either a pre-trial conference as requested in [our] application . . . or, if the Court prefers, a status conference."[26] The letter emphasized Polaroid's circumstance: "the passage of time continues to advantage Eastman Kodak and to prejudice Polaroid." This time Judge Zobel apparently got the message. Within a couple of days, counsel received notification from the judge that she had set a "status and preliminary scheduling conference" for Monday, July 14, 1980.[27]

As the lawyers sought to get on track with their new judge, momentous events were taking place at Polaroid. The year 1979 had been a very difficult one, arguably the worst in its forty-two-year history. The company saw its earnings plummet seventy percent, from $118 million in 1978 to a mere $36 million in 1979. Its stock, always on a bit of a roller-coaster ride, lost more than half its value, falling to just over twenty dollars a share in early March 1980 from its January 1979 high of fifty-four dollars a share.[28] These figures were particularly dramatic in view of the fact that Polaroid had sold 7.2 million cameras in 1979, the second-highest figure in company history, and that its total sales for the year were only down by one percent worldwide from 1978's record level.[29] There were several factors, both internal and external, that overwhelmed these sales figures and contributed to the company's poor financial results. The aftermath of Polavision's disastrous performance in the marketplace was a lingering and major contributor.

Polaroid's financial decline in this period took place in the context of a weak U.S. economy and a national crisis of confidence that pervaded the country. Things were not going well for Jimmy Carter in the last years of what would be his one-term presidency. As of late 1979, forty-nine Americans were being held hostage by the followers of the Ayatollah Khomeini in Iran. On the economic front, inflation was out of control, with the cost of living rising at a double-digit rate, fueled in large part by interest rates that had reached twenty percent. There was nothing that Carter seemed able to do about either problem, other than complain about a "national malaise." In any event, the recession the country was experiencing created a very difficult environment for a luxury or recreational industry like instant photography. Although the OneStep remained the best-selling camera in America for the third year in a row, the overall impact of the nation's doldrums cost Polaroid sales growth that might have improved its fiscal condition.

Another contributing factor was the effect of what appeared to be market saturation for instant cameras. Instant cameras had seen extraordinary sales over the previous few years, a pace that was apparently unsustainable. A Polaroid executive admitted to the press that the company had been "surprised at how quickly the market soured."[30] Having increased its production capacity based upon unrealistically optimistic projections, Polaroid was faced with the necessity of cutting back on personnel and facilities it apparently no longer needed. Business had just not been good in 1979, and the prospects for 1980 were not encouraging. The company faced significant challenges, and, as a result, a sentiment developed within Polaroid's boardroom and upper management that change was needed at the helm of the company. Perhaps a fresh perspective might help Polaroid adapt to an evolving business environment. Events had been moving in that direction for years, but it seemed that the time had finally come. Polaroid needed a shake-up.

Polaroid made the announcement on March 6, 1980. Effective with the shareholders meeting on April 22, Land would step down as the company's chief executive officer, turning that additional title over to William McCune, the man who had already succeeded him as president and chief operating officer five years earlier.[31] (See Fig. 17-2.) After more than forty years, Land would not be in charge of the company he had created and built. While he retained his position as chairman of the board of directors, and entered into a ten-year agreement with the company to continue his role as "consulting director of basic research," Polaroid's patriarch was relinquishing control.[32] This was a difficult move for Land. He did not agree to this transition voluntarily, and the view of those close to him at the time was that he was pushed out.[33] An unnamed director was quoted as saying that "it happened quickly" and Land "accepted the thinking . . . although he didn't volunteer."[34]

If a change was necessary, McCune offered continuity, a technological orientation, and the solid business experience he had garnered at the company in recent years. Frustrated by this turn of events, Land briefly sought independent legal advice to see if he could fight off the forces arrayed against him, but quickly capitulated.[35] In the end, he likely realized that this development was inevitable. His work would continue in the private research facility, known as the Rowland Institute, that he was establishing on the bank of the Charles River in Cambridge, just down the road from Polaroid's headquarters and the campus of MIT.

Anticipation about this historic changing of the guard turned Polaroid's 1980 annual shareholders meeting into a special occasion. While Kerr had a previous commitment, Schwartz attended representing Fish & Neave.[36] For the first time in its history, Boston's Symphony Hall, home to the Boston Symphony and the Boston Pops Orchestra, made its ornate facility available to a corporation for this occasion. "In a setting of orchestrated adulation," the *Wall Street Journal* reported, "Polaroid Corp. heaped praise and applause on outgoing chief executive officer Edwin H. Land."[37] The Empire Brass Quintet punctuated the proceedings with performances of Pezel, De Prez, and Handel.[38] During his remarks, Land received no less than nine ovations.[39] He talked about Polaroid's accomplishments and looked forward to his continuing role in the company:

> I enter this phase of my own life with a good sense of contentment and fulfillment. No initial fantasy in 1944 about perfect instant color pictures or of a magic camera in which they would be taken could have been courageous enough . . . to foretell the beauty to be found in our [newly introduced] Time-Zero . . . pictures, nor the exquisite sense of technological perfection in our self-focusing SX-70 Sonar cameras.
>
> ***
>
> These . . . [evolutionary products] could not have occurred without the remarkable philosophic, scientific, social, aesthetic environment provided by the research-engineering-manufacturing continuum at Polaroid. To help maintain this environment over the years when I shall be associated with it and to help insure its permanence when I am no longer part of it is as happy an assignment as a man can have.[40]

As he later told the *New York Times* about the move, "I expect to be making just as many contributions—perhaps more basic ones—than I have been," Land promised.[41] (See Fig. 17-3.)

In part, this reassurance was a calculated message aimed at the financial community who continued to look to Land to provide Polaroid with the technological leadership and creativity only he was capable of furnishing. He was widely recognized as the company's biggest asset and still, despite this management reshuffling and his advancing age, its biggest hope for the future. Not unexpectedly, Land's impending departure had some observers expressing concern over Polaroid's prospects. It was clearly a company that relied on introducing new and exciting products in technologies it pioneered. At that time, the only hint that something might be in the Polaroid product pipeline was the disclosure that Polaroid's expenditures for research and development had increased twenty-seven percent in 1979, the most ever.[42] The pressure was clearly on the company to deliver given that, as one journalist expressed it, "the hope that a new marvel will soon emerge from the company's labs and instantly capture the public's fancy has now become an expectation."[43]

This expectation was seen by some as a continuing challenge for Land, something he always embraced. "Though he will be seventy-one years old in May," noted *Fortune*, "there is always the chance that the autocratic Land, who is always eager to prove the rest of the world wrong, will come up, once again, with inventions to bolster the fortunes of the enterprise he created."[44] After all, its fundamental financial condition, with no long-term debt and a largely self-sufficient manufacturing operation, remained strong. And as long as Land remained associated with the company, albeit in a new role, miracles were yet possible.

Immediate concern about Polaroid quickly faded, however, when, in a sudden turnabout, Polaroid's 1980 first-quarter sales surged to "exceed even the 'most bullish of street estimates.'"[45] In the second quarter, Polaroid's net income increased thirteen percent despite a one percent drop in sales, a drop attributed to the effects of the country's continuing woes, economic and otherwise. With the financial picture of the company improving, and with the drama and the emotion of the executive transition having ebbed, Land's focus quickly returned to the lawsuit. He sat for the final four days of his deposition in mid-May. By now, completely comfortable with the process, and thoroughly educated on the technical and legal issues, he acquitted himself beautifully, from Polaroid's point of view, effectively frustrating Carr on every line of inquiry.

But Schwartz was puzzled: why did Kodak keep up these exhaustive (and exhausting) examinations of Land and other key Polaroid witnesses? To Kerr and Schwartz, there were at least two major mistakes in this approach. First, there was Kerr's conviction that a good trial lawyer should never waste his or her good stuff in depositions, unless you thought that by doing so you could blow your opponent's case out of the water. The better strategy was to save it for trial, he believed.[46] In contrast, Carr's approach revealed most, if not all, of Kodak's best arguments to Polaroid's attorneys, who were then able to analyze them and prepare a response.

Second, Carr's approach resulted in helping Polaroid's key witnesses prepare for giving testimony at trial and, in particular, for their inevitable cross-examination by Carr.[47] Polaroid's most important witnesses were forced during depositions to learn every aspect of the technical and legal case that Kodak was asserting, ensuring that there would be no surprises if and when they took the witness stand. Carr was also acclimating them to the otherwise uncomfortable experience of being a witness. By spending such long periods in deposition engaged in detailed interrogation—Rogers was deposed for twenty-one days and Land for twelve days—there was every chance that at trial the witnesses would be neither uncomfortable nor unprepared. Kerr and Schwartz believed strongly that this gave Polaroid an important strategic advantage.

Largely as a result of this puzzling strategy, Polaroid's counsel continued to wonder whether Kodak ever really expected the case to reach the courtroom. They were certainly aware that Carr, in his prior experience, had been very effective in using discovery to effectuate favorable settlements for his client.[48] He may have had good reason to assume that this case would be no different. Or perhaps he truly believed that Kodak had the facts to win at trial, no matter what advantage his discovery approach might impart to Polaroid. Given that he had studied the Polaroid patents so carefully, and had concluded that they were flawed, Carr surely believed that he had sound defenses against each and every one of them.[49]

In the larger context, Carr's confidence was likely bolstered by the fact that this case occurred in what was considered an anti-patent era.[50] In the 1960s and 1970s, when Kodak had developed its instant photography system, and when Carr first evaluated Polaroid's patent portfolio for his client, the protection ostensibly offered from the ownership of a patent was not such a sure thing. Although philosophical reservations about the benefits of a strong patent system can be traced back to Thomas Jefferson, the genesis of the most recent incarnation of an anti-patent era was in 1938. At that time, President Franklin Delano Roosevelt asked Congress to examine the patent laws as one of many possible factors contributing to the concentration of economic power in the United States, a trend he believed was adversely affecting the country's fiscal health. In response, Congress set up a committee, known as the Temporary National Economic Committee, to conduct a broad review. In its enabling legislation, Congress gave it the task of considering an "amendment of the patent laws to prevent their use to suppress inventions, and to create industrial monopolies."[51]

Even though most of the recommendations of the committee were not adopted at the time, as one scholar describes it, the "deliberations and proposals had a greater impact on the patent climate in the courts . . . dramatically altering patent owners' expectations about both the validity and infringement of their patents."[52] This shift in attitude led to an environment in which the percentage of patents upheld in court challenges steadily dropped, dipping below ten percent in 1952.[53] As Supreme Court Justice Robert Jackson wrote in a dissent at the time, "the only patent that is valid is one which this Court has not been able to get its hands on."[54] With the help of pro-patent voices like Land's, who had assured the legal community in his landmark 1959 speech to the Boston Patent Law Association that "the effective operation of the anti-trust laws has brought us to a period when we no longer need be concerned that an aggregation of patents in the hands of a single company might be considered as a monopolistic threat to our society," a patent's survivability rate increased over the years from that nadir.[55] Yet, it was only around thirty-five percent during the 1970s.[56] Perhaps Carr and his client believed that, given the risk of having Polaroid's patents adjudicated invalid and the technical arguments against each one Carr had devised, Polaroid would back down rather than risk a complete breakdown in the patent portfolio it had worked so hard to create around its technology.

With the costs and burdens of discovery mounting, and given the business risks of holding out for a complete victory for either side, a negotiated settlement seemed to make sense for both parties. Most litigators will tell you that in a major litigation, with so much at stake, a reasonable settlement is often preferable to the uncertainty of risking everything at trial. Yet neither party seemed especially eager to make it happen. Polaroid feared it would have to license its precious and exclusive technology, even though it would have benefited from the significant revenue flow it would have received. There was certainly a risk of losing further market share to Kodak and others if the field was opened up, but this risk may have been overblown. Polaroid's experience during the last few years of competition had demonstrated that although Kodak had captured roughly one-third of the instant photography market, Polaroid's sales had, save a few temporary blips, continually increased due to the expansion of the overall market. Polaroid had demonstrated that it really did have the best products and thus could compete effectively. It had the reasonable expectation of being able to maintain a technological lead in the field, an advantage that had already translated into a larger market share for the smaller company against its otherwise dominating rival.

Instead, Polaroid's new management seemed prepared to face the more threatening risk at trial of having its patent shield vitiated, a disastrous outcome that would allow rivals to compete in its field without paying its innovator. In early June 1980, Kerr and Schwartz met with Julius Silver, still a Polaroid board member, and brought him completely and comprehensively up to date on the case. He was provided with a detailed chronology and a set of "helpful Kodak documents correlated to . . . the patents in suit."[57] At the meeting, Silver asked for materials concerning the determination of reasonable royalties in patent licensing and patent infringement suits, and Schwartz followed up by sending the information along the following week.[58] Schwartz kept Peck and Ford, the in-house Polaroid legal team, fully informed on the information being transmitted to the Polaroid board through Silver. Perhaps confident about the outcome, or still fearful of opening its field to other competitors through a settlement, Polaroid management initiated no discussions, let alone actions, on settlement.

Land's perspective was quite different. He clearly did not look at the case solely from a business point of view. Ultimately, this was a personal battle for him. On a visceral level, Land could not help but react emotionally to the basic thrust of Kodak's "invalidity" defense, which, in essence, asserted that the so-called inventions disclosed in the Polaroid patents were not worthy of protection because they either had been previously discovered or, more insultingly, were so trivial that they would have been obvious to any reasonable person skilled in the art of instant photography. He couldn't help but view this line of attack by Kodak as a denigration of the work he and his colleagues had done. It was bad enough that Kodak had challenged him and his company in the field they had created by, in his view, using patented and protected technology. But the more Kodak assailed Polaroid's inventions, the angrier Land became. He believed in the righteousness of Polaroid's position and clearly looked forward eagerly to having his day in court.

For Kodak, a negotiated settlement would have ensured its ability to remain in the instant photography business and would have allowed it to compete with Polaroid for a larger share of what was at that time still a growing field. If it was looking at the situation objectively, the certainty of royalty payments had to have some appeal over the abject uncertainty and risk of having its future in the market foreclosed by the courts. But Kodak had always believed that there was no need to license the use of Polaroid's technology. When Fallon had visited Land at Polaroid's Cambridge headquarters in October 1974, he had left the impression that Kodak did not think any of Polaroid's patents were valid and enforceable and thus did not place any significant value on a license under them.

Now, almost six years later, there seemed to be little change in Kodak's attitude. Either Kodak was counting on Polaroid backing down rather than risk its patent portfolio, or it was truly determined to break, once and for all, what it perceived to be the Polaroid patent blockade. Apparently, Kodak wished to end what it considered to be Polaroid's improper and illegal utilization of the patent system to perpetuate its patent monopoly. Carr and his Kenyon & Kenyon legal team seemed to encourage Kodak in this approach by providing strong reassurance that the Polaroid patents were indeed invalid and thus would not withstand the scrutiny of a trial.[59] Although this was a huge risk to take just to teach Polaroid a lesson, Kodak continued to believe Polaroid was not entitled to any meaningful compensation and that Kodak would be successful one way or the other.

Despite Kodak's apparent intransigence, Polaroid made one last attempt to ensure that its opponent had an objective view of the case. Concerned that Kodak executives were having their judgment clouded by either overly optimistic assessments provided by its trial counsel or its isolated and insular existence in Rochester, Polaroid lawyers were determined to make sure that Kodak management understood the nature of the case Polaroid was assembling and the magnitude of the risk Kodak was taking by heading to trial. The idea was to convey the notion that a total victory for Kodak was anything but a sure thing.

During the course of discovery, Fish & Neave had uncovered and assembled a series of internal Kodak documents that painted a fairly convincing story helpful to the Polaroid case. The documents detailed how Kodak had tried several times unsuccessfully to develop its own instant photography system and had finally adopted a particular approach. But when faced with the elegance of the SX-70 system upon its commercial introduction, and the realization that the system it had in the works was inferior, Kodak made a dramatic about-face in its program in an effort to develop a more competitive product. From its own documents, it further appeared that Kodak had done all this with little regard for Polaroid's patent portfolio. Schwartz made the assembly of this "Kodak Story" one of his priorities for the discovery process because he knew it would be a critical part of Polaroid's case at trial.[60] The potential of this story coming to light, in a presumably sympathetic home field Boston courtroom, would be a prospect that one might expect any prudent Kodak executive to be interested in avoiding.

Accordingly, to provide a dose of reality to its leadership and perhaps to move Kodak off the dime on the concept of a settlement, Schwartz traveled to Rochester in early 1980 to depose, for the second time, Walter Fallon, Kodak's chief executive. Schwartz was determined to show Kodak's leader the strength of the case Polaroid would be able to make at trial based on Kodak's own internal documents and, as Schwartz put it, to "rub his nose in" some of the harmful Kodak documents Polaroid had uncovered during discovery.[61] As the sessions unfolded, document after document was put in front of Fallon, many of which it appeared he had never seen before. Time after time, Fallon's dismay was evident when Schwartz pointed out a particularly damaging passage. His was not a poker player's face or demeanor, at least on this occasion. There was virtually no meaningful substantive information that Fallon was able to provide in response to Schwartz's questions, but that was not the purpose of the exercise. The education of Kodak's CEO having been emphatically delivered, the deposition was adjourned and Schwartz returned to New York to await its impact. Not unexpectedly, nothing seemed to come of it.

As the summer of 1980 rolled around, Kerr and Schwartz waited anxiously for the July 14 conference with Judge Zobel, in the hope that she would take control of the case and finally set a firm schedule for the completion of discovery and the commencement of trial. Carr continued to move forward with the depositions of Polaroid witnesses, and wrote to Schwartz seeking to extend the schedule well into November, beyond the September 30 discovery cutoff date proposed by Kerr. The Polaroid side did not respond immediately. But on July 11, Schwartz advised Carr: "Polaroid is of the view that unless and until the Court cuts off discovery, Kodak is willing and able to, and plans to, conduct discovery interminably. We see no reason to schedule any depositions beyond September 30 until that conference has occurred."[62]

On July 14, Polaroid and Kodak, in the persons of their respective counsel, had their first opportunity to appear before Judge Zobel. Diminutive, with short-cropped dark hair, professorial glasses, and a faint German accent, she welcomed the parties to her courtroom. Kerr began by giving a brief history of the case, including a recitation of the succession of judges who had preceded her. He told the judge that "pretrial discovery activity on both sides has been vigorous and extensive" and cited impressive figures to illustrate how much had occurred.[63] He summarized the motions still pending for the court's decision, as well as those awaiting action by the Special Master, but he pointed out, "like all good things, discovery in this action should not go on in perpetuity. It must reach an end," Kerr insisted. He underlined Polaroid's need and right to move to trial and cited Kodak's delaying tactics. "Polaroid needs this Court's help to abate Kodak's infringement and to stop Kodak's expanding penetration into the field created by Polaroid."[64]

Carr focused on the work that remained to be done before discovery could be completed. He handed Judge Zobel a detailed five-page list of the various motions that remained to be decided, motions that, as Kodak had argued in its opposition to Polaroid's application for a conference, could change the substantive scope of discovery and thus create considerable additional discovery.[65] Judge Zobel reacted immediately. Instead of taking the matter under advisement to issue a decision at a later date, she set a schedule for the case right then, something she had obviously decided to do before coming into the courtroom. Her "presumptive" date for a cutoff of discovery was set for December 31, 1980, with a possible trial date in July of 1981.[66] She also set September 19 as the date for oral argument on Kodak's partial summary judgment motion on the Excedrin film unit patents. Although these were not the dates Polaroid had requested, Polaroid, Kerr, and Schwartz were elated.[67] For the first time, a judge had taken full control of a messy and seemingly endless pretrial process. Zobel was clearly not intimidated by the enormity or the gravity of the case and seemed determined to actually try it. The Polaroid side at last could be bullish that something was going to be done.

Following Zobel's action, there was suddenly a certain irritable tone that seemed to appear in Carr's daily correspondence. Carr reverted to complaining about the depositions Schwartz had been reluctant to schedule before the conference with Judge Zobel.[68] Carr was not happy about the firm deadlines set by the judge. Nonetheless, the setting of this very tentative schedule would not deter him from continuing to make efforts to prolong discovery and to avoid going, finally, to trial.

In actuality, there was a tremendous amount of ongoing discovery activity yet to be completed. Judge McLaughlin worked diligently to decide disputes and to keep the process moving along. Fish & Neave did everything it could to keep Kodak's lawyers on schedule with respect to its discovery obligations.[69] The Polaroid lawyers refused to engage in legal civilities like granting what would otherwise be considered routine extensions and hammered away at the main point: Kodak had done everything it could to avoid what it was supposed to do. Kerr himself, who had for months stayed out of the discovery fray, wrote Carr when Kodak's counsel claimed his firm and client could not meet its deadline "because of other discovery requirements." This is not acceptable," Kerr insisted.[70] "You should know now that Polaroid will oppose any attempt by Kodak to delay unreasonably Kodak's compliance with the Master's order." Carr immediately brought the matter to Judge McLaughlin, who gave Kodak counsel the relief they sought on the grounds that it was a "reasonable . . . extension of the time set for compliance."[71] Yet, the message had been delivered, and for the ensuing months, Fish & Neave stayed on Kodak's case as best it could in an effort to tie up all the discovery loose ends and nudge the case towards trial.

CHAPTER 18

THE FINAL TURN

With at least a "presumptive" trial date in place, the next major obstacle on the path to the courthouse was Kodak's request for partial summary judgment made in April 1979. This was critical because both parties knew that if Kodak was successful, Polaroid would seriously have to question the viability of its case at trial on the remaining patents, making a speedy resolution of the litigation on terms favorable to Kodak very likely. The motion was finally going to be argued to the court on September 19, 1980.

The essence of any summary judgment motion is that there are no genuine issues of fact material to an adjudication of a claim that need to be determined at a trial and that on the basis of those undisputed facts the law requires a finding for the party seeking summary judgment. Kodak's argument was that Rogers' invention merely involved putting together elements from two earlier Polaroid patents, both issued in the name of Land. Therefore, the court could rule that the Rogers patents were invalid because they did not meet the patentability standard of being a discovery that would not have been obvious to one of ordinary skill in the art.

In his rebuttal before Judge Zobel, Kerr had urged on behalf of Polaroid that "there are two separate, distinct and fundamental reasons why Kodak's motion for partial summary judgment should be denied." First, Kerr asserted that there were genuine issues of material fact that could be decided "only by a full-dress trial on the merits, during which the court will have the benefit of live expert testimony, subject to thorough cross-examination. Indeed," noted Kerr, "a full array of such issues virtually leaps out of the voluminous papers filed on this motion." Second, Kerr urged the position that even if no genuine issues of material fact existed, Kodak's arguments "do not come close to showing that the Rogers patents are invalid. On the contrary," Kerr continued, "the totality of the papers before the court constitute a highly persuasive demonstration of the merit and validity of the patents to which Kodak's motion is directed." Furthermore, Kodak's claim that the Rogers invention was obvious "is idle, and comes with ill-grace," argued Kerr. "Kodak itself attempted to acquire a virtual worldwide patent monopoly on these very same inventions," declared Kerr in reference to the patent applications filed by Kodak before it learned of the prior Rogers applications.[1] This was perhaps the single most important argument that Kerr hoped to bring before Zobel, and it was surely delivered with all the incredulity he could muster.[2]

With the oral arguments concluded, the parties went back to their ongoing discovery efforts with no idea of when Judge Zobel might issue an opinion on the motion. Kerr wrote Julius Silver that he was "confident, or at least as confident as counsel can be in a contested matter, that Judge Zobel will deny the Kodak motion. The open question is when this will happen," he confided. "A show of alacrity would be a pleasant and welcome change."[3] Kodak, for its part, was anxious to keep the pressure on Polaroid by continuing to take pot shots at its case. While Carr and Kerr were in the courtroom arguing the summary judgment motion on the Rogers film units, Kodak surprised Polaroid by filing with the court a second motion for partial summary judgment.

This motion was directed at the Trap patent issued to Polaroid in the name of John Campbell. It covered a feature of the film unit designed to contain and neutralize excess processing fluid. As Kodak described it, this invention was nothing more than "an empty area—a void—within the film unit, which serves as a sink for collecting and retaining [i.e., "trapping"] excess [left-over] processing fluid. . . . It is, so to speak, a built-in chemical dump site."[4] Again, Kodak contended that this technique, although used in an integral film unit by Campbell, had been taught in earlier Polaroid patents on peel-apart film configurations. Polaroid saw this motion as yet another desperate stratagem by Kodak to create a diversion and perhaps delay the court schedule. Kerr told Silver, and through him Polaroid's board and, most importantly, Land, that Kodak's "second motion has even less merit than Kodak's first motion."[5] The obligatory opposition papers were filed, but Polaroid pressed on, focused on the main target: completing discovery by the December 31 date set by Judge Zobel so that they could get the case to trial.

It would not be easy. Kodak tried to schedule every deposition it sought before time ran out. Schwartz, however, was clearly trying, in sports parlance, to run out the clock and schedule as few depositions as possible before the deadline. Carr was clearly frustrated and, in heated correspondence with Schwartz, tried to document his "concern," already submitted in a letter to the judge, that Kodak would be unable to meet the deadline.[6] In his plea for an extension of the discovery deadline, Carr had told the judge that "his first inclination" was to ask for a ninety-day delay until March 31 but that he was "hesitant" to do so because of all the pending discovery matters yet to be completed.[7] Perhaps he was going to need even more time than that.

Kerr responded to Carr's request in his own letter to the judge and did not dispute Kodak's assertion that some kind of extension of the deadline might be inevitable. But, noting that Carr had "mention[ed] but [did] not endorse" a March 31 cutoff, he suggested the end of February "in the spirit of compromise."[8] The judge responded promptly by holding a meeting with both counsel on November 25, at which she set the firm dates Kerr had requested: March 31 as the discovery cutoff date, and June 29, 1981, as the date for the trial to commence.[9]

At that conference, the judge also inquired about the possibility of arriving at a settlement, asking that both lawyers have their clients meet to discuss a negotiated resolution and report back to her. Charles Mikulka, Polaroid's senior in-house legal executive, was present at the hearing and undertook to organize the meeting from Polaroid's side.[10] On December 10, Judge Zobel followed up by requesting that counsel for each side appear before her on December 30, accompanied by a member of management, to report on the status of their discussions.[11] The meeting between Kodak and Polaroid was held on December 22. McCune, Polaroid's president and CEO, attended together with Mikulka.[12] Walter Fallon, Kodak's chief executive, did not attend. Instead, Fallon again sent his lieutenant, David Greenlaw, whose title at that point was vice president and director of corporate affairs.[13]

Fallon may well have been indicating his lack of interest by sending a subordinate. Yet Greenlaw was the one Fallon had sent to Polaroid back in April 1977 to conduct the very first exploratory talks about reaching some sort of deal to head off litigation. Now, Greenlaw once again served as an emissary, acting under the direction of Fallon and Cecil Quillen, the Kodak lawyer supervising the case. Kodak's position, however, had not changed, and the judge's initiative came to nothing. Schwartz never thought the case had any genuine prospect for settling anyway.[14] Judge Zobel may have been serious in her attempt to explore the possibility, but he believed it was very unlikely. Kodak had never given any indication that it was willing to make a serious deal that involved a licensing payment to Polaroid of any significance. Schwartz knew that without some kind of meaningful compensation, Polaroid was not prepared to settle.

After two meetings were postponed because of bad weather, the settlement conference with Judge Zobel finally took place in mid-January 1981. It "turned out to be a substantial dud," was how Kerr characterized it to Julius Silver.[15] Apparently, the judge wanted only a general discussion of what had transpired and requested that "no figures or other specifics [be] mentioned by either side." Although to Kerr "the conference . . . amounted to a waste of time," he counseled Silver that it did "not mean that the Judge will not press the parties further."[16]

With the discovery cutoff looming, Carr did everything and anything he could to create more issues, more controversies, and more work that would be difficult to complete even by the new March 31 deadline. The question now was how much more time, if any, Carr could buy. McLaughlin, in trying to mediate a dispute over Kodak's request to "bypass the formalistic procedures" used up to that point in resolving discovery disputes,[17] had already indicated that he "would have no hesitancy in recommending to Judge Zobel that the March 31 cutoff date is unrealistic and should be extended" if either party could demonstrate to him that it could not "comfortably complete pretrial discovery."[18] Of course, as the Special Master pointed out, he had no idea of how the judge would react to such a recommendation, but even the suggestion of another extension had to appear as a gaping escape route to Kodak, which Kerr and Schwartz were determined to block.

By this point, both sides were getting a bit ragged from sheer fatigue and unremitting pressure, so patience had worn very thin. On December 19, Kodak served on Polaroid a new motion seeking to compel its compliance with some aspect of discovery.[19] This would have required Polaroid's lawyers to work through Christmas. Though this demand was in keeping with Carr's delaying tactics, its timing had a certain malicious edge to it. Schwartz wrote to the Special Master the following Monday morning, December 22, asking for an extension of time to respond until January 12.[20] "Kodak has recently made plain to you that so far as their witnesses and lawyers are concerned, discovery should not be had over the Christmas holiday," Schwartz wrote, merely seeking the same accommodation for his people.

Surprisingly, Carr actually objected on behalf of Kodak, characterizing Schwartz's request as "excessive" in a letter dated December 23 but not received at Fish & Neave's office until after it had closed for business on Christmas Eve.[21] Schwartz was incensed. "It is hard for us to comprehend how the same Kodak counsel who was sufficiently 'outraged' at the prospect of his attorneys conducting discovery the week prior to Christmas . . . should, in the next breath, seriously urge that Polaroid's attorneys should not be entitled to the same treatment," Schwartz complained incredulously.[22] The Special Master, trying to respond to both parties' concerns without betraying his real reaction to the tone and the nature of the dispute, granted Polaroid an extension to January 9, thus giving it a full week after the holiday to prepare its brief.[23] It was a reasonable compromise, but one that, under most any other circumstances, would have been achieved out of professional deference and courtesy, without the necessity of the Special Master's intervention.

Meanwhile, Land was trying to adjust to his new situation. By February 1981, it had been almost a year since he had stepped down as Polaroid's chief executive. Bill McCune had reorganized top management by appointing four senior vice presidents to oversee the diversification of Polaroid's activities beyond instant photography for the amateur market.[24] This strategic move did not make Land happy—it was one he had long opposed. At one shareholders meeting he had vehemently insisted that the company had no plans to diversify. "It would be madness," he said, "now that we are on the 90-yard line with the other guy 30 yards behind to make an end run around nothing."[25] Now, out of sheer frustration, he took the unusual (for him) step of inviting a journalist from *Business Week* to his Cambridge office for a series of interviews. Taping the conversations himself to ensure that he would be quoted accurately, Land paced his office floor as he sought to put his position on the record.

Land's central message was to make it "clear that he does not intend to allow anyone at Polaroid to devote too much of its resources to diversification," read the published account of the session.[26] While he conceded his successors' right to pursue "new opportunities," Land wanted to make sure that they did not lose sight of what he believed to be the company's primary mission, to make instant photography the photographic system of the masses. "They are not entitled to any neglect whatsoever of making the most out of the great initial opportunity, and I honestly believe that the basic amateur field is in its infancy and not its maturity," said Land, pledging to put his still considerable weight and "ultimate power" at Polaroid behind his views.[27] Citing his ownership of 12.5 percent of the company, his "total knowledge of the phenomenology of the company," and even a veiled threat to terminate his contract with Polaroid so as to enable him to contribute his technological prowess to another company, the message Land delivered, as reported in *Business Week*, was that "he threatens to derail the diversification plan if the company's new managers stray too far from his original vision."[28]

If Land were simply trying to get the attention of Polaroid's new management, he surely succeeded. But what really alarmed them were reports, perhaps leaked by Land himself, that he was exploring the sale of his Polaroid stock to Irving Shapiro, then outgoing chairman of DuPont. If true, it was assumed by the media that any purchaser of Land's block of stock "would approach the acquisition as the first step in an eventual takeover."[29] Both Polaroid and DuPont denied that any discussions were going on. Yet Shapiro, who confirmed that no talks were taking place between the two companies, was "uncommonly reticent" and refused to respond to inquiries about whether he personally had engaged in conversations with Land. (See Fig. 18-1.)

McCune and some of Land's allies on the board confronted Land. What was said during their meeting is not known, but in the end, Land was persuaded to publish a joint statement with McCune aimed at putting an end to the highly disruptive speculation. The letter, issued in late February 1981 to members of the Polaroid Company, first denied the stories about the stock sale and then refuted the reports that Land was "personally dissatisfied with Polaroid's top management team. Both of us have high respect and deep appreciation for each other; both of us share high respect and deep appreciation for our management team," the joint statement read.[30] The letter went on to issue a direct rebuttal to the *Business Week* article, a clear signal, they hoped, that the two men had reached a meeting of the minds, or at least an accommodation for the good of the company. The statement denied any breach between McCune and Land and reaffirmed the company's primary focus on instant photography, but also said that it would be a mistake to pass up opportunities to use Polaroid's vast resources and experience "to enter new and exciting fields."[31]

This face-off had put a real strain on a forty-year relationship between McCune and Land—and left open the question of how it would evolve. Even though the laboratory had always been his primary domain, the transition for Land out of the executive suite was clearly not an easy one. Despite his supportive public pronouncements, he was still bitterly disappointed at having been pushed, as he saw it, out of the top spot in the company he had created.[32] His interview may have been a ploy to reconnect some lines of communication, if not influence, with McCune and his team. Whatever his true motivation, Land had at least received assurances that Polaroid would continue to treat instant photography as its main priority.

McCune had a perfect opportunity to smooth over the breach publicly just weeks later when he introduced Land at the 1981 Polaroid shareholder's meeting:

> In the 43 years from Polaroid's founding in 1937 to the present, Polaroid has changed profoundly in size and shape, but certain things about the company have remained immutable. As its scientific, intellectual and entrepreneurial leader, Dr. Land established very early a company character that has remained strong and well defined. He has fostered a creative environment, one in which idealism and humanism find a ready combination with science and technology. In addition, it is an environment in which aesthetic values are recognized to have an important role. We have tried to create an industrial organization responsive to the needs of the communities in which we function. These and other personality traits, if you will, have exerted a strong attraction for the host of creative, determined, competent people who have helped Polaroid grow from a company once small and local to one now large and international. It gives me great pleasure to return the chair to my friend and colleague of 42 years, Dr. Edwin H. Land.[33]

Eager to move on and to put this incident behind him, Land was able, once again, to focus on winning the litigation with Kodak. He knew by this time that he was to be Polaroid's first witness at the trial, and so he got back to working with his trial counsel and Nan Chequer to prepare. There was still considerable work to be done if Polaroid's legal team was ever going to wrestle Kodak to the courtroom mat.

Carr seemed to have endless ways of delaying the trial. He was apparently not winding down his deposition program despite his representations to the court when he sought an extension of the deadline to March 31. In a letter to Carr, Schwartz complained that since November, instead of completing the depositions still in progress, Kodak had noticed the depositions of twelve additional Polaroid personnel.[34] Before he would agree to schedule any of these new witnesses, Schwartz insisted that Carr finally declare where the end was in the seemingly inexhaustible list of Polaroid deponents.

The depositions continued to be as contentious as ever. Kerr himself was defending the deposition of Richard Wareham, a co-inventor of the light shield / deflector, one of the camera patents in suit. He decided that

John Fogarty, one of Carr's partners, was abusing the witness by having him interpret prior art references that he had not written. Kerr continued to believe that this kind of examination was not appropriate. In his, and Schwartz's, view, and despite the Special Master's previous ruling to the contrary, the witness was there to testify about facts within his knowledge and not as an expert whose job it is to render an opinion on unfamiliar material. Accordingly, Kerr ordered Wareham not to answer the questions and terminated the examination. Fogarty immediately wrote a long and detailed letter to the Special Master seeking relief.[35] When Kerr received his copy, he took his "BULLSHIT" stamp out of his drawer and proceeded to adorn the letter with his personal seal of disapproval. He also added, for good measure, a handwritten note above the stamp, which read "Quintessential Fogarty," and proceeded to circulate the letter to the twelve or more Fish & Neave lawyers working on the Polaroid case. It was a very undignified, un-Kerr thing to do, but a clear indication that the patience of counsel and client was running out.

Meanwhile, there was still no word from Judge Zobel on either of Kodak's two motions for partial summary judgment, motions that had the potential to influence greatly the outcome of the case. The more important motion, involving the Rogers "Excedrin" film unit structure, had already been argued five months earlier in September 1980 but was still awaiting a decision. On February 6, Carr wrote the judge, ostensibly asking that she schedule oral argument on the second of Kodak's motions. His real purpose was to get some idea of when the judge intended to rule on both. As Kerr had pointed out to Julius Silver, "it would not be unprecedented for the Court to hold [both of] these motions, and not to decide them, electing to have the issues decided after trial."[36] Carr tactfully suggested to Judge Zobel that her "disposition of one or both of the Summary Judgment Motions would be useful as a guideline in our trial preparations."[37] Kerr responded to Carr's letter by advising Judge Zobel that he continued "to believe that Eastman Kodak's Second Motion for Summary Judgment is lacking in merit, and that oral argument on it is not justified" but that he would be "guided by the Court's wishes."[38] Within a week, Judge Zobel responded by scheduling oral argument on the Trap patent motion for April 10.[39] Just one week after setting that date, she dropped the first judicial bombshell on the case.

In a decision issued February 25, 1981, Judge Zobel denied Kodak's first motion for partial summary judgment involving the Rogers film unit patents on the ground that there existed material issues of fact that had to

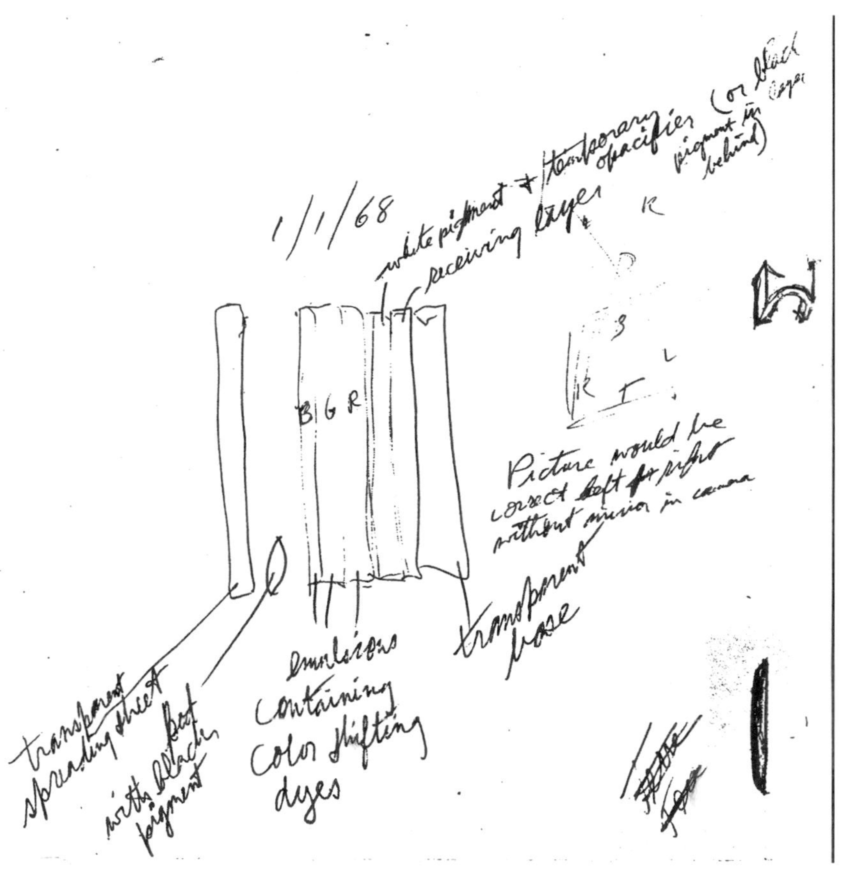

Figure 6-4: Howard Rogers' sketch of the "Excedrin" integral film unit structure, later used by Kodak in its PR-10 instant film, as drawn on New Year's Day, 1968.

Polaroid Corporation v. Eastman Kodak Company, United States District Court, District of Massachusetts, Civil Action 76-1634-Z, Trial Exhibit DF-449

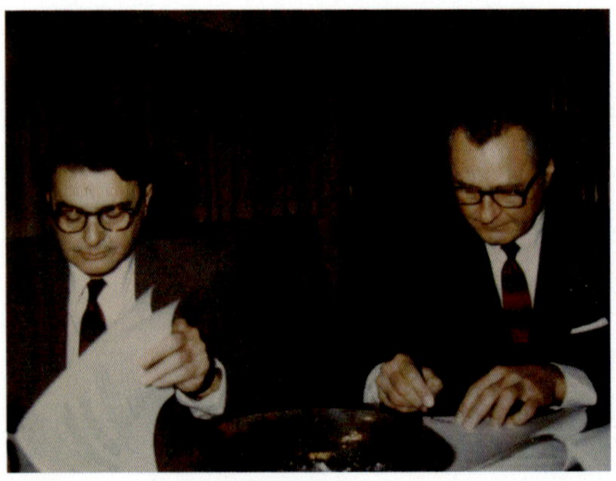

Figure 7-1: Land and Louis Eilers executing supply and license agreements between Polaroid and Kodak in New York City, December 10, 1969.

Polaroid Corporation Legal and Patent Records, Box II.227, f. 1, Eastman Kodak Supply and License Agreement, December 1969, Courtesy of Baker Library Historical Collections, Harvard Business School

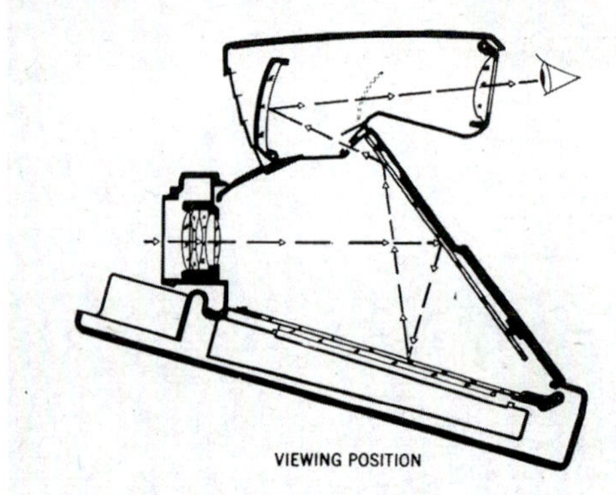

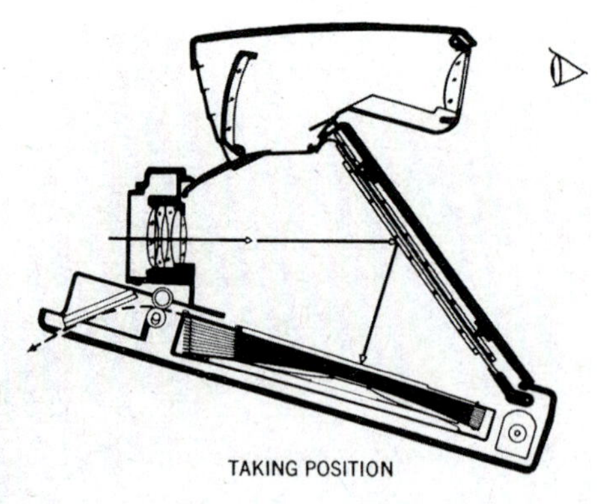

Figure 7-2: These two diagrams depict the path of light through the folding SLR SX-70 camera.

(Top) In viewing position, light from the lens is reflected off the permanent mirror on the folding camera back and down onto a special hinged Fresnel mirror, which reflects it up into the viewfinder.

(Bottom) Taking the picture: when the shutter button is pressed, the hinged Fresnel mirror springs up to the folding back of the camera body, leaving the film uncovered and ready for exposure. The light coming through the lens is reflected down onto the film by an optical mirror mounted back to back with the Fresnel.

Both diagrams: Reprinted from Land, Edwin H., "Absolute One-Step Photography," The Photographic Journal, *July 1974, courtesy of The Royal Photographic Society, UK/www.rps.org*

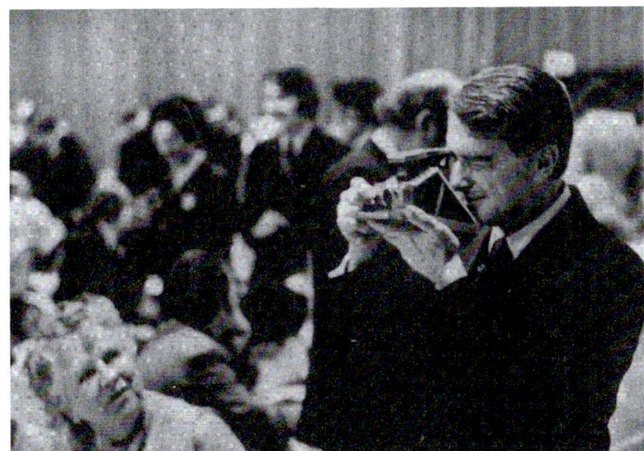

Figure 8-1: Land demonstrating SX-70 at the April 1972 Polaroid shareholders meeting.

Courtesy of Polaroid Corporation Archives

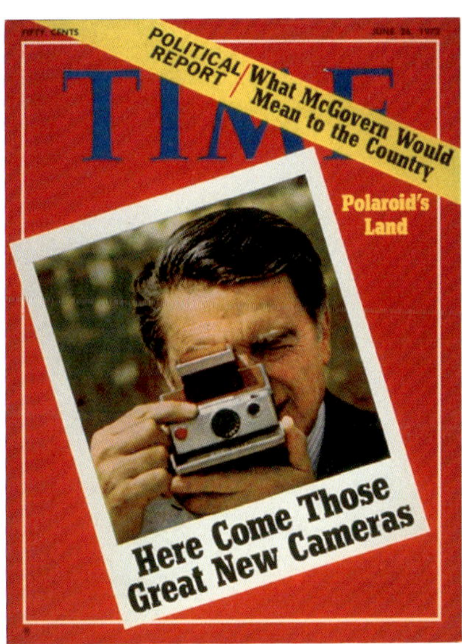

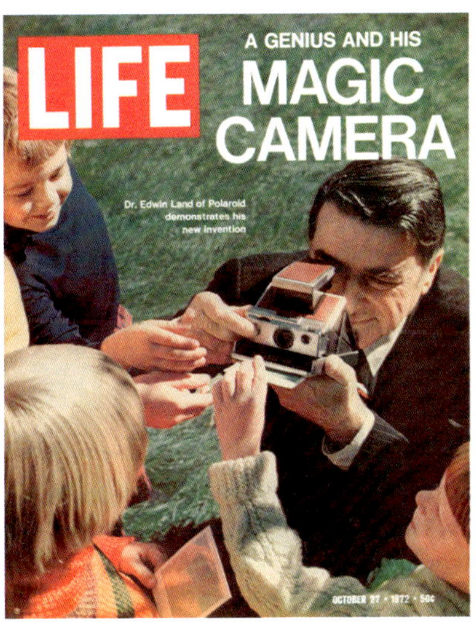

Figure 8-2: Cover, *Time*, June 26, 1972.

Courtesy of PARS International

Figure 8-3: Cover, *Life*, October 27, 1972.

Courtesy of Getty Images

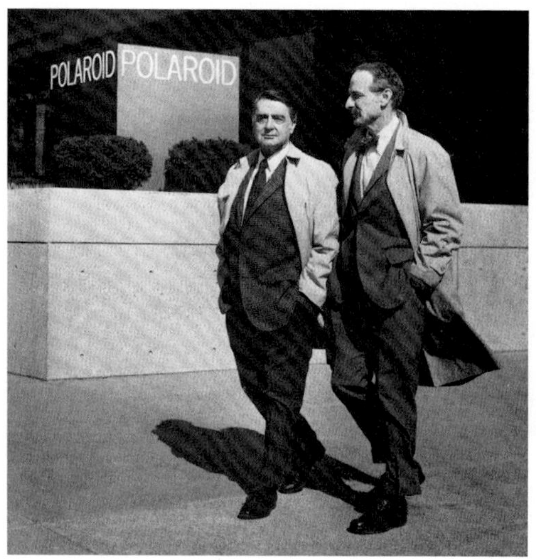

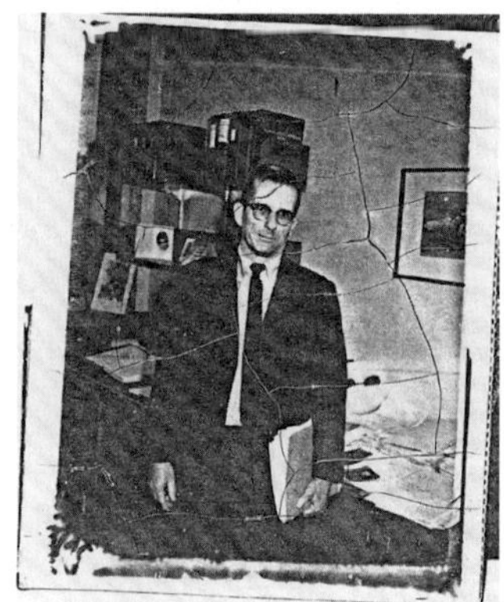

(Top left) Figure 9-1: Land (left) and McCune (right) in 1974 walking outside Polaroid's Technology Square headquarters in Cambridge.

Courtesy of Polaroid Corporation Archives

(Top right) Figure 9-3: Bob Ford, the Polaroid patent counsel who worked closely with Schwartz throughout the Berkey and Kodak litigations as an invaluable guide to the technical issues presented.

Author's collection

(Right) Figure 9-2: Cartoonist's depiction of Polaroid's organizational chart before McCune's ascension to the presidency.

Attachment to Letter, Robert Peck to H.F. Schwartz, December 23, 1977, Fish & Neave archives

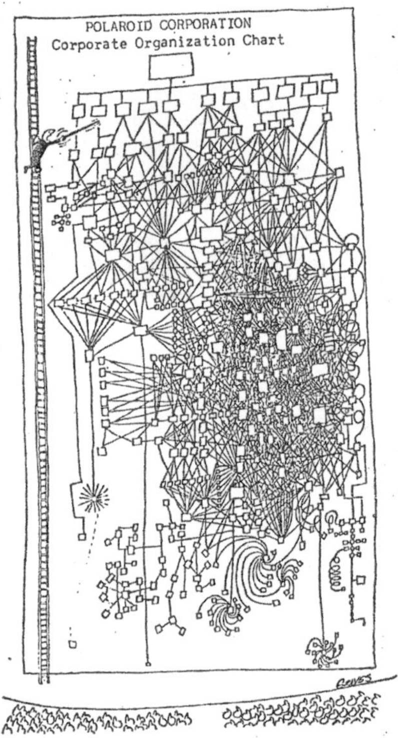

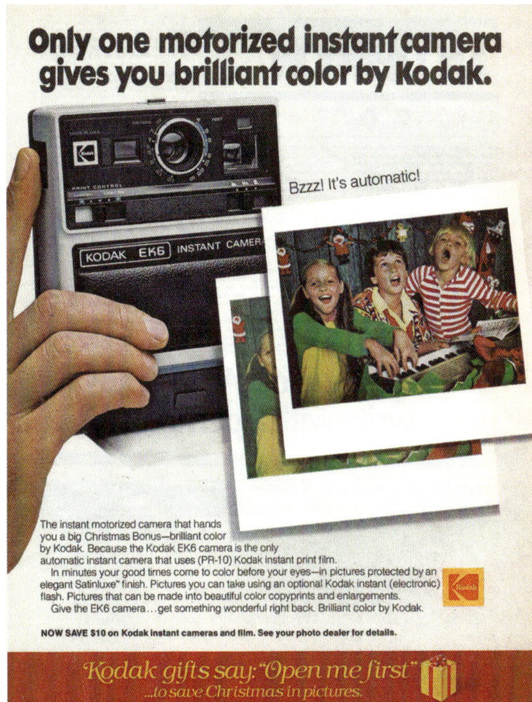

(Left) Figure 10-2: Full-page ad for EK6 camera and PR-10 instant film system.

Author's collection, Courtesy of Eastman Kodak Company

(Below) Figure 10-1: Walter Fallon, Kodak's chief executive, holding an EK6 camera at the company's April 20, 1976, launch event.

Photograph by Neal Boenzi, Courtesy The New York Times/*Redux*

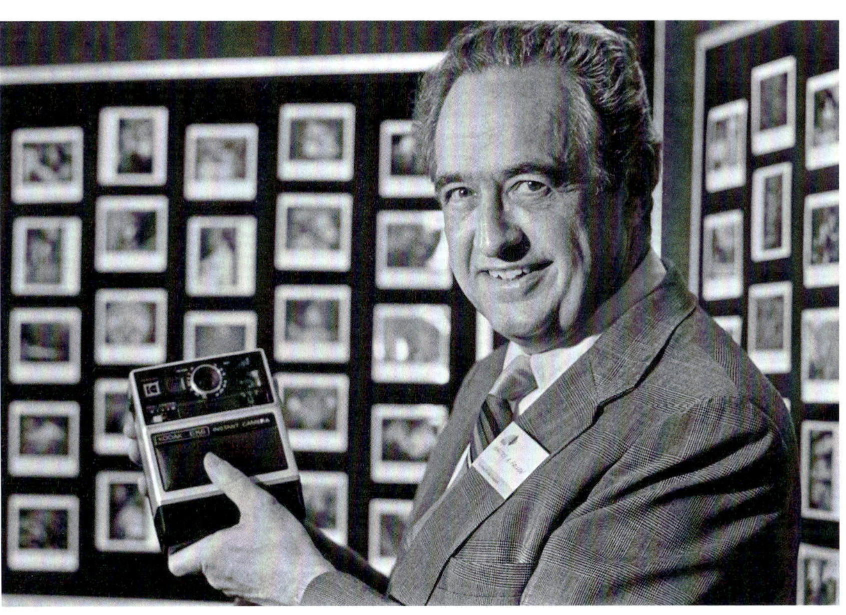

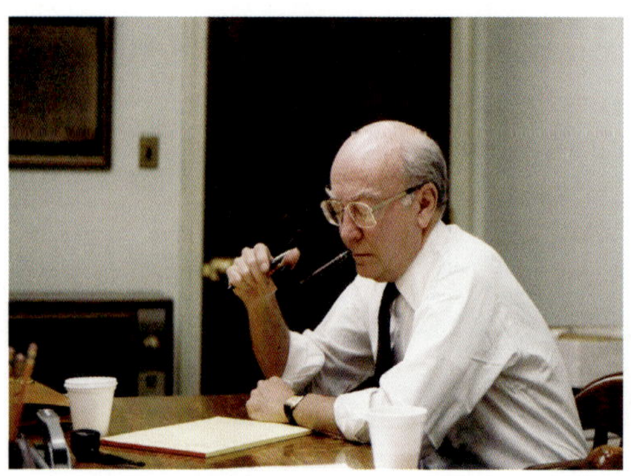

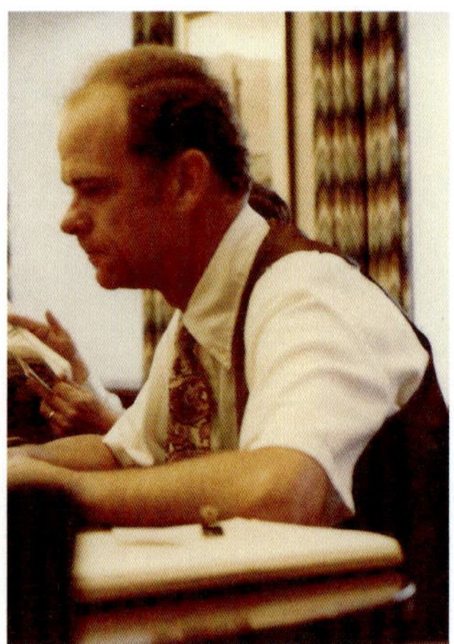

(Top left) Figure 11-1: Francis T. Carr, Kodak's lead trial counsel, with his ever-present pipe.

Courtesy of Cecil Quillen

(Top right) Figure 12-3: Cecil Quillen, the senior Kodak legal executive charged with supervising the Polaroid litigation.

Courtesy of Cecil Quillen

(Below) Figure 12-2: Land demonstrating the Polavision camera and monitor at the Polaroid shareholders meeting, April 26, 1977.

Courtesy of the Associated Press

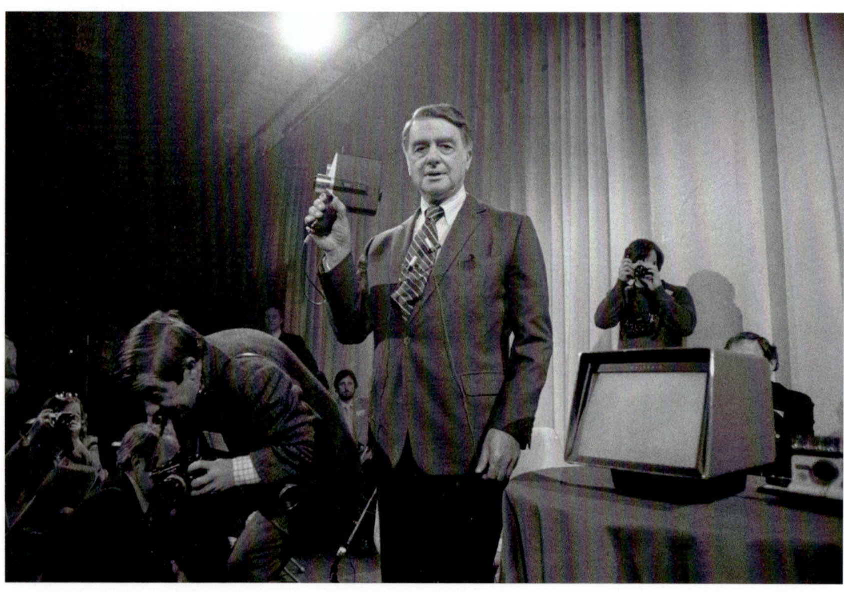

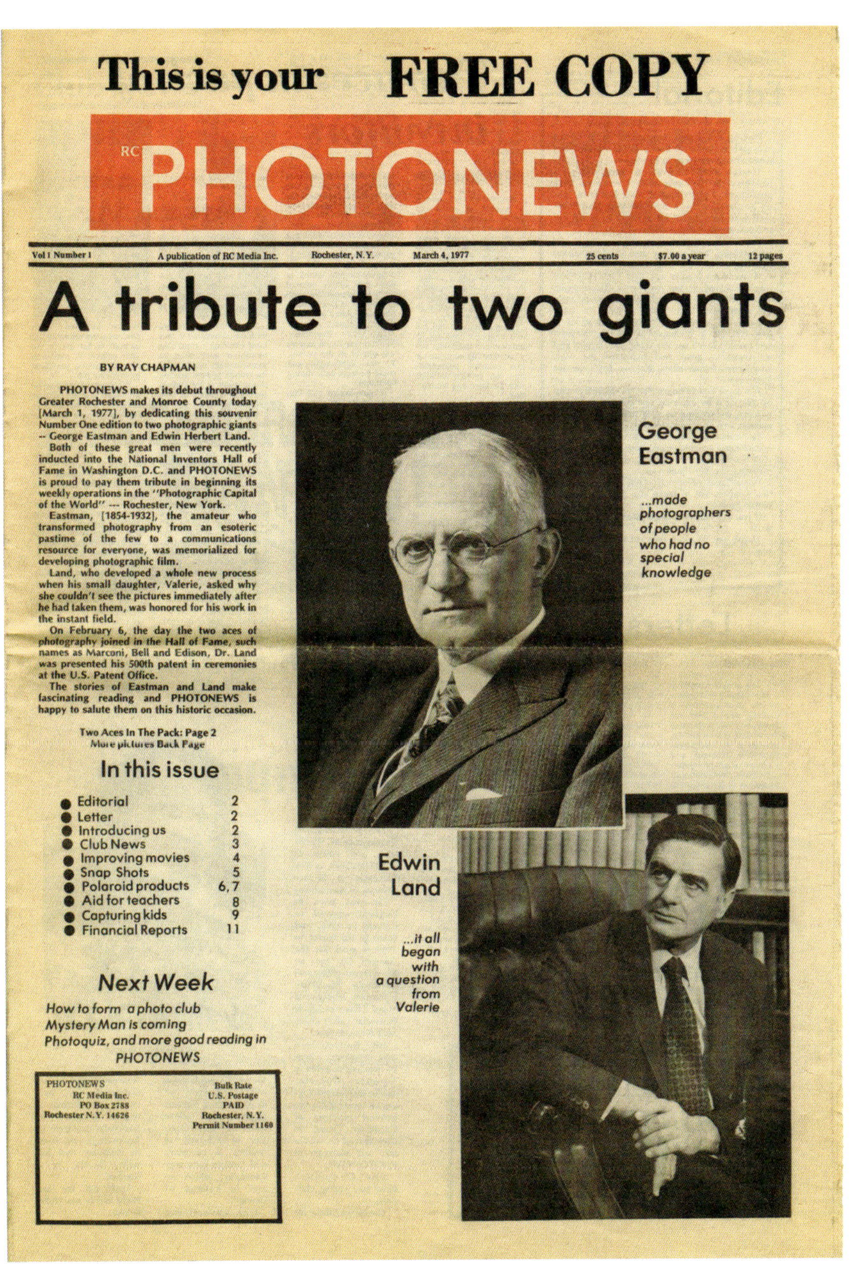

This is your FREE COPY

RC PHOTONEWS

Vol 1 Number 1 | A publication of RC Media Inc. | Rochester, N.Y. | March 4, 1977 | 25 cents | $7.00 a year | 12 pages

A tribute to two giants

BY RAY CHAPMAN

PHOTONEWS makes its debut throughout Greater Rochester and Monroe County today [March 1, 1977], by dedicating this souvenir Number One edition to two photographic giants -- George Eastman and Edwin Herbert Land.

Both of these great men were recently inducted into the National Inventors Hall of Fame in Washington D.C. and PHOTONEWS is proud to pay them tribute in beginning its weekly operations in the "Photographic Capital of the World" --- Rochester, New York.

Eastman, [1854-1932], the amateur who transformed photography from an esoteric pastime of the few to a communications resource for everyone, was memorialized for developing photographic film.

Land, who developed a whole new process when his small daughter, Valerie, asked why she couldn't see the pictures immediately after he had taken them, was honored for his work in the instant field.

On February 6, the day the two aces of photography joined in the Hall of Fame, such names as Marconi, Bell and Edison, Dr. Land was presented his 500th patent in ceremonies at the U.S. Patent Office.

The stories of Eastman and Land make fascinating reading and PHOTONEWS is happy to salute them on this historic occasion.

Two Aces In The Pack: Page 2
More pictures Back Page

George Eastman

...made photographers of people who had no special knowledge

Edwin Land

...it all began with a question from Valerie

In this issue

- Editorial 2
- Letter 2
- Introducing us 2
- Club News 3
- Improving movies 4
- Snap Shots 5
- Polaroid products 6, 7
- Aid for teachers 8
- Capturing kids 9
- Financial Reports 11

Next Week

How to form a photo club
Mystery Man is coming
Photoquiz, and more good reading in
PHOTONEWS

PHOTONEWS
RC Media Inc.
PO Box 2788
Rochester N.Y. 14626

Bulk Rate
U.S. Postage
PAID
Rochester, N.Y.
Permit Number 1160

Figure 12-1: The simultaneous induction of George Eastman and Edwin Land into the Inventors Hall of Fame was front-page news in the photographic industry trade paper, *Photonews*, March 4, 1977.

Polaroid Corporation Legal and Patent Records, Box II.178, f. 1, Magazine and News Articles, 1968–1977, Courtesy of Baker Library Historical Collections, Harvard Business School

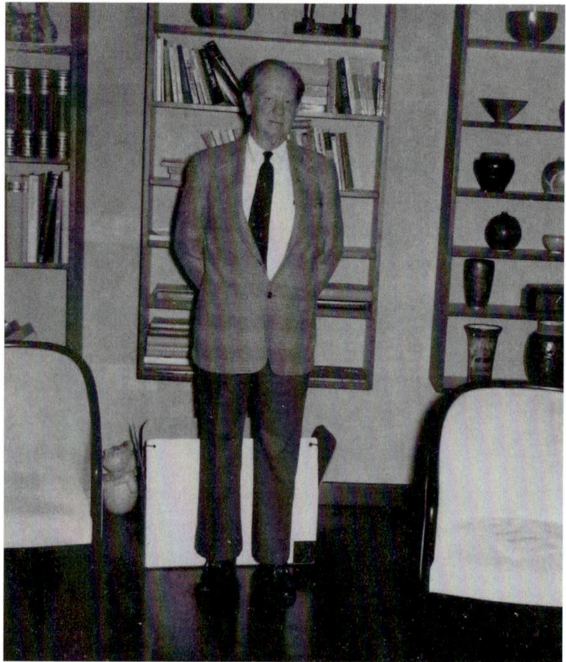

Figure 15-1: William K. Kerr, Polaroid's lead trial counsel at the start of the trial, posing for Land in his Cambridge office during a pre-trial preparation session.

Polaroid Corporation Legal and Patent Records, Box II.177, f. 2, Edwin H. Land Cross-Examination Notes, Courtesy of Baker Library Historical Collections, Harvard Business School

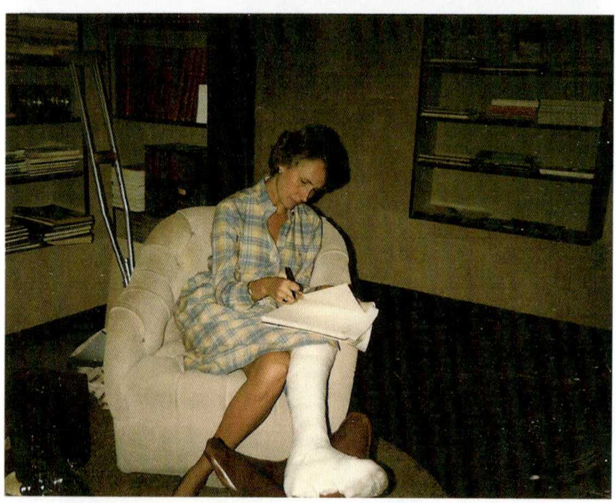

Figure 16-1: Land assistant Nan Chequer, who worked closely with Polaroid trial counsel as a conduit to Land throughout discovery, the trial, and then the appellate process. The cast is protecting a badly broken leg Chequer suffered in a fall.

Author's collection

be determined at trial. In her written opinion, she pointed out that "however persuasive defendant's arguments may be with the benefit of hindsight, whether the patents at issue here were anticipated by the prior art must be judged in terms of obviousness to one skilled in the art at the time the invention was made."[40] With regard to this consideration, she concluded that "in view of the contradictory . . . [expert witness testimony] before the Court it is impossible to conclude that no issue of fact is presented regarding obviousness." Zobel pointed out that for this reason courts had generally been "reluctant to grant summary judgment in patent cases."[41] This was a major victory for Polaroid, mostly in the sense that it avoided the potential disaster of having two of its fundamental patents knocked out of the lawsuit before trial. Kerr sent copies of Judge Zobel's decision to Land and to Julius Silver, and through him to the Polaroid board of directors, with a modest note: "This result is not surprising—but it is gratifying."[42] In actuality, Schwartz was not overly encouraged by winning this motion.[43] To him it merely confirmed that there were difficult issues that had to be determined at trial.

Kodak had to be disappointed that its major initiative to blow a hole in the side of Polaroid's case had failed. Calvert Crary reported to his readers that the outlook for Polaroid in the case was improved and that "Kodak is probably feeling a little uneasy about the situation."[44] Yet the Kodak team saw many positives in Judge Zobel's decision and believed that they had come close to winning the motion.[45] Confidence was high that they would be able to tip the scales, at least on these patents, in Kodak's favor at trial.

In the meantime, Kodak seemed content to continue stringing out discovery as best it could. At a March 10 hearing on discovery issues, the Special Master did Kodak a good turn by announcing that he had told Judge Zobel that he believed discovery should be extended until the end of April. He also informed both sides that the judge said that this would mean the trial date would have to be delayed until early September.[46] Both parties understood that this was a clear signal for them to cooperate in resetting the dates along those lines. As a result, a joint application was submitted to Judge Zobel on March 20, resetting the discovery cutoff to May 1 and the trial to the "earliest available date in September 1981."[47] On March 24, 1981, Judge Zobel acted on this application and set what were, finally, to become the ultimate dates for the case. She adopted the May 1 date for the end of discovery but set October 5, 1981, as the first day of trial.[48] Kerr informed Julius Silver of the new dates, explaining that they "reflect[ed] the wishes of the Special Master and Judge Zobel."[49]

With things seeming at last to be going Polaroid's way, the question of a settlement was raised once more. After consulting management, Bob Peck called Cecil Quillen to find out if Kodak had given any further consideration to a possible settlement in view of Zobel's decision on the Excedrin film unit patents.[50] Once again, this initiative went nowhere. Kodak was still not willing to entertain any kind of meaningful payment to Polaroid, even as it did its best to avoid the imminent inevitability of going to trial. Kodak's position was hard to figure. Perhaps it thought it might do better on its second summary judgment motion, but that remained a long shot.

Oral argument on Kodak's second summary judgment motion was held on April 10, 1981. Carr's argument, as laid out in Kodak's briefs, was straightforward. All patents have very specific language at the end of the document that defines the precise scope of the invention, the metes and bounds of the invention protected by the patent similar to the recitation one finds in the deed on a parcel of land. This section is known as the "claims" of the patent. During a patent application prosecution, the inventor's attorney often engages in a to-and-fro with the patent examiner over the language of the claims, in an exercise meant to define the invention as narrowly as necessary so as to find it patentable over the prior art.

Based on the history of the Trap patent prosecution, Carr averred that the invention covered was limited to the inclusion of an acid material in the spacer element of the film unit to neutralize excess processing fluid.[51] This was not a new technique, he urged, but was clearly taught in prior art references, particularly in earlier patents issued to Polaroid. Ironically, as occurred in the Rogers film unit situation, one of these prior patents belonged to Land. Thus, the Kodak argument went, the solution allegedly invented by Campbell, and covered in the Trap patent, would have been obvious to one skilled in the art of designing instant photography film units at the time Campbell had his discovery because the technique had been taught in other Polaroid patents. Kerr vehemently disagreed. In essence, he took Zobel's position on the first motion—the facts necessary to adjudicate the patent could only be determined through a full and fair exposition at trial.

A few days later, Carr tried an end run around Kerr, writing to Judge Zobel to provide some additional information about prior art patents he had sent to her just days before and that he had referenced during the oral argument.[52] Kerr could not contain his annoyance over what he perceived as an improper attempt by Carr to save a hopeless motion. Kerr wrote to the judge, objecting to the tactic and reiterating his arguments on the motion:

> Polaroid's position remains precisely as it was stated during oral argument, namely, that irrespective of the legal interpretation accorded to the Campbell patent claims (the scope of the invention), Kodak's motion is doomed to failure because (1) it is replete with contested issues of fact, and (2) the evidence before the Court demonstrates that the Campbell invention would not have been obvious.
>
> Kodak's motion should be denied forthwith.[53]

At the oral argument on April 10, Judge Zobel had set May 8 as the date for a pretrial conference to start the process of getting ready for trial. During this phase, the judge works with both parties to set ground rules and procedures necessary for the orderly conduct of a complicated proceeding. Some of the items addressed include defining and narrowing the issues that actually need to be tried, the sharing of information such as lists of witnesses and exhibits, and the handling of evidence gathered during discovery. Different kinds of cases require different sets of such rules, and judges themselves have personal preferences on how many items should be handled. Judge Zobel asked both parties to submit a proposed agenda for that conference within a week to ten days. As interpreted by Kerr, this meant that Zobel expected the parties to confer and to agree on an agenda, at least, if not on a set of procedures for the trial.

On April 20, Carr sent the court a letter unilaterally outlining Kodak's proposals, together with a large pile of supporting documentation.[54] The submission had some novel ideas. First, he proposed adopting procedures used in the New York federal court trial of a major antitrust case between SCM and Xerox in the late 1970s. As big a case as the *Polaroid v. Kodak* litigation was, that one was even bigger, having taken a full year to try before a jury. There was to be no jury in this case, so that seemed an inapposite choice of a model. Second, Carr suggested that the trial be split up, with all of the film patents being fully tried before the camera patents. Finally, he proposed that the court appoint "neutral experts" to hear the detailed technical testimony and to interpret it for the judge.

Kerr was livid when he received his copy of Carr's submission. Controlling his outrage, he diplomatically "confess[ed]" to Judge Zobel "some surprise upon receiving [Carr's] voluminous submission . . . I had understood Your Honor to suggest . . . that the parties should endeavor to prepare a joint agenda of matter to be discussed with the Court on May 8," explained Kerr. He pointed out that Carr's ex parte submission was contrary to his understanding of Judge Zobel's usual pretrial procedures and "not in conformity with pretrial procedures that I have encountered in other District Courts. . . . I can not refrain from saying," continued Kerr, "that I regard Mr. Carr's submission as high-handed and one-sided and designed to arrogate to defendant the customary prerogatives of a plaintiff with respect to pretrial procedures."[55] He then went on to detail his objection to virtually all of Carr's proposals. Carr responded to Kerr's letter by advising Judge Zobel that he would address the substance of Kerr's letter at the hearing but explained that he had submitted the material on the tenth day following her request after having "awaited the initiative of plaintiff . . . [but] having received nothing from [them]."[56]

When the conference was held in open court at three p.m. on Friday, May 8, Judge Zobel announced that she had reviewed both parties' submissions and then proceeded to hear from both sides. Over the course of the next two hours, she set a specific day-by-day schedule for all pretrial activities, leading up to the trial date of October 5 that she had previously set. These rulings were then formalized in a pretrial order issued on May 11.[57] In the end, the calendar she set and the procedures built into it were not at all based on the model Carr had suggested from the *SCM v. Xerox* trial. Also rejected were each of Kodak's other novel proposals over which Carr and Kerr had argued. She refused to bifurcate the trial into film and camera segments and denied Kodak's request that she should appoint a neutral expert. She tasked Judge McLaughlin with resolving any outstanding discovery disputes that had arisen prior to the May 1 discovery cutoff and asked him to "take all steps necessary to close discovery as soon as reasonably practicable."[58] She also advised that the motions filed back in 1976 relating to Kodak's counterclaims of patent misuse directed at the entire Polaroid patent portfolio would remain undecided until the case had been heard and decided.[59] Additional pretrial conferences were scheduled for September 10 and 28.

Carr had seemingly gotten off on the wrong foot with the judge. At least one observer used the word "rude" in describing Judge Zobel's treatment of Kodak's counsel in court that day.[60] Schwartz, however, took little comfort from the episode. He believed that it was impossible to truly tell whether a judge's comportment toward an attorney signaled anything with respect to the ultimate adjudication of the case. He had learned that only when you see the decision do you really know what the judge thought.[61]

Kerr, however, was elated with the results of the conference and told Land in a letter that, "things went well before the Court," advising him that the trial date had held, that discovery "is virtually concluded" and that

"Kodak had failed to get some of the major things it requested, such as separate trials on the film and the camera patents, and the appointment by the court of a 'neutral' expert."[62] Kerr, in his cheerful frame of mind, for the first time employed the salutation "Dear Din." But this use of Land's nickname, allowed only to close friends and colleagues, was probably another misstep in terms of his ongoing relationship with his star witness.[63] It was an impolite, if not arrogant, thing for Kerr to do. There was no immediate negative reaction by Land, but it surely did not go unnoticed.

CHAPTER 19

OVER THE PRECIPICE

With summer approaching and the discovery phase of the lawsuit finally coming to an end after five contentious and exhausting years, both sides began in earnest their preparations for trial. For the lawyers expected to live in Boston while trying the case, Fish & Neave reserved thirteen rooms from mid-September to mid-January at the comfortable Ritz Carlton hotel. From the Ritz, it was a pleasant half-mile walk to the courthouse along the Boston Common. An extensive amount of space was also reserved at the Park Plaza Hotel, diagonally down and across the street from the Ritz, for the support staff Fish & Neave would be bringing up from New York for the duration. More than twenty rooms were reserved for paralegals, who would handle the documents and trial exhibits, and for secretaries, who would process all the words generated by the lawyers in this era before the personal computer. In addition, a huge suite was reserved to house the more than 100 file cabinets it was anticipated Polaroid would need to store all the paper generated by the litigation.

As plans were being finalized, the Park Plaza's engineers advised Fish & Neave that the weight of all this material would be too heavy for the floor to bear, so a plan was devised to split the filing cabinets between two floors in the hotel. Office furniture and the latest technology in word-processing equipment were leased, and two station wagons were rented from Avis for use by the team. All in all, it was a huge logistical undertaking. Kenyon & Kenyon faced the same challenge of trying a lawsuit on the road, away from its home office, and set up its temporary operation in office space provided by Kodak at 141 Milk Street in Boston, near the courthouse, with its lawyers housed in apartments rented for the duration.

For Polaroid, Schwartz immersed himself in the logistics as well as a series of required pretrial submissions to the court—a joint statement of issues to be tried, contentions of the parties, lists of expert and fact witnesses and supporting information, lists of exhibits to be introduced at trial, designations of deposition testimony to be offered into evidence, and trial briefs to support the parties' contentions—all due on a precise schedule. With Schwartz heading up the troops, Kerr, who was rapidly approaching his sixty-fifth birthday, focused on one very important piece of the preparation, one he knew would be his "last hurrah" in a courtroom: Land's testimony. Once again, Land had arranged for his trusted associate, Nan Chequer, to come in from California sometime in July, but Kerr began sending her "formidable looking volumes" of preparation materials just after the May 11 pretrial order was issued confirming the trial date of October 5.[1]

Early on in the process, in addition to collections of patents and technical materials relating to the specific issues in the case, Kerr provided Chequer with a memorandum he had prepared entitled Polaroid Patent Policy.[2] This document, drawn from speeches and publications Land had issued over the years, was an expression, mostly in his own words, of Land's philosophy of business and the patent system and the fundamental principles embodied in Polaroid, the company he had built. "Polaroid has always been a research-dedicated organization," the memo began. "I have always believed that the patent system . . . should be utilized by such research organizations to ensure that worthwhile inventions are protected, for limited periods of time, from scrupulous or unscrupulous competitors," it went on. Emphasizing Polaroid's underdog status as compared with Kodak, Kerr reminded Land of his previously published view "that patents are more important to newly-formed research enterprises than they are to larger, more established organizations. . . . This has been true with respect to Polaroid and to the field of instant photography."

In the memo, Kerr also brought together Land's observations on his role in directing a team of researchers, an important theme in the case, given that Land was the inventor on only three of the twelve asserted patents. "Creation of a new technology, such as one-step photography," Kerr wrote in paraphrasing Land, "requires that a single individual have in mind the objective to be reached. . . . This master plan must be supported by the efforts of many others . . . but the single dominant individual must constantly assure himself that the individual efforts complement one another and create support for an integrated system."[3]

This reminder to Land fit in with the strategy Kerr and Schwartz had devised for the trial. They believed it critical to provide to Judge Zobel an overall context in which the specific issues of the lawsuit should be considered.[4] Their focus, despite all the technical issues involved, would be on Kodak's relationship with Polaroid—its early mentoring and support of Polaroid when it considered instant photography a curiosity and a niche market, its complete change of heart when it determined that Land's one-step technology might actually become a real threat to Kodak's dominance of amateur photography, and finally, its determined invasion of Polaroid's domain without regard to the pioneering company's patent property. During discovery, Schwartz had concentrated on developing evidence from Kodak's internal documents that he could use to tell this story.

Land's view of the relationship between the patent system and scientific innovation, in the real-world context of an entrepreneurial businessman, was a related and equally important subject. Simply put, the hope was that when presented at trial, these two themes would, in essence, brand Polaroid as the innovative, underdog good guy and Kodak as the monopolistic and ruthless transgressor. In his cover letter, Kerr told Chequer that this material was likely outside the strict scope of testimony Land was entitled to provide as a fact witness in the trial—but suggested to Chequer that Land "should not be reluctant to express these views, or at least some of them, to the Court as the occasion arises."[5]

Before May was over, Kerr also began to outline the actual testimony Land would present to open Polaroid's presentation of its case, probably the most important phase of the entire trial. Having Land make a good impression on the judge would be critical in establishing his credibility and his expertise as he explained both Polaroid's position on the technical issues of his patents in the lawsuit, as well as the history of instant photography. It was critical to set a tone with Land for everything else to come in Polaroid's direct case. On May 27, Kerr wrote another "Dear Din" letter to Land, conveying his views on the subject:

> I have been giving further thought to your direct testimony from the standpoint of both content and sequence.
>
> As I explained during our recent meetings, the conventional way to proceed is to open the direct examination of a witness with testimony concerning his or her background, so that the Court [or jury] may acquire some understanding of who is doing the testifying.

> I am now seriously considering whether or not we should depart from this conventional procedure in your direct examination. The necessity of introducing you to Judge Zobel is, to coin a phrase, like carrying coals to Newcastle. Also, I am influenced in making this decision by Judge Zobel's no-nonsense approach and her desire to get into the guts of this case as expeditiously as possible.
>
> As it now lies in mind, I believe that we should start your direct examination with a leap into your first work relative to one-step photography, your continuing dedication and manifold contributions to the field, culminating in the creation of absolute one-step photography. There will be many stops along the way, but when those aspects are amply dealt with, we can move immediately into the inventions of your patents in suit and the prior art relied upon by Kodak against them. As I now visualize it, the conclusion of your direct examination would include such matters as the recognition of your signal accomplishments by the scientific community, as exemplified by your awards, honorary degrees, etc.[6]

This all came down to giving Land, as well as the Polaroid lawyers and executives anxiously following developments, a preview of the script for what was sure to be a momentous event.

The path to the courthouse proved to have some further bumps, however. On Friday, July 10, Judge Zobel delivered some unexpected and worrisome news to Polaroid. In a decision that clearly adopted, almost verbatim, most of Kodak's arguments, she ruled that "the alleged invention of the . . . [Trap] patent would have been obvious to one skilled in the art at the time the invention was made, and that the patent is, therefore, invalid."[7] Rather than take the cautious route, as she had on the previous summary judgment motion, of denying Kodak's motion by simply acknowledging that there were contested issues of material fact that needed to be determined at trial, Judge Zobel cited prior case law suggesting that summary judgment might be appropriate even in patent cases where the technology is "readily understandable by the lay mind. This is just such a case," she concluded. Analyzing the two Polaroid prior art patents proffered by Kodak as teaching the Trap invention, Zobel found that their "solution" to the problem addressed by the Trap patent was identical.[8]

In a finding most troubling to Polaroid, and most encouraging to Kodak, Zobel declared "immaterial" Polaroid's argument that the prior art patents related to peel-apart film units, whereas Campbell's Trap invention addressed the problem in an integral film unit that stayed together forever. Zobel decided that "the problem itself remains the same—to contain the liquid and to neutralize it . . . [and the Trap] patent solves the problem in the same manner as do the two prior art patents."[9] Employing the techniques taught in the prior Polaroid patents, including Land's, in an integral film unit "is not the type of 'advance' upon which patentable monopolies were intended to be predicated," Zobel ruled.[10] This theory—that the solution claimed as an invention in a color integral-film unit could be found in earlier generations of one-step technology, for example, peel-apart film units or black-and-white processes—was the foundation for many of Kodak's attacks upon the Polaroid patents in suit. To have Zobel embrace the theory provided Kodak the affirmation it had long sought that its case had merit, and it certainly emboldened Kodak's attitude toward the outcome.[11]

For Polaroid this was an unnerving blow, with Judge Zobel apparently sending a new and different message to the parties. As Kerr explained to Julius Silver, "speculation concerning the reason for any Court decision is probably not fruitful. . . . My guess is, however, that this decision is calculated by the Court to bring home to the parties that this case is not a sure winner on either side, and that the parties had better get back to the bargaining table."[12] When Land heard of the decision, he asked to see a copy, and Bob Ford sent him one the following Monday.[13]

Within weeks of issuing the decision, Judge Zobel took further steps that confirmed Kerr's interpretation. On Friday, August 7, she held a meeting in her chambers with Schwartz and Carr, and suggested that settlement talks be held with Judge McLaughlin, in an effort to bring the parties together.[14] Apparently, it was Zobel's normal pretrial practice to push for a settlement. She had done this in another complex litigation, one involving antitrust and patent issues that she had on her upcoming docket for trial. In that case, using her background in tax law, she had even suggested a method for structuring a proposed settlement payment to minimize tax liability for the party receiving it.[15] Kerr responded promptly to Judge Zobel, accepting her proposal but, intent on avoiding any delay or distraction with respect to the preparations for trial, urging that any conferences with Judge McLaughlin "be scheduled promptly and . . . proceed expeditiously."[16]

Kodak had a completely different view of Zobel's suggestion. Carr wrote to the judge on August 12 expressing his "judgment [that] the problem with compromise here lies in the vast difference in the relative values placed upon this litigation by the parties. With such a difference," Carr advised, "Kodak believes that mediation is very unlikely of success." Carr argued that it would take far too much time to acquaint Judge McLaughlin with the positions of the parties on the merits of the lawsuit. Instead, he told Zobel, "Kodak had repeatedly proposed to Polaroid that the General Managements of the two companies and their counsel sit down in a small meeting for a frank, off-the-record discussion of the relative strengths and weaknesses of their positions. It is our belief," concluded Carr, "that such a meeting could move the parties closer to a reasonable settlement."[17] There was no immediate response from the judge.

Having ostensibly rebuffed Judge Zobel's attempt to work toward a settlement, Kodak instead launched another potential torpedo at Polaroid, perhaps in an effort to weaken further its resolve for whatever discussions might take place. Just two days later, on August 14, Kodak filed yet a third motion for partial summary judgment.[18] This time, Kodak's assault was on Land's cherished Rear Pick patent. In essence, Kodak asserted that Land's patent should be declared invalid because "every feature of the claimed invention save one is found in the prior art . . . and . . . that single feature on which patentability was predicated is plainly disclosed, in the same or similar environments, in at least three prior art patents that were not cited during the prosecution of the . . . [Rear Pick] patent."[19]

This was a risky move for Kodak. Since the trial was scheduled to commence in just seven weeks, it was awfully late in the game for this kind of motion. Given how long it had taken to deal with the two previous summary judgment motions, Kodak must have realized that it would be virtually impossible for this motion to be adjudicated before trial—given the current schedule—and that Judge Zobel might not be pleased with this tactic, especially since Carr had just frustrated her initiative toward finding a settlement. From a strategic perspective, the motion was a complete diversion. No matter how it was handled, it would not eliminate the necessity of moving forward at trial on all the remaining patents. In retrospect, it's still difficult to figure out why Kodak risked annoying Judge Zobel. Perhaps the motion was merely intended to influence the Judge's view on this patent by suggesting that even Land's inventions were not beyond reproach. Perhaps it was intended as some kind of message for Polaroid. Perhaps it was both.

In a vigorous response, Kerr first addressed Carr's reply to Judge Zobel's settlement initiative. He reported to the judge on August 18 that he had considered Carr's letter and its proposal with his client and reiterated Polaroid's "willingness to explore the possibility of settlement of this case on a fair and reasonable basis."[20] He then threw a small bombshell of his own, questioning the veracity of Carr's characterization of Kodak's previous settlement efforts. Kerr informed Zobel that "Polaroid was surprised by Mr. Carr's statement to the Court" concerning Kodak's repeated proposals for a meeting of the companies' managements. "We know of no repeated proposals; we know of no proposal of such a meeting of 'General Managements,' which we assume was intended to refer to 'Top Managements,'" Kerr continued. "As your Honor will recall," he reminded her, "no member of top management of Kodak appeared at the settlement conference before your Honor, at which Mr. McCune appeared as Chief Executive Officer on behalf of Polaroid." Nonetheless, Kerr advised Judge Zobel that McCune would be contacting his counterpart, Walter Fallon, "in an endeavor to arrange such a meeting."[21]

An incredulous McCune wrote to Fallon and told him that he had seen a copy of Carr's letter about Kodak's supposed repeated settlement attempts. "I am astonished by this statement," McCune wrote. "Nevertheless," McCune continued, "we understand from Mr. Carr's statement that you are interested in participating in a small meeting . . . [and] if this is so, we would be happy to meet with you and your colleagues for this purpose . . . keeping in mind that the trial date is set for early October."[22]

On the other front, Polaroid moved to strike Kodak's summary judgment motion and moved to extend its time to respond on the merits of the motion until thirty days after the court ruled on Polaroid's motion to strike. This was a clever maneuver by Kerr and Schwartz; no matter how the judge ruled, it gave them time to continue to prepare for the trial without wasting time responding to the motion—and it also gave her a way to ignore Kodak's attack on Land's patent indefinitely. It worked; on August 26, Judge Zobel allowed Polaroid's request to postpone a response until she ruled on the motion to strike. Since Zobel never did, in fact, rule on Polaroid's motion to strike, Polaroid never had to deal with Kodak's summary judgment motion at all. Nonetheless, in its submission, Polaroid did not miss its opportunity to point out that Kodak's timing in filing this motion on the same day that it maintained "that it would be too 'time consuming' to involve Judge McLaughlin in settlement negotiations" represented Kodak's "latest . . . attempt to distract this Court and counsel from the task at hand, *i.e.* to get this case pre-tried and tried as quickly and as simply as possible."[23]

As it turned out, Kodak's bag of trial-avoidance tricks had not yet run out. Emboldened by its success on the Trap patent, Cecil Quillen came up with a novel idea, one that Carr resisted but ultimately was persuaded to pursue.[24] On August 27, the very day after Zobel, in effect, tabled the Rear

Pick motion, and just five weeks before the trial was set to begin, Kodak topped all previous ploys. It suggested to Judge Zobel that she postpone the trial until March 1982 so as to allow Kodak to file eleven separate summary judgment motions, one for each of the remaining patents in suit. In his submission, Carr had the effrontery to describe this "program" as an attempt to "expedite a resolution of this matter. We expect that the trial of this case will be arduous and time-consuming," explained Carr, "[so] . . . to the extent that the summary judgment proceedings are dispositive, a full-scale trial with respect to the affected patents will be completely obviated and the resources, time and energy required of all, substantially reduced." Carr told Judge Zobel that his "proposal is put forth solely as a mechanism for shortening and simplifying the length and complicated task faced by the parties and the Court."[25] To the contrary, Polaroid saw this "astonishing proposal" as an outrageous attempt to complicate the proceedings so as to avoid or at least postpone a trial.[26]

There is some evidence that the judge herself may have had the same impression about the real motive behind Carr's ongoing campaign. A Polaroid local counsel who called the judge's clerk, Nina Singer, regarding the timing of a response to Carr's proposal reported that she told him that the court had not yet received the letter but "expressed surprise and stated something to the effect of 'what could they be saying now to postpone the trial?'"[27] This was no way to gain the favor of a judge who had worked so hard to put the case on an orderly track. Again, the motive for this maneuver is hard to fathom, since Carr had to know that it stood little chance of succeeding and had more of a chance of exasperating the judge than resolving the case. Judge Zobel did nothing to respond to Carr's proposal when she read his letter. She was apparently prepared to wait two weeks until September 10, when the next pretrial conference was already set to occur.

In the meantime, following up on McCune's letter, McCune and Fallon scheduled their meeting of "General Management" for Friday, August 28, at a neutral site in Boston.[28] The day before, McCune, Land, and the rest of the Polaroid contingent met with Kerr and Schwartz in Polaroid's boardroom to discuss the status of the case, the outlook for the trial, and the posture they should adopt in their meeting with Fallon. The session with Kodak the following day was held in two adjoining suites at the Ritz Carlton.

Trial counsel was not invited to the meeting between McCune and Fallon, but both executives had their in-house lawyers and negotiators present: Charles Mikulka and Robert Peck, for Polaroid, and Cecil Quillen and David Greenlaw, for Kodak. Also present for Polaroid was Land. Fallon, still displaying the resentment some within Kodak had toward

Polaroid's patriarch, later quipped to Quillen that he "didn't know Polaroid had two chief executives."[29] Once the meeting began, Quillen made a full presentation of Kodak's case on the merits with respect to each of the patents in suit. The Polaroid contingent listened respectfully, but when Quillen was finished, they offered no rebuttal on the merits of the case. The only comment made was by Peck, who asked Fallon and Quillen whether in evaluating its prospects in the litigation Kodak had taken into account the possibility of its being enjoined from infringing the Polaroid patents, should Kodak lose on one or more of them. Such an injunction would, as a practical matter, force Kodak out of the instant photography business altogether unless it could redesign its products to avoid the infringement, which would be an expensive and impractical enterprise.

The Kodak executives did not respond, and the meeting quickly came to an end. Privately, there was a divergence of opinion at Kodak about the disastrous possibility of an injunction issuing from a defeat in the trial. Greenlaw and others held the view that no matter the outcome, there was no way a federal court would enjoin Eastman Kodak and force it out of the instant photography business. To them, the worst that could happen would be that the company would have to pay some indeterminate sum in damages for infringing any patents Polaroid might prevail on. But on the plane trip back to Rochester, Quillen felt the need to express a contrary view to Fallon. "Come on guys, let's get real," Quillen advised his colleagues. "No matter how godly the Kodak Company is, if we lose, we will suffer an injunction."[30]

In the end, no settlement discussions occurred at that meeting of top management. Quillen commented later: "We didn't persuade them, and they didn't persuade us, and no offer came directly out of the meeting."[31] However, McCune followed up with a phone call to Fallon a few days later on September 3, the Thursday before the long Labor Day weekend. During that phone call, McCune offered to settle the case for a five percent royalty on Kodak's sales of instant photography cameras and film, retrospectively and prospectively.[32] The five percent royalty rate McCune offered was not without precedent. It was the very same royalty agreed to by Kodak for licensing Polaroid's patented technology under the provisions of the 1957 agreement between the two companies regarding the development of Polaroid's original color negative.[33]

Fallon did not respond immediately. He told McCune that conversations about the Boston meeting were still ongoing with his colleagues, and he promised to get back to him after the weekend. It was clearly an awkward conversation for the two men, who had worked together so closely for decades, and who knew each other well. McCune told Fallon that "except for this" he was doing a "great job."[34] Fallon responded by telling McCune that he "could be proud" of Polaroid's SX-70 system, "coming from someone who understands the technological accomplishments."[35] When McCune asked Fallon how he knew about SX-70, Fallon responded, "What makes you think I don't have one?" Well, "use lots of film," McCune said, as the two hung up.[36] Fallon did not call McCune back after the weekend, as he had promised.

At the pretrial conference on September 10, Zobel ignored Carr's proposal to substitute eleven summary judgment motions for a normal trial and instead opened a discussion with the parties about some procedures for the trial that she thought might simplify the process. She made a suggestion about the order in which the parties would present aspects of their cases that Kerr said he was prepared to accept on behalf of Polaroid.[37] Carr told the judge he'd consider it and discuss it with his client but a week later wrote to her rejecting the idea as not in their best interest.[38] Instead, he once again pushed his multiple-summary-judgment proposal, an initiative that was never seriously discussed again and thus fell by the wayside.

Before the September 10 conference broke up, Zobel asked both parties about the status of the settlement discussions. Kerr told the judge about the August 28 meeting of McCune and Fallon and their discussion of the parties' respective views on the merits of the case, as well as the subsequent phone conversation in which McCune had offered "to settle this controversy on the basis of a reasonable royalty arrangement."[39] At that time, Kerr did not mention the specific figure that was discussed but told Zobel that as of yet, McCune had received no response from Fallon. Zobel asked both lawyers to consult with their clients and to report to her on further developments. That September 10 conference took place on a Thursday. Fallon called McCune the following Tuesday, September 15, and declined Polaroid's offer. Fallon reiterated that the only settlement they would consider would be a "fee" covering the "cost of the trial" and the "value of the litigation," and that a "conventional royalty of [even] four percent of total sales [was] unrealistic."[40]

Frank Carr wrote to Judge Zobel on September 17 informing her that Fallon had called McCune but that "no agreement was reached, and no definite arrangement was made for further discussion, although further discussion was not ruled out."[41] As might be expected, Kerr was not happy with Carr's report. In his letter the very next day, Kerr reminded the judge of what he had told her during the September 10 conference about Polaroid's offer to settle the case on the basis of a "reasonable royalty" and advised the judge that during the September 15 telephone conversation "Mr. Fallon stated that Kodak had no interest in settling . . . on the basis of a conventional royalty arrangement. Mr. Fallon also informed Mr. McCune that Kodak's position continued to be that settlement should involve no payment to Polaroid other than a payment commensurate with the value of the litigation," Kerr reported.[42] Apparently, nothing had changed in Fallon's view of the case since his original meetings with Land in 1974, almost two years before Kodak released its instant photography system. Kerr reminded Judge Zobel that "this was [also] the position taken by Mr. Greenlaw on behalf of Kodak at the settlement conference before the Court in January."[43]

Carr, for some reason, felt the need to question the accuracy of Kerr's version of the conversation. Carr wrote to the judge on September 23 advising her that he had consulted with his client about the substance of Kerr's letter. "I have . . . been advised that in the parties' conference there was no such discussion of a 'reasonable' royalty arrangement. . . . [S]uch a characterization of Polaroid's royalty proposal is entirely that of Mr. Kerr or Polaroid," Carr contended.[44] As Carr explained it, in Kodak's view the royalty proposed by Polaroid would exceed any amount Polaroid could expect to recover if it won the entire litigation. "I am impelled to make this correction by the implication in Mr. Kerr's letter that Kodak had refused to settle on a 'reasonable' basis."[45]

There's little doubt that Carr's letter raised the Scottish ire in Kerr, since it effectively questioned the veracity of his report to the court. This was a matter of his integrity before Judge Zobel, and thus of supreme importance to him. Given that the dispute came down to a mere difference in perception between the parties of what could be characterized as "reasonable," Carr certainly could have crafted his letter differently, avoiding the suggestion that Kerr had somehow misrepresented the situation. But, for whatever reason, he had chosen not to. Kerr shot back on September 24 to clarify the matter, and to defend his reputation:

> Because of Mr. Carr's statement that my letter of September 18, 1981 implies "that Kodak has refused to settle on a reasonable basis," I see no alternative but to advise the Court that Polaroid is willing to settle, for the past and for the future, at a royalty rate of 5% of Kodak's net sales of infringing cameras and films. I submit that this rate cannot properly be claimed as anything other than "reasonable," and that it is not, as Mr. Carr suggests, "significantly in excess" of any amounts Polaroid could recover if it prevailed in this litigation.[46]

Carr was clearly shaken by Kerr's letter disclosing the details of the settlement discussions, in particular the specific royalty rate that McCune had offered to Fallon. Carr responded to Kerr's letter the very same day and told Judge Zobel that he preferred "to discontinue this correspondence."[47] Carr offered to discuss the matter "on an off-the-record basis, if you desire." Perhaps he had fallen directly into a trap laid by Kerr; Judge Zobel now knew exactly what had been discussed between McCune and Fallon and that Kodak had turned down the offer and had not even made a counteroffer. Perhaps Carr had never been told of the substance of the conversation between Fallon and McCune. Worst of all, Carr's desire to keep this episode confidential was for naught, as Kerr's August 18 letter reporting on Polaroid's "willingness to explore the possibility of settlement of this case on a fair and reasonable basis" wound up on the court's public docket, and within days was discovered and published by the *Boston Globe*.[48] Now, in addition to Judge Zobel, the world knew.

That marked the end of any attempt at settlement. It was only eleven days until the trial, and with both companies hurtling toward a risky collision in the courtroom, Polaroid had blinked, but Kodak had ignored the opportunity. The paradox seemed to be that Kodak's counsel was doing everything possible to postpone the trial, while their client showed no real interest in settling the case, which understandably perplexed Polaroid and its counsel. Settlement now seemed, after all these years and attempts, to be a mirage.

Whatever factors went into Kodak's intransigence, it apparently was in denial about the reality that it needed to win on each and every patent in order to prevail. Winning on some but losing on others was not going to result in a positive outcome for the company. Negotiating some kind of royalty with Polaroid would have legitimized Kodak's presence in the instant photography field and allowed it to compete fully for market share without any fear of throwing additional resources into an area of its business plagued by legal uncertainty. The royalty could have been factored into its financial plan as a normal cost of doing business, a regular occurrence when companies license technologies from third parties.

Kodak also failed to appreciate the public relations impact. To some extent, the decision to enter the instant photography field had already damaged Kodak's relationship with its industrial customers in the conventional photofinishing business, none of whom were exactly enthusiastic about a technology that severely impacted their revenues.[49] But what about Kodak's image generally? Given its status as the preeminent photography company in the world, how would it look if a court adjudged it to have ripped off the intellectual property of one of its customers and competitors? And although it was a long shot in Kodak's view, what if it lost and was forced to abandon the business altogether? All in all, these were indeed huge risks, and it would seem that at least exploring a negotiated settlement would have been the prudent thing to do.

Perhaps Kodak was completely confident in the merits of its case, especially after its success on the Trap patent motion.[50] Notwithstanding Carr's opinion letters declaring each of the remaining Polaroid patents invalid, not infringed, or both, it is hard to understand how Fallon could have been so convinced that Kodak would successfully defeat all eleven of the remaining patents during a trial held on Polaroid's home turf, in which its formidable founder might be an ominously effective star witness. And even if Fallon assumed that Carr would be able to knock out *most* of Polaroid's patents, it would have been virtually impossible for him to calculate before the fact that the damages from infringing the surviving patents would be significantly less than what a negotiated royalty might have amounted to. In the end, perhaps it was just Kodak's isolation and arrogance, catalyzed by its resentment of Land and his company, which impelled it to risk all in a trial.[51]

By moving ahead, what Kodak was certain to get was an Edwin Land committed to fighting and to winning. He remained resolute in his belief in the importance of the patent system to the advancement of entrepreneurial science in general and to the company he had built upon it in particular. On the other hand, Polaroid's new generation of management was more pragmatic and perhaps more fully cognizant of the potentially disastrous consequences for Polaroid of a defeat in the courtroom. Better to negotiate a royalty and let competitors into the instant photography business, rather than have its patent portfolio shot to pieces. This was an eminently sensible position and, the closer the trial date, the more it seemed to take hold in Polaroid's Cambridge headquarters. At this point, even Kerr and Schwartz were concerned about the outcome and would have been happy to avoid having to risk everything in the courtroom. They were, despite their confidence, realistic about the true risks of a trial.[52] But, like the tango, it takes two to settle and the photography giant from Rochester was just not willing to dance. Kodak did not seem to share Polaroid's circumspection. Instead, it disregarded the lifeline Polaroid had offered and, seemingly eager to take the risk, took both companies over the precipice.

CHAPTER 20

LAND'S DAY IN COURT

Monday, October 5, 1981, was a sunny, cool, and crisp New England fall day. The beloved Red Sox of sports-obsessed Boston had just completed their sixty-third straight season without winning a World Series, but the basketball Celtics, led by Larry Bird, were preparing to start another campaign as reigning NBA champions. Ronald Reagan, the third president to work in the Oval Office since the *Polaroid v. Kodak* lawsuit began, and the sixth president to serve since Land first showed a prototype of SX-70 to his colleagues at Kodak in 1968, was only in his eighth month as the fortieth American president. He had narrowly survived an assassination attempt in March, had passed a twenty-five percent tax cut through Congress in July, and had fired 12,000 federal air traffic controllers in August when they struck in violation of a court order. His first Supreme Court nominee, Sandra Day O'Connor, had been sworn in as the court's first female justice just ten days earlier.

Despite the nip in the air, Edwin H. Land emerged without a topcoat from a nondescript Polaroid company van and walked up the steps and into the federal courthouse wearing his customary three-piece suit.[1] The moment had finally come for Polaroid to have its day in court. The trial was expected "to be one of the toughest corporate courtroom battles of the 1980s."[2] It had become Land's personal crusade. He was angry with Kodak not just because he believed it had wrongfully appropriated Polaroid's technology but also because it sought to defend its conduct by denigrating the work Land and his Polaroid colleagues had done.[3] (See Fig. 20-1.)

The scene was set to open in courtroom three of the Boston federal courthouse. The case had been reduced to ten patents a few days earlier when Polaroid, at a pretrial hearing on September 28, announced its intention to withdraw its claims relating to the Crank patent.[4] This decision was made based as much on the fact that Kodak had ceased manufacturing the cameras that employed this device as on the sober reality that Polaroid had little faith in the viability of this particular patent and didn't want to taint the strength of the rest of its case by pursuing a patent that might have insurmountable flaws.[5]

The trial was a monumental logistical and intellectual effort that was carried out with superlative professionalism by both law firms. Nothing was overlooked or taken for granted. Kerr had issued a memorandum instructing all his witnesses and courtroom attendees to conduct themselves "quietly and with dignity," to "remain impassive at all times," and to "not show any sign of excitement if Polaroid appears to make a point, and do not show concern or distress if a point appears to have been made against Polaroid."[6] Among other things, reading materials, dark glasses, gum, and candy or eating of any kind were prohibited.

Judge Zobel had moved the case steadily through the end of discovery and to trial with the same decisiveness and fearlessness she had already exhibited in other cases that had come before her. Most notable was a dispute concerning the strike of Boston school bus drivers the previous year, which she grappled with by sending several drivers to jail for contempt of court while declaring that the strike "was akin to anarchy and [it] cannot and will not be tolerated."[7] Both sides had learned to not test her patience.

As sole arbiter of the case, Judge Zobel had little, if any, technical education or background that would be useful in patent litigation. A science columnist for the *New York Times* predicted that the trial would "be a difficult one" for her, given that the "technical questions to be answered are horrendously complex."[8] But both judges and juries were (and still are) expected to render a verdict despite having little or no previous knowledge of the technology involved. It is up to the lawyers to explain that technology in terms a reasonably intelligent person can understand, while avoiding deluging the court with unnecessary and confusing detail. Such is the challenge and the true art of the patent litigator. Judge Zobel entered the courtroom precisely at nine a.m., a diminutive figure with a large white bow draped over the top of her black robe. She faced a formidable congregation of lawyers, company executives, and reporters, each of them with different expectations and concerns.

William K. Kerr knew this would be not only his last but perhaps his most monumental appearance as a litigator. Kerr's part in the trial was to be limited since, once he had presented Polaroid's opening statement and led Land through his direct testimony, he had agreed to turn over the reins of Polaroid's legal team to Schwartz, something that was initially viewed with apprehension by Polaroid management.[9] Given the magnitude of the case, such an arrangement was unusual to say the least. Yet, Kerr had long ago planned his retirement, while Schwartz, who had been increasingly setting the strategy, was coming into his prime as a big-time litigator.

For Francis T. Carr, this trial was the culmination of more than a decade's effort to help Kodak steer a course through the Polaroid patent blockade and into the instant photography market. He had been enlisted to review Polaroid's patents at the inception of Kodak's development program in 1969.[10] He had never wavered in his conviction that Kodak would prevail on the merits, even as he knew that hundreds of millions of dollars, as well as his own professional reputation, were on the line.[11]

Land's reputation as one of the most innovative figures in technology was also on the line. He would, at last, be able to fight for the vindication of his life's work and the very survival of the company he had founded, built, and led for more than forty years. Despite his determination and months of preparation, the gravity of this moment was clearly weighing on the seventy-two-year-old Land. When approached by an admiring spectator in the courtroom before the proceedings began, Land rebuffed him, nervously blurting out, "This is no show. I'm not allowed to talk to you."[12]

After years of legal skirmishing, Kerr and Carr, in their opening statements, would finally present to Judge Zobel the basic themes of their respective cases. Then, for many weeks to come, a parade of witnesses for both sides would testify. Polaroid, the plaintiff, was to go first. The judge, having settled onto her perch, nodded to the court clerk, who called the case: "Polaroid Corporation versus Eastman Kodak Company, Civil Action 76-1634-Z." Following some brief introductions to the court of the lawyers sitting at the respective counsel tables, Judge Zobel surprised everyone with an announcement. "Before we begin," the judge said, "I should tell you that I have a mother-in-law problem. . . . It is not the sort Ann Landers would be interested in. But I learned yesterday that my mother-in-law owns a thousand shares of Kodak stock. . . . She has been a stockholder of Kodak for some 50 years." Before the lawyers and company executives could even digest this news, she added, "I do not regard this as disqualifying, and at this point will proceed to hear the case."[13]

Kerr, whose wife and daughters were present in the courtroom for this special occasion, got to his feet. Speaking from an outline printed double-spaced on yellow paper, he began by characterizing "this action for patent infringement . . . as conspicuous and noteworthy for at least two reasons: First," Kerr explained, "the unusually large number of patents involved and, second, the background and history of the parties in the field to which the litigation relates." Previewing the strategy he and Schwartz had decided on years earlier during discovery, Kerr began to articulate the context of Polaroid's case: the "big picture" story of the history of the technology and the relationship between the parties as it evolved in the commercial marketplace. He went on:

> The evidence will demonstrate that instant photography, which is a shorthand way of describing broadly the field of technology involved here, was created by Polaroid, largely through the inventive and pioneering efforts of Dr. Edwin H. Land. The evidence will demonstrate, also, that instant photography was a radical and revolutionary departure from conventional amateur photography. Dr. Land will testify about the origin of instant photography and the history of its development at Polaroid. Dr. Land's testimony will establish that instant photography can be divided conveniently into two categories: one-step photography, which was introduced to the world by Polaroid in the late 1940s; and absolute one-step photography, which was introduced to the world by Polaroid in 1972 and exemplified by Polaroid's SX-70 system. In one-step photography, the film unit had to be peeled apart after development of the picture was completed. In absolute one-step photography, the camera delivers to the photographer a complete dry film unit containing the image photographed. The photographer has nothing more to do except to watch the picture as it emerges and materializes in the film unit. It is absolute one-step photography which is the subject of this infringement action before your honor.
>
> The evidence will show that the relationship between Polaroid and Eastman Kodak, as far as instant photography is concerned, goes back to the late 1940s. At that time, the relationship was one in which Kodak manufactured and supplied to Polaroid a conventional photographic negative. Polaroid added this negative to its positive to produce Polaroid's one-step film.
>
> As time passed, Kodak began to recognize one-step photography as an attractive segment of the field of amateur photography. Pretrial discovery has established that this recognition continued to grow

> within the Kodak organization and that Kodak eventually decided that it was necessary for Kodak to enter the domain of instant photography, the domain created and pioneered by Polaroid.[14]

Kerr then proceeded to describe to the court in "a quiet, steady voice" the rest of the "Kodak Story" Schwartz had focused on so intensely during discovery—its effort to develop a peel-apart film for Polaroid cameras that was abandoned, after the expenditure of $94 million, as "essentially . . . obsolete" once SX-70 was introduced and the ups and downs of its attempts to develop an integral, "absolute one-step" system to compete with SX-70.[15] By quoting directly from Kodak internal documents, Kerr underlined how even Kodak's engineers hailed SX-70 "as a masterpiece of engineering" that "set the standard and level of expectation for all future products of similar type," while conceding that the product it had under development at that moment "had lost its impact from a marketing point of view" and would be "no more than equal . . . [and in] some ways . . . less than equal" to Polaroid's products. After years of further work to come up with a more competitive product, including the "continuous study of Polaroid products, Polaroid patents in suit, other Polaroid patents and prior art, Kodak arrived at a system which capitalized upon and bore a marked resemblance to Polaroid's commercial SX-70 products and which, morc importantly to thc issues here before your Honor, infringed the ten Polaroid patents in suit."[16]

Kerr acknowledged that there were differences between the Kodak and Polaroid commercial products. "But notwithstanding these differences," he continued, "Kodak, with its large resources, and after an intensive research program came on to the market with a system that is a remarkable parallel to the system commercialized by Polaroid and which parallels the system described and claimed in Polaroid's patents."[17] Kerr then made his ultimate accusation in words that would be reprinted in newspapers like the *New York Times* the next day. He told the court that during its research, "Kodak had unsuccessfully tried other non-infringing approaches, but Kodak repeatedly bumped its head against a series of stone walls and was ultimately driven to appropriate Polaroid's patented inventions."[18]

In his conclusion, Kerr stressed how Kodak's misappropriation of Polaroid's technology "strikes at the very heart of Polaroid's business. On the other hand, Kodak is and has been for years the dominant force worldwide in amateur photography," he explained. "The absolute one-step photography field so vital to Polaroid is only a very small part of Kodak's overall business. . . . It is for these reasons that we have been and are eager to prove to this Court the soundness of Polaroid's position and to convince the Court that Polaroid's patents are valid, enforceable, and infringed."[19]

When Carr took the floor for Kodak, he decided to digress from his prepared presentation and to, instead, according to one observer, "ravage Polaroid's case . . . [by] jump[ing] on Kerr" for his characterization of the dispute:[20]

> May it please the Court, I believe in the very end of Mr. Kerr's remarks, he put his finger on what the case is all about as far as Polaroid is concerned. He said that this involve[s] an infringing invasion of Polaroid's market. This is hardly the case. And I didn't know that there was a Polaroid market . . . or a Kodak market, [or] anybody else's market. The market is there for anyone who [can] manufacture a product properly, legally, and can sell it. That's the market. Polaroid views this as a domain which no one can enter, or enter certainly at terrible risk.[21]

With that said, Carr quickly returned to his script and began to outline the strategy Kodak had followed throughout discovery and clearly intended to pursue at trial. Its case was focused on the technical issues of each patent in suit individually, on its own terms within the confines of its own "specification," the portion of the patent document that explains the invention, how it works, and how it differs from what has come before it. From Kodak's perspective, the trial was more about judging the merits of each invention in isolation against the standards of patentability or infringement and less about the overall story surrounding the development of the technology in question. For the most part, Kodak considered this "real world" context an irrelevant diversion. In each and every case, with respect to each and every patent asserted against it, Kodak believed that it had the best of the argument. This confidence was based on the theories originally developed in Carr's opinion letters and then honed over the years during patent proceedings abroad and through discovery in the case. Now, Carr's strategy was to concentrate the court's attention on the particulars of the patents and the prior art, while diverting it away, for the most part, from the personalities or the circumstances involved in the underlying story.

To that end, Carr set out for the judge the framework for Kodak's approach to the case, "scoffing," as one journalist characterized it, at Polaroid's attitude:[22]

> I thought we would look first at what the case is not about. Kodak has not and will not contest the fact that Polaroid has been a very innovative company over the years, and Dr. Land assuredly has been at the forefront of that innovation. On the other hand, Polaroid cannot challenge the fact that over these years, Polaroid and Dr. Land have been vastly rewarded financially by acclaim and otherwise for that innovation. To the extent they have had an absolute monopoly in the field from 1946 to 1976, when Kodak entered the market.
>
> But the focus of this case is not, in our view, upon the pioneering of instant photography or the question [of] who should be the perpetual pioneer in the field. . . . I submit that this is a straightforward patent case which now involves ten patents, where the questions simply are: Are the patents valid? And, if they are, are they infringed and are they enforceable, one by one, an invention and a patent at a time. That is this lawsuit.[23]

Carr explained how Kodak had a differing view of the standard for judging inventions than that of its adversary and proceeded to set forth that difference in language reminiscent of that used by Judge Zobel in granting summary judgment on the Trap patent. "Polaroid believes that patentable invention is involved in adapting to a color system known solutions to the same problems arising in black and white systems," Carr explained. "That's wrong," he declared. "Polaroid believes that patentable invention is involved in adapting to an integral, non peel-apart system known solutions to the same problems arising in peel-apart systems," he continued. "That's wrong," Carr repeated. "Polaroid believes that patentable invention resides in appreciation of some advantage in an otherwise old structure or process." Again Carr intoned, "That's wrong."[24]

Carr then laid out the philosophical argument behind Kodak's case—that the framers of the Constitution had sanctioned patent protection as conferring exclusivity for only a limited period of time. In his view, Polaroid had abused the system to perpetuate its monopoly in instant photography beyond what the law allowed. "Polaroid's penchant for patenting every slight improvement, rather than only true innovations, had led to a vast inventory of look-alike patents, ponderous in size, bewildering in complexity, but differing only in trivial and predictable respects from each other and from the prior art. The ten patents in suit fall squarely in this category." In conclusion, Carr again did his best to focus the court on Kodak's narrow view of the case:

> The real question here is whether Kodak achieved its objectives by means validly patented by Polaroid or not. That is where we are. That is where we start, and I submit to you that is where we will end. Are these patents valid? If they are valid, are they infringed by Kodak? If infringed, are they enforceable? That is what the evidence should focus upon, and not the myriad of secondary factors which will occupy much of the time in this case. Thank you.[25]

His presentation complete, Carr gathered his notes and returned to Kodak's counsel table. It was time for the trial to begin. "If you would please call Dr. Land forward to be sworn, we'll proceed," declared Judge Zobel.[26] From his aisle seat in the audience, Land moved forward to the witness chair. As he watched Polaroid's "guiding light" approach the front of the courtroom, one journalist observed that Land did not "look his age. . . . His hair is still full and dark, his frame heavy but without an old man's paunch."[27] Nor did Land convey the "manner of a corporate chieftain . . . walking almost diffidently to the stand." Land walked in almost a shuffle, gliding rather than striding, to the front of the courtroom. After the oath was administered, Kerr began Land's direct examination.

In trial procedure, an attorney is not permitted to "lead" his or her witness during direct testimony. That is, the questions are not supposed to suggest responses. For example, "What did you have for breakfast?" would be proper form. But "You had eggs for breakfast, didn't you?" would not. Still, despite this constraint, nothing had been left to chance, and Land's testimony was anything but a spontaneous affair. Land's direct examination had been prepared by one of Kerr's associates and reviewed and refined over the months leading up to trial. The outline itself was never shown to Land because, under the rules, doing so would have made it subject to being turned over to Kodak. But the direct examination was rehearsed over and over by having either a young lawyer or Land's assistant, Nan Chequer, ask him the questions in sequence so that he could formulate the precise answers he would want to give on the witness stand. The object was to help him frame those answers within the specific issues of the lawsuit and to keep the material focused, avoiding irrelevant tangents.

To this end, a draft of Land's direct examination had been provided to Chequer as early as July 27. She conducted countless preparation sessions with Land, and his attorneys met with them for at least eight days prior to trial to review all of the material. Even after these review sessions, when

Kerr provided to Chequer a "final" draft of the questions he intended to ask on September 29, just a week before the trial commenced, he noted, "I am confident that there will be additions, deletions and emendation."[28] Indeed, the direct examination was reviewed and polished by his associate late into the night before the trial began, and reviewed for the final time at Kerr's daily five a.m. pretrial preparation session.

Kerr began with a brief review of Land's curriculum vitae, an exercise that is standard procedure but was clearly unnecessary in this particular case, and thus promptly disposed of. As illustrators sketched Land's face and the courtroom scene for use on television broadcasts, Kerr began the exacting process of having Land lay out the basics of Polaroid's case: the story of the development of instant photography, the history of Polaroid's relationship with Kodak, and the details involving the patents in suit on which Land was the named inventor. He quickly got into the heart of the matter and asked: "Is the term One-Step photography a term which is familiar to you, Dr. Land?" "Yes," answered the witness. "Would you please define for the Court what this term means to you?" Kerr followed. "It means a system of photography in which the photographer, having taken the picture, has the picture made available to him with one single mechanical photochemical step, so that he can look at it promptly after taking it," explained Land. "Is this a term which you originated?" asked Kerr. "Yes," Land stated for the record.[29] Land seemed a bit uneasy at first, not yet having settled into this strange and very public setting. Indeed, as reported by another journalist, his testimony at first was delivered "in a quiet voice that was difficult to hear in the back of [the] Courtroom."[30] Even the judge, who was sitting close by, had to advise him, "Dr. Land, I have trouble hearing you."[31]

Kerr's next objective was to demonstrate just how remarkable an advance instant photography had been. He had Land contrast his one-step process with the complicated multistep processes used in conventional photography to take exposed film from a camera and then to produce a picture. In conventional photography, these steps were all done in a carefully controlled environment with respect to light and temperature, using a series of specialized chemicals to develop, rinse, and stabilize both the negative and later, in a separate process, the positive print. As Land described it, in his one-step process, the picture was taken and the photograph developed in the photographer's hands.

Judge Zobel listened intently and showed no reluctance to intervene in the questioning. "I'm sorry," said the judge early in Land's explanation of how an image is formed on a film negative by exposure to light. "If I am to understand this, I don't yet. You'll have to explain it in greater detail than that," she said.[32] Quickly becoming more comfortable on the witness stand as his role evolved into, in effect, the court's personal tutor on his favorite subject, Land obliged the judge, "speaking slowly and quietly" to explain the process to her satisfaction.[33] To tie up this line of questions, Kerr asked Land whether it would be "accurate to say that from the standpoint of the photographer, one-step photography replaces the long sequence of steps which you have described in connection with conventional black and white photography?" "Yes," replied Land.[34]

After describing how the peel-apart process of Polaroid's first commercial instant photography system worked, Land, as the *Boston Globe* described it, "broke the law in federal court. . . . He took a picture."[35] "I'll waive the ban against cameras in the courtroom," Judge Zobel announced as Kerr handed Land a 1948 vintage Polaroid Model 95 camera and a roll of black-and-white film.[36] Land loaded the camera, took a photograph of Kerr, and then demonstrated the process for manually pulling the combined negative and positive out of the camera, timing the processing, peeling apart the two elements to expose the print, and finally treating the print with a chemical swab to stabilize the image before mounting it on a cardboard frame in order to prevent it from curling. Land then performed the same process using the Polacolor film first introduced in 1963. (See Fig. 20-2.)

Minutes later, when Land demonstrated an SX-70 camera system, with the film automatically ejecting from the camera after exposure to develop in the light with no further manipulation by the photographer—no timing or peeling apart, no coating with a chemical stabilizer, no mounting of the print—the evolution and dramatic improvement from peel-apart OneStep photography to the Absolute OneStep technology of the integral SX-70 system (and the patents in suit) was apparent. It was indeed the instant photography system that Land had envisioned as an "idealized camera and film" at the outset but in reality had taken decades of intensive investigation to realize.[37] Land explained how he had coined the term "Absolute One-Step Photography" to describe "the camera and the film that enabled the photographer merely to press the button and to have the finished, dry package emitted from the camera; not sticky, not to be peeled apart, just to be watched as the image materialized . . . it was the camera and film which achieved my original purpose of letting the photographer merely look and then have to do nothing else . . . look at the subject, look through the finder at the subject, compose it and press the button and then have a photograph."[38]

Kerr then led Land through a basic explanation of the chemical processes involved in making a color photograph in conventional photography, having him explain that "it's more difficult to make a colored photo in the conventional fashion than it is to make a black and white photo, and requires many more processing steps."[39] He then briefly described the work done at Polaroid to come up with a color system for instant photography. Land recalled that he had started working on processes employing the basic color coupler chemistry used in conventional color film but that Howard Rogers, who was simultaneously working on the problem, invented a new class of photographic chemicals called dye developers, which proved to "have tremendous appeal" in the one-step color system.

Land explained that, after Rogers' discovery of the dye developer, "the most startling . . . realization [was] that although it seemed unlikely, you could indeed control all the three layers of . . . photosensitive emulsion and the three layers of dyes so even though the dyes were passing through each other and passing through all the emulsions, you nevertheless could have each layer do its job in the way it ought to be done. . . . What you'd expect a priori—and I think every good person did—it would be extraordinarily difficult to have all those reactions going on at the same time and the transfer process going on at the same time, or shortly after each other, without each one modifying and deforming what was happening in the next layer."[40] Since one of Rogers' basic dye developer patents was asserted against Kodak, Rogers himself would provide the expert testimony regarding it later in the trial. But Land had set the stage by describing the historical context for the work that had led to Rogers' invention.

Having thus established through Land some of the basic groundwork for Polaroid's case, Kerr then moved his witness into the subject matters of his patents in suit. First up was Land's eponymous L-Coat patent, which covered a technique for automatically stabilizing the image within the integral film unit with no intervention on the part of the photographer. Land began by explaining that stabilization of the image is the third basic operation in making a photograph. Step one is the exposure, in which you form the latent image in a silver halide layer in the negative. Step two is the development, which includes all of the sub-steps involved in "converting the latent image either to a black and white or a colored image." "The third step of photography," explained Land, "is stopping all those processes, preventing new processes from occurring so that the image you then have at that stage will stay there for many years, you hope forever." Land went on to testify about how stabilization is "a very special and unique problem in one-step photography because you don't have either the equipment, the baths or the trays, and you must generate a whole new technique for stopping the processes and keeping the image where it was. . . . And for each kind of film," Land went on to emphasize, "sepia, black and white, and color, techniques involved were markedly different."[41]

Land described for the judge how the essential method for controlling an instant photographic process is to control the chemical environment within the film unit, in terms of the level of acidity. The term "pH" is a standard method for measuring and describing the relative position of a substance on the scale between acidity on one end and alkalinity on the other. The pH scale ranges from one to fourteen, with one being the most acidic and fourteen the most alkaline. Water is considered to be neutral at seven. In instant photography, adjusting the pH controls the process.

He explained how his L-Coat patent taught "the technique, the solving of problems of how to maintain the system alkaline long enough for the negative silver halide grains to be developed, for the dye to be controlled in the emulsion, for the dye to migrate to the positive player, for the three dyes to dye the positive layer, and then and only then to have the alkalinity drop rapidly to start to slow up the process to a point where it was safe and the image will not be damaged or altered by further dye transfer or by decomposition of the dyes or by return of the dyes from the positive back to the negative."[42] The invention of his patent, Land testified, was to accomplish this using "not an ordinary acid but . . . a high polymeric acid . . . [which is essentially] a plastic with acidic properties." In addition, his invention taught the way this polymeric acid layer effectively prevented the alkali from doing any harm to the image: "In a sense it purges the system of the alkali by sequestering it outside of the system and then leaving the system free both of the alkaline and of the salts that would be formed if you neutralized it with ordinary acid."[43]

Next, Land told the court how his patent in suit also disclosed "the location in which that plastic layer is placed in the system." Although Land discovered two possible locations, this particular patent covered the placement of the L-Coat in the photosensitive element "adjacent the support, i.e., between the support and the innermost layer of dye developer," whether that film was a peel-apart unit, like Polacolor, or an integral one, like SX-70 or Kodak's PR-10. Finally, Land explained how a "spacer layer" was necessary to help time the action of the polymeric acid layer.[44] His testimony for this momentous first day of trial wound up with Land standing in front of the courtroom, using blown-up charts on large easels, to demonstrate how the location of the polymeric acid layer, as depicted in the drawings of film units in his patent, was the same as in Kodak's

PR-10 film. The chart that depicted PR-10's structure was a schematic diagram of Kodak's film, prepared and published by Bunny Hanson, "one of Kodak's distinguished scientists," according to Land, and its former head of research.[45] (See Figs. 20-3 and 20-4.)

As the second day of Land's direct testimony began, he reviewed in greater detail the differences between the problems of stabilization in each of the one-step films—sepia and black-and-white—and the methods employed to solve them. On this second day, Land was clearly feeling more comfortable, and his answers were long and detailed, prompted by short cues in Kerr's questions. The very public setting was proving not to be the kind of humiliating or unpleasant experience he may have dreaded. In fact, Land was being accorded all of the respect that he had become accustomed to. A Rochester reporter even noted that in the courtroom "he was called Dr. Land with deference not granted to other witnesses with PhD degrees."[46]

Significantly, Land never lost sight of the fact that Judge Zobel was his audience, and each time she interrupted to ask for some clarification, he patiently reviewed the material until she was satisfied that she understood his testimony. Whenever the judge showed initiative to learn more than might have been intended for his direct testimony, Land obliged. For example, Land testified that the stabilization system used for sepia did not work for black-and-white film even though "one would have thought so," and that swabbing the picture with a print coater appeared to be necessary.[47] The judge asked, "How does the print coater work? How does the print coater perform the stabilizing function? Or doesn't it matter?" Land pounced on his student's curiosity and replied, "It matters because you're interested" and "because we're showing . . . that stabilizing is one-third of the fundamental process. It's peculiar to each process, part and parcel of the personality of each process," Land declared as he proceeded to explain in greater detail how stabilization worked in black-and-white. Minutes later, when Judge Zobel asked another question concerning the operation of his invention, Land complimented her: "I think you have a shrewd insight and I'll have to give an intuitive answer."[48] Clearly, a productive teacher and pupil relationship was forming between the two of them. Polaroid's lawyers were delighted as they observed Land performing effectively in a role he relished, with Judge Zobel handling her "intense education . . . from Dr. Land . . . with great tact and fine humor."[49]

As he continued to tell the story of his L-Coat invention, Land explained how the original plan was to provide a print coater with the color film Polaroid was set to introduce in 1963. He admitted that he was dissatisfied with the prospect of having to do that and so, in the summer of 1962, "dropped everything else I was doing . . . and with a small group of people we went at it day and night . . . [until] I arrived at the theories that we've been talking about today."[50] Land explained that the L-Coat invention was incorporated into the new color peel-apart film, although in a different location—within the positive—as covered in a related patent. This last-moment breakthrough avoided the need for a print coater when Polacolor was finally released to the public. Land also testified that, following his discovery, he immediately shared the information over the telephone with one or two of the key Kodak scientists with whom he had been working for more than a decade on adapting Polaroid's technology to Kodak's film production facilities, in particular Henry Yutzy. Land made the call, he explained, out of concern that, "in spite of the suddenness and quickness with which this had become possible to do and the lateness in the program, that we wouldn't be able to get into production in time with it. . . . We were sharing our enthusiasm with them," explained Land, clearly harkening back to the days when his relationship with Kodak had a very different feel to it.[51]

When Kerr handed Land a memorandum written by Yutzy referring to Land's phone call, Carr objected on the grounds that this was an internal Kodak memorandum and that since Yutzy had "been dead for quite some time," it could not be used to establish what Yutzy may have said, since the Kodak executive was not available to testify about it.[52] Nonetheless, Judge Zobel allowed Land to review the document to refresh his recollection about what he had told Yutzy, and Land went on to testify that he had told him that the new polymeric acid layer produced "prints [that] were more stable than the ones we got by swabbing and . . . that I thought the swabbing operation could be omitted." Despite all the testimony that Land provided on this second day, and the marginal relevance the phone call represented, the newspapers seemed to pick up on this particular interchange as being the most dramatic of the day.[53]

To complete his testimony on the L-Coat patent, Kerr had Land carefully, yet dismissively, distinguish each of the prior art references cited by Kodak as either teaching his invention or rendering it obvious to one skilled in the art. One by one, he explained how the prior art taught the use of different materials and techniques or applied only to either sepia or black-and-white systems. He did so from the unique perspective of someone who had struggled mightily to overcome what might seem like minor obstacles to second-guessers. When Kerr finally asked whether Land believed that any of the references cited by Kodak "teaches the invention" of his L-Coat patent, Land promptly responded, "I think they do not."[54]

Then, leaving the witness box to use a large chart displayed for the court, Land demonstrated how, in his opinion, the process used in Kodak's PR-10 film to stabilize its image infringed the process taught and claimed in his L-Coat patent. (See Figs. 20-5 and 20-6.)

Kerr then turned the examination to the next major subject, Land's Symmetrical Supports patent. This was the patent that Schwartz had originally left off the list of patents to assert against Kodak because he, and lawyers in Polaroid's patent department, had believed the invention claimed would not stand up to the rigorous scrutiny a vigorous litigation would subject it to. It had reluctantly been added to the lawsuit only after Kodak had included it in its counterclaim seeking to have a host of unasserted Polaroid patents held invalid and unenforceable. Of course, circumstances had changed mightily since Schwartz had made that original assessment. Given Land's lack of involvement in the litigation at that point, he did not have the benefit of the inventor's input, especially his explanation of the merits of the invention. But now the Polaroid patriarch was determined and, in fact, eager to explain to all just how improbable and, in legal terms, how novel and nonobvious his concept truly was. In response to Kerr's request that he explain his invention, Land delivered the description he had settled on over the preceding months of preparation, all the while keeping in mind the legal and technical arguments Kodak had leveled against it:

> The invention has to do with solving the problem presented by the fact that most of the structure that . . . forms [an integral] film unit with plastic exterior surfaces . . . will curl or deform seriously after a length of time, varying from hours to weeks or sometimes even months. The invention is to use external sheets which have good mechanical stability. But most importantly have extremely low permeability to water and water vapor. Consequently, with a very low water vapor transmission through the external sheets, the water content from the pod that was spread between them is retained for days or weeks, a typical period of drying out being about two weeks. And that very slow permeation of the water vapor and transmission of the water vapor is a necessary condition for keeping the film . . . unit flat both promptly and for extended periods in nearly all environments under most conditions. And that is true for years.[55]

Land explained how the prevailing wisdom at the time he did his work was counter to his discovery that relatively impermeable supports should be used on both sides of the film unit. To the contrary, it was believed that a film unit needed to dry out quickly, both to minimize the potential of curling and to create a stable system within it that would not affect the image. But Land testified how he "had tremendous faith in the power of the L-Coat and its ability to keep the system stable," so they tried a strong and impermeable plastic material, polyester, on both sides of the film unit. He recalled with apparent satisfaction how they were rewarded "with very prompt success so the pictures made right from the beginning, to the present day, I'd say, are . . . as flat as anything anyone has ever seen in the photographic art over many decades."[56]

Having described his invention, and the break with conventional thinking that was necessary to achieve it, Land was then once again led through the prior art references that Kodak had cited as teaching the Symmetrical Supports invention. As he had done with the L-Coat patent, Land patiently explained each one away. One was concerned with peel-apart film units and taught nothing with respect to the construction of a flat integral film unit.[57] Others taught the use of cellulose acetate and triacetate, more conventional and relatively permeable plastic materials that had long been used as the backing for conventional as well as peel-apart instant film but demonstrated a tendency to produce severe curl when used in integral film units. According to Land, each and every prior art reference was flawed in some way, and none disclosed his invention or rendered it obvious. Some of the patents cited by Kodak were actually in his name. One covered what Land admitted was "the heart of the structure" for what became the SX-70 film unit, but Land explained how this earlier patent did not address the problem of achieving a flat film unit, and "did not describe or recognize any special benefits or advantages which would accrue from the use of film supports having the characteristics specified in . . . [the Symmetrical Supports] patent in suit."[58]

The examination on this patent continued into the third day of trial, at which point Kerr and Land promptly made the headlines for that day's newspaper accounts of the trial. Kerr questioned Land about his familiarity with Bimat, a system that was developed at Kodak in the mid-1960s for use in aerial and satellite reconnaissance. This system made a sepia negative that would depict the terrain being surveilled and used two sheets of impermeable plastic on large rolls that were brought into contact after exposure and then kept together on another roll. Land testified that he acquired a "general knowledge of the process" as part of his work within the national defense intelligence community. He explained to Judge Zobel that he "was a science advisor on the Science Advisory Committee to Presidents Eisenhower,

Kennedy, Johnson, Nixon and Ford, and also on the Board of Foreign Intelligence Advisors." "A part of my particular responsibility was to advise on the progress and potential of domains to be supported in the photographic and electronic and optical aspects of overhead reconnaissance," Land continued.[59] He emphasized, though, that because of similarities to Polaroid's sepia system, Land "took care not to learn more" about Bimat so as to avoid any potential conflict of interest.[60]

As Land continued explaining at great length about his reticence to acquire knowledge of the Kodak system, Judge Zobel interrupted to ask him if he "wasn't getting a little far afield." She questioned whether his knowledge of Bimat was pertinent to its use as prior art by Kodak. Kerr explained, over Carr's objection, that it was important because Kodak had "contended that Dr. Land got confidential information about Bimat as a result of his activity on these national [security programs] . . . in some unauthorized way."[61] This fact was not really germane to Bimat's relevance as prior art, because as long as the reference was available to the public, it didn't necessarily matter if the inventor was aware of it or not. Kodak had described its Bimat system in brochures used to promote it for sale, and so there was no real question as to whether it qualified as prior art against Land's patent.

But clearly this was a sensitive allegation by Kodak that Land took personally. Perhaps it was raised as just another barb from the Kodak contingent, which held on to its resentment of Land. In any event, Kerr was determined to make his point. Judge Zobel quickly grasped the situation, assured Kerr that she understood Land's testimony, and the direct examination moved on. But it was one of those testy moments that stood out to the casual observer, including the press, from the hours and hours of detailed technical testimony that consumed most of the day. As essentially irrelevant as it may have been, the *New York Times* headline the next day read, "Land Knew of Kodak's Film,"[62] while the *Boston Globe* took a more defensive posture in its headline: "Land: No 'conflict' on Kodak Film."[63]

Land completed his testimony on the Symmetrical Supports patent with another line-by-line comparison of Kodak's PR-10 film with the claims of his patent, testifying that the Kodak product infringed because it met each and every element of his invention as defined by the patent. When Kerr's questioning brought out the fact that Kodak's film used polyester supports of differing thickness on the front and the back of its PR-10 film, Land testified that the thickness "makes no difference in any measurable sense, and perhaps in any sense."[64] Judge Zobel jumped in at this point to clarify this testimony. "Do I understand, Dr. Land, the important part of this . . . [is that the] polyester is not permeable by water?" she asked. "That is right," Land responded. "Thickness doesn't make any difference?" asked the judge. "That is right," Land declared, going on to testify about scientific research done at Kodak on various thicknesses of their film support material, demonstrating that any difference in their relative permeability was "so small that the range would be almost immeasurable."[65] In sum, the point of this testimony was to establish that the term "symmetrical" as used in the informal title given to Land's patent referred to symmetry as measured in impermeability to water vapor as opposed to other physical characteristics such as thickness.

Two down, one to go. Kerr then turned Land's attention to his third and final patent in suit, the Rear Pick patent. This was the Land patent against which Kodak had belatedly filed a summary judgment motion just days before the trial was to start, unsuccessfully seeking either to knock it out of the lawsuit, as it had done with the Campbell Trap patent, or somehow to delay the trial. Following the pattern established in his testimony on the first two patents, Land began by describing, in his own words, the essence of his invention:

> The invention has to do with the mechanism for handling, after exposure, the individual film elements in the cassette of films. The task, after exposure through the front of the cassette, is to advance one and only one of the elements in the cassette, the top one, since the exposure is from the hole in the top of the cassette . . . [and] advance that through a slot in the leading edge of the cassette and into rollers which are running with the circumferential velocity that matches the speed of the film as it is being pushed towards them. So that they continue to draw the film, first, bursting the pod within the unit and then spreading the reagent between the . . . sheets, and then ejecting the finished film out through the aperture in the front of the camera. This must be done as one. . . . The cassette has an opening in the rear corner of the cassette so that the pick, as we call it, can engage the rear corner of the film unit and pull that forward.[66]

Once again, Land left the witness stand to use large charts and then an actual SX-70 film cassette and camera to point out to the judge the various elements of the rear pick mechanism and their operation. After the elements of the patent were explained, Kerr led Land through a description of how the invention was made, with particular focus on all of the other methods for advancing film units out of a cassette that he and his colleagues had tried before he conceived the rear pick. Land described several mechanisms developed at Polaroid but ultimately found unsatisfactory, including devices that pulled film units from the front by small holes in their sides and others based on "the permanent intuition [that] comes from handling a deck of cards, put your thumb on the top . . . [card] and push it forward."[67] From that concept came "a great many proposals and some models of fingers with rubber ends on them that would push the film . . . and some having two rubber rollers pushing on the edge of the film."[68]

In the long run, Land testified, none of these other approaches proved reliable as transport mechanisms, and each presented other issues to complicate the process. Some raised the potential of damage to the pod, while holes in the border of the film unit seemed to have an impact upon the rate at which the film unit dried out. The rear pick arrangement, allowing the use of a film unit without holes, had the slot for the pick located in the rear of the film cassette, the opposite end from the film unit pods. This location also turned out to have the benefit of decreasing the possibility that light could sneak into the film cassette to ruin the film before it was exposed. Land testified how, in contrast to all of the other techniques they had tried, the rear pick proved to be incredibly reliable. He proudly pointed out that it has "taken out billions of pictures, a billion this year, and we never hear of a failure."[69]

Day three of Land's direct examination ended with the start of the now familiar parade through the prior art references cited by Kodak. The exercise spilled over into the next day, with Kerr putting before the witness one after another of the patents Kodak claimed revealed his invention. As he had done for his previous two patents, Land batted each one away, spelling out for the judge the distinctions between each from the mechanism taught and claimed by his patent. Some of the references were other Polaroid patents in the names of other Polaroid researchers, covering alternative approaches tried during the SX-70 development project but rejected as unreliable. They used rollers or rubber fingers or pads to advance a film unit or involved manual as opposed to automated transport mechanisms. Kodak also cited references related to systems other than instant photography. These included devices as disparate as a street vendor's camera that produced postcards or a device for moving computer cards. Once all of the prior art patents were disposed of, Kerr again led Land through the task of matching Kodak's product, this time the EK-6 camera, against the claims of his patent to demonstrate that it did indeed infringe.[70]

Before concluding his direct examination, Kerr asked Land to describe for the court his and Polaroid's relationship with Eastman Kodak through the years. With some evident nostalgia, Land explained how he had developed working relationships with several people in the Kodak organization through governmental and scientific society work prior to 1944, when he became interested in one-step photography. He recalled how, early on in his research, he decided he needed to build the process around a conventional black-and-white photographic emulsion in the negative and, "based on a general friendship and long period of cooperativeness, I was able to bring myself to ask" one of the Kodak scientists he knew, Cyril J. Staud, for a small supply.[71] Staud agreed and "kindly and generously provided the samples of emulsion without knowing what they were for." Once Land's research had progressed to the point where he "had quite handsome looking pictures," he decided it was time to tell Kodak what he was up to, and to see whether it would be willing to manufacture the negative to be incorporated into Polaroid's new film.

That disclosure was made to a group of Kodak scientists led by another longtime colleague, Kenneth Mees. This "amiable and happy occasion," Land explained, led to a long period of cooperation between the two companies with regard to the manufacture by Kodak of negatives for Polaroid film. The supply relationship worked well for both companies. "We were a very good customer as well as friends," Land testified.[72] This joint effort necessitated an extensive exchange of information between the companies. Accordingly, Staud and others at Kodak were aware that Land and his Polaroid team were also working on a color one-step process and even supplied standard Kodak color photographic emulsions, as they had done at the outset with black-and-white.

Once a color one-step product became viable, principally because of Howard Rogers' invention of dye developer chemistry, Land told the court how he conducted a demonstration in 1957 for Yutzy, with whom he had served on the Science Advisory Board to the Air Force. Yutzy "was very enthusiastic," and he said, according to Land, "'that looks commercial.'"[73] According to Land, this was a pivotal event that sparked an even deeper relationship between the two companies. Kodak agreed to manufacture the negative for Polaroid's color film, using Polaroid's new color-imaging technology. The two companies embarked on a five-year cooperative effort to adapt Polaroid's new color dye developer system to Kodak's manufacturing equipment. The venture was celebrated many years later in Polaroid's 1974 annual report, a copy of which was offered into evidence. The report reproduced on its cover the very color print from 1957 that had been shown to Yutzy.

As a result of this new cooperative undertaking, Kodak supplied color negative to Polaroid for more than a decade, starting in 1963 when Polacolor film was released to the public. A contemporaneous Polaroid press release, introduced into evidence, trumpeted both Rogers' invention of the dye developer imaging chemistry and Land's L-Coat, which obviated the necessity of using a print coater to stabilize the picture, as major developments that had made Polacolor possible. Ironically, a Kodak press release of the same era, also introduced by Kerr at this point, acknowledged its cooperation with Polaroid but clearly expressed its lack of interest in competing with its customer: "The company has no current plans, however, for entering the field Polaroid has pioneered."[74]

The final chapter in the Polaroid—Kodak relationship began, Land testified, in 1968 when he wrote to then president of Kodak Wren Gabel, informing him that Polaroid had developed "a new film that we thought was very important and that would have a deep effect on the future of instant photography."[75] This "new film" was what ultimately became SX-70, Land explained. His disclosure led to a series of meetings during which representatives of both companies, including Land and McCune for Polaroid, and Gabel and Louis Eilers for Kodak, discussed the possibility of Kodak's assuming the same role of negative supplier for the new film as it had done in black-and-white and then color peel-apart films for more than two decades. Land recalled that during one of those meetings in early November 1968, he had shown some samples of the new film to the Kodak executives, and they "were impressed and asked if they could take the pictures back to show others in Rochester."[76] Land agreed, and they did so, returning them to him shortly afterwards. Those very samples were then produced and entered into evidence.

Finally, after explaining that those discussions did not "arrive at a definitive or satisfactory delineation of a potential, modified relationship for the new film," Land told the court that Kodak decided in April 1969 to terminate its production agreement with Polaroid.[77] Nonetheless, he continued, he was able to negotiate an extension of the production agreement for "a limited period."[78] At the same time, Land revealed, Polaroid granted to Kodak an option to license Polaroid technology sufficient for it to manufacture its own color, peel-apart film for use in Polaroid's cameras. However, Land purposefully did not mention Kodak's desire for a license to compete with its new integral product nor any connection between Polaroid's refusal to do so and Kodak's refusal to continue its manufacturing relationship.

At this point, Kerr turned towards the judge and announced that his direct examination of the witness was complete. He gathered his papers from the lectern and returned to his seat at the Polaroid counsel table, satisfied that he had accomplished what he had intended: to lay the first row of bricks in the foundation Polaroid planned to construct for the judge over the coming months. Land, the inventor, had made the case for each of his inventions. Land, the pioneer of one-step photography and Polaroid executive, had provided the context Kerr and Schwartz very much wanted the judge to deliberate within—the real-life challenges confronting the development of Polaroid's sequential instant photography systems and the long history and relationship between the parties that had changed them from allies and partners to competitors in the marketplace and vigorous opponents in the courtroom.

Whether or not the judge received the essential message behind this presentation, the journalists in attendance seemed to. The *Boston Herald* reported on how Land "voluntarily showed . . . Kodak a new instant color film eight years before Kodak allegedly copied the product and brought out its own self-developing camera system."[79] The *Boston Globe* described how "the long, friendly commercial relationship between . . . Kodak and Polaroid . . . broke down in May 1968 when Kodak refused to produce the negatives for the color film used in SX-70 cameras."[80]

One thing was for certain from the perspective of the Polaroid side: Land had performed well so far. As one commentator later observed, Land "painted a picture of Polaroid as a small company that used its brainpower to reach elegant solutions. . . . Kodak, by contrast, had assigned thousands of researchers and spent tens of millions of dollars to arrive at a copycat solution."[81] The real test, however, was still to come. It was now Carr's turn to cross-examine the man at the heart of the lawsuit. After years of studying the issues and writing detailed opinion letters with respect to the patents in suit, it was finally his chance to chip away at, if not topple, those very bricks that Kerr had laid.

CHAPTER 21

THE CROSS-EXAMINATION

As Frank Carr rose from his chair to begin the cross-examination of Land, his objective was clear. At the very least, he needed to poke as many holes as possible in Land's highly effective testimony. He could also hope to trap Land into making a damaging admission or perhaps provoke a loss of temper or composure that might tarnish Land's likability or credibility. Kodak's trial counsel was certainly well prepared for this moment. Carr had studied instant photography, including Polaroid's patents, for more than twelve years. He had written opinion letters for Kodak, laying out in detail his arguments as to why each of the patents now being litigated was either invalid or not infringed by Kodak's products or both. He had also supervised or consulted on litigations in several foreign jurisdictions seeking either to invalidate or cancel the overseas counterparts to several of the patents in this suit. And he had conducted long, detailed, and contentious depositions of each of Polaroid's key inventors, including twelve full days with Land.

Land's experience from those protracted depositions, however, gave him an edge; he knew how Carr worked and thought, and he had a strong sense of Carr's relatively limited (as compared with Land's) understanding of the technology, as well as his positions on the issues. Given the comprehensive nature of Land's deposition examination, it was unlikely that Carr could spring any nasty surprises on him. The two men were prepared for a battle that might well determine the outcome of the years of litigation.

Carr began by advancing Kodak's core arguments—that the inventions covered by Polaroid's patents, including Land's, were merely old and known techniques applied to new circumstances and that Polaroid had developed a practice of patenting every idea or concept that arose in its labs, regardless of its merit or the quantum of invention involved. He began by showing Land the first paper he had authored on the subject of one-step photography back in 1947, believing it covered several of the basic elements of instant photography that would later appear in every instant system released.

Carr asked: "Do I understand it correctly that in visualizing the concepts, as you did in that first paper in 1947, you were utilizing mentally and otherwise in the laboratory principles and materials which were then already known in photography . . . ?"[1] Land answered philosophically: "As it has been said of many great processes, they certainly stand on the shoulders of processes that preceded them." Carr was not satisfied with Land's response and tried to get a definitive answer. "Your answer is 'yes'?" he shot back. "No. No," Land repeated, "because the word 'using' is an ambiguous term. . . . You asked if I were using known processes. I was employing some old processes as the basis of generating new ones," he explained. "But as far as the basic principles of photography, you weren't creating any new ones, were you?" Carr persisted, still sparring gently. "I think it depends on how you define basic principles," Land began, as he offered a detailed response, concluding with the observation that "without some new principles, [one-step photography] couldn't have come about."[2]

Carr then went after Polaroid's alleged profligate patent policy, asking Land whether in the early days of his research, "as you came upon a useful thought, was it your practice to get a patent application on file?"[3] "You can't file a patent on a thought," Land responded. "It has to be an operating mechanism," he told the court. Carr then offered into evidence a collection of seven early Land patents, each of which had expired before the lawsuit had commenced. He took Land through each of them, trying to pin him down on what the covered invention was. They all apparently related to the basic instant film unit structure of using a pod to contain the processing composition, with some relating to various aspects of stabilizing a sepia image. Carr could not have been unhappy when it eventually became clear that, even for the inventor, distinguishing between the various patents in terms of the specific invention they actually covered and claimed was no easy task.

Next, the cross-examination focused on Kodak's claim that many of the steps in various instant processes were similar to those in conventional photography and that even between the various one-step systems—sepia, black-and-white, and color imaging, or peel-apart and integral film units—the problems confronted were the same or similar. Specifically, he took Land through some diffusion transfer processes that existed as early as the mid-nineteenth century. He tried to equate them to the early work Land did at Polaroid and some of the early patents he had secured in the period leading up to the introduction of the first commercial one-step system. As one newspaper account described it: "Carr was trying to get on the record that Land's sepia photography process, invented in the 1940s, was not an amazing new and patentable invention but a photographic process based on prior art that was generally known at the time."[4]

But Land vigorously resisted this superficial reconstruction of his work. Robert Lenzner, the *Boston Globe* journalist who had long covered Polaroid, was scrupulous in showing that Land was absolutely resolute in the face of Carr's attempt to equate the unequatable and, in his view, gloss over the kinds of crucial subtleties that make all the difference when a researcher confronts a problem in the real world. Lenzner's headline proclaimed: "The Kodak, Polaroid Trial a Picture of Splitting Hairs." "Land's job was to carefully slice hairs and obfuscate Carr's proddings, a mission for which the 71-year-old [sic] inventor is a natural," wrote Lenzner. "Land does not like to answer simply yes or no. He gives ground grudgingly when he gives ground, which is not very often. Then, he tries to be exquisitely precise and sometimes ends up opaque. It is all in the interest of Polaroid's case . . . one that could be worth many tens of millions of dollars to Polaroid."[5]

For hour after hour, Carr and Land sparred back and forth, with the witness parsing each question into oblivion. Land resisted providing simple and straightforward answers to even simple questions. Kodak's hometown *Rochester Democrat and Chronicle* reported, "Carr was having a tough time getting Land to admit to anything. . . . His questions were frequently answered by Land with lengthy analyses of why Land thought the questions were based on unwarranted assumptions."[6] After sparring for about four hours on some subject, with the witness refusing to give an inch, Carr tried again to rephrase a particular question and pleaded, "It's a simple question. . . . Do they or don't they?"[7] Land rejoined instantaneously, "I think there is no simple answer to your simple question." "Then give me a complex answer," snapped the now red-faced Carr, clearly showing his frustration.

Moments later, when Carr showed Land a diagram Kodak had prepared to depict the film unit and process disclosed in one of his L-Coat patents, Land refused to acknowledge that the diagram was accurate because it was lacking a key element. "This is not an illustration of what is in the patent. . . . [It is] as different one from the other as the buzz of the bee is from the sting. . . . When you do it right, the results are positive, absolute and violently correct," Land maintained. "That's what this patent is about, how to do it right. And that drawing is in general the way to do it wrong."[8]

Throughout this exercise, Judge Zobel seemed to maintain her good humor, or, as one observer put it, her "dry wit," and interceded often in an attempt to keep the examination moving.[9] At one point, Carr asked Land if, in the sepia process, the "image-receiving layer was permeable to the alkaline processing composition."[10] Land began to respond, "That was a very remarkable surface, which was made by," when the judge interrupted him, stating, "I think the question simply was: whether it was or it wasn't . . . I think the answer is either yes or no." "I'm afraid the answer would have to be somewhat," said Land. "I am not being evasive," he assured the judge. "I know," she responded with a small but somewhat weary smile.

On another occasion, Land had objections to another of the charts Kodak had prepared. He believed this one inaccurately depicted one of the prior art references. "I am having trouble . . . those aren't the film units in the process," Land explained.[11] "They are highly idealized statements. . . . I always have the fear that if I agree with your simple questions . . . I am by implication agreeing that these charts are indeed a description of the product, and they are not. . . . I am worried about generalizations implicit in the question that makes me a witness to something . . . that I don't believe," he protested. "I don't know if it is my place to object," Land wondered, turning to the judge in frustration, "but I object to being such a witness." Judge Zobel immediately jumped in to respond, "You are putting Mr. Kerr out of a job, Dr. Land." All eyes turned to Polaroid's counsel, but implicit in his response was an acknowledgment that his witness had the proceedings well under control. "I will be very quiet, your Honor," said Kerr.

The questioning directed at Land's L-Coat patent continued into its third day. Carr's main thrust continued to be an attempt to find the invention in Land's earlier work, particularly in his basic patents on the sepia process. In one instance, an earlier patent discussed techniques for slowly lowering the pH. Carr tried to equate those teachings to the timing layer used in the L-Coat patent. Although Land provided a detailed distinction, at least on a superficial layman's reading of the language of the two technical documents, the processes did sound an awful lot alike, which was Carr's main point. But would Zobel, in considering Land's careful parsing of his responses, find his testimony conclusive and convincing? Or did she regard them as obfuscations designed to mask infirmities in Polaroid's positions? In any event, it was clear to all present that the judge was totally engaged in the proceedings, following every question and answer while referring to the piles of loose-leaf notebooks, binders, and other references that covered her desk. One account described how "she has to stand up to reach all the exhibits and reference material on the bench, and [thus] seems to spend as much time standing up as sitting down."[12]

The *Rochester Democrat and Chronicle* followed the trial closely, keeping the thousands of Kodak employees back home abreast of the proceedings. Phil Ebersole, the reporter posted to Boston to follow the trial, described the scene, pointing out how "Kodak lawyers and staff dominate the proceedings, in numbers. . . . There's not enough room for all of them at the lawyers' tables so they fill up half the jury box, which they share with reporters." He also described the woman presiding over Kodak's fate, noting that Judge Zobel "speaks with a faint central European accent."[13] On this occasion, Ebersole apparently thought Carr had scored some points when he questioned Land about some discussions he had with two Kodak scientists. Although Land rejected the notion that the mention by a Kodak scientist of using an "immobile acid" in the film unit gave him the idea for his L-Coat invention, the Rochester paper's headline the next day read, "Land Asked about Kodak 'Help' on Film."[14]

Thursday, October 15, was the seventh day of the trial and the fourth day of Land's cross-examination. Carr's questioning had now reached the same length as the entirety of Kerr's direct examination but remained mired in the details of only the first of Land's three patents in suit. Carr was not making much progress, and the newspaper report back in Rochester advised readers, "Land . . . continued to be a tough witness."[15] The back-and-forth continued when Judge Zobel suddenly asked Carr, "How much longer do you expect to be on this patent? . . . Just in general."[16] After Carr responded that it could take the rest of the day, the judge asked whether he was planning to be as long on the other patents. "On one of them, I don't believe [as] long . . . no, your honor. . . . But I have considerable more examination, if that is your question, yes," Carr advised her. "Weeks, months?" she asked. "No. I hope for both our sakes, no," Carr replied. Land spoke up: "Can you repeat the answer? I can't hear you. You are telling them how long you expect me to be here, and I would like to hear what you say." "He is not sure, Dr. Land," Zobel advised. "I don't know," concurred Carr, as he resumed his examination with no indication that he sensed the judge's patience wearing thin.

Instead, Carr turned to an issue key to Kodak's insistence that its PR-10 film did not infringe the claims of the L-Coat patent. Land's patent called for the polymeric acid layer to be located during processing in the "photosensitive element" of the film unit. Under the terms of the patent, that meant that it should be located adjacent to the layers of photosensitive chemicals in the negative, and on the opposite side of them from the image-receiving layer. In the manufacture of PR-10 film, virtually all of the chemical layers, starting with the image-receiving layer and progressing through the photosensitive layers, were coated on one plastic support. The polymeric acid layer was coated on the other plastic support—a clear one through which the film would be exposed—before the two were attached during manufacture to form the sandwich of the film unit. In Kodak's film, the processing solution was spread *in between* these two supports. Thus, the polymeric acid layer was on one side attached to one support, and the photosensitive layers were coated onto the other support. Kodak's argument was that since its polymeric acid layer was not coated on the same support with the photosensitive layers, it was not within the "photosensitive element" and for that reason, Kodak maintained, its film did not infringe the claims of Land's patent.

In an effort to establish this, Carr took Land through a series of other Polaroid patents he thought would support his case. One was a relatively recent patent of Land's that, according to Kodak, covered a polymeric acid layer coated on a clear "spreader sheet." Superficially, the structure seemed to resemble that of Kodak's PR-10 film. Kerr objected to having Land questioned on this document because, as a later reference, it did not qualify as prior art to Land's invention. But the judge overruled him in the face of Carr's argument that "it's our position they can't have it both ways. Clearly," he explained, "the . . . [L-Coat] patent . . . didn't describe the use of such an acid in such a position for such a purpose because they got another patent on it later, in the face of the . . . [L-Coat] patent [as prior art]."[17] Polaroid "should have sued for infringement on this patent instead?" joked the judge. "I will face that when the time comes, your Honor. At the moment, I am dealing with the [L-Coat]."[18]

Carr then moved on to other patents, including the Rogers "Excedrin" film unit patent asserted against Kodak in the case. His aim was to try to establish a definition of "photosensitive element" that would exclude Kodak's PR-10 film unit because of the way it was constructed. As usual, Land struggled with Carr's exercise, unwilling to agree that a certain diagram depicted a photosensitive element. Judge Zobel interceded to ask Carr, "Are you suggesting that Mr. Rogers describes this as a photosensitive element?" "Yes, your Honor," Carr replied. "I understand that's what you're trying to do," she acknowledged. "The witness disagrees," she pointed out. "Why don't we go on to the next question."[19]

The rest of that day's session continued along the same vein, with one journalist describing it as "semantics . . . [as] a battleground."[20] The following morning, the argument over the definition of "photosensitive element" picked up just where it had left off, bouncing back and forth from reference to reference in which the term was employed in different and specific contexts and on charts created by Kodak to depict film structures that Land seldom agreed were accurate. When Carr asked Land if he was "aware that Kodak's PR-10 film unit is comprised of separate elements, each having the relatively thin support on which are coated relatively thin layers," Land immediately objected. "I think I don't subscribe to the idea of 'separate elements.' The Kodak integral structure has in it many strata bound together to make an integral film. And I regard that totality as a single unit."[21]

But this was a critical point for Carr, and he was not about to give up. He got Land to admit that he knew that the PR-10 film was manufactured as separate elements, with the photosensitive emulsions coated on one base and the acid polymer and timing layer coated on the other. At that point, Carr pulled out a PR-10 film unit and had one of his colleagues cut around the edges with a pair of scissors so that it came apart. He then proposed to question Land about it. Kerr repeatedly objected, particularly to Carr's persistent attempts to have Land agree that that the parts were "separate elements." With Carr's approval, Judge Zobel asked Land if he "recognized . . . [the exhibit] as a PR-10 film unit after surgery?"[22] But the witness was not prepared to capitulate. "I regard it as a product of taking a unitary structure [in] which the essential concept and commercial significance is having a unitary structure and arbitrarily separating them and giving them separate characterizations," he protested. "What you are cutting in half and trimming off is the structure of a whole field of achievement." Jumping on the metaphor the judge had introduced, and in a line reported widely the next day, Land went on to complain that "after surgery, I recommend that from this body you have taken the head and left the foot."[23]

The exercise continued for some time longer, when Carr finally appeared to be getting ready to move on to the next patent. Just before doing so, seemingly out of left field, he asked Land to confirm that he did not keep laboratory notebooks, a fact Kodak had learned during discovery. Since Land had testified to a long succession of laboratory failures in trying to use acids in connection with his sepia process, Carr asked whether "it was fair to say that your testimony [regarding the laboratory failures] was based principally on your memory?" Land acknowledged that he did not keep notebooks but pounced on the implication that because he couldn't produce written records of the failed experiments then perhaps they had not occurred. "No," Land proclaimed, as he tried to put the research process into some real-life context for Carr, but more importantly, for the judge who would, effectively, decide the fate of Polaroid:

> It is based on my memory and the dynamics of research situations in which things that work are carried out into the future and utilized. And the things that are not working yet remain where they are to be utilized at some later date. . . . The fact is as you so well know, the research of science is nothing but failures. You fail and fail and fail, and when you succeed you stop. So the record of science is experiments that didn't work and that are then the basis of one that does work. . . . The reason I know is because what we offer and what we are selling is the embodiment of all those failures, it is the record.[24]

When Land finished, Carr couldn't resist making an embarrassed quip, perhaps realizing too late that he had again opened the door for more testimony from the witness than he had sought, testimony that did not necessarily help his case. He acknowledged, "I think I got an answer at the outset of the answer, and I won't pursue that any further" and announced that he had finally reached the end of his examination with respect to the L-Coat patent.

At this point, Carr was already into his fifth day of cross-examination of a witness who had testified for only three days on direct. His strategy, right or wrong, was starting to become apparent. Carr seemed determined to lay out for the judge Kodak's best case through the inventor seated in the witness box, whether or not he could get Land to agree with any of the theories he was presenting. In other circumstances, a litigator might circumscribe a cross-examination to cover only essential and strategic areas. The idea is to ask only targeted questions and to demand equally targeted responses. The best way to achieve this is to ask only questions to which you know the answers, preferably answers that can be established from another source, such as a document, a publication, or prior testimony in another proceeding, like a deposition. The testimony of the witness on cross-examination can thus be reined in by this independent source, should he or she give anything but the expected response.

Thus, the conventional wisdom is that it is dangerous to take cross-examination any further than the strict limitations of the documents to which one can tie the witness. To try to accomplish more only provides the witness with the opportunity to expound on theories and explanations that bolster the case for the witness's side, thus making the cross-examination counter-productive. To the seasoned trial lawyer, when it comes to the dangerous terrain of cross-examination, less is more. The common axiom is that you never want to ask that one additional question that could open the floodgates to testimony that might damage your case. Clearly, Carr was feeling no such inhibition in his approach to Land as a witness. No matter how stubborn the witness might be, Carr seemed prepared to battle Land over the theories he had long ago developed on these patents. He presumably hoped that the judge would be educated to Kodak's positions as a result. It remained to be seen how this strategy would work. Could he trip up Land, causing him to make a fatal confession or admission on a pivotal fact? Were the lengths to which the witness was going to explain the errors in Carr's assertions likely to make the judge begin to see points of weakness in Polaroid's position? Only time would tell.

Next up for Carr was Land's Symmetrical Support patent. Carr started his examination on this patent by taking Land through some scientific articles showing that the properties of polyester as a film base, in terms of "durability and dimensional stability," were well known as far back as the early 1950s.[25] He tried to pin Land down about his own knowledge of those properties prior to 1958, including its relative impermeability to water vapor, but Land resisted. "I can't remember, in sitting here, how my own awareness of a dramatic difference in water vapor permeability came into being, or just when it did," he insisted.[26] Carr then began a line of examination concerning passages in one of the basic Polaroid patents, Rogers' patent on the positive dye developer chemistry utilized in all Polaroid color instant film. Carr's objective was to demonstrate that this seminal patent already taught what Land and Polaroid asserted to be his Symmetrical Supports invention. The tussle over words and concepts that resulted was by now all too familiar to everyone in the courtroom.

When Carr tried to pair some text from one part of the Rogers patent together with a drawing from another section of that same patent to come up with a structure that, as Carr phrased it, "would be basically useful in carrying out the invention of the [Symmetrical Supports] patent," Land erupted.[27] "Not at all one whit," he declared as he proceeded to draw the distinctions. On and on the testimony continued for the rest of that day, and into the next, with Carr and Land going round and round concerning the teachings of hypothetical film units created from various passages of the earlier Polaroid patent as they might relate to Land's invention. It was an excruciating exercise, with Kerr objecting often along the way. But the judge indulged Carr and merely reminded Kerr that both parties could "make whatever argument you wish to make from" the hypothetical exercise. And so Carr plodded forward, with the examination getting seemingly more and more confused, finally prompting the judge to "confess [that] I am at a loss as to what the question was." It was now late during the Friday session after a grueling trial week. "I have a suggestion, your Honor," Carr said, "to break through this haze." "You are going to make a chart?" guessed the judge correctly, and with that she adjourned for the weekend.[28]

As the trial resumed the following week, the chart promised by Carr appeared but seemed to have little, if any, of the effect he had hoped for. Kerr objected to the chart as "misdescriptive," while Carr disagreed with Kerr's characterization of it as being "hypothetical," given that he believed it came "directly out of the [Rogers] patent."[29] Zobel did her best to keep the examination moving by having everyone avoid the distracting word play. Eventually, Carr either believed he had obtained the testimony he sought or he finally gave up. Then he moved on to another procession of Polaroid patents, again mostly in Land's name. Some of these dated as far back as the 1940s. Carr began to ask Land about each one's disclosure of the use of polyester and "relatively impermeable" materials as a base or support for one-step film units. With each wave of these questions, however, Land just got more and more intransigent and irascible. The cross-examination went completely off course for Carr when he began to question Land about the film structures described in one of Land's early patents.

In that instance, Carr asked Land to agree with what seemed a fairly obvious concept: that an integral film unit like SX-70 was "simpler to handle" than one that had to be physically laminated together by the user at some point during processing. "I don't think I want to take that position," replied Land. "I'd like you to take a position about that, Dr. Land," Carr said. "I don't think I have to," Land responded, and Carr turned to the judge for help. "If you can answer the question, please answer it," the judge advised Land. "If you cannot, say so." "I cannot answer the question," was Land's response. Carr inquired, "Dr. Land, tell me, is it your position that you can't answer the question or that you won't answer the question?" "Ask the question again and I'll tell you which of those it is," responded the witness.[30] And so Carr complied, asking, "Is it your opinion that a film unit which is laminated before exposure and processing is simpler to handle than one which is . . . in separate parts which have to be brought together?" The response that followed was quintessential Land, and certainly not unexpected, given what had transpired over the past six days of cross-examination:

> I suppose it has to do with what the word "handle" means, and in that case, we know of the infinite number of chemical complexities, impermanences, difficulties that are a consequence upon having them together in the first place, problems which were not solved for decades afterwards. Is that handling or not? I don't know.
>
> If you make a primitive question: Is one piece of something easier to hold in your hand than two pieces, perhaps it is.
>
> If you ask the question: What's the most direct way to handle films for getting a good and usable picture—it turned out this was not the most direct way. It took years and years of research and study to be able to do anything that remotely resembled it, and it then had in it all sorts of things that aren't in this [early patent] at all.
>
> I think it's a perfectly proper statement to say it's an object of the invention [of the early patent] to provide a composite film unit which is simple to handle. That is an object. But ordinarily, I think when you talk about object in a patent . . . you indicate goals that you're heading for. So this did turn out to be an indication of a very, very distant goal for an entirely different process in color, and so on.
>
> This doesn't show any of the techniques for keeping it together or handling extra reagent, doesn't show what you do to spread the closed system or to take care of the excess. So I think the best answer to your question is that that is not a question that can be answered by anybody.[31]

Unperturbed, Carr apparently still believed at that moment that his line of questioning was going to accomplish something, and so he continued to try to get Land to agree with the idea that an SX-70 integral unit is simpler to handle than a Polacolor peel-apart unit. The judge, though, had had enough. "Why do we have to get into that, Mr. Carr?" she asked. "The witness just explained at some length that the term 'simple to handle' could have any number of different meanings. . . . It may be simple for the person who takes the photograph but it may be very difficult to handle for the chemist who makes up the various parts of it. I understood that that is what the witness was trying to explain," Zobel concluded, clearly displaying some exasperation over Carr's seemingly ineffectual relentlessness. Message apparently received, Kodak's trial counsel moved on to another subject, but not another approach.

The judge once again showed what seemed to be her growing impatience when some questions concerning the structure of a film unit in another patent again got twisted into a hypothetical. "As I understand the witness' testimony throughout, Mr. Carr, he will not admit—and you will not get from this witness—any admission that you can simply take one thing away or add one thing and that is has no significance. I don't think you will get very far with the kind of questioning you have just put."[32] But Carr pressed on, and the judge allowed him the leeway while doing her best to keep the process moving forward. She remained determined to avoid limiting Kodak's opportunity to conduct a full cross-examination in a way that might later be used as grounds for appeal, should it lose.

Nonetheless, even before Kerr could rise from his chair to do so, it wasn't long before she began to object to what she thought were Carr's repetitive questions. "We now get into the same thing," said Zobel.[33] "The sheet is dimensionally stable, but in combination with something else, it may not be. I believe that is what the witness has repeated over and over again, and I understand that to be his testimony," she concluded. When Carr disagreed, the judge asked the witness to concur, and he did, beginning to explain why. "It's all right, Dr. Land," said Zobel. "I think I understand the testimony. If you're asking the question that I understood you to be asking, the objection that Mr. Kerr was about to make to it is sustained because it is repetitive, and I think the witness has answered it." All that the now chastened Carr could muster as a response was "Thank you."

When that day's session finally ended, the judge asked counsel for both parties to approach the bench. She asked Carr for an estimate of how much additional time he was going to need to complete his cross-examination. Carr stated that he planned to use as many as ten additional days.[34] The judge did not react immediately but asked the lawyers to return before the trial resumed the next day, October 21, for a conference. Schwartz immediately sent two of his young lawyers to the law library to look, late into the night, for cases that he could cite to Zobel for her to rely on in limiting the length of Carr's cross-examination of Land.[35] When counsel for both sides reconvened in conference before the trial the next day, Judge Zobel was still unwilling to put an arbitrary limitation on the length of Carr's cross-examination. She did inquire, though, as to whether there were any other means of expediting the process, including having Carr give some advance notice to Kerr on the subject matters to be covered the following day.

Kerr and Schwartz came away from the conference believing that Carr had agreed to this approach and memorialized it in a letter to him that was hand-delivered to his firm's temporary office. In it, Kerr noted, "Naturally, we would like to have this information as early as possible on the preceding day."[36] Carr, however, was not about to surrender any more of the advantage he thought he had than necessary. Grudgingly, he immediately wrote back to Kerr, advising that he was willing to provide the information "no later than the late afternoon of the day before the examination" but reserving "the right, however, to examine on patents and references which, for whatever reason, are not included in such advance notice."[37] In short, he was willing to give virtually nothing at all.

Carr was focused on his mission and seemed unwilling to cooperate in some of the professional niceties one might expect litigators to accord each other. He was apparently unconcerned about whether this attitude, and his whole approach to the cross-examination of Polaroid's star witness, was going to have any negative impact on the judge. He had displayed a similar lack of sensitivity in the weeks leading up to trial with all of the transparent and ineffective attempts Kodak had made to delay or to derail Polaroid's day in court. Although it was reasonable to assume that Judge Zobel had taken notice of these tactics, it was impossible to know what impact, if any, they would have on her decision-making process. Yet Carr and his Kodak colleagues had clearly decided it was worth the chance, and so on they went with this tack. But Kerr and Schwartz, egged on by Land as well as their Polaroid clients, determined that the matter needed further action on their part, especially after receiving Carr's letter eviscerating any benefit from the procedure the judge had tried to work out. They readied themselves to act when the time was right.

After the conference with the judge on Wednesday, October 21, Carr worked his way through another day of seemingly endless inquiry into the Symmetrical Supports patent. Carr's primary agenda for the day seemed to be to get Land to agree that the invention of this patent was merely "the substitution of polyester for cellulose acetate in both of the supports."[38] Suffice it to say, this was not a proposition that Land was disposed to agree with. After more sparring over the meaning of terms like "dimensional stability" and a lengthy dispute over whether glass would qualify as a support within the meaning of the patent, the day ended with the judge adjourning the case for a week because of other court obligations. With the end of Land's cross-examination nowhere in sight, Kerr and Schwartz put their plan into motion.

On the next day, October 23, they filed a motion with the court on behalf of Polaroid seeking an order limiting the time Carr would have for further cross-examination. The motion sought a limit of three additional days for Land, and the establishment of a rule "limiting all future cross-examinations in the case by either side to approximately the same time as that consumed by the direct examination of the witness in question."[39] In its brief, Polaroid pointed out that without intervention by the court, the original estimate for the duration of the trial would be rendered meaningless "because of the needless, wasteful and cumulative cross-examination" Carr was conducting. They noted that Land's direct examination had taken fourteen hours of court time, but that the still incomplete cross-examination had already consumed twenty-two hours.

> There is no justification for the tremendous disparity between the length of direct and the length of cross-examination. Only two months ago, Kodak insisted that all ten patents in this lawsuit were simple and ripe for summary judgment. Now Kodak is bound and determined to plod through literally hundreds of items of prior art and thousands of exhibits as if discovery had never occurred.[40]

The brief went on to point out that Kodak had deposed forty-nine Polaroid witnesses for more than 300 days, and that the parties had already agreed that this material could be offered into evidence in the case. Thus, covering the same ground in court was a waste of everyone's time. In sum, Polaroid urged, "Kodak's dilatory tactics on cross-examination are but the latest Kodak attempt to delay the ultimate resolution of this case. This Court has previously rejected Kodak's attempts to spin out discovery, institute inordinately complex pretrial procedures, and substitute summary judgment for the trial. Kodak's latest stratagem for delay . . . should be similarly rejected."[41]

Kodak did not immediately respond to Polaroid's motion, nor did Judge Zobel ask for an opposing brief. She did not hold a hearing on the subject, nor did she issue a formal decision. Instead, she somehow was able to convey to both parties that she was sympathetic to the argument Polaroid was making, at least in terms of its contention that the length of any cross-examination should bear some reasonable relationship to the length of the direct examination of a given witness. Zobel's sub rosa approach to dealing with the matter seemed to achieve the desired result. According to the estimate Carr had provided to the court on October 20, he was going to need an additional nine days to complete his cross-examination of Land when court resumed. Instead, he actually used less than three days, completing his examination in precisely the time period Polaroid had requested in its motion.

No one can be sure how, quite unexpectedly and almost miraculously, this came about. Some young, less experienced Fish & Neave lawyers even wondered whether Zobel may have had a private conversation with Carr, but that would have been an improper ex parte communication and thus likely never happened.[42] But perhaps Carr, simply and accurately reading Zobel's body language, decided that an extensive cross-examination was becoming counterproductive, for one or both of two reasons: one, because it was annoying and thus possibly alienating the judge, and two, because it was only providing the uncontrollable Land with more and more opportunity to further Polaroid's, rather than Kodak's, case.

When the trial recommenced on Wednesday, October 28, Carr spent virtually the entire session questioning Land in a manner designed to reduce the Symmetrical Supports invention to a level of simplicity, or even absurdity, that Schwartz had originally feared when he passed over the patent in his initial selection process. Kodak asserted that all of the high-tech legalese in the patent came down to a simple choice of materials; in essence, you want a flat film unit so you select thick plastic sheets for both sides. In trying to overcome Kerr's objection to a particular question, Carr made his goal clear: "I would like to narrow the invention down to what I believe it actually is, and that is using a relatively inflexible support member on both sides so the whole unit stays flat."[43] "I didn't understand them to claim anything beyond that," answered Judge Zobel, much to the chagrin of the entire Polaroid contingent. Kerr promptly broke in to point out that, contrary to this simplistic view, the invention is what is described and claimed in the patent—that is, "supports [that] have specifically defined characteristics" in order to achieve the objective of a flat and stable film unit.

Carr also attempted to establish that the choice of polyester was in fact caused by factors other than those discussed in the patent, such as its superior performance in the manufacturing setting because of its mechanical strength. In the face of this relentless barrage, Land did his best to highlight the complications inherent in the research process and the fallacies behind the overt simplifications and hypotheticals that Carr used repeatedly in his cross-examination. Kerr objected frequently about repetitive or unartfully composed questions. Nonetheless, when Carr finally announced, close to the end of the day, that he was moving on to the final Land patent, one had to wonder if he had indeed successfully exposed the emperor clothed in the Symmetrical Supports patent.

Last for Land's cross-examination was his Rear Pick patent. Carr flew through this subject in less than two days. But it was more of the same. He painfully took Land through other Polaroid camera patents that described a variety of means for transporting a film unit out of a cassette within the camera into processing rollers that would burst the pod and eject the film out of the camera. As with other features of its instant photography system, Polaroid had tried many approaches and patented many alternatives beyond those actually used in its commercial products. Now Land was confronted with explaining how they all differed and how none of them solved the problem in the elegant and reliable fashion that his Rear Pick apparatus did.

At one point, in phrasing a question, Carr gave a lengthy description of the way a particular prior art apparatus advanced the film unit. Kerr objected that Carr's recitation amounted to a legal argument, something he could do later. "That is a long argument, and I do not think it is appropriate for counsel to argue what that patent shows now," complained Kerr.[44] When the judge asked Land to explain why Carr's description was incorrect, Kerr again objected that this was an unfair question because it assumed that Land had everything Carr had said in mind. Zobel turned to the witness and asked, "Do you think it's unfair?" "Yes," Land replied, "but I am used to it," as he launched into a lecture for the next several minutes on the flaws in Carr's description of the apparatus.[45]

On the tenth day of Land's cross-examination, Carr announced early in the session that he intended to conclude "shortly."[46] Instead of returning to the subject of the Rear Pick patent, he asked some questions about the history of Polaroid's relationship with Kodak, a subject that Land had touched on during his direct examination. Then, without warning, he surprisingly returned to the subject of the Symmetrical Supports patent. Carr seemed to be taking a run at getting Land to admit that the description of this invention that he and Polaroid were now proffering was a revisionist account—that the principles on which Land had expounded in his direct examination were not appreciated at the time the patent was obtained. It was a long shot for Carr to take and a convoluted argument in any event. In the end, besides failing in his attempt to get Land to make a fatal misstep, he simply gave the witness another opportunity to explain how unexpected and counterintuitive his discovery had been. Land explained, in part:

> It was a general attitude that a picture could not be left wet. That was offensive intuitively to most people with chemical backgrounds, they felt that to make an image that would not be vulnerable in many ways, the image needed to be dried out as soon as possible as a general principle, and that was a dominant intellectual principle until I persuaded the group with the passage of time that the acid polymer alone [his L-Coat invention], with appropriate timing controls, would create an environment in which, although the water content might take a week or two to be lost, the image would nevertheless be a useful and stable image and that the benefits to be gained mechanically . . . by using the impermeable layers far outweighed any imagined hazard due to the continued presence of water for the 10 or 15 days, as the case might be.[47]

It's not hard to imagine Kerr and Schwartz looking at each other with bemused expressions as Land orated eloquently on the merits of his invention. They surely must have appreciated how their witness preparation had paid off in a succinct description for the court and his recounting of the kind of "aha" moment one might expect an inventor to have when a discovery is made. It was the kind of insight they knew a judge without a sophisticated technical background could appreciate and, in this case, it was being presented in its most effective setting, on cross-examination. It was another blunder by Carr in trial tactics, but whether it would be enough to save what, at one point at least, they considered their weakest patent, the Polaroid lawyers could not be certain. In any event, they certainly were not unhappy that Carr's cross-examination had provided the witness with yet another chance to make his best pitch. Carr pressed on with some questions, trying to limit the damage, but he likely soon realized that he was doing more damage than good and so finally uttered the words everyone had been waiting for: "I have no further questions at this time on cross-examination, your Honor."[48]

With Land's cross-examination complete, it was clear that an important phase of the trial was in the books, but it was not immediately clear what its impact on the outcome would be. Some litigators might have opted for a very targeted, surgical cross-examination of a difficult witness like Land. The more conventional approach would have been to go after a few easily demonstrable points that might at least inject a few wisps of doubt into the witness's overall credibility. But Carr opted for a different strategy, one of going through pretty much every aspect of Kodak's arguments on each of Land's patents in what turned out to be a seemingly futile attempt to get a damaging admission from the inventor on a key point. After weeks spent deposing him, Carr had to know that he was going to be faced with a consummately prepared and combative witness, willing and able to debate every point he tried to make. Yet he stuck with his approach, despite its potential for providing an uncontrollable witness like Land an opportunity to propound upon his own, contrary views. In the end, Carr surely did not get his Perry Mason moment, and, more importantly, it was unclear whether he had helped or hurt Kodak's case with his cross-examination technique.

The conclusion of this phase of the trial represented another major milestone for the Polaroid side—the end of the active roles for the two senior members of its cast, the star witness and his counsel. Land's work was done, the many months of anticipation and preparation having resulted in what he and his closest colleagues thought was a vintage performance in defense of his work and the intellectual property of his company. For Kerr, this marked perhaps an even more consequential event. He was, virtually immediately upon the end of Land's testimony, retired from the practice of law. Although he would stick around the courtroom until his formal retirement at the end of the calendar year, Schwartz would handle the trial from this point on. When the trial took a break for the holidays, Kerr was set to return to New York to close up his office and repair to his beloved home in Bermuda. He would watch the rest of the drama play out from afar. (See Fig. 21-1.)

To commemorate the dual occasion, Land arranged a special dinner to be held in a small, private dining room in the Boston Ritz Carlton hotel. The Fish & Neave legal team was living in the hotel during the trial, but it was also a regular stomping ground for Land, who was particularly fond of its famous lobster bisque. Present for this intimate celebration were just Land, Nan Chequer, Kerr, Schwartz, and the young Fish & Neave lawyer who had been assigned to work directly with Land throughout the litigation. It turned out to be a very stiff and uncomfortable meal. The room was small, much too small to contain the two huge egos of Land and Kerr, both of whom felt compelled, as usual, to try to outdo the other.

The primary source of contention on this occasion was the choice of wine. Given the menu, it was clearly going to be white, and given the two men deliberating over the selection, it was clearly going to be the most elite of the world's white wines, a Bourgogne from the famous Côte de Beaune. The one-upmanship continued until Land ordered the single most expensive bottle on the astronomically expensive Ritz wine list, a rare and exquisite Puligny Montrachet. Everyone enjoyed the libation, but the process of ordering it was, in some ways, just another demonstration of Land's tenacity. He never gave in to Carr on the witness stand and, in this final meeting with the legal brain trust, he was not about to give in to William K. Kerr.

CHAPTER 22

AFTER THE MAIN EVENT

With Herb Schwartz now at the reins of the Polaroid trial team, Polaroid's obligatory technical case was presented, but always with a focus on telling the real-life story of the making of the inventions at issue, and the history between the two companies that had led to this litigation. Schwartz knew that virtually any patent, dissected and analyzed in isolation, would have warts and blemishes exposed that might serve to undermine it. By forcing the judge to consider each patent within the framework of actual events as they took place, taking as much hindsight and assumption as possible out of the analysis, he believed he was creating the best possible environment for the judge to conduct the technical analysis of the patents required to adjudicate the case.

The next witness for Polaroid was Robert Duncan. Land had hired Duncan in September 1968 to serve as the program manager for the SX-70 camera and film system then secretly under development. As Duncan described it, the project at that time included only "a small group of engineering personnel sequestered from the rest of the company."[1] The group had built a series of prototype cameras, but a full year later, in the fall of 1969, they were still experiencing problems with the latest model, particularly its film transport and processing system.

The pressure was on because at that time, Land was already eager to go public with the new system, something he would now be restrained from doing until Polaroid's 1970 shareholders meeting. Duncan was concerned: "my view was that we had so many problems and the rate of progress that we were making in solving these problems was such that we were not going to meet the scheduled introduction date of the system." Duncan told the court that he brought his concerns directly to Land, who promptly created a special group of researchers to address these problems.

Land joined them in a special effort ensconced in a room near his office. Duncan recalled: “During the course of these intensive meetings, Land invented the new film transport and processing system, the very heart of which was rear picking.”[2]

Over the next two weeks, Fish & Neave presented the balance of Polaroid’s case on the other camera patents in suit through two Polaroid engineers who testified as expert witnesses, John Driscoll and Milton Dietz. Dietz was a last-minute replacement for Otto Wolff, an independent camera authority and crony of Land’s who had become ill and was hospitalized just days before he was scheduled to testify. The witnesses rendered their expert opinions that the prior art references cited by Kodak did not teach the Polaroid inventions, nor would the discoveries have been obvious to one of ordinary skill in the art when they were made. They also testified how the Kodak commercial cameras infringed the claims of each of the other three Polaroid camera patents in the lawsuit, the Gear Train patent, the Light Shield / Deflector patent and, finally, the Detachable Spread Housing patent.

Schwartz’s examination of Driscoll took considerably less than one court session, broken up between two days. John Fogarty, Carr’s partner who had focused on the camera patents throughout discovery, handled Driscoll at trial. When his cross-examination of Driscoll extended into a third day, Judge Zobel began to show some impatience. “How much more do you have of this witness?” she asked Fogarty.

“I think perhaps an hour, your Honor,” he replied.

“I thought you told me [the previous court session] that you thought you had another hour?”

“I was wrong, your Honor,” admitted Fogarty.

With a perfunctory, “Yes,” Zobel called a recess.[3]

When the trial resumed, Schwartz reminded the judge that Polaroid had filed a motion to limit cross-examination, and that the motion was still pending before her. He urged the judge to act on it. “After hearing two more witnesses and two more witnesses crossed, we believe that the motion is appropriate and timely and we would like . . . a ruling.”[4]

All that the judge would commit to at that moment was that she would think about it. However, she quickly turned to Carr and warned, “That isn’t designed to encourage Kodak to extend its cross-examination.”

“We so understand,” Carr acknowledged. Clearly, Zobel did not want to have to take a formal position limiting cross-examination if she didn’t have to. She seemingly hoped, as Polaroid counsel did, that the mere pendency of the motion and the possibility of her imposing formal limitations if Kodak abused the process would be all the encouragement Carr and his team would need to keep their cross-examinations to a reasonable length. This approach seemed to shorten Carr's cross-examination of Land, but now the same excessively long questioning by Kodak counsel had reemerged.

As the trial moved on through the month of October, an otherwise innocuous, yet strange, development outside the courtroom raised the stakes for Polaroid. On October 12, the Fuji Photo Film Company of Japan announced that it would be introducing its own instant camera and film products.[5] Although initially the sale of these products was slated only for Japan, the industry consensus was that a U.S. introduction was inevitable. According to press reports, Fuji's film and camera system was to be compatible with Kodak's. It was reported in the *Wall Street Journal* that Kodak had issued licenses to Fuji in 1979 that enabled the products of the two companies to be interchangeable.[6] This was a continuation of Kodak's overall practice of licensing other manufacturers to make cameras that could use Kodak film, and film that could be used in the Kodak cameras.[7]

In the short term, this meant that Kodak would immediately have a new Japanese market for its film. Long term, once it entered the U.S. and other international markets, Fuji would be able to sell its film to everyone who owned a Kodak instant camera. Kodak, by enabling Fuji's entrance into the field, had to have made a determination that the Japanese photo giant's participation in the market would be a net benefit for the company, despite any competition it might create. But what would be the impact on Polaroid of having Fuji join the market? Clearly, a positive outcome in this trial was going to be the only effective way of keeping this new competitor from further eroding Polaroid's critical market share.

The curious aspect of this announcement was its timing. Why had Fuji chosen this very moment in the Polaroid litigation (at a time when Land was still on the witness stand) to make a statement that seemed somewhat pointless, since no date was given for the availability of this new product line? Industry followers pointed out that it might have made more sense if Fuji had waited until some early rulings in the case could "clarify the strength of Polaroid's basic patents in the field, an issue that could be no less important to Fuji's worldwide prospects in the field than to Kodak's."[8] Could Kodak have convinced its licensee that making such an announcement at this juncture would help Kodak clear the road into the instant photography market for both companies?

Kodak had previously announced at a pretrial conference on September 28 that, upon the close of Polaroid's direct case, it would follow the normal procedure of making a motion to dismiss Polaroid's complaint with respect to one or more of its asserted patents. This maneuver was known as a "Rule 41 motion," referring to the relevant section of the procedural rules governing federal trials in effect at the time. According to those rules, the ground for seeking a dismissal at this early stage, before the defense has put in its case, was that "upon the facts and law the plaintiff has shown [in its case] no right to relief."[9] Although a long shot, this stratagem would provide the judge with the opportunity, if she thought Polaroid's direct case was weak, to force Polaroid's hand by granting Kodak's motion to dismiss on one or more patents. Fuji could easily have waited until this occurred before going public with its incomplete plans.

There is no way to know whether there was any communication between Kodak and Fuji on this subject. Fuji may simply have been acting on its own initiative. If it had any role in convincing Fuji to act, was Kodak attempting to affect the real-world environment in which the judge would be deliberating so as to improve its chances of success? Was Kodak trying to send a message, indirectly, that there was more at stake in the case before Judge Zobel than just a dispute between two companies? Was it trying to make sure she was cognizant of the fact that Polaroid's intractability with regard to licensing its portfolio, which Kodak contended was based on a collection of flawed and improperly obtained patents, was impeding progress across the industry, in fact, across the world?

After all, if Judge Zobel, after hearing Polaroid's case and Kodak's cross-examinations, had felt that Polaroid's defense of its patents was less than compelling, she might well take the opportunity of the Rule 41 motions to knock some of them out of the lawsuit, a clear message to Polaroid that perhaps a negotiated accommodation with Kodak and the rest of the photography industry was advisable. If this is what Kodak was hoping for, some collaboration between the two companies to influence the proceedings at this critical juncture cannot be ruled out.

As the calendar marched toward Thanksgiving, Fish & Neave completed its presentation on the camera patents in suit. On November 18, Lloyd Taylor, a longtime Polaroid chemist, took the stand to testify about his Mordant patent. The mordant is the chemical layer in the film unit to which the migrating dyes attach and form the image. It attracts the dyes out of the processing solution in which they are soluble and then holds them in place so the picture is sharp. Taylor's patent was on a particular mordant comprised of a specific polymer designed for an integral instant film unit.

Taylor testified how Land personally asked him to get involved in the SX-70 project in December 1970 to help develop a mordant that would work in the new integral film. Up to that point, they had been using the mordant employed in the Polacolor peel-apart film. Although that mordant had proven reliable, Land concluded that the new SX-70 film required a new mordant specifically designed to work with the new "metallized" dye developers it used. Taylor's new mordant was not included in the initial SX-70 film but was later incorporated into the improved Time Zero instant film Polaroid introduced in 1979.[10]

Yet Polaroid contended that Kodak used a mordant within the class developed by Taylor and patented by Polaroid. The testimony on this patent involved complex organic chemistry. Kodak's primary contention was that the compound used in its mordant did not infringe the specific claims of Taylor's patent, claims drawn in complex "chemicalese." As a result, the going was slow, with repeated interruptions from the bench for further explanation. The direct examination of Taylor covered one full day and parts of two others, while his cross-examination spilled into a fourth day. When that was complete, both sides were unsatisfied with the status of the record and so redirect and recross were both conducted on Taylor, taking the trial right up to the long weekend for the Thanksgiving holiday.

Polaroid's final witness, Howard Rogers, took the stand to begin his testimony on the following Monday, day twenty-seven of the trial. Rogers was, perhaps, the most important witness after Land. He was the inventor of the crucial patents in Polaroid's case in the sense that they covered fundamental structures and processes rather than specific components or features. According to Polaroid, his "Excedrin" film unit patents covered the basic geometry of the Kodak film units. While there were two separate patents asserted, one with and one without the use of Land's L-Coat layer, for purposes of the entire litigation, they had been considered as one. Rogers' other patent covered the Negative Dye Developer imaging chemistry that, Polaroid alleged, Kodak employed in its PR-10 film. In many ways, they stood as the two primary obstacles around which Kodak researchers had labored for years. During discovery, Kodak counsel had confirmed their importance by spending far more time deposing Rogers than any other witness, even Land. Further, Kodak moved for summary judgment with respect to the Excedrin patents, and although the motion was denied, Kodak hoped its effort represented the first salvo in its barrage against the foundation of the patents' validity.

Tackling first the Negative Dye Developer patent, Schwartz led Rogers through an overview of his working history at Polaroid. Rogers had already been with the company for a decade when, in 1947, about the time Polaroid introduced its first instant system, which took sepia-toned pictures, he volunteered to start an investigation into making color images with the new one-step technology, and Land agreed to give him that charter.[11] Rogers had started from scratch back then, learning all he could about the color chemistry used in conventional photography. On this day at trial, using a series of color charts created to illustrate the various imaging chemistries at issue, Rogers began teaching Judge Zobel about that chemistry—known as "color coupler chemistry"—and the basics of the process for making a conventional color image. He explained how Land had tried to adapt color coupler chemistry to the diffusion transfer process that made one-step photography possible and how those attempts had failed. Rogers' subsequent efforts had also failed. In the end, he had concluded that the various color coupler processes they had worked on were all impractical for this new application.[12]

Throughout this testimony, the judge was as engaged and active as ever. Rogers, being a low-key and amiable fellow, gentle and gentlemanly, was another professor perfect for educating the inquisitive judge. She clearly wanted to understand this subject matter thoroughly, and for long periods during Rogers' testimony, the proceedings turned into a direct discussion between the two of them, with Rogers patiently and quite effectively explaining whatever it was she wanted to know. The rapport between them was obviously excellent.

As the day proceeded, Rogers described how, in the early 1950s, he had proposed the use of a new type of photographic compound, a dye developer. This was a molecule in which a developer and a dye were joined together, with the developer controlling whether or not the dye became part of the image. He explained how he had conceived two basic processes using these compounds. In the "positive" process, dye developers in exposed regions were developed and designed to stay in place, while those in areas that were not exposed were not developed and were thus free to migrate away. This left the dyes in the developed molecules located in exposed areas to form the image. In the "negative" process, the exact opposite occurred. The dye developers in exposed regions of the film unit were developed and rendered free to migrate to the positive layer of the film unit to form the image. Those in unexposed and undeveloped areas just remained in the negative. (See Figs. 5-1 and 5-2.)[13]

Rogers received patents on both systems, but only one—the Positive Dye Developer system—was adopted for Polaroid's commercial products, beginning with its Polacolor peel-apart film introduced in 1963. However, it was Rogers' patented Negative Dye Developer process that, Polaroid alleged, covered the imaging chemistry utilized in Kodak's PR-10 film. Rogers explained in detail to the judge why, in his expert opinion, the chemistry used by Kodak fit within the scope of the process he had patented.[14]

Rogers completed his basic testimony about the Dye Developer patent by distinguishing all of the prior art Kodak had cited against his invention. However, before leaving the subject, Schwartz had one more special item left on his agenda to attack Kodak's contention that PR-10 film did not infringe Rogers' patent because its imaging chemistry did not use dye developers. He handed Rogers a copy of a document from Kodak's research labs that had been produced during discovery. The memorandum had two references to "sulfonamidophenols as *dye developers*." (emphasis added) Rogers explained how this Kodak document described a negative dye developer process, and how the sulfonamidophenol compounds it discussed were closely related to the sulfonamidonaphtols used in PR-10 film. Judge Zobel quickly caught on. "Are you saying that what is in this memorandum describes the PR-10 film unit?" she asked Rogers. "Yes," Rogers replied, his bushy eyebrows upraised in his modest, yet authoritative, way.[15]

The judge sat back in her chair—the point had been made and noted. Schwartz quickly moved on and very shortly announced that he had completed direct examination on that patent. The judge immediately called a recess, leaving this dramatic moment lingering over the proceeding. This document and its description, innocent at the time it was made in a Kodak laboratory, yet a devastating admission in the context of this trial, had become a real threat to Kodak's defense.

After the recess, Schwartz and Rogers dove directly into the Excedrin film unit patents and began to explain for the judge once again why they were going to be treated as one patent for purposes of the testimony. But before Schwartz could get very far, the judge interrupted to ask, "[Are] these the patents that deal with carbon in one of the layers to make the darkroom?" "Yes," Schwartz confirmed, acknowledging that they had been the subjects of Kodak's first summary judgment motion, a fact she obviously already had in mind.[16] Although Judge Zobel had denied

Kodak's attempt to defeat these patents on summary judgment before the trial, her characterization of the invention at this juncture was very troubling for Polaroid. This was just the kind of simplistic view of the patent that Kodak had long urged.

Nonetheless, Schwartz moved on, using the patent as a reference point for Rogers to describe the invention for the court in more detail—particularly, the arrangement of chemical layers, including their relationship to the two outside supports that were sealed together to make it an integral film unit.[17] He explained how the film unit worked, including the opacification system that called for the inclusion of carbon black in the processing solution to form the dark curtain over the photosensitive part of the film necessary to allow it to be ejected from the camera and to develop in the light. Rogers also described for the judge how, because his film unit was exposed from one side and the image later viewed from the other, it had the advantage over the SX-70 system of not requiring the image to be reversed by the use of a mirror in the camera. In mid-answer, Judge Zobel interrupted him. "Is this film made commercially, the film of these patents?" she asked. Rogers didn't hesitate. "Yes," he quickly replied, "it's PR-10 film."[18] (See Fig. 22-1.)

As Rogers described the operation of his film unit, which was as depicted in a chart based on the text of the patent, the judge again jumped in to ask, "What would a film unit like this look like when the picture has been developed?"[19] Rogers responded that it would "look like a normal picture on one side." The other side, he explained was the transparent support through which it had been exposed. "After processing," continued Rogers, "where the goo is spread . . . pardon me, the viscous reagent. . . ." But before Rogers could continue, Zobel broke in, "I can understand language like 'goo,' it's the rest of it I have problems with." Schwartz seized the opportunity to help out—and simultaneously help make Polaroid's infringement case—by offering to the judge the PR-10 film unit that had previously been admitted into evidence during Land's testimony. Over objections from the Kodak side, Zobel took the Kodak film unit, and asked Rogers, "All of the black that one sees on the back, does all of that come from the carbon black in the processing goo?" "Yes," confirmed the witness. "All right, that part I can understand. Thank you," said the judge.[20] In effect, by jumping on this serendipitous moment, Schwartz had been given the chance to let Zobel equate a chart depicting the patent in suit with the actual Kodak film unit in terms of its structure and operation, as the witness had described them. (Compare Fig. 22-1 with Fig. 20-4.)

When Rogers had completed his testimony on the structure and operation of his Excedrin film unit, Schwartz asked him to describe the circumstances under which he had proposed it. He recalled making a sketch on New Year's Day of 1968 of "an alternative way of making an instant film unit."[21] He explained that he meant an alternative to the SX-70 film unit that, at the time, "was very well along and nearly done." Rogers described the essential differences between his new film unit and the one, as invented by Land, Polaroid was developing as SX-70. He enumerated some of the potential advantages of his design.

The Excedrin format allowed for a very simple opacification system, as opposed to the extremely sophisticated system developed for SX-70. Because the image was going to be viewed from the other side of the film unit, the processing composition itself could be made opaque through the inclusion of a common substance like carbon black. Another advantage of the Excedrin film unit, as Rogers testified, was that it did not require an image reversal because it was exposed and viewed from opposite sides. The SX-70 camera used a complicated mirror arrangement to reverse the image. Also, because of the position within the film unit that the processing solution was spread, the migrating dyes did not have to pass through the "goo" to reach the image-forming layer, resulting in greater tolerance in spread thickness as well as a system that could even tolerate some bubbles, "which would be troublesome in SX-70."[22] Further, the white-pigmented layer (titanium dioxide, or TiO_2) against which the image would be viewed could be preformed during manufacturing of the unit. In SX-70 the opacifying layer formed by the spread processing solution had to become bleached out after processing to become the white background for the image.

Despite all these advantages, ones that Kodak certainly recognized when it selected this format for its film, Polaroid stuck with the SX-70 design rather than switch over to Rogers' new concept. Rogers explained, "The SX-70 system was nearly ready to go and there had been heavy investment made in research and engineering and production machinery to produce it . . . [so] there was insufficient incentive to change those plans."[23]

On the morning of December 2, 1981, day twenty-nine of the trial, Polaroid was nearing the end of its case. Schwartz had already advised the judge that he would likely complete Rogers' direct examination on that day. As soon as Kodak completed its cross-examination, Polaroid's case would be complete, and it would rest. Accordingly, the time was right for Judge Zobel to address the question of the motions to dismiss that Kodak had said it would file at the end of Polaroid's case. The previous week,

Kerr had written a letter to the judge about Kodak's intentions. While he realized that the court could not prevent Kodak from pursuing this tactic, he wanted to do his best to ensure that the motions would not otherwise impede the progress of the trial.

First, Kerr advised the judge that Polaroid's lawyers had "conducted extensive research . . . establish[ing] that Rule 41 dismissals at the close of Plaintiff-patentee's non-jury case on the grounds of patent invalidity or non-infringement are virtually non-existent."[24] He then explained, "[We will] share the results of this research with the court and Kodak at this time, before any such motion is tendered, because we believe that a Rule 41 motion in this case is foredoomed to failure and that it will serve no purpose other than to waste the time of the Court and the parties." After reciting all the relevant case law, Kerr pointed out that Kodak's motions were not "a satisfactory substitute for proceeding in the time-honored way, namely to resolve the complex issues presented here on the basis of a complete record and with the benefit of full post-trial written submission." Finally, he acknowledged that while the judge could not prevent Kodak from making its motions, she did "have the power to regulate the manner in which any such motion is argued and disposed of." He urged her to address those procedures in a way that ensured that the trial would "not be delayed for consideration of any such motion."[25]

What Judge Zobel had to say on this particular morning undoubtedly brought a mixture of comfort and angst to both sides. First, she granted Kerr's wish that the trial not be delayed if Kodak filed its motions, by announcing that she would "consider them even as the evidence goes on."[26] Accordingly, to avoid the potential for wasted time should she grant one or more of the motions, she "invited Kodak, if it wishes, to structure the beginning of its case in such a way that it addresses primarily the patents that are not subject to the Rule 41 motion." Carr immediately protested because, "of course, the defense is already structured, your Honor," but before he could further explain, she cut him off. "Well, I don't know, Mr. Carr, how much of a problem this really is," and explained that she wanted to proceed this way because of the approaching holidays and because she just did not want to interrupt the trial. She was willing to accept the risk and advised Carr, "you can go ahead and present your case in any way, and if, indeed, any of the Rule 41 motions are allowed, then . . . [some of Kodak's case] may have gone in for naught."[27]

Kerr and Schwartz were delighted with the judge's approach, perhaps until she added one additional comment directed to Kodak's counsel.

"Mr. Carr, I have some thoughts now as to problems I have with either infringement or validity as to some patents. I may be wrong about that, but at the moment, I have some thoughts about that [and] I will tell both of you what those thoughts are"—after the motions are filed.[28] Although Zobel never did follow up on her promise to share her thoughts on those "problems," her comment couldn't help but deepen any feelings of vulnerability that might have already existed among the Polaroid contingent.

This business complete, Rogers returned to the witness stand. Schwartz completed the process of having him distinguish the various prior art references Kodak had cited from the invention of his Excedrin film unit. Then, repeating the same dramatic pattern he had used with Rogers on the dye developer patent, Schwartz handed to the witness a copy of a Kodak patent application and noted to the judge that the document was not prior art. "What is the purpose of this?" Zobel asked. Having been cued up perfectly, Schwartz explained that "this is the patent application filed on behalf of Kodak which we believe represents Kodak's aborted attempt to patent the subject matter of . . . [Rogers' Excedrin Film Unit] patents."[29] Polaroid had already used Kodak's attempt to patent the Rogers film unit structure for itself as part of its argument in defending Kodak's summary judgment motion. So, theoretically, this was not news to Zobel. But Schwartz milked it for all it was worth, slowly and carefully taking Rogers through the Kodak application so that he could demonstrate to the judge that Kodak had indeed tried to patent for itself the very same invention it was now claiming was unworthy of patent protection. Not only that, but the Kodak application also distinguished away three of the film units in Land's name that Kodak was now asserting taught the Rogers invention and thus invalidated his patents.

As Schwartz next took Rogers through his testimony on infringement, describing how each and every element of the claims in his patent were satisfied by Kodak's PR-10 film unit, the judge had a sudden and critical epiphany. While Rogers was responding to a series of questions equating the various layers of the patented film unit to those in PR-10 film, Zobel interjected. "Do I understand Kodak's position to be that these claims only claim the inclusion of carbon as an opacifying agent in the processing fluid?" she asked.[30] "My recollection of the summary judgment motion is that . . . Mr. Carr argued . . . that this patent deals with the inclusion of carbon in the processing fluid for opacifying . . . [and] your argument at that time was that the patent is invalid because the use of carbon as an opacifying agent was well known prior thereto?" "Precisely," was the quick but perhaps unfortunate response from Kenneth Madsen, Carr's partner handling Rogers' cross-examination.

"But your position, Mr. Schwartz," continued the judge as a realization dawned on her, perhaps for the first time, "is that the patent doesn't deal only with that but also with the arrangement of the layers?" she asked. "Yes," confirmed Schwartz, "the arrangement of the layers and the other features which Mr. Rogers has testified about, where you put the fluid in it, how it is used, and so forth." "All right," said Zobel. Message received. Her earlier simplistic characterization of the patent that had alarmed Polaroid was now hopefully redefined in her mind forever. And with that, Schwartz sat down, Polaroid's direct case complete.

Although Carr had taken the deposition of Rogers for twenty-one days and was clearly the architect of Kodak's arguments, his partner and primary deputy, Kenneth Madsen, stood up to handle the cross-examination. Rogers had already demonstrated that he was a tough nut to crack as a witness. Even though Rogers' direct examination had taken less than two days, Madsen would continue to hammer away for all or part of six trial days. It's hard to imagine what Kodak counsel thought they were going to accomplish. The length of Rogers' deposition had given him the opportunity to focus on and to master all of the documentary material he would be confronted with, and by this point he had been well schooled by Polaroid's lawyers to understand all of the legal issues regarding his patents, and Kodak's arguments, inside and out. There was simply no room for surprise at this point, and none occurred.

Also, Rogers had demonstrated a temperament as a witness different from Land's. While Rogers was patient and willing to explain gently for Judge Zobel anything she asked, he was not prone to go off on extended discourses. He was, in life and on the witness stand, a man of few words. Seemingly egoless, he was not enamored of the sound of his own voice. Further, he had another trait deemed admirable for any witness in a legal proceeding: He was not at all reluctant to admit that he did not recall something, if indeed he did not recall it. Nor was he unwilling to admit that his recollection was not refreshed when he was shown some document intended to do so. These characteristics helped to keep brief or, in some instances, to short-circuit many lines of cross-examination.

Notwithstanding Rogers' effectiveness as a witness, Madsen was able to demonstrate for the court several of Kodak's key arguments against his patents. Even though Rogers seldom helped in the exercise, by pursuing well-honed lines of examination, Madsen was at least able to make sure

Judge Zobel understood what Kodak's arguments were. One such subject was its contention that Rogers' Excedrin film unit was obvious in light of a series of patents issued to Land on various structures for integral instant film units. According to Kodak, one of these film units disclosed Rogers' film structure except that, instead of spreading a layer of black processing solution under the transparent outer support of the film unit to opacify it for ejection out of the camera, the Land patent taught the use of an opaque superposable flap instead. Polaroid's position was that the use of a flap made any such film unit not an "integral" one—that is, one that stays together before, during, and after processing. Madsen worked mightily to get Rogers to agree that Polaroid had considered such film units "integral" in various patents and documents filed with the U.S. Patent Office, including some in his own name, but he got nowhere with the witness.

Zobel finally asked Madsen what conclusion she was to draw from his line of examination. "Your Honor, we have heard testimony about the fact that a film unit, which has a superposable flap, instead of opaque goo, is a different animal in the sense it is not integral in the way they have used the term integral," Madsen explained. "This [examination] is to point out, first, that [in] the same application which Mr. Rogers filed on the black goo in the pod, he also thought of the idea of using the opaque flap and does not characterize the end result any more or less integral irrespective of which one he chooses and, secondly, to verify or supplement our argument that the prior art looks at these as interchangeable mechanisms of opacifying a film unit on the exposure side."[31] Just to make sure she understood what he was saying, Zobel asked, "You mean the flap and the black goo?" "Right," confirmed Madsen, having in effect testified in lieu of Rogers. "All right," said the judge, "I understand where you are trying to go."

For three more days, Madsen continued in the same vein. Rogers did little to help him, but Madsen persisted in his strategy of educating the judge by laying out again and again Kodak's key arguments through the questions he posed to the witness, whether or not any helpful testimony was actually secured in the process. Finally, on December 9, 1981, the thirty-fourth day of trial, after another series of "I don't remember[s]" from Rogers, Madsen announced, to the relief of all, "We are prepared to finish our cross-examination."[32] "Does that mean that it is finished or you are being ready?" asked a bemused Zobel. "It is an interesting notion," the judge could not help observing. After just a few more inconsequential questions, Madsen then announced, "We are concluding." The judge, showing she had not lost her sense of humor, as well as revealing perhaps an expression of her pleasure at reaching this milestone in the case, asked, "The preparation to finish is concluded and the finish has begun?"[33] "Yes," Madsen replied, and sat down. The judge excused Rogers from the witness stand and asked Schwartz if the plaintiff had any other witnesses. "No, your Honor," said Schwartz, and he then uttered those magic words, "The plaintiff rests."[34]

CHAPTER 23

KODAK'S DEFENSE

With Polaroid's presentation complete, Judge Zobel turned to Frank Carr to begin Kodak's case. "The first order of business of course, your Honor, is our filing of Rule 41 motions," said Carr, announcing that Kodak had five of them, each directed to a different Polaroid patent.[1] Interestingly, all three of Land's patents were included in the group. Given that Kodak was seeking to eliminate them from the case on the ground that Polaroid had not established its right to relief during its direct case, the motions were, at least in part, capable of being construed as a direct swipe at the adequacy and effectiveness of Land's testimony. In addition, Kodak challenged two of the camera patents, those covering the gear train and the detachable spread roller housing. For whatever reason, Howard Rogers' patents were not included.

Actually, it was hard to say what Kodak's strategy was, given the long odds against its success on the motions. Were the motions just another personal tweak of Polaroid's patriarch? Perhaps. More likely, they were simply another wave in Kodak's cumulative effort to erode the strength of Polaroid's patents by giving the judge repeated opportunities to focus on Kodak's arguments against them. Judge Zobel set a schedule to handle the motions while the trial continued. Polaroid was given nine days to respond in writing, and then oral argument was set for December 23, just prior to the Christmas and New Year holidays.

Kodak's Rule 41 motion papers set out its now familiar arguments. It contended that PR-10 film did not infringe Land's L-Coat patent because its polymeric acid layer was not coated on the same support as the rest of the photosensitive layers and thus not located within the "photosensitive element" of the film unit, as required by the patent.[2] Similarly, and somewhat surprisingly, Kodak's attack on Land's Symmetrical Support patent was based solely on a very technical noninfringement argument. Neither motion put forth Kodak's main contentions that the patents were invalid because the inventions claimed were previously known or were obvious and thus did not rise to the level required for patentability.[3] Such was not the case regarding Land's Rear Pick patent. Here, Kodak did not, at least at this point, contest infringement. Instead, it argued only that the patent should be declared invalid because "the evidence adduced during Polaroid's main case allows only one conclusion, that the 'invention' of the . . . [Rear Pick] patent would have been obvious to a person of ordinary skill in the art of camera design at the time it was made."[4] On the other patents, Kodak contended that Polaroid had failed to prove infringement with respect to the Gear Train patent, and maintained that the Detachable Spread Housing patent was neither valid nor infringed.[5]

When Polaroid filed its papers in opposition the following week, its basic position was that it had indeed discharged its burden of proof and demonstrated its right to relief. It asserted: "Kodak's motions either ignore or contradict the [trial] record, are unsupported by evidence or the law, and should be denied in all respects."[6] The Polaroid brief went on to reiterate the legal principles Kerr had set forth in his earlier letter on the subject—that successful motions to dismiss in patent cases were extremely rare and "should be granted only in clear cases." Although Polaroid put forth its position in great detail, the tone was clearly one of disdain for what it obviously believed was Kodak's audacious attempt to short-circuit the full trial on the merits that the parties were already almost halfway through. Yet, given the judge's off-hand comments a few weeks earlier that she had "problems" with one or more patents, Polaroid could not afford to treat Kodak's motions lightly. The briefs were filed on schedule, and with the matter submitted to Judge Zobel for her to consider over the holiday break, Kodak returned to the task at hand of presenting its defense.

Kodak's first witness was Leo Thomas, its current director of research. Thomas was not presented as an expert witness to opine on the Polaroid patents in suit but as a fact witness who could describe generally the developmental project that went on at Kodak and led to the introduction of its instant photography camera and film system in April 1976. Responding to Carr's direct examination, Thomas recalled how in late 1968 and early 1969, Kodak embarked on a program known as "PL-976 . . . to put an instant photographic product on the market." He explained that the "'PL' was the designation for laboratory project, and '976' referred to the presumed or estimated date of appearance on the market of the product—1976."[7] His understanding was that their objective was "to produce a very high-quality color print in a camera and that it would have to be free of the encumbrance of existing patents, because we didn't think we would get any licenses," clearly referring to Polaroid.

According to Thomas, what Kodak was attempting to do was "to develop what we called an 'integral product,' a picture unit which would not require timing and peeling apart." By mid-1969, according to Thomas, the project had been renamed "PL-974" as "a prod" from management, which was impatient with the 1976-projected date.[8] Of course, none of this testimony was inconsistent with the story Schwartz and Polaroid had developed. Land had testified that he had shown Kodak executives the prototype for what would become SX-70 in November 1968, and Schwartz was prepared to demonstrate through Kodak's own documents how the company had reacted dramatically to that disclosure. Carr, however, wanted to lay out in detail the research effort Kodak had undertaken to enter the market in an attempt to bolster its case that it had labored hard to develop its own system so as to avoid Polaroid's patents. Getting the testimony of Thomas on record was the start of that effort, and in the next several days other Kodak executives and researchers took the stand and helped to tell the story.

As Kodak's version of these events was retold for the court, it became clear that, in several key instances, a certain amount of revisionism was taking place. At least, it appeared to Schwartz that some of the explanations for events provided by the witnesses were at odds with the black-and-white record memorialized in Kodak's internal documents collected by Polaroid during discovery. This provided him with material for the kind of traditional cross-examination he intended to conduct—one that is very focused and seeks to impeach the witness by tying him or her to contradictory documents or previous testimony.

One such opportunity for Schwartz involved Thomas' explanation of why Kodak's P-129 program, aimed at developing a peel-apart film for use in Polaroid's existing cameras, was abandoned in November 1972, just months after Polaroid had introduced its integral SX-70 system. On direct examination, Thomas suggested that the project had been abandoned because the research team had been unable to solve a technical problem associated with containing the excess processing fluid in the film unit.[9] While such problems may have existed and may have been the target of ongoing research efforts in the Kodak program, Schwartz knew that this was not the whole truth.

On cross-examination, Schwartz asked Thomas if, when he gave his explanation, he had been aware that the decision to abandon the project had been made by the Kodak board of directors? "Well, no," conceded the witness. Schwartz then showed Thomas a copy of the minutes of the board held on November 16, 1972, and directed him to a section reporting that Walter Fallon, Kodak's president, "reviewed at length the Company's progress in the development of its own products for in-camera processing and explained the *business* [emphasis added] consideration for management's conclusion. . . . The Board was advised and concurred in management's present intention to discontinue further efforts to market peel-apart film packs for Polaroid cameras," Schwartz continued to read. Schwartz asked if Thomas had that board decision in mind when he gave his testimony concerning his understanding of the reason for abandoning P-129. "Mr. Schwartz, it was not the habit of the Board of Directors to send me a copy of their minutes," Thomas admitted. When Schwartz pressed for a direct answer, Thomas acknowledged that he did not, indeed, have the board's business rationale in mind when he presented his version of the event.[10]

But Schwartz was not done. Referring to Fallon's reference in the minutes to "the business considerations for management's conclusion," he asked Thomas whether he was aware of any testimony Fallon had provided during depositions relating to those business considerations.[11] When Thomas admitted that he wasn't, Schwartz showed him a Kodak memorandum dated November 6, 1972, just ten days before the board meeting. He pointed to a passage that read, "It is recommended that further effort on Project 129 be discontinued." Then he directed Thomas to the reason why: "In light of these facts, it would appear that in 1975, we would be introducing an essentially obsolete product, in a dying phase of this market." Schwartz asked Thomas whether he found that statement in the memorandum. Flustered, Thomas started to offer some kind of response, but the judge cut him off. "There is no question except do you find that, and I guess the answer is yes," Zobel said. Schwartz quickly followed up asking whether he had taken that report into account in connection with his testimony on the reasons P-129 was abandoned, and when the witness started to explain, Schwartz jumped in again mid-answer. "My only question is: did you take into account this memorandum in giving your answer?" hammered Schwartz. "No," Thomas was forced to admit.[12]

With that, Schwartz moved quickly on to the next topic with the same precision, with both of his objectives accomplished. It was classic cross-examination technique. First, the judge had been educated on the story Polaroid wanted to tell, and thus put on alert that Kodak's version of the history behind the case might not be as innocent as presented. Second, Schwartz had somewhat undermined the credibility of this particular witness by suggesting he was either less informed than he had been portrayed to be, or was perhaps a willing party to the spin Kodak was attempting to put on the events in question. Both of these were useful goals. Most importantly, though, Schwartz's skillful cross-examination of this first Kodak witness almost certainly raised the possibility that Judge Zobel might become increasingly skeptical as Kodak continued to lay out its case.

As Kodak's defense case continued, it wasn't long before Schwartz again felt compelled to inject some reality, at least as perceived by the Polaroid side, into the story being spun by Carr and his Kodak witnesses. Following Thomas, Kodak's next witness was the director of the P-130 project, Albert Sieg, who testified about Kodak's instant photography development program during the period starting in 1971. Another of Carr's partners, Robert Fier, handled Sieg's testimony. Among other things, Sieg described the "Lanyard Camera" and film unit that were the prototypes of choice in early 1972, just before Polaroid introduced SX-70.[13]

This system was operated manually; a lanyard pulled the film unit out of the camera using two metal tow bars in the front of the camera. The camera was designed to strip away from the film unit the pod that had contained the processing solution, as well as the trap that collected the excess. As Sieg got into his detailed description of the Lanyard system, Judge Zobel asked Fier, "Why are you getting into this in such detail?" "Your Honor," Fier explained, "Polaroid has made quite a point about the fact that we were, at Kodak, sort of stumbling along until Dr. Land showed the world his SX-70 camera and that we then dropped everything we were doing and ran to copy it. I will show you through this witness and subsequent witnesses that that is not the case at all; that we were started on our own path and did our own thing . . . you will see the connection between this camera and the camera that is now on the market, and I think your Honor will perceive immediately the similarity."[14]

Satisfied that she understood the purpose, Judge Zobel allowed the testimony to continue. Fier led Sieg through the rest of his description of the Lanyard system. He readily acknowledged that he and others at Kodak watched Polaroid's launch with great interest and even admitted to having attended the Polaroid shareholders meeting in April 1972 at which Land first demonstrated SX-70 publicly.[15] Sieg also conceded that Kodak had made changes to its product after having analyzed Polaroid's, principally because Kodak realized that its system would ultimately have to compete with what Polaroid had on the market. According to Sieg, the principal realization Kodak took from its examination of the SX-70 was that it did not have to strip away the pod and the trap and that a film unit could be designed to keep those components intact within the film unit when it was ejected from the camera. Eliminating these operations gave the Kodak engineers the opportunity to make the film cartridge, and thus the camera, smaller. In conceding this, Carr and Fier were very much aware that, not coincidentally, these particular aspects of the system design were not related to any of the Polaroid patents Kodak was being charged with infringing. Thus, they were trying to cast Kodak's reaction to SX-70 in a relatively benign light from a legal perspective.

Schwartz, however, determined that a more thorough explanation of these events and their implications for the case was necessary. Given the statement Fier had made, he knew it was crucial to demonstrate for the judge right there and then that there was more to know about the events that occurred at Kodak after the introduction of SX-70 than this witness was disclosing.

As soon as Fier sat down, Schwartz got right to the point and asked Sieg to demonstrate for the judge how the Lanyard camera operated, using the model that had been introduced into evidence.[16] Schwartz's mission here was to highlight each of the inventions in suit that the Lanyard camera lacked, but that Kodak's EK-6 camera ultimately had. He had Sieg confirm that, in the Lanyard camera, the film exited straight out without being deflected, that it did not have a gear train extending along the side of the camera from the rollers to the back to coordinate its mechanical functions, that the film units were transported out of the camera by two little metal pieces at the forward end of the film unit that moved a tow bar, and that there was no detachable spread housing. He then showed Sieg a Kodak EK-6 camera and asked him to show how the film cartridge was loaded and where the film unit exited the camera. "Is that deflected at an angle, approximately 15 degrees?" Schwartz asked. "In this camera . . . yes," replied Sieg. "Does this camera have a gear train which extends along one side?" Schwartz continued. "Yes," the witness replied. "Does it have a pick in the rear of the cartridge which picks the film unit one at a time?" asked Schwartz. "Yes," Sieg again responded, falling into the rhythm of an effective cross-examination. "Does it have a separate spread assembly?" Schwartz continued. "It is removable by the camera technician, yes," acknowledged Sieg. Schwartz went in for the kill. "Will you agree that the features we have just gone through for the EK-6 do not appear in the Lanyard camera?" he asked the witness. "Yes," conceded Sieg.[17] So much for Fier's representation about the similarities between the Lanyard camera and the EK-6.

Schwartz then turned to Sieg's testimony concerning camera size. He asked Sieg if he recalled Kodak's marketing people informing him in March 1972 "whether or not they considered the Lanyard camera to be sufficiently small in size."[18] As Sieg hesitated to respond, Schwartz said, "Maybe I can help you," and handed the witness a copy of a memorandum Sieg had written dated March 17, 1972. Schwartz directed him to the last paragraph and read aloud: "It is desirable to have the smallest possible camera consistent with good design and engineering. . . . The Marketing Organization is of the opinion that the camera size as presently known meets this goal and that its present size is not too large," Schwartz continued reading. "Is that a reference to the Lanyard camera which you have testified about . . . ?" he asked. "Yes, that would be the camera," Sieg replied. "At least in March of 1972, so far as you were concerned, the Marketing Organization was content with the size of the Lanyard camera?" Schwartz reiterated. "Yes," Sieg confirmed.

Schwartz then showed Sieg another memorandum he had written, this dated May 11, 1972, at which the comparative size of Polaroid's SX-70 was addressed, using "styled wood block models of the present Aladdin camera" made at Kodak after the April 1972 Polaroid shareholders meeting. Sieg confirmed that the "Aladdin camera" was a reference to the SX-70 camera. Then Schwartz pointed him to a section of the memo reporting that "the committee members were in agreement that the P-130 camera as presently conceived appears too large."

Schwartz asked Sieg, "Could [you] explain what transpired between March 1972 and May 1972 that caused your committee to believe that the cameras presently conceived appeared too large?"[19] Sieg resisted the obvious connection and launched into a reiteration of his testimony regarding the development path Kodak was already on and how its realization that the pod and trap could be retained had influenced considerations about the size of the camera. Schwartz tried to rein him in. "Can you point to any other event between March of 1972 and May 5, 1972 that caused any change in the camera size other than the introduction of the SX-70 in April 1972?" Sieg resisted again and started talking about "a continuing dialogue between designers and marketing people" when Schwartz cut him off. "Can you point to any other event in that time period besides the introduction of the SX-70?" he insisted. "I don't believe there is any single event," Sieg said. "Other than that, is that correct?" Schwartz pressed. After the witness did not respond, Schwartz moved on, considering the point had been made, whether the witness would agree or not.

When Schwartz was done with his cross-examination of Sieg later that morning, Carr called his next witness. Lee J. Fleckenstein was the Kodak chemist responsible for the sulfonamidophenol imaging chemistry used in its PR-10 film and accused in this case of infringing Rogers' Negative Dye Developer patent. Kodak believed it had Rogers' patent caught between a rock and a hard place. On the one hand, it asserted that the sulfonamidophenol compounds used in PR-10 film were not dye developers and therefore did not infringe the patent. Thus, Fleckenstein's job at trial was to try to convince the judge that his chemistry operated more like Kodak's conventional coupler chemistry and was not a dye developer within the meaning of Rogers' patent. On the other hand, if one accepted a definition of negative dye developers broad enough to include Fleckenstein's sulfonamidophenol compounds, as Rogers had testified, then Kodak believed that the dye developer process as broadly defined would be anticipated by a process disclosed in a 1960 Australian patent issued to two scientists by the names of Whitmore and Mader. Under this broad definition, Polaroid would *not* have been the first to develop the dye developer process after all.

Rogers had testified that the Whitmore-Mader chemistry was a two-step process that didn't work and thus could not be considered as effective prior art against his patent. Following Fleckenstein, Kodak put on the stand Charles Schallhorn, an organic chemist from the research labs in its Color Instant Photography Division. Just before the start of the trial, Schallhorn had been asked to conduct experiments to evaluate the Australian patent in connection with the ongoing litigation. He testified that as a result of those experiments, he concluded that the compounds disclosed by Whitmore and Mader did, in fact, release their dyes in the areas where there had been exposure, just the way one would expect in the Negative Dye Developer process.[20]

According to Schallhorn, his experiment was done in a single procedural step. However, he acknowledged that he did not use the dye described in the 1960 Australian patent because it had subsequently been determined to be carcinogenic. He chose other available dyes instead. On cross-examination by Edward Mullowney, a young Fish & Neave partner, Schallhorn also acknowledged that other elements of his experiments were not available in 1960, including the mordant, despite the fact that the Whitmore-Mader patent disclosed a number of mordants that could have been used in his experiment.[21] He also admitted that he did not ensure that the formulation of chemicals of the processing composition he used was known in 1960. The question thus left for the judge to consider was whether or not those departures from the strict teachings of the Whitmore-Mader patent invalidated the conclusions Schallhorn reached from his experiments. She asked Mullowney, as he proceeded through his cross-examination, "Are you suggesting that . . . the experiment was not done according to the technology of 1960?" Acting the innocent, Mullowney responded, "I just want to find out . . . I am just probing . . . to see which actual things which were used were available in 1960."[22]

R. Frederick Porter followed Schallhorn to the stand. Porter was an MIT-trained chemist who had worked at Kodak since 1958 in various capacities involving photographic imaging chemistry. He was currently the assistant director of Kodak Research Laboratories and a corporate vice president.[23] Carr examined Porter as an expert on the Rogers Negative Dye Developer patent, as well as a fact witness with respect to the program Kodak had undertaken to develop its PR-10 film. With just a week remaining before the court would recess for the holidays, Carr was working to complete most of Kodak's case on the film patents before the break. Porter's testimony took about two days of trial time and, consistent with the pattern that had developed during Kodak's case, Schwartz's targeted cross-examinations took only a fraction of that time.

Accordingly, in the next few days Carr was able to call two additional Kodak chemists as witnesses concerning Kodak's development of its instant film, Judith Schwan and Kin Kwong Lum. By this point, everyone working on the case was profoundly exhausted from the incredible pressure of the trial and the 24/7 effort it required of each team member, from the secretaries to the paralegals to the young lawyers, and all the way up to Schwartz and Carr. Neither of Carr's witnesses seemed especially critical, and one couldn't help getting the impression that both sides were happy to run out the clock until the much-anticipated holiday break. The cross-examination of Lum was completed on Monday, December 21, 1981, whereupon the trial was adjourned until January. The oral argument on Kodak's Rule 41 motions, scheduled for December 23, was never held, having been waived by mutual consent of the parties.

When the trial reconvened on Monday, January 4, 1982, Judge Zobel got right to the subject everyone was awaiting. "The first matter to be dealt with today—and I see this air of anticipation," she noted, "are the pending Rule 41 motions."[24] Attorneys and executives for both sides held their collective breaths. "Because the factual questions in the case are very complex," she explained, "and because any judgment by this Court is most likely to be appealed, I deem it appropriate and in the long run more efficient to make findings and rulings as to all of the issues that are raised so as to facilitate a complete review by the Court of Appeals in the first instance. Accordingly, I deny without prejudice," she continued, "that is, not on the merits" each of the motions filed. Although even Kodak knew that it had been a long shot to expect the judge to grant any of its motions, the Polaroid side felt a palpable sense of relief. Judge Zobel went on, "[I] assure counsel that I do not regard the filing of the motions as having been in any way an exercise in futility. . . . I appreciate being focused on the issues and having, at least in retrospect . . . a good deal of the cross examination . . . which was obscure at the time, made clear and obvious to me now."[25] Thus, for Kodak counsel, even though they did not score the upset win they had hoped for, the motions were at least partially successful. To the extent part of their strategy was to educate the judge on Kodak's case by putting its arguments before her over and over in different forums, they had succeeded, even by the judge's admission.

For Polaroid, there was some relief that it had avoided a result that would have been as unexpected, given precedent, as it would have been devastating. It also got the benefit, for whatever it was worth, of a public relations boost from the judge's ruling. The headline of the Associated Press report in the next day's papers proclaimed, "Court Denies Kodak's Bid to Water Down Polaroid Suit."[26] Calvert Crary, the Bear Stearns analyst, reported the denial of Kodak's motions in his *Monthly Newsletter on Corporate Litigation*, but admitted "Judge Zobel's decision doesn't tell us much about how she will ultimately decide the case."[27] He acknowledged that he "thought Kodak's arguments against the patents were pretty good" but recognized that they were all "subjective and debatable" and thus could go either way. "Under these circumstances," he concluded, "it must be assumed that Kodak's likelihood of prevailing on all 10 of the patents at issue in the case is fairly small."

Thus, the failure of Kodak's long-shot motions sounded something like a Polaroid victory in the ears of the public and especially Polaroid supporters, even if the reality was that it only amounted to a "no contest" in an aborted skirmish. If there was any kind of emotional let-down for Kodak's counsel from the failure to have even a single patent eliminated, Judge Zobel left no time for anyone to dwell on the ruling she had just made, and promptly said to Carr, "I will hear your next witness."[28]

Prior to the holiday break, Kodak had the floor for only ten trial days, so there was still a considerable amount of material to cover. It would take almost an additional two months. Next up for Kodak was Donald Smith, the assistant director of the Chemistry Division in its Research Laboratories. Smith's role was to testify as a witness about the Taylor Mordant patent. He was assigned a double duty: to testify as an expert, opining on the patent's validity and infringement, and as a fact witness, regarding the development of the mordant used in PR-10 film. Kenneth Madsen handled this witness for Kodak and spent his first morning on the stand having the witness describe the chemical structure of the mordant claimed in Taylor's patent, as well as the mordant used in Kodak's PR-10 film. It was a class in organic chemistry, complete with diagrams of carbon rings and their attachments, and Judge Zobel, as usual, was engaged and participated actively.

When Smith concluded his comparison of the two chemical structures, Madsen asked whether the attachments in the PR-10 compound were within the description of those described and claimed in the Taylor patent. "Well, two of [the three attachments] are," Smith replied, but the third "does not fall within the description of the polymers" in the Taylor patent claims. Zobel was right on top of it, and jumped in to ask, "Because it has got more than six carbons?" "Yes," confirmed the witness, as the judge sat back in her chair, and Smith explained why.[29] Satisfied that Kodak's noninfringement case was now squarely on the record, and perhaps in the mind of the judge, Madsen announced that he was then moving on to the prior art and a recess was called. Finally everyone had a moment to allow the event of the morning—the denial of Kodak's motions—to sink in. For two more days, Smith testified about the mordant patent, undergoing a rigorous cross-examination by Edward Mullowney. But Smith proved to be adept at parrying Mullowney's attacks, and the impact of Judge Zobel's earlier apparent epiphany persisted.

The next target for Kodak's case was Land's Symmetrical Support patent. John Otto, a Kodak technician who specialized in its aerial photography products, testified about the Bimat system that Kodak had developed and sold for in-flight photography. Kodak contended that Land had learned about this system through his work with the intelligence community, but Land had testified earlier in the trial that he had cut off any detailed exposure when he became aware of it. As Otto began his description of the Bimat film, explaining that it had a polyester support, Judge Zobel perked up and interrupted him. "Excuse me," she said. "It had the . . . [polyester] supports in the 1960s?" "Yes, your honor," Otto replied. As Otto continued to describe how the process worked, the judge asked some questions to confirm that when the two polyester supports are laminated together, "the wetness stays inside?"[30] "That is correct," Otto responded. He did note, however, that after exposure the laminated supports would then be wound back onto a roll and not used or handled as an individual film unit, as occurred with an instant photograph. While it was clear that Bimat did not operate like SX-70 or Kodak's PR-10 films, Carr believed that it undercut the notion that there was something revolutionary, or even new, about Land's supposed discovery that it could be viable to keep a photograph "forever wet" by using two impermeable supports without negatively affecting the image or the flatness of the film unit.

Schwartz went right after this notion on cross-examination. First, he had Otto emphasize the essential difference between Bimat and any kind of instant photograph by asking if when "the actual process is carried out, it was always done on a reel-to-reel basis?" "Usually in rolls, yes," conceded the witness.[31] When Schwartz asked him how large the rolls were, Otto admitted that for "a military application, it could be rolls that were a thousand feet long." Schwartz then moved quickly to the issue of image stability in the Bimat system. Again, the huge collection of internal Kodak documents Polaroid had amassed during discovery provided him with the material he needed to make his point with little risk that the witness could stray from the testimony Schwartz intended to elicit. He showed Otto a memorandum he had written in March 1965 concerning the "cover sheeted Bimat material."[32] In the memorandum, Otto had written that the material "does not have an extended life" and that "after about a week or 10 days, the image begins to fade."

Using another textbook cross-examination technique, Schwartz merely asked him if that statement accurately reflected his views at the time he wrote the memorandum, and Otto could do nothing but agree. Schwartz followed by showing him a manufacturing approval document from November 1966, more than a year and a half later. This document stated that the "image in the laminated state has a useful life of [only] approximately three days to a week." Schwartz asked again if that was an accurate statement at the time it was written, but Otto decided to try to equivocate, claiming: "Probably some purists wrote that statement."

Otto offered that he had informed his supervisors that he did not agree with that description of the image stability. Schwartz was not going to let that stand, asking Otto if the manufacturing approval was ever changed in light of his objections. The witness answered, "No." Schwartz then showed him a Kodak brochure on the Bimat system and referred Otto to a statement that "the image keeping . . . on the Bimat Transfer film is reasonable . . . about a week or so." "Did you ever make any attempt to get that brochure changed in that regard?" Schwartz asked Otto. "No," was the response. "As far as you know, was it ever changed?" "I don't think it was ever changed," Otto conceded.[33]

Following Otto, Kodak presented its expert on the Symmetrical Support patent, Peter Adelstein, the supervisor of Kodak's laboratory titled Physical Performance of Photographic Film. Adelstein testified extensively about various film support materials and how they behave in use. He opined that, concerning distortion of a film unit from changes in humidity, "it would have been evident from anyone with any knowledge in this field that the use of a polyester film base would give lower curl than any other material."[34] "Furthermore," Adelstein continued, "he would have been aware of the teachings . . . about a balanced film structure reducing curl." At this point, Judge Zobel sought some clarification, asking whether his "statement [would] apply also to the realities of the film unit which is exposed not only to humidity but to moisture from within itself?"[35] Adelstein agreed. "Yes, it does, but I suspect that will come out in later testimony." Obviously getting the point Kodak was trying to make, Judge Zobel then perhaps showed some of her growing impatience with the deliberate nature of Kodak's presentation. "Maybe we could move a little faster," she suggested.

The following morning, as she took her position on the bench, Judge Zobel addressed the courtroom. "Before we begin, I would like to have some idea from counsel as to their present estimate of how much longer this case is likely to take."[36] It was the end of a long week, January 8, 1982, the forty-seventh day of testimony. Carr admitted that while he had previously told the judge that Kodak could finish its direct case by the end of the third week in January, it now looked as if because of "absolute unavoidable slippage . . . we will finish our case before the end of the month, and as much before the end of the month as I can possibly arrange."[37] It was clear that the judge wanted to move things along. A moment later, when discussing an evidentiary matter pending her decision, she asked Carr when he needed her to rule. When he replied that it wouldn't be for another week or ten days, Judge Zobel quipped, "If I hurry up, will you?" "I promise you," said Carr.[38]

Nonetheless, the deliberate pace of Kodak's presentation continued. As Judge Zobel had demonstrated repeatedly—in refusing to officially limit Kodak's cross-examinations, in denying its Rule 41 motions—she was determined to have a full exposition of the case so as to create a complete record for appeal. However, she was clearly a quick study despite the complexity of the technology, and she may have felt she needed less explanation and elemental testimony than the trial lawyers had anticipated when preparing their cases. But they were stuck with the scripts they had written, Carr in particular.

As Adelstein resumed his testimony, he continued to describe how instant film units react to various environmental forces and how a person of even ordinary skill in the art would have known what to do to alleviate the problem of distortion in an integral film unit. "It would have been very obvious," Adelstein contended. "Of course, there would be only two supports he could have used [and] one is so far superior to the other, he would have chosen the superior one," referring to the polyester disclosed in Land's patent.[39] Pressing his point, Adelstein added that, with respect to the problem of distortion and curl, "someone with a rather rudimentary nature understanding of the problem would have still come up with the right answer."

In what was clearly becoming a direct attack on Land, as well as on the validity of his patent, Adelstein was shown two Polaroid experimental film units from the 1960s that used asymmetrical (different) supports—that is, cellulose triacetate on one side and polyester on the other. They were badly curled. During his direct testimony, Land had used them to show how badly film units would curl and become distorted from having asymmetrical supports. But when Carr had asked Land on cross-examination to explain why the film units curled in one particular direction—that is, towards the triacetate support rather than toward the polyester support—Land had become cautious. "I just want to be careful in answering that," he had said. "I suspect the answer is very intricate, and I don't want to give it without thinking about it," he had explained as he proceeded to describe some of the factors that would have to be considered in any such analysis. Carr had pressed on anyway, and Land had refused. "I don't want to discuss anything that important so casually," he had insisted.[40]

When handed the same film units now, Adelstein apparently did not suffer from the same reluctance that Land had shown when he had declined to provide a simple answer to what at least he perceived to be a complex question. "In his testimony," Carr noted, "Dr. Land declined to offer an opinion as to the reason the film units curled towards the cellulose triacetate support. In your opinion," he asked, "what has caused the curl?"[41] Adelstein launched into a lengthy explanation of the curling phenomena evident in the old Polaroid experimental film units without hesitation. Judge Zobel was intrigued by the testimony and conducted her own examination of the witness as she sought to understand his theory. When the examination moved on to another aspect of Land's testimony, Adelstein was apparently eager to take Land on directly yet again. This time he challenged Land's assessment that different supports react differently to chemicals in the film unit's processing solution and that this might cause deformation of a film unit having asymmetrical supports. "Do you agree with Dr. Land's assessment . . . ?" Carr asked his expert. "No, I do not," replied Adelstein, "I can't accept . . . [his] logic."[42]

Although Land was not physically present to hear his expertise challenged by Adelstein, it was not long before he learned what happened in court that day. Once his own testimony had been completed, Land had stopped coming to the courtroom. He was still completely engaged in the day-to-day events of the trial, but he believed his presence might create a distraction. Instead, he sent Nan Chequer to observe and to report back to him every evening. Land also reviewed the trial transcripts every day as soon as they were delivered by the court reporter. As Schwartz sat at Polaroid's counsel table listening to Adelstein attack Land, he knew that Polaroid's star witness was not going to take this criticism well. Schwartz understood that he was going to have to cross-examine this witness with even more gusto than usual, both for the sake of Polaroid's case, as well as for the sake of Land's credibility as the paramount expert in the field of instant photography, a status upon which so much of Polaroid's case depended.

Schwartz started right in on Adelstein's qualifications. Adelstein admitted that he had been working in the same job within Kodak's film manufacturing operation since 1958 and that during that time he had never done any research or development work on either peel-apart or integral instant film units, nor had he ever constructed either a peel-apart or integral instant film unit. He confirmed that he was not involved in any way with the development of Kodak's PR-10, nor were any of the items listed on his curriculum vitae as evidence of his expertise in the field—articles he had authored or lectures he had given—concerned with the field of instant photography. Finally, Adelstein acknowledged that he had no patents issued to his name and accordingly "had never made any inventions in the field of instant photography."[43] So much for Adelstein's hands-on experience with the kinds of instant film units he had testified about for days in an "expert" capacity.

At this point, Schwartz started in on the prior art references that Adelstein had testified taught Land's invention. One by one he focused Adelstein on the key distinctions. He got Adelstein to agree that the Bimat system did not use an integral film unit of the type to which the Symmetrical Support patent was directed—that is, one that stays together during exposure, processing, viewing, and storage. He went on to other prior patents cited by Kodak, emphasizing that they mostly involved peel-apart film units or, to the extent they described an integral film unit, they did not teach the concept of using relatively impermeable supports on both sides.

Then Schwartz took on the subject of the experimental SX-70 film units, and the explanation for their curl that Land had been reluctant to provide. He asked Adelstein whether he knew the composition of the different layers of chemicals within those film units. Adelstein admitted, "I know very little about the composition of the specific layers." Schwartz asked if he took into account the specific effects of their differential drying properties when he provided his explanation, and Adelstein stated that he "did not think that was pertinent to explain the magnitude of curl I observed." Schwartz pressed on, "Did you take into account the differential rates of drying through those layers due to the different moisture permeabilities of those two supports in giving your testimony?" Again, Adelstein admitted he had not. One by one, on and on, Schwartz put Adelstein through a list of items he had not considered in forming his opinion. Schwartz had a very good source for this material. Land had provided the list to him through Nan Chequer overnight. They happened to be the very factors that had given Polaroid's chief witness pause on the witness stand and had led him to decline to provide a simplistic answer, when Carr had asked him why the film units curled as they did.[44] Schwartz's goal was to demonstrate that Adelstein simply did not evaluate the issue as thoroughly as Land had.

Finally, Schwartz took on Adelstein's testimony that the choice of two polyester supports would have been obvious because of polyester's known characteristics and his contention that there were really only two support materials to choose from in any event. The day before, Schwartz had shown Adelstein internal Kodak documents establishing that its choice of two polyester supports was made in 1973 only after it had done extensive laboratory analysis of Polaroid SX-70 film units. Although there was little the witness could add to what the document said, clearly Schwartz couldn't have cared less. His sole purpose in bringing them out was to educate the judge on the events that surrounded Kodak's choice of supports for its PR-10 film unit, events he was convinced would demonstrate that Land's invention was not as obvious as Kodak and its expert witness were now claiming.

On this occasion, Schwartz showed Adelstein a series of Kodak laboratory documents reporting on all sorts of testing done at Kodak with many different support materials. James Galbraith, the young Kenyon lawyer handling the Adelstein examination, sought to object to this line of questioning because Adelstein's name was not on these documents, and he had already testified that he was not involved in the PR-10 research program. However, the judge allowed Schwartz to proceed since the witness had expressed an opinion on why Kodak had chosen its supports. She agreed with Schwartz's contention that he was entitled to see if Adelstein was familiar with actual work done at Kodak in making that choice.

After painfully slogging along with the resisting witness through two documents from 1971, before Polaroid had released its SX-70 system, Schwartz skipped to the bottom line and asked Adelstein whether he was aware that Kodak had tested more than 600 supports during its development program. "I am not personally aware," Adelstein responded, "but I have been told of that figure," he admitted.[45] Then Schwartz showed Adelstein another Kodak laboratory document. This one reported on Kodak's search for a moisture-permeable support as an essential part of its film unit. Schwartz directed Adelstein to a passage stating that "the present most favored concept for . . . [PR-10] requires a new film support which is highly permeable to moisture."[46] Schwartz's objective was clear: if choosing two impermeable supports was such an obvious thing to do, why in the days prior to the introduction of SX-70 did Kodak conduct all of this research in *precisely the opposite direction*?

After almost three days of cross-examination, Schwartz finally wrapped it up, hoping that he had done enough to defend not only the validity of this patent but also the credibility of Polaroid's star witness. Of course, the great irony in all this was the simple fact that the Symmetrical Supports patent was the one that Schwartz had originally passed over as not being strong enough to assert against Kodak. But now, belatedly educated by the inventor as to certain merits of the invention he might have missed in his original assessment, Schwartz was using his estimable legal skills to fight for the patent's survival. There was no doubt that his cross-examination had been masterful and effective at undermining a good portion of Kodak's expert testimony. But would it be enough? Would Kodak's simplistic description of the Symmetrical Supports patent prevail, despite all of the sophisticated expert testimony on both sides? Would the judge accept Kodak's view that the "invention" amounted to no more than merely the selection of polyester as the supports for both sides of the film unit? Was it all truly obvious? As the trial continued, that question remained a topic of much consternation among members of the Polaroid legal team.

CHAPTER 24

THE BATTLE OF THE EXPERTS

It was now mid-January, and Frank Carr still had seven witnesses on his list. Most were directed to specific pieces of the case and so could be disposed of in a day or two of testimony for each. But also remaining to be presented were two outside photography authorities that Kodak had decided to call as its principal experts. Franz Trautweiler was a Swiss scientist who was enlisted to testify regarding the primary film patents in the suit: Rogers' Negative Dye Developer and "Excedrin" film unit and Land's L-Coat. While he had been a Senior Research Physicist at Eastman Kodak for four years in the mid-1960s, Trautweiler had spent most of his career in Europe, working for another technology company active in the photographic field, the Ilford Group of the conglomerate, Ciba-Geigy.[1]

Trautweiler began his testimony with a professorial lecture on how the eye sees color and how that relates to the reproduction of color in photographic processes. Judge Zobel was clearly fascinated by this exposition and, as had happened on so many occasions, Trautweiler's testimony turned into a dialectic between judge and witness with counsel standing by as an observer. After some time, as their to-and-fro came to a conclusion, Carr apparently felt slightly embarrassed that the discussion had gone on so long. He assured Zobel that "it [was] not intended to be a day-long discourse, your Honor." "Why not?" responded the judge, "I might even learn something."[2]

Carr picked up his lead of the examination and moved Trautweiler into a discussion of the Negative Dye Developer patent. This testimony went on for the rest of the day and the entirety of the next day's court session. Kodak's expert reiterated its now familiar arguments. First, he testified that the Rogers process, as disclosed in the patent, was inoperable without the use of a ballast to anchor the negative dye developer in place. He alleged that the patent did not teach the use of that kind of expedient. If true, this would be a ground for invalidation, as patent law requires that a patent teach "any person skilled in the art" how to make and to use the disclosed invention.[3] Second, he recited and explained at length Kodak's contention that if Rogers' Negative Dye Developer patent were interpreted broadly enough to cover the photographic process in Kodak's PR-10 film, then it would have to be rendered invalid due to having been taught by the prior Australian patent issued to Whitmore and Mader.[4]

As Carr seemingly retraced territory that had been covered with previous witnesses, Schwartz finally stood up to object. He grumbled that the testimony was cumulative and thus a waste of time. Carr tried to justify his approach by pointing out that Trautweiler "is a totally disinterested witness testifying as an expert on behalf of the defendant. The other witnesses who have testified," he argued, "clearly were employees."[5] Judge Zobel jumped right on Carr's rationale, pointing out that Trautweiler had been a Kodak employee at one time, albeit many years earlier. However, she once again was determined to err on the side of completeness and to avoid any criticism later that she somehow prejudiced either party by limiting the presentation it intended to make.

Over Schwartz's protestations, she gave some insight into her position. "Well, I don't wish to have cumulative testimony," she explained, "but a certain amount of repetition does indeed help. It is a little like listening to a Mozart symphony in which the melody is repeated over and over again, and you remember it after the fourth or fifth or sixth or eighth repetition. I don't think we need to go that far," she cautioned, asking Carr if "perhaps you can cut short this testimony."[6] Yet this was not what Schwartz wanted to hear, knowing that Judge Zobel's remark signified a certain acceptance of, if not vulnerability to, one of Kodak's key strategies. From its various procedural maneuvers, as well as the design of the testimony in its direct case, Kodak had sought to bring its arguments before the judge as often and in as many contexts as possible. Its apparent hope was that this repetition would have precisely the impact on Judge Zobel that it was appearing to have, assuming, of course, that she accepted the validity of its positions.

Over the next two days, Trautweiler completed his testimony on the Negative Dye Developer patent and moved to Rogers' Excedrin film unit patent. As soon as Carr asked Trautweiler to provide a general explanation of what the patent covered, Schwartz stood up to raise his objection anew. "Your Honor," he said, "I would like to repress my objection . . . on cumulative [testimony], and at this point I think there is an additional factor, namely, this is a patent that Kodak moved for summary judgment on. They told your Honor," he argued, "that it was so simple that no explanation of the patent was needed, no explanation of the prior art was needed, [and so] your Honor could decide it simply from reading it. Since the time we've been here, Polaroid has put on a witness and explained it. Kodak put on Ms. Schwan [and] she explained the patent, she explained the prior art and she gave her ultimate opinion, in her view, on what she said the obviousness of the invention was," he reminded the judge. "It seems to me to go through this with a second witness on that set of facts is improper, unsound and unfair."[7]

Judge Zobel looked at Schwartz and asked if he was "suggesting that I'm going to be influenced by the number of witnesses who talk about a particular patent?" "No," Schwartz assured her but repeated his contention that it was "blatantly cumulative to put on a witness to do the very same thing" as a previous Kodak witness had done.[8] The judge remained unwilling to concede. "My recollection is I denied summary judgment on the ground there was a difference in the views of the experts about this and that I was not in a position, on summary judgment, to decide the question where experts might differ. . . ." Schwartz pleaded, "I'm not suggesting that Kodak is not entitled to explain it. . . . What I'm saying is [that] it comes from Kodak's mouth . . . that no testimony was necessary, and to do it twice . . ." is, by definition, cumulative.

Finally, Judge Zobel seemed moved to intervene. "I won't hold them to their position that no testimony was needed because I said specifically that it was," she declared. "But I would like there not to be cumulative testimony; that is, I would not like Dr. Trautweiler to be asked the same question and give the same answers as did Ms. Schwan. I assume," she told Carr, "Dr. Trautweiler is here to either expand on what she said or elucidate what she said without repeating what she said." "Most assuredly," responded Carr. "He's not going to repeat what she said and he's not going to be asked the same questions she was asked."

With this assurance provided, Carr resumed his line of questioning with the witness. After the very first question, Schwartz stood up immediately and pointed out, "Just for the record, that question was asked and answered by Ms. Schwan on [page] 4646 [of the trial transcript]. . . ."[9] Carr turned to Judge Zobel for help and asked, "May Mr. Schwartz have a continuing objection to this . . . ?" Zobel turned to Schwartz and told him she was "allowing introductory questions to set the stage for new materials over and above what Ms. Schwan gave us" but refused to prohibit him from objecting if "we get beyond the introduction and there is cumulative evidence." Carr resumed with a few more basic questions. Within moments, Schwartz was on his feet again. Judge Zobel looked at him and said, "We are still on the introduction." "May be the introduction," Schwartz complained, "but it is also this witness' definition of the invention, which was also the subject of discussion by Ms. Schwan. . . . I think that is a little more than introduction." Judge Zobel overruled him and allowed Carr to repeat his question, but it was clear that the witness had been confused by the colloquy. "You have lost the thread?" Carr asked him and Trautweiler agreed. It appeared that the Kodak side had been shaken by Schwartz's repeated interruptions. The witness had lost his train of thought, and Carr seemed to realize that he had to pare down his examination. "Would it be useful to take a recess?" Judge Zobel offered. "Yes," responded Carr, "because I would like to minimize harassment. Not harassment by the Court," he quickly added. "We will take a recess so that you may eliminate all of the harassment materials," Judge Zobel announced, continuing to display her sense of humor.

By this point in the trial, everyone involved was being pushed to their limits of endurance, Carr and Schwartz above all. Each personally bore the ultimate responsibility for his client's case. Schwartz clearly felt he had to be as challenging at every turn as the situation allowed. As a result, his relationship with Carr was showing evidence of strain outside the courtroom as well. For example, while Trautweiler's examination went on during the day, in the evenings, Carr and Schwartz were attempting to work out evidentiary issues relating to some camera models Kodak wanted to introduce into the case. This was relatively routine stuff, but Schwartz was dragging out the process, being as recalcitrant and combative as he could be.

Finally, in frustration, Carr wrote to Schwartz, accusing him of violating an agreement he had made long ago with Bill Kerr that Polaroid would not challenge the authenticity of these cameras.[10] Schwartz was unmoved, and perhaps relished the chance to tweak by after-hours correspondence the opponent with whom he was engaged by day in the courtroom struggle over Trautweiler's testimony. In a terse reply, Schwartz wrote, "I have your letter . . . the tone . . . [of which] can only be described as intemperate. I do not see," Schwartz concluded, "how we can continue to discuss anything with respect to cameras . . . unless you are willing to conduct those discussions in a rational fashion which is consistent with all of our dealings to date during the many years of pretrial discovery in this case and during the trial."[11]

When his testimony resumed, Carr moved Trautweiler quickly into his interpretation of the prior art. He provided his expert opinion that the Rogers film unit structure could be found in various combinations of earlier patents on various instant film units, all of which were from Polaroid's portfolio. Again and again, Schwartz interjected his objection when the testimony mirrored that of Kodak's earlier expert. Sometimes the judge allowed the question, sometimes she did not. At one point, when Carr put on the easel a chart that had been covered previously, Schwartz rose, but the judge cut him off. "I recognize the chart," she admitted, "but I haven't heard the question yet." Schwartz was undaunted, and with the faintest of smiles said, "I recognize the entire camel in the tent, your Honor."[12] Although more often than not, Judge Zobel allowed Carr great latitude in his examination, the overall effect of Schwartz's repeated objections was undeniably unnerving for Carr and made it extremely difficult for him to proceed with his case in any kind of smooth manner. It was getting toward the end of another grueling week of trial. Finally, Carr relented, and in the face of another Schwartz objection, he agreed to "defer" the line of testimony until the following Monday, promising to eliminate any cumulative material before then.[13]

On that following Monday, January 18, 1982, the fifty-second day of the trial, Trautweiler finally completed his testimony on the Rogers patents and moved on to Land's L-Coat patent. For once, the examination seemed streamlined and did not even take the full day to complete. Carr took Trautweiler efficiently through each of Kodak's now familiar arguments on the L-Coat patent. He provided his explanation of how, in his opinion, Land's old patent on the sepia process taught the same process as claimed in his L-Coat patent, that of using a polymeric acid layer to lower the pH of the processing composition. He also stated his view that one skilled in the art would use the teachings of that sepia patent in a color system. "If you leave out . . . the words 'color' and 'dye,'" Trautweiler contended, the process of Land's L-Coat patent "describes the process of the . . . [sepia] patent."[14]

Furthermore, Trautweiler testified that, based on his calculations, the polymeric acid layers disclosed in Land's patents did not have enough acid to effect the change in acidity claimed by the patent.[15] This assertion, if accepted by Judge Zobel, might provide grounds for declaring the patent invalid for inoperability. Finally, Trautweiler provided his expert opinion that Kodak's PR-10 film unit did not infringe the L-Coat patent because its polymeric acid layer was not located in its photosensitive element since it was not coated on the same support as the photosensitive layers of the film unit during its manufacture.[16]

Carr completed his direct examination with little time left in the day's schedule. Schwartz decided to fill the rest of the day with an exploration of Trautweiler's expertise in the field he was testifying about. He showed him a copy of a 1972 article he had written that categorized color photography into three basic groups: dye-coupling, silver dye bleaching, and diffusion transfer. Trautweiler confirmed that the first method was associated with conventional color processing, as used in Kodak products like Kodachrome and Kodacolor. He then confirmed that the second method was one also used commercially to make prints and transparencies and that he had conducted research in this area from 1968 to 1980. Finally, he agreed that his third major category, diffusion transfer, was the process he associated with Polaroid, both its Polacolor peel-apart film and its integral SX-70 film. His foundation in place, Schwartz moved in to make his point. When asked, Trautweiler admitted that he had never done any research in the area of diffusion transfer photography, that all he knew was from reading about it, and that the only analysis he had ever done of either Polaroid's or Kodak's instant products was such analysis as he may have done while taking pictures.[17]

The following morning, Schwartz picked up on the same theme, having Trautweiler admit that none of his publications or patents in the photographic field related to the diffusion transfer technology used in instant photography. Judge Zobel turned to Schwartz and asked, "You're not questioning the witness' qualifications, are you?" She noted that the parties had stipulated prior to the trial that each of its designated expert witnesses had the requisite qualifications to testify. Schwartz acknowledged the stipulation but started to explain, "I think the questions go to—" when Zobel jumped in to finish his sentence for him, "the weight of his testimony," she added.[18]

With that clarification made, and his point clearly registered with the judge, Schwartz moved on to the same general subject that Carr had used to begin his direct examination of Trautweiler. He inquired into the basis for the witness's explanation for how the eye sees color. In short order, Trautweiler conceded that his knowledge was based on his "education at the university" and that he had done no independent research on the subject and was not aware of work or publications by others on that subject.[19] Given that this subject matter had no direct bearing on any issues in the case, the fact that Schwartz addressed it was some indication to anyone paying attention that at least some of the material for his cross-examination had a genesis other than with the Polaroid legal team. It was coming directly from Land's Osborn Street office.

In fact, color vision had been one of Land's lifelong interests. He had personally conducted extensive research in the area and had published several important papers on the subject. As he reviewed each day's courtroom proceedings, Land became more and more incensed about what he viewed as the inaccuracy of the testimony of yet another Kodak expert on another subject he knew so well. Since color vision was not a topic of particular relevance to the lawsuit, the witness had little motivation to skew his views for a strategic reason. He was simply uninformed and wrong. To Land, it was a matter of scientific integrity and principle. To Schwartz, it was a valid means of impeaching the witness's credibility and persuasiveness. If Trautweiler could be so wrong about this generic subject of photographic science, one not in dispute in a legal sense, then how could his expert opinion have any weight in the area of instant photography generally, one he had already admitted he had little experience in?

Even though Schwartz had established that Trautweiler may have had a limited basis for his understanding of color vision, he still needed a way to demonstrate that the man put forward by Kodak as its outside expert might be plain wrong about the substance of his explanation. Throughout Kodak's case, the Polaroid legal team had been working behind the scenes to figure out how the end of the trial would play out. In normal trial procedure, the plaintiff presents its case first, and then the defense has its turn. Afterwards, the plaintiff is given the opportunity, if it elects, to present a rebuttal case addressing issues or facts presented during the defense case that it did not cover in its initial presentation.

The Polaroid team was constantly assessing Kodak's case to determine if it even needed to present a rebuttal case and, if so, which witnesses it would call. While it had been pretty clear to Schwartz from the outset that he would indeed present a rebuttal case of some extent, the big question remained: would Schwartz call Land to the stand again? Carr, still in the dark about what Polaroid counsel might be planning, was anxious to have Polaroid disclose its list of rebuttal witnesses, but Schwartz had resisted, agreeing only to follow the rule, which had been in effect throughout the trial, of three days' advance notice. Schwartz's questions to Trautweiler on color vision were, perhaps, the first indication that a plan to include Land in Polaroid's rebuttal was actually in the works, but at this point any such plan was a closely guarded secret.

With the preliminary material covered, Schwartz launched into a comprehensive cross-examination of Trautweiler on each of the film patents he had testified about. There would be no room for exposition by the Kodak witness of the type Carr had allowed Polaroid witnesses like Land to engage in during their cross-examinations. For example, when discussing the Excedrin film unit, Schwartz read Trautweiler a portion of the Rogers patent claim that defined his invention and asked the witness if an earlier Land patent disclosed that element. He already knew the answer. "No," Trautweiler agreed. Methodically, Schwartz repeated the process five times with five different aspects of the Rogers film unit not disclosed in the Land film unit.[20] Of particular importance was the "photosensitive laminate" of the Rogers film unit. Schwartz read that portion of the Rogers patent and asked Trautweiler, "Now, you would agree, would you not, that the . . . [Land] patent does not disclose such a photosensitive laminate?" Again, Schwartz knew the only answer the witness could give. "It discloses a different photosensitive laminate," conceded Trautweiler.

When he moved on to the L-Coat patent, Schwartz knew that, once again, he had an especially interested audience looking over his shoulder: Land himself. It certainly raised the stakes, but it also provided Schwartz with a unique source of material to pursue on cross-examination. He began by pointing out to Trautweiler specific passages from Land's patent, time and again forcing the witness to concede that the patent taught other advantages of the L-Coat beyond merely lowering the pH of the processing composition. Foremost among these was the teaching that the L-Coat provided an improved image by serving as a "mordant for alkali," thus "washing" the image layer of the ions that could potentially harm the image.[21] Using the same methodical technique, he then got Trautweiler to acknowledge that Land's patent taught several techniques for timing the action of the L-Coat to allow sufficient time for development and then the formation of an image before shutting down the process.

Turning to another point that was surely rankling Land, Schwartz addressed Trautweiler's testimony that the polymeric acid layers described in the patent did not contain enough acid to lower the pH the requisite amount. First, he had Trautweiler confirm that his opinions on the efficacy of the examples provided in the patent were solely based on his calculations. Schwartz then, one by one, asked the witness about other aspects of the L-Coat process that might consume alkali and therefore contribute to the reduction in pH. In each case, Trautweiler had to confirm that he had not included those factors in his analysis. Finally, Schwartz had Kodak's expert confirm that his testimony was not based on any experiments he had conducted. "I take it," Schwartz inquired, "all of that testimony is based strictly on opinion and not on actual tests you have run?" Trautweiler resisted. "I tried to find an explanation why the pH was lower than it should be according to . . . [my] calculations and the data given in the patent," he said. "[Your opinion] is based on your hypothesis as to what you think?" Schwartz rejoined. "That's correct," Trautweiler confirmed.[22] Schwartz was done.

With the Trautweiler chapter of the trial finally complete after seven days on the witness stand, Judge Zobel excused him and wished him a good trip back to Switzerland. For the next two weeks of trial time, Kodak presented a series of fact witnesses, each with a specific story to tell concerning Kodak's development of its instant camera and film system. All of this was a prelude to its final witness, Edward R. Kaprelian. Kaprelian was to serve as Kodak's outside expert on all of the camera patents, in the same capacity as Trautweiler had served with respect to the film patents. He had worked as an engineer in the photographic field for almost 50 years. His extensive experience included stints working in the U.S. Patent Office, research and engineering assignments in the U.S. Army and other military entities, and consulting work in the private sector. "Mr. Kaprelian is an author of scientific and engineering papers and encyclopedia articles, and is a patentee of more than 65 patents in the field of his expertise," read Kodak's submission to the Court describing his qualifications.[23] Kaprelian was also a camera collector and historian of great repute and testified that he had loaned Kodak several cameras from his collection to be used as prior art in this trial.

Once again, Carr's partner John Fogarty handled the examination in the camera portion of the case. For the next five days, Fogarty led Kaprelian through his testimony on each of the four camera patents in the suit, laying out all of Kodak's theories on why the patents were invalid because of what was already known in the art of camera design at the time their claimed inventions were made. In contrast to Kodak's case on the film patents, its emphasis on the camera side was much more directed to arguments of invalidity as opposed to noninfringement. Over and over, after describing the similarities between the Polaroid patents and various preexisting camera models and other references, Kaprelian opined that the mechanical arrangement constituting the invention of each of the Polaroid patents would have been obvious to one skilled in the art at the time the alleged invention was made.

When Fogarty was finally done, Schwartz embarked on another of his systematic and incisive cross-examinations, taking each prior art camera and reference one by one and getting Kaprelian to acknowledge whatever distinctions existed between it and the Polaroid invention in question. He was determined to get each and every concession by Kodak's expert on the trial record, and so this process consumed more than three days of trial time. But Schwartz had saved the best for last. He began his ultimate line of cross-examination by having Kaprelian agree that the inventions in each of the four patents he had testified about were embodied in Polaroid's SX-70 camera. The examination moved on from there:

> Q: Now, you have been studying and collecting cameras for at least 35 years, is that correct?
> A: Yes.
>
> Q: And you have about 2,000 cameras in your collection?
> A: Yes.
>
> Q: And a number of the prior art cameras which you have testified about come from that collection?
> A: Yes.
>
> . . .
>
> Q: And you also have a large collection of literature relating to cameras and camera design?
> A: I do.
>
> Q: And I take it that during this period of time you tried to keep abreast of camera models as they were announced?
> A: As much as I could.
>
> Q: And there came a time, I take it, that you became aware of the SX-70?
> A: Yes.
>
> Q: And did there come a time that you became convinced that the SX-70 was the most innovative and advanced mass-produced amateur camera ever made?
> A: Yes.

. . .

Q: And did you also conclude at some point in time that the SX-70 performed its many functions through such wholly novel, optical, mechanical, chemical and electronic designs as to justify Polaroid's statement that it probably includes more technology than any other consumer product in the world today?
A: I did.[24]

There was a stunned silence in the courtroom. Schwartz walked over to the witness box and handed Kaprelian a copy of an article he had written in 1976 reviewing the history of American cameras since 1840 for a special edition of the magazine *Photomethods* entitled "A Tribute to American Ingenuity."[25] The witness agreed that during this period, hundreds of American cameras were marketed, including approximately 500 by Kodak itself. Yet from all those cameras, Kaprelian acknowledged that it was the SX-70 that he had singled out and written about in such glowing terms. The incongruity was inescapable between the testimony this expert had just given for almost two weeks, belittling so many mechanical features of the SX-70, and his previous views on the same camera, published before he had been retained as a hired gun to testify for Kodak.

Kaprelian had clearly anticipated that this moment was coming. He knew his article was out there, yet there was little he could do during his cross-examination but agree with each of Schwartz's precisely drawn questions. Kaprelian tried, however, to mitigate the damage. As the cross-examination was coming to an end, Schwartz pointed the witness to another statement in his paper—that the SX-70 "performs . . . many functions through a wholly novel mechanical design." This time, Kaprelian tried to qualify the statement, and minimize the impact of what had just transpired, by saying that he was referring to features other than those in the patents in suit when he wrote his statements. Schwartz asked, "Where do you find that in your paper?" "I find it in the camera," Kaprelian said. "Where do you find that in your paper," Schwartz repeated. "It doesn't say that, does it?" Kaprelian could only offer: "it doesn't say it isn't."[26]

It was impossible to know whether Kaprelian's attempt to save the moment had any effect. Cecil Quillen looked shaken as he watched this unfold from the jury box, where several overflow observers sat every day.[27] Schwartz thought he saw Judge Zobel with a wry smile on her face as she observed it all from her perch atop the bench.[28] Her expression, at least in part, was likely one of professional admiration for the classic example of witness impeachment that had just been expertly conducted by Schwartz. Whether or not the exercise would have any impact on the merits of the case, only she knew. In any event, with his point firmly made, Polaroid's trial counsel announced, "I have no further questions."

After some brief redirect from Fogarty and recross from Schwartz, Kaprelian was finally excused from the witness stand. At that point, the court recessed for a previously scheduled winter vacation. When it reconvened ten days later, Kodak called three witnesses to testify regarding some rather peripheral issues in the lawsuit. It was a truly anticlimactic end to Kodak's case, but finally, after some evidentiary matters were resolved, the defense rested on Tuesday, February 23, 1982, on the seventy-second day of trial.

All that remained was Polaroid's rebuttal case. Although Schwartz had announced in court that Polaroid might need up to two weeks for its rebuttal, he knew that he would need nowhere near that amount of time. He had been trained at Fish & Neave as a trial lawyer by some of the best in the history of the patent litigation bar, and had been taught to believe that one should not put on a rebuttal case unless one had to.[29] Nonetheless, with Land insisting on having an opportunity to have the last word, he was determined to play out this final chapter with whatever tactical advantage he could muster.

Carr had asked as early as January 26 for help from Judge Zobel in getting Schwartz to disclose his rebuttal witnesses, to no avail. On February 5, Schwartz advised Carr that Polaroid intended to call at least three witnesses, including two of its inventors of patents in suit, Howard Rogers and Lloyd Taylor, who had previously testified.[30] The third witness was a longtime Polaroid researcher and known associate of Land's, Vivian Walworth. There was no mention of Land. Carr followed with a letter to Schwartz a few days later, again seeking the identity of Polaroid's rebuttal witnesses, together with information concerning the substance of their expected testimony, so that Kodak could prepare.[31]

Schwartz responded the following day, repeating his mantra that Polaroid would comply with the three-day notice rule both parties had followed throughout the trial.[32] He was not about to give an inch more than he had to and was determined to do his best to obfuscate the possibility of Land's appearance. Finally, on February 19, precisely three days before Polaroid's rebuttal case could be expected to begin, Schwartz sent a letter to Carr informing him that "at the present time we are considering calling the following rebuttal witnesses."[33] The letter listed eleven potential witnesses. Land sat squarely in the middle at number five down the list. Schwartz conceded that "it is possible that we may not call all of these witnesses," but beyond that, he offered no insight into Polaroid's plans.

In the end, Polaroid presented only seven of the eleven identified rebuttal witnesses. The first six were completed in two days of trial time. The seventy-fourth and last day of the trial, Thursday, February 25, 1982, was devoted completely to a return to the stand of Polaroid's first and star witness, Edwin H. Land. It was truly a dramatic moment. Land resumed his position in the witness box some four months after his original testimony had been completed. Until the moment he walked to the front of the courtroom, no one but Schwartz and his colleagues knew that Land would return as Polaroid's final witness. A lot had transpired in the interim. Once again, Land was there to defend his company's case. But now he was also there to fight on a more personal level. Land had insisted on this opportunity to defend what he perceived as the personal attacks some Kodak experts had leveled against his work, as well as on his scientific opinions, and thus his ultimate credibility as an expert in what had become, not surprisingly, a battle of experts. The rebuttal presentation was as thoroughly planned as Land's earlier direct examination, but this time it was as much intended as a refutation of Kodak's experts as it was a strategic exercise driven to advance the legal and factual issues of Polaroid's case.

First up was Land's Rear Pick patent. Kaprelian had testified as Kodak's expert about conventional single-frame cameras and motion picture cameras that used claws to advance rolls or cartridges of film with perforated edges. Kodak had cited several of these as teaching the technique of Land's patent. Serving as the legal straight man, Schwartz asked Land to "explain the relationship, if any, that the film-advancing mechanism of those cameras has to solving the problems of designing a film transport and processing system for an integral instant camera."[34] Cued up like an eager volcano ready to spout, Land launched into his response. "The systems, the function of the system, the problems, the technique of handling the problem, are as antithetical as one could imagine," he declared. With great enthusiasm, back in his role as professor in residence on the witness stand, Land proceeded to explain the differences in great detail, providing a historical perspective on the development of these camera mechanisms going back to the days of Thomas Edison. Land then addressed Kaprelian's testimony that his patent did not solve any problems that were not already solved by the prior art and how a person of ordinary skill in the art would have approached this aspect of the camera development. He systematically ticked off more than half a dozen considerations that Kaprelian, who had conceded he had never designed an instant camera, had not taken into account in his testimony.[35]

Next to find himself in Land's crosshairs was Trautweiler, who had testified as Kodak's expert about Land's L-Coat patent. In this instance, Land took on Trautweiler's explanation of how he believed the L-Coat worked and his testimony that he had concluded on the basis of some calculations he had made that there was insufficient acid in the polymeric acid layers disclosed in the patent to make the invention operative. Schwartz asked Land if he was aware of the testimony. Land responded with no hesitation. "Yes, I think the mistake he made was in ignoring the whole structure of this system and, hence, not reading the . . . [patent] carefully enough." What Trautweiler had failed to understand, according to Land, was that there are several elements of his invention—not simply the polymeric acid alone—that operated together to control and finally to reduce the pH of the processing composition.

In his explanation, Land highlighted each of the mechanisms Trautweiler had confessed during his cross-examination to having omitted from his calculations. Then, using a chart he had created, Land explained mathematically how his L-Coat functions to lower the pH by the amount, and in the manner, disclosed in the patent. "If one were to follow Dr. Trautweiler's recommendation and have enough acid . . . it would be an unnecessarily large amount, in some cases a dangerous amount," Land warned. He declared that his patent "describes precisely and accurately what is necessary."[36] Finally, under questioning from Schwartz, Land then disclosed that both the Polaroid and the Kodak commercial instant film units used an amount of polymeric acid within the range taught in his patent.

Land's rebuttal now moved to the subject that really stuck in his craw: Trautweiler's explanation of color vision. Once again, Schwartz acknowledged that the subject did not have any direct bearing on issues in the lawsuit. To the court, he justified bringing it up in rebuttal on the basis that Kodak had thought it important enough to have the subject covered by its expert. Carr did not resist. "If Dr. Land would like to carry on about this, I have no objections," Carr told Judge Zobel. The judge acknowledged, "I admit it is very confusing," when Land jumped in: "For the sake of science and truth in this court, let me say just two or three sentences." "For the sake of my edification," offered the judge.[37] For the next few minutes, Land offered more than just a couple of sentences. It was more of an introductory lecture in a field of tremendous fascination for Land, one he would continue to do research in for years to come. In explaining the differences between how the eye sees color and how a color film must be designed to duplicate that color image, Land exposed what he contended was the profound inaccuracy in Trautweiler's explanation. To sum up his refutation of Kodak's rival expert, Land noted finally that the common misunderstanding Trautweiler espoused was "a creationist's view of color, but it is not the way it works."[38]

To this point, Land's rebuttal of Kodak's experts, however dismissive, had been intellectual in substance, yet polite in tone. It turned out that he had saved his real animus for Adelstein's testimony about the Symmetrical Supports patent. Schwartz began with a question about the curled Polaroid experimental film units that Land had declined to opine on when Carr originally asked him about them during his cross-examination. These units used two different (asymmetrical) support materials on either side and, at the time, Land had demurred because of the complexity of the answer he believed necessary to what Carr had characterized as a simple question. Adelstein, however, as Kodak's expert, had unhesitatingly offered his explanation of the curl when asked by Carr during his direct examination. Although Schwartz had effectively cross-examined Adelstein about his opinion, undermining it as best he could, Land was livid when he read the transcript.

According to Land, not only was Adelstein's opinion based on what he considered to be an inadequate scientific understanding, but Adelstein had felt free to declare Land wrong despite having admittedly zero actual experience in the field about which he was testifying—the design and construction of integral instant film units. Considering his lack of experience, the air of confidence with which Adelstein had offered his baseless opinions offended Land's scientific sensibilities profoundly. Land was determined to refute Adelstein completely and dramatically—and insisted to Schwartz that he be given that opportunity.

Schwartz began to ask Land about the experimental film units, but before he could complete his first question, Carr jumped to his feet and objected. "I submit, your Honor, if Dr. Land declined to discuss such an important matter so casually during his cross-examination on plaintiff's direct case, he should not be permitted to discuss the same subject matter in this rebuttal case."[39] Schwartz pointed out that Kodak had brought the subject up in its case. "I might agree with Mr. Carr," Schwartz said, "if he didn't choose to put on a witness who spent many hours giving his own explanation of why that was so and why Dr. Land was in error in not discussing it." After some additional argument, Judge Zobel decided to allow the testimony, and Schwartz again asked Land why the film units had curled. Land could not resist beginning his response with a direct attack that only began to unleash his utter disdain for Adelstein. "Let me say that the Kodak witness in this case was wrong," Land proclaimed.[40]

In his best professorial presentation, Land then explained that the film units had curled "because of the unsymmetrical penetration of water into the [two different supports]." He described how, in use, the two supports were hit with moisture on only one side, the inside. Adelstein had provided a theory that had assumed wrongly, according to Land, that the plastic supports became uniformly wet through their thickness. According to Land, since one of the materials absorbed "a considerable amount of water" while the other one absorbed "scarcely any," the surface that was wet expanded and caused the plastic sheet to curl. In his explanation, Kodak's expert had assumed that the two outside plastic supports of the film unit were fixed together, a construction that he reasoned contributed to the curl. Land testified that, contrary to Adelstein's testimony, the two sheets of the experimental film units were not bonded together but could slip against each other. As Land explained, Adelstein's explanation for the failed experiment was flawed because in order to end up in the shape they assumed, it was vital that the supports slip on each other so that they could curl. "If they were rigidly bonded," he said, "they couldn't bend."[41]

Land, however, was not content to dispute Kodak's expert merely on the basis of his words. Ever the scientist, Land told Schwartz early in their conversations about a possible rebuttal appearance that he wanted to conduct some demonstrations in the courtroom to illustrate visually that Adelstein had it all wrong. Schwartz was less than enthusiastic when he first heard about this idea. It was a basic tenet of litigation that one should avoid, if at all possible, conducting an experiment or a demonstration in a courtroom.[42] The litigation annals include a litany of such experiments gone wrong, the most infamous in Fish & Neave lore being the demonstration of the unbreakable light bulb, which shattered right on cue when it was dropped in court.[43] But Land was adamant. Testifying in a courtroom may have been a new experience for the Polaroid founder, but conducting scientific demonstrations in public was something he had done enthusiastically, year after year, at Polaroid's annual shareholders meetings.

Ultimately, Schwartz determined that, especially for the Symmetrical Supports patent, it was worth the risk.[44] The demonstrations were discussed and then previewed by Land in his office in the days preceding his rebuttal testimony, where counsel and witness struggled over what Land could and could not do in the courtroom.[45] The aim was to reduce as much as possible the chance of anything going amiss. In the end, despite his reservations, Schwartz had to have confidence in the indefatigable Land. With Judge Zobel watching intently from her bench above the witness box, Land began a series of simple, yet illuminating, presentations. First, he took pieces of the two plastics that had been used in the old film unit and began to show the effect of wetting one surface of each of the two sheets, causing them to curl. "Mr. Carr, would you like to come and watch?" Judge Zobel inquired.[46] "Why not?" responded Kodak's counsel dismissively, as he walked over to stand with Schwartz.

As Land embarked on his next demonstration, Schwartz handed him a stack of white paper. In a moment of rare levity, Carr smiled as he noted out loud that he now understood why Kodak had received a ream of blank paper together with the copies of other exhibits Polaroid intended to use in rebuttal. When Schwartz met with Land during their preparation, Land had initially told him that he intended to conduct this particular demonstration with a stack of $100 bills.[47] Schwartz advised his client that this was not his best idea, and the irrepressible Land admitted almost as much to Judge Zobel. "I wanted to do this with dollar bills," was what Land told the judge as he began his presentation, "but my lawyer said it would be in bad taste."[48]

His mission this time was to show the difference between bonding the sheets and allowing them to slide. Manipulating the stack of paper like "a banker" might, Land showed "that when you bend them, they slide. . . . Now, if I hold it rigidly, if they were cemented [along the edges]," he continued, "then, it is a beam, it can't bend because they must slide." Judge Zobel was watching intently, and noted, "But that assumes they don't stretch," recalling that Adelstein had testified, "that the top sheet also stretches as it bends." Land pounced on cue. "Yes," Land declared, "and he is wrong. That is one of the things he is wrong about. It takes enormous forces to make it stretch," Land declared as he went on to explain the fine distinctions between bending and stretching.[49]

So far everything was playing out as planned, but the witness had saved his most sensational demonstration for last. It was as daring a piece of scientific showmanship as that fish in a fishbowl on a brightly lit windowsill in a Boston hotel room decades ago. When Land had first suggested it to Schwartz, the anxious, somewhat disbelieving Polaroid counsel insisted that, like the others, he show it to him first.[50] Sure enough, the rehearsal went precisely as Land had predicted, and so, with the viability of his patent at stake, Land was about to perform it for Judge Zobel. Examining one of the old, curled film units, Land again posited his explanation that because the two sheets were not bonded, one slipped over the other as the one that took in more water from the inside curled. He theorized that the unit got stuck in the curled position because the chemicals within ultimately dried out and acted as a glue to keep the unit in that distorted shape. Adelstein had testified that the plastic sheets had reacted with the processing composition and were thus deformed, causing them to stay in the curled position.

Land knew that this simply was not the case. To prove that he, not Adelstein, was right, Land took one of the old curled film units, showed it to the judge, and predicted that "if one were to cut this out and peel it apart . . . the sheet would become flat. . . . In other words," he predicted, "they are not significantly deformed or stretched," as Adelstein had testified.[51] It was a bold statement. After all, the film units had been curled like the letter C since the late 1960s. With that, Land took a pair of scissors, cut around the edges of the film unit, and peeled the two plastic support sheets apart. Precisely as he had predicted, they immediately gave up their curl and lay flat on the table.[52] It was as close to a jaw-dropping moment as likely ever occurs in a patent trial.

As dramatic as the demonstration was, Schwartz and Land were out to ice the cake. Schwartz asked Land if he was aware that Adelstein had offered his own explanation for the curl in the film units. "Yes," replied Land. "Do you agree with his explanation?" Schwartz inquired. Land responded with no attempt to hide his indignation. "I am sorry to say that I can remember no part of it that was correct," he said.[53] Finally, when Schwartz asked if he was aware of Adelstein's testimony on the properties of some various plastic supports, Land again could not restrain himself. "Well . . . I was surprised in the testimony," he acknowledged. "It isn't really my view, it is just a well known fact in the literature. . . . I don't understand how that testimony could have been given."

With Land having completed his best attempt at demolishing Kodak's expert, Schwartz announced that he had no further questions. After a brief recess, Carr rose to cross-examine Land, but it was clear that his heart was not in it. Perhaps he had decided that there was little to gain strategically from trying to challenge Land on his rebuttal presentation. Carr asked just a couple of questions about the curled film units and then a few about the patent examples in the L-Coat patent on which Trautweiler had done his calculations. It was over in minutes. Judge Zobel excused the witness, and Land left the witness box and returned to his seat. "Any other witnesses?" Zobel asked Schwartz. "No, your Honor," he replied, and with that, the contest was finally at an end.

In over seventy-four days of testimony covering almost five months, both sides had presented their cases with the same thoroughness and intensity that had marked the litigation from the outset and all the way through pretrial discovery. Judge Zobel had proven to be up to the challenge; completely engaged in the process, she took notes, asked questions, and made sure that she understood all the testimony each day before leaving the bench. She would have mountains of written material to refer to in making her decision, but it was clear that she was determined not to wait until later to sort it all out. Given that this was to be the final formal session, Judge Zobel chose to conclude the proceedings with some final comments about the trial:

> It seems to me that the quality and, indeed, the fairness of the trial depends in very large measure on the quality of counsel. Pretentious, unprepared, unskilled lawyers cause not only error, but sometimes even injustice. On the other hand, lawyers who cooperate with each other and the Court, who know their case factually and legally, who are familiar with and abide by the rules of the courtroom, unquestionably serve not only the Court, but most directly, their clients.
>
> I must say in this trial both parties have been extremely well prepared. I have been impressed daily with your extraordinary command of that huge mass of materials and the details of a very complicated technology, and by your extraordinary feat of organization of that material to make it both accessible and understandable.
>
> I have reference not only to the availability of the materials when you needed them, but to your . . . graphic presentation of some of the materials.
>
> I have been equally impressed by your, if you will, pedagogic techniques. Of course, the proof will be in the pudding.

> It is said that patent cases are interminable and deadly dull. I confess that contrary to that perceived learning—although, some might . . . consider it interminable—and the predictions of my colleagues, I did not find this case at all dull. In fact, I found it interesting and enjoyable. I attribute that to your cooperation and your professionalism, and I thank you all for that.[54]

It was 11:30 a.m. on Thursday, February 25, 1982. The gavel fell, the court adjourned, and the *Polaroid v. Kodak* trial was over.

CHAPTER 25

THE WAITING GAME

After nearly five months of testimony and courtroom drama, the trial had finally come to a close. It had been the fourth-longest trial concluded in any U.S. federal court that year.[1] Like a campaign headquarters on the day after an election, both sides quickly dismantled the huge operations each had set up in nearby hotels and office buildings. Documents and equipment followed the weary lawyers and support personnel back to New York. Most had scarcely seen their families for the past six months, and like soldiers back from overseas duty, they decompressed and gradually eased back into their pretrial lives. Upon returning to his office, Schwartz took the time to reach out to Land to express to him in writing his appreciation for the extraordinary help he had provided during the trial. He had become Polaroid's lead witness and, behind the scenes, a source of tremendous insight and technological expertise upon which Schwartz had regularly relied.

"I would like to take this opportunity to write some of the thoughts I expressed orally to you at the close of the trial," wrote Schwartz. "Thank you, on behalf of all of us at Fish & Neave, for all of your time, energy and effort which you devoted to the trial of this lawsuit. Not only was it extremely significant and beneficial to us in connection with your own direct and rebuttal testimony, but it was extremely useful as an aid in the cross-examination of many Kodak witnesses and in the direct and rebuttal testimony of many Polaroid witnesses."[2] The effective working relationship between counsel and witness had been greatly facilitated by Nan Chequer, yet a comparable personal relationship between the two men never developed, despite their shared and intense experience.

In large measure this was due to Land's innate shyness and his penchant for privacy. Also, while he seemed to appreciate his lawyers' efforts, Land had always had a problem in praising colleagues directly and personally.[3] Land did manage to convey his appreciation to Schwartz in other characteristic ways. For example, with Chequer along, Land gave his lawyer a personal tour of the Rowland Institute—the new research facility Land was building in Cambridge. For Schwartz, it was a grand gesture, a "surreal" experience. This was the closest to an expression of friendship and appreciation as one could expect from the reserved and complex Land.[4] Before leaving Boston, Schwartz's young associate who had worked most directly with Land throughout the litigation asked Chequer whether Land might agree to autograph a copy of his L-Coat patent as a souvenir of their experience together. He was surprised and delighted when, a few days later, she produced an envelope containing the patent copy with a small "EH Land" inscribed in red ink near the top.

Although the trial-weary litigators on both sides could have used a vacation at this point, there was, in fact, little time to waste. Judge Zobel had set a deadline of May 14 for the submission of posttrial briefs. It was a chance to present their cases as clear and coherent stories drawn from the mass of material that had been accumulated over the years throughout discovery and introduced into evidence at trial and to apply those facts to the governing law as they interpreted it. Judge Zobel had requested that each side present a separate brief for each of the patents in suit. These were to be accompanied by a general brief that addressed legal issues and factual contentions common to all of the patents. Following these initial submissions, each side would then have until July 12 to submit reply briefs addressing the contentions made by each in the first set. With strict page limitations having been set for all of these submissions, the challenge was not only to sort through and then synthesize the vast amount of factual material but also to present it as tersely, and yet as persuasively, as possible.

Polaroid's general brief, consistent with the strategy adopted by Schwartz early in the litigation, attempted to put the case, and thus the detailed technical arguments regarding the specific patents, in some real-world perspective. "This is no ordinary patent infringement suit," it began. "Polaroid's remarkable achievements in instant photography have been properly accorded protection by the patent system. Kodak, given its vast economic and technical resources, and its dominance of conventional photography, had an opportunity to make a unique contribution to this field. Instead, it consciously chose to utilize Polaroid's patented technology."[5] The brief then went on to tell the "Kodak Story" that Schwartz had worked so hard to assemble during discovery. It was a narrative that detailed, largely through the use of Kodak's internal documents, the photography giant's "decision that Polaroid had become a significant economic threat to a segment of Kodak's business and that Kodak needed an SX-70 type system to counter that threat." The story of Kodak's cooperation with Polaroid during the early days of instant photography going back to the 1940s was retold, as was the pivotal event in 1968 when Land showed Kodak executives a prototype of what was to become SX-70. The brief described how this revelation had such a significant impact at Kodak's Rochester headquarters that the company immediately committed to developing an instant photography system of its own.

The final chapter, according to the Polaroid brief, occurred in 1972 when Polaroid introduced SX-70 commercially. Kodak realized that the Lanyard camera it had under development at the time was not competitive with the new Polaroid camera, which a Kodak engineer hailed in a memo as "a masterpiece of engineering." According to Polaroid, Kodak then changed course in its development program and moved towards what its own internal documents termed a "me too" program, an effort that inexorably resulted in its having to use the inventions of the patents asserted in the lawsuit. Polaroid pointed out how Kodak's twenty-year effort to enter the instant photography field was further evidence of the merit of Polaroid's inventions and thus the validity of its patents. "Despite its vast resources, unparalleled expertise in conventional photography, knowledge of the prior art, prior work with Polaroid and its own instant photography research dating back to the 1950s, Kodak could not commercialize its instant photographic system until after examining Polaroid's SX-70 camera and film." Quoting a 1921 Supreme Court patent case, Polaroid's brief submitted that "this 'page of history is worth a volume of logic' with respect to the significance of the inventions of the patents in suit."[6]

Polaroid also presented a fundamental argument relating to the relative qualifications of the experts presented by the two sides, and thus the relative weight the court should accord to their testimony. It noted how Kodak had used witnesses who, while experts in photography in general, had "no research or development or 'hands-on' experience in instant photography." Kaprelian, Trautweiler, and Adelstein were all cited as examples. Polaroid pointed out that "there are hundreds of individuals within the Kodak organization with substantial practical experience in instant photography," individuals who, according to Kodak, had supposedly developed the company's own novel instant photographic film and camera system. While Polaroid acknowledged that a few of Kodak's experienced scientists were called as fact witnesses, "the instances where Kodak rejected logical choices for experts from within its own ranks are legion." In contrast, Polaroid had educated the court on the technology of instant photography using the very scientists who had done the work, including each of the inventors named on the patents at suit. "Why is it that Kodak chose experts with no 'hands-on' experience in instant photography to meet the testimony of Dr. Land, Mr. Rogers and others? . . . Polaroid submits that Kodak's choice of expert witnesses reflects Kodak's recognition that its experienced internal scientists could not or would not persuasively testify about the patents in suit."[7]

Polaroid's individual briefs recounted the "real world" problems confronted by the inventor, and how the discovery claimed in each patent solved those problems. Also addressed was the development history at Kodak on each feature and how its efforts and failures underscored the merit of Polaroid's inventions. For example, the brief described in detail how Land's L-Coat patent emerged when his attempts to adapt stabilization techniques that had been used in prior sepia and then black-and-white instant systems failed to work with Polaroid's new color-imaging technology and how it was the product of his intense effort to avoid using a print coater with Polacolor film. Polaroid also noted the fact that during Kodak's development of its PR-10 film, one of the scientists working on its stabilization approach, Eugene Wolfarth, had warned in a written report that its proposed use of a polymeric acid layer "may run into a patent conflict" but that Kodak was unable to find an alternative approach despite a two-year effort to do so.[8]

In its posttrial submission, Kodak stressed that the controversy involved "nine separate disputes," each of which was to be considered on its own merits. "The apparent complexity of the present lawsuit derives more from the sheer number of patents involved than from the technical aspects of any one of them. In its blunderbuss approach to patent litigation, Polaroid apparently is pursuing the theory that the greater the number of patents asserted, no matter how weak, the better are the odds that one will survive unharmed and found to have been infringed. However," Kodak wrote, "there should be no room here for such theory. . . . Each patent in suit . . . must measure up to the demanding standard of the Patent Laws, as if each were the only patent in suit."[9] Kodak declared its confidence that if the decision were to be approached in that manner, each of Polaroid's patents would be found "either to have fallen far short of the special quality needed to justify a monopoly grant, or not to have been infringed, or both." Although its overall assault on Polaroid's patent portfolio had been effectively tabled since the early days of the litigation,

Kodak could not resist the chance to chastise Polaroid for securing "patents by the carload." Once again Kodak criticized what it considered to be Polaroid's patent policy of making "efforts . . . to patent not just the significant technical achievements, but every feature, apparently no matter how trivial."

Kodak also addressed the notion that Polaroid had tried somehow to portray itself as a righteous victim in defense of its proprietary turf. Kodak's counsel was determined to pierce whatever aura of sanctity Polaroid and its inventors may have created by focusing the court on its pioneering history in instant photography. To do so, they invoked some patent history surrounding another famous inventor, the pioneer of radio, Guglielmo Marconi. This was a critical theme that Kodak had urged on the court even in its pretrial briefs. At that time, trying to set the stage for the trial, it reminded Judge Zobel that "each patent must stand on its own, unaided by the glow of the inventor's reputation for past accomplishments."[10]

With Land clearly in mind, Kodak reminded the judge that "courts have uniformly rejected attempts to rub the gloss of a general reputation onto specific patents." It quoted from a 1943 Supreme Court case holding one of Marconi's radio patents invalid: "Marconi's reputation as the man who first achieved successful radio transmission rests on his original patent . . . which is not here in question," wrote Chief Justice Harlan Stone in the Court's opinion. "That reputation, however well-deserved, does not entitle him to a patent for every later improvement which he claims in the radio field. Patent cases, like others, must be decided not by weighing the reputations of the litigants, but by careful study of the merits of their respective contentions and proofs."[11] Now, Kodak urged Judge Zobel to remember those tenets while evaluating the merits of this case:

> Polaroid, as stated by Mr. Kerr on the first day of the trial, regards instant photography as its "domain," as if pioneering a particular field gives perpetual, exclusive rights to it. Polaroid's pioneering role in instant photography is not disputed and is not to be denigrated. However, it does not establish a perpetual domain and it does not, in itself, justify later patented improvements any more than Marconi's early work established a domain over radio or saved Marconi's secondary "improvement" patents. Each patent must be separately examined, and must stand or fall on its own merits.[12]

When it came to the individual patents, Kodak put forth each of its positions on how the inventions in Polaroid's patents were invalid or not infringed by Kodak's products or both. These were long-held and well-honed arguments. Their genesis could be traced back to Frank Carr's original opinion letters for Kodak management, as refined over the years as each deposition or trial witness was exposed to them and offered the opportunity to challenge them intellectually or factually. Invalidity was generally urged on the grounds that the inventions were either already known in the prior art literature or were merely incremental advances in the technology not worthy of patent protection on their own. For instance, Kodak argued that Land's L-Coat patent represented "an example of Polaroid's penchant for repatenting the prior art . . . [in that it] simply applies to color instant photography techniques long known in black-and-white instant photography—with predictable results." Kodak also set forth at length its contentions regarding infringement. As expected, Kodak insisted that its PR-10 film did not infringe Land's L-Coat patent because, during manufacture, its polymeric acid layer was not coated on the same support as the photosensitive layers of the film and thus was not located within the film's "photosensitive element," as required by the patent.[13]

With regard to Land's Rear Pick patent, Kodak similarly argued that there was no invention behind the complicated language of the patent document. "There is no question but that the asserted claims of the . . . patent are to a combination of old elements, at most representing the substitution of a rear 'edge' pick for a front pick or a rear 'end' pick in an otherwise known camera," it contended.[14] Kodak described several prior art patents, cameras, and other devices that it believed employed a rear pick like that allegedly invented by Land. They included a scanning camera developed at Polaroid, two different versions of cameras designed to take photographs that could be used as the front of postal cards, a number of commercial camera models, and even an IBM computer card feeder.

Over and over, Kodak cited the testimony of its camera expert, Edward Kaprelian, that these references and devices rendered Land's Rear Pick invention obvious to anyone skilled in the art of camera design. Moreover, in yet another direct slap at Land, Kodak contended that the rear pick mechanism adopted at Polaroid was not even invented by him at all but that its origin could be traced back to work done as early as 1967, two years before Land's claimed date of invention, by another Polaroid researcher, Richard Wareham, on a "crochet hook" mechanism.[15] "Apart from Dr. Land's *ipse dixit*, there is no evidence establishing that he had any connection with the design of the specific rear pick structure described in . . . the . . . patent. . . . The evidence reveals that Wareham did."

Kodak's position on Land's Symmetrical Supports patent was even more disdainful. It characterized "the invention . . . [as] the 'discovery' that polyester is a useful material for the external supports of an integral film unit, and that it allegedly gives improved properties of flatness compared to the earlier use of [other plastics] for the same purpose. In fact," the Kodak brief contended, "the undisputed facts present a classic case of non-invention: the substitution of a newer [but known] material for an older material to take advantage of the known improved properties of the newer material. . . . The result may be useful, indeed an improvement—but it is definitely not a patentable invention."[16]

Kodak's brief went on to explain how polyester had been a known material for use in photographic applications since 1955, and its use in instant photography had even been described by Land and other Polaroid researchers in patents dating as far back as 1958. It cited as prior art Kodak's Bimat aerial reconnaissance film system that, it declared, "utilized the essential feature of the alleged [Land] invention . . . that a laminate having two external polyester supports will resist distortion caused by moisture within the laminate."[17] Kodak argued that the Symmetrical Supports patent should be declared invalid on the ground that "applying known principles to known products using known materials is not patentable," citing a case from 1850 in which a court invalidated a patent on a door knob assembly in which the only advance was the substitution of a new material—porcelain instead of wood—for the handle.[18] Taking its argument even further, Kodak invoked no less a historical figure than Thomas Jefferson, a member of the original U.S. Patent Board, noting that Jefferson had been quoted in the landmark Supreme Court patent case of *Graham v. John Deere Company* as saying that "a change of material should not give title to a patent." Finally, directing its attack at Land's professorial courtroom testimony, Kodak warned that "cloaking the obvious in mystical and pliable phrases like 'symmetry' does not make it less obvious."

The ferocity of Kodak's attack was not reserved for Land's patents. It contended that Howard Rogers' Negative Dye Developer process "is an essentially useless one," citing the "dubious results" Rogers had obtained in his original experiments with the chemistry.[19] It also set out its now familiar argument that, because of what Kodak urged were similarities between its PR-10 process and the prior art Whitmore-Mader process, if the Rogers patent was interpreted broadly enough to be "found to be infringed by the Kodak process, the patent would be invalid in light of the prior art." Kodak sought to draw a strong distinction between Rogers' process and the imaging chemistry used in its PR-10, "a successful process, extensively and originally developed by Kodak scientists, which has found wide-spread application in commercial multi-color film." It also pointed out that Kodak's process had been "separately patented."

In contrast, Kodak asserted that Rogers' Negative Dye Developer patent was "based on very limited single-color experimentation in peel apart units and exhibited pitiful results . . . did not enrich the art, and was understandably laid aside at Polaroid. . . . Given the very narrow and lame nature of the technical disclosure, the lack of commercial interest . . . and a history underlying the invention which smacks more of opportunism than meritorious discovery," Kodak urged the court to give the Rogers patent, if valid at all, "a very narrow interpretation."[20]

When it came to Rogers' "Excedrin" film unit patents, Kodak reminded the court that its summary judgment motion seeking to invalidate the pair for obviousness was denied only because material issues of fact needed to be decided at trial. "Kodak now submits that the record fully supports its original contentions that there is simply nothing unusual, unexpected or unobvious in the alleged invention . . . that justifies the monopoly they represent."[21] Kodak's brief contended that Rogers conceded at trial "that every element in . . . [his Excedrin] film unit is old." It then presented Judge Zobel with a laundry list of what it contended Rogers' patent "is not": "It is not original in providing for an 'integral' film unit; it is not original in providing such a unit which is designed to be exposed from one side and viewed from the other; it is not original in using an opacifying agent in the goo to provide a light shield for out-of-camera processing; it is not original in providing a preformed permeable opaque layer as a light shield; it is not original in permitting the user to see the picture emerge; and it is not original in providing for introduction of the processing liquid on the side of the photosensitive layers opposite the receiver. All of these 'features' are taught in the prior art," Kodak declared, proceeding to set forth in detail where each could be found in other Polaroid patents, most of which were attributable to Land.[22]

In an attempt to belittle the patent even further, Kodak quoted a 1968 Polaroid internal memorandum that described Rogers' concept for this film unit as "Howie's idea of a carbon black pod."[23] This reference uncomfortably (for Polaroid) echoed the shorthand description Judge Zobel had used for the patent during the trial. The real crux of Kodak's argument was that the only difference between Rogers' film unit structure and Land's prior structure was that Rogers substituted an opaque processing composition for the opaque flap used in the Land film unit to prevent further exposure of the photosensitive layers once the film was ejected by the camera into the light. Kodak argued that this "obvious substitution" was unworthy of patent protection. "It has long been the law," urged Kodak, "that no patentable invention results from substituting one known element for another one in an otherwise old combination, no matter how useful the result, where the 'new' combination performs precisely as one would anticipate based on the known functionality and interrelationship of its component parts."[24] In sum, Kodak urged the court to declare the Rogers patents invalid because, among other reasons, they "represent an effort to foreclose continued use of available teachings of the prior art . . . they subtract rather than add to the sum of knowledge."[25]

Kodak handled each of the other patents in suit in a similar fashion and filed its briefs with the court, as required, on May 14. Both sides then had until July 12 to submit reply briefs, and each took full advantage of that opportunity, adding several hundred additional pages of argument to the pile of materials on Judge Zobel's desk, from which she was to make her decision. While each party rebutted the other's arguments on the individual patents in great detail, each continued to push its thematic overview of the case. Polaroid, for its part, tried to convince the judge that Kodak's approach was to ignore much of went on at trial in favor of the positions it had espoused long before the litigation even started:

> Kodak's Briefs After Trial open the door to a new and uncharted world where history can be rewritten and reality can be suspended at a moment's notice. Kodak's approach submerges the five-month trial of this action so as to suggest that it never happened. Kodak ignores that it spent weeks at unsuccessful cross-examination of Polaroid's experts and that its own experts, on cross-examination, agreed with Polaroid's experts about many aspects of the technology at issue, the disclosures of the prior art, the scope of the claims of the patents in suit and the construction and operation of Kodak's products at issue.
>
> Kodak's . . . [briefs] subvert the very point of having a trial. Our system of jurisprudence places great emphasis on live testimony of witnesses and the value of cross-examination. Briefs After Trial are meant to assist the Court in weighing and interpreting that evidence. Kodak makes no attempt to persuade this Court that the weight of the evidence adduced at the trial entitles it to

> prevail here. Kodak's Briefs seek to replace or distort that evidence. Kodak's mainstays are lawyer's arguments, hyperbole and unsupportable allegations directed at Polaroid, its witnesses and its patent attorneys.
>
> Kodak's strategy at trial was to make the record as cumbersome as possible. Kodak's Briefs augment that strategy by creating an intricate web of fantasy to obscure the evidence of the validity and infringement of the patents in suit. When this web of fantasy is replaced by the realities of the evidence, Polaroid's claims of infringement should be sustained.[26]

The Polaroid reply brief recounted once more the highlights of the history between the two companies that led to Kodak's decision to pursue its own instant photography system, and the events that occurred during its development process as they bore on the inventions of the patents in suit.[27] It responded to Kodak's "irrelevant" complaint about Polaroid's propensity for securing patents by quoting Kodak's own published patent practice: "'It is the policy of . . . [Kodak] to apply for patents on all worthwhile and patentable inventions in any field of its present or prospective business.' Kodak has advanced no reason why Polaroid is not entitled to do the same," argued Polaroid.[28]

Kodak, in its reply brief, continued to pound on the same theme it had offered from the outset of the litigation:

> A review of Polaroid's General Brief confirms the fact that Polaroid still seeks to bootstrap what are at most minor specific improvements in the art into patentable invention by reiterating, *ad nauseam*, the pioneering role that Polaroid and Dr. Land played some 35 years ago in establishing instant photography as a commercially acceptable approach for amateur photography. However, being a pioneer conveys no special status to every subsequent improvement. Courts have uniformly rejected attempts to rub the gloss of general reputation onto specific patents.[29] (emphasis in original)

The Marconi case was once again cited and once again quoted as the ultimate precedent for this proposition. It downplayed Polaroid's account of the history of instant photography as "largely irrelevant" and merely reflective of "Polaroid's 'domain' mentality and the implication that it built instant photography and therefore Kodak's entry is *per se* wrongful." However, there was one specific piece of the story that it now tried to undermine directly. Again, this salvo seemed squarely aimed at Land personally. Kodak challenged Land's account of the meeting in his Cambridge office in November 1968 when he showed prototypes of his new SX-70 film to a Kodak contingent. Land had stated that a Kodak executive, who was no longer alive, had actually taken a sample with him to Rochester. Kodak conceded that Land had shown the prototypes but that they were "carefully mounted in frames" so that little about them, except for the image, could be observed. More importantly, Kodak disputed Land's recollection that a prototype had been taken back to Rochester. It pointed out that no other witness had confirmed Land's account and thus contended that the lack of corroboration "strongly suggests that Dr. Land's memory is faulty."[30]

Although the posttrial briefing was now complete, there remained one more major task for the weary lawyers on both sides. Judge Zobel requested that each party submit proposed findings of fact. These documents were to be compilations of factual conclusions on a patent-by-patent basis, drawn from all of the evidence presented in light of the posttrial briefs. In his memorandum organizing this process for the dozen Fish & Neave lawyers still working on the case, Schwartz tried some levity to spur his fatigued troops on for this final task. Supplied to him by Patricia Martone, a colleague who was more of a baseball fan than Schwartz, he borrowed the famous Yogi Berra quotation, "It ain't over 'til it's over."[31]

When the proposed findings of fact were filed on July 30, 1982, the posttrial process was, indeed, finally over, and everyone hunkered down nervously to await the decision. Over *six years* had passed since Kodak had introduced its instant photography system, prompting Polaroid to react immediately by filing suit. An Associated Press report carried across the country predicted that Judge Zobel "would take the next few months to review 74 days of testimony, volumes of evidence and at least 900 pages of final briefs before deciding if Kodak infringed on patents protecting Polaroid's instant photography empire."[32] Schwartz sent copies of all the briefs and findings to Land, who had been continuously involved in contributing ideas and information to Schwartz during the briefing effort. Not all of Land's suggestions were actually used though. According to one published report, he was apparently frustrated, and perhaps amused, by the pedagogical rigidity of the legal process when Schwartz would not adopt for Polaroid's submissions his description of Kodak's case as "phantasmagoria."[33]

Schwartz also shared the posttrial submissions with Bill Kerr, who by this time was comfortably ensconced in his Bermuda home, "Singora," on the grounds of the Mid Ocean Golf Club. With sincere collegiality, Schwartz offered that Kerr could "read, sprinkle with Cointreau, or do anything else to them that you like."[34] But in an equally sincere tribute to the man who had mentored him and taught him much about how to practice the art of patent litigation, Schwartz advised Kerr, "I think you will find throughout our briefs authorities, phrases and thoughts which you are very familiar with."

Walking out of the courthouse on the last day of the trial, Frank Carr was sure that he had won.[35] As the Kodak team disbanded, Cecil Quillen returned to Kodak headquarters in Rochester also feeling very confident about the outcome of the case. When asked by colleagues how he thought Kodak would do, he acknowledged that, while logically it might be hard to expect a victory on each and every patent, when he worked his way through the case patent by patent, he could not pick out any that he expected Kodak to lose on.[36] Close observers of the case gave Kodak a good chance of prevailing on a majority of the patents in suit. Wall Street's Calvert Crary reviewed the trial patent by patent in great detail and forecast a quick decision, "within the next two or three months," in which Polaroid "would likely win on only three of the ten patents in suit," a result that would all but vitiate any claim of victory.[37]

While Judge Zobel had on her desk the question of liability, the issue of damages, should Polaroid prevail, was not. It had been severed at the start of the litigation by agreement of both parties, to be determined only when and if Kodak's liability was established. Nonetheless, "how much?" was a question on the minds of everyone discussing the case. Richard DeLima, one of Polaroid's in-house lawyers, was quoted as stating, "Hundreds of millions of dollars is in the proper order of magnitude."[38] Ty Govatos, an analyst with the firm of Donaldson, Lufkin & Jenrette, put forth publicly his own, more conservative calculation: "If we were to assume they decided Polaroid deserved $5 a camera . . . that would be $100 million," based on the fact that Kodak had sold 20 million instant cameras to date.[39] According to Govatos, such an award would not be "significant enough" to help Polaroid, whose profits had declined significantly in 1981.

Although its sales were only down 2.1 percent for the year, from $1.45 billion to $1.42 billion, Polaroid's bottom line was hurt by adverse foreign currency translations, startup costs for a new camera and film line, and an inventory write-down for its older models.[40] Net profits for the full year of 1981 had fallen to just ninety-five cents per share from

$2.60 a share in 1980, in large part due to a very weak fourth quarter in which earnings dropped 94.7 percent.[41] The future was unclear for Polaroid on several levels, but the importance of winning the lawsuit remained a critical component of whatever its future would be. A loss in the courtroom that would open the instant photography market to all comers would clearly exacerbate the company's problems.

With final submissions in the case en route to the Boston federal courthouse, the legal teams decompressing from their marathon efforts, and observers prognosticating over the result, another landmark event with profound implications for Polaroid occurred. On July 27, 1982, Edwin H. Land stepped down from his last formal post at Polaroid, as chairman of the board of directors. As reported in the *Boston Globe*, at the age of seventy-three, "the company founder and progenitor of instant photography severed his corporate connection as Polaroid chairman to free his energies for another passion, pure scientific research."[42] The headline in the *New York Times* succinctly read: "Era Ends as Land Leaves Polaroid."[43] "In his typically reticent style, Mr. Land refused all interviews," the *Times* reported and instead quoted Land's statement as released by the company. "I look forward to a new period of creative freedom for myself and to a generation of industrial grandeur for Polaroid." (See Fig. 25-1.)

The Polaroid board immediately elected William McCune, who was already serving as its president and chief executive, to replace Land. Although the *Wall Street Journal* reported that Land had left "abruptly," in reality this was but the last chapter of a long and gradual transition that had been initiated in 1975 when McCune assumed the role of Polaroid president.[44] The metamorphosis in Polaroid's leadership accelerated in 1978 and 1979 in light of the Polavision disaster that led to McCune's assumption in 1980 of Land's role as chief executive. Now, with the Kodak litigation complete, it was time to sever the last cord.

Land's gradational dis-enthronement as head of the company he had created had clearly not been easy for him. His disputes with McCune and others regarding the direction the company should pursue had surfaced publicly and embarrassingly from time to time. Executives "closest to Land" reportedly speculated that the Polaroid board had reached its breaking point and had told Land "that he would have to let McCune run the company his own way."[45] There was also clearly a sentiment within at least some parts of Polaroid's executive ranks that it was time to turn the page. For example, when McCune prepared a draft of his letter to shareholders for inclusion in the forthcoming Polaroid annual report, he opened with a tribute to Land, noting how "we all have been personally affected by his decision to leave the company."[46] When McCune circulated the draft to some of his top associates for comment, he received back a long impassioned note from Richard DeLima, urging him to move the tribute to the end of the letter. "To put these paragraphs in the preeminent position suggests that they are the most important thing you have to say to the owners of the company *about this year's operations* (emphasis in original). I think, and I'm sorry to say so, that you do yourself a disservice."[47] McCune, nonetheless, ignored the advice.[48]

Brenda Lee Landry, an industry analyst for Morgan Stanley, shared her view that Polaroid's management had been suffering "big conflict . . . for some time" and pointed to Land's "tearful plea" at the recent April 1982 Polaroid shareholders meeting for patience and understanding as the company worked to right itself.[49] On that occasion, as he completed his extemporaneous remarks, and apparently aware that his days at the company were numbered, Land asked the shareholders for their support: "I don't mean in terms of money, I mean in terms of enjoyment of the creative life."[50] His voice broke as he uttered what were in fact to be his last public words as chairman of Polaroid, and the shareholders erupted into enthusiastic applause as he sat down. Reportedly, there was not a dry eye in the room.[51]

Now, three months later, this was indeed the perfect opportunity for Land to "finally put behind him the pesky annoyances of prying security analysts, litigious shareholders, management squabbles and muckraking journalists," as Robert Lenzner wrote.[52] A spokesman for Land confirmed, "He really wants to devote full time to the research he's been planning. . . . For him to remain on the board would be a major commitment of intellectual energy and time."[53] Land himself offered that his departure might be a good thing for Polaroid, while at the same time issuing a subtle challenge to his successors. "[Now that I am] not present there to stir up the pot, the cream will rise to the top."[54] He had always hoped that he would be able to leave behind a talented and motivated research force that could carry on the work necessary to propel Polaroid to future horizons. "One would hope that one has left an environment in which people with my kind of pace and competence will sprout," he had told an interviewer from *Forbes* in 1975. "That is the best I can do."[55]

In fact, Land was more than ready for this day and had long planned an elegant escape route. Using personal funds, variously reported to be somewhere between $20 million and $38 million (between $58 million and $110 million in 2014 dollars), he had endowed in 1980 the Rowland

Institute for Science, a private nonprofit center for research, initially focused on vision and the human brain.[56] Property was secured in Cambridge, along the banks of the Charles River just down the street from his old Polaroid laboratory, and a state-of-the-art research facility was built and completed in the fall of 1981. By the spring of 1982, there was already a staff of twenty at work in the institute, ready for the arrival of its founder and benefactor. The facility itself was impressive, designed by Boston architect Hugh Stubbins with careful insight and input from Land. Deceptively described by one observer as a "simple, elegant red brick building," the facility was positioned "to give those inside inspiring and magnificent views of the river, the Longfellow Bridge and Beacon Hill."[57] Land described his new working environment as having the "tranquility of a distant mountain resort." Reportedly, his attention to detail extended to a semicircular bench in the building's Japanese garden that had been tailor-fit for him and his key colleagues by having cardboard bent to the contours of their backs.

Land's enthusiasm for his new endeavor was genuine and obvious. "Like most men my age who are in good health," Land explained, "I seek to make application of my enthusiasm and accumulated experience—in my case to science, to education and to sustaining and teaching those historic values which have counted for us."[58] When he granted an interview to Robert Lenzner to show off his new digs, Lenzner was impressed with Land's "infectious energy" as he bounded up the stairs two at a time and with his "playfulness, the willingness to see things in a new relationship and to see science as fun."[59] Land seemed to revel in the freedom he and his colleagues would enjoy in their new facility to pursue whatever interested them. "Here we will not be under industrial pressure or academic pressure or student pressure," he exclaimed.[60] His informal motto for the Rowland Institute showed that he still lived by the mantra developed in his youth: "Never go to bed with a hypothesis untested." He was intent on creating "an environment [in which] the scientists will achieve frequent and significant results and scientific insights."[61] Land's aspirations for his new professional home were apparently realized. A few years later, immersed in pure research, and having authored a stack of scholarly scientific publications, Land proclaimed that at the Rowland Institute he was "now living out the dream I have had since I was seventeen years old."[62]

The real question for many observers of Polaroid, however, was how the company would fare with its founder and prime visionary physically gone from the premises. There were clear challenges for the company on the horizon. The field of instant photography, effectively Polaroid's sole business, was starting to show some signs of trouble. Polaroid's camera sales in 1980 had dropped to six million units from a high of almost ten million in 1978. Film sales were correspondingly off. Uncertainty over the future of the technology had affected the company's stock price as well. Other new imaging technologies—a still video camera developed by Sony, a 35-millimeter three-dimensional camera from Britain called the Nimslo that was about to be released in the United States by Timex, and Kodak's recently introduced film disc technology—were the topics that now fueled excitement on Wall Street. One Boston analyst went so far as to be quoted anonymously declaring instant photography "a dead industry."[63] Despite this dire pronouncement, it was clear that the challenge was to regain the pace of growth seen in earlier years, rather than sound an immediate death knell for the company. As summed up in one analysis, "no one doubts that Polaroid . . . will continue to sell millions of its instant cameras. The issue is how many millions will it sell—and whether, in the long run, its technology will be passed by."[64]

In fact, notwithstanding its "erratic earnings" in the short term, Land was leaving his company in solid financial shape. According to an analysis published by *Barrons*, which noted that Polaroid was not "approaching fossil status," the company had more than $330 million on hand in cash and securities when Land left, and shareholders' equity of $952 million.[65] Remarkably, the company had but $124 million in long-term debt. "Polaroid's woes are not financial," it was acknowledged; rather, the problem was about the direction the company would take going forward.[66] After all, Land was a hard act to follow. Michael E. Porter, a professor at the Harvard Business School predicted that Land would "be remembered as one of the great technical and creative visionaries in American business in the last 30 or 40 years."[67] Without Land to provide a creative vision, and with more fiscally circumspect management in place, where would the next frontier for Polaroid be? "It's a delicate trade-off," noted *Barrons*. "How adept the company proves at striking the right balance between inspiration and pragmatism remains to be seen."[68]

In the meantime, competition in the worldwide instant photography market remained as fierce as ever. In March, word spread that Fuji was "accelerating its assault on the U.S. market."[69] At that point, its instant camera, designed to use Kodak instant film, was on sale only in Japan. Fuji executives were quoted as saying that Fuji "is very, very committed to the U.S. market," raising the expectation that a U.S. introduction of its products would be announced before the end of the year. Notwithstanding Fuji's collaboration with Kodak to enter the Japanese instant photography market, Polaroid's business in Japan remained strong. In fact, Japan was Polaroid's best overseas market, boasting sales in excess of $100 million in 1981, a growth rate of 10.5 percent from 1980 in a year when Polaroid's worldwide sales had dipped 2.2 percent.[70] However, the prospect that Fuji might make the U.S. market a three-way competition was another matter. For its part, Kodak continued its pressure on Polaroid, introducing new camera and film models in an effort to keep pace with developments from Cambridge.

Polaroid's most recent entries, in May 1981, were its high-speed 600 Instant Film and the Sun camera line that featured a sophisticated built-in flash. High-speed film is more sensitive film, meaning that less light is necessary to take successful pictures. As a result, consumers would have more success with photos taken indoors and in dim light. Kodak countered in early 1982 with four new instant camera models of its own—and faster film to go with them. Although analysts acknowledged that the Kodak offerings were improvements in its product lines, they were not "revolutions in instant photography."[71] Accordingly, there was a real split among industry experts as to whether Kodak's new products were likely to help the com pany get beyond the thirty to thirty-five percent share of the instant market it had held for several years. For one thing, the Kodak film, although faster than its previous versions, was not nearly as fast as Polaroid's new entry. The cameras also seemed merely to echo features that had previously been available on Polaroid models. "The net is that this new line is also a me-too thing," pronounced George Elling, a Bear Stearns analyst.[72] "I see nothing superior to Polaroid's Sun cameras."

As the instant photography competition continued in the marketplace, a ruling in the courtroom battle remained a waiting game, as the spring of 1982 became summer and then fall. After the trial, each side had taken Judge Zobel's copies of its trial exhibits, to be delivered to her chambers at a later date upon her request. As November rolled around, Polaroid's trial exhibits remained piled up in the office of its local Boston counsel. When it was learned that Kodak had delivered its five boxes of exhibits to the judge in late summer, some concern developed on the Polaroid side. William Cheeseman, one of Polaroid's local counsel wrote to the judge's clerk to inquire. The object was to make sure that Polaroid's exhibits, a meager two boxes, would also be available to the judge as she worked on her decision. "Exhibits on hand may be more readily examined than exhibits which need to be sent for," reasoned Cheeseman in his letter.[73]

Judge Zobel's clerk called within a couple of days inviting Polaroid to send over its exhibits, and Cheeseman was asked by Fish & Neave to deliver the materials personally.[74] He did so and reported that Judge Zobel "made a few tongue-in-cheek remarks about reading the exhibits, but gave no indication about when she was planning on doing that."[75] Cheeseman also reported that he noticed Kodak's exhibits in the judge's chambers, and "that they appeared to be in some disarray," an observation that could only further ratchet up the anxiety already felt by Polaroid lawyers and executives in that the judge had possibly been working on her decision with only the Kodak exhibits as references.

Even though Judge Zobel had in her possession all of the materials she would need to decide the case by the end of 1982, no news emerged from her chambers for month after month after month, stretching through 1983 and into the spring of 1984. The attorneys had moved on to other matters, but through this period both Polaroid and Kodak were mired in difficult business circumstances, and both were anxious for a decision. Like Polaroid before it, Kodak went through a leadership succession. In May 1983, Walter Fallon, the combative chief executive who had steadfastly maintained Kodak's position against any kind of compromise in the Polaroid litigation, announced his retirement. The Kodak board elected Colby Chandler to succeed him as chairman and CEO and Kay Whitmore as president. Whereas Fallon was a hard-nosed executive who often challenged his associates and engendered considerable intimidation among some of his subordinates, Chandler, an engineer who had grown up on a dairy farm in Maine, was mild and affable, a more gentlemanly manager. But Chandler had inherited challenges facing Kodak in just about every aspect of its amateur photography business, challenges that, as one observer wrote, have "already disrupted the admirable esprit that once prevailed throughout the company."[76]

Fuji, in its increasingly recognizable green box, was making significant inroads into Kodak's formerly "impregnable marketing bastion" of conventional amateur film, including landing the highly sought-after sponsorship of the Summer Olympic Games to be held in Los Angeles in 1984.[77] This was a critical marketing platform Kodak missed out on because of its corporate "arrogance," according to Peter Ueberroth, president of the Olympic Organizing Committee. Moreover, the disc system Kodak had introduced with great fanfare in 1983 was not living up to its potential, and was already being called "a flop—a humiliating Edsel of a product."[78] Finally, there was instant photography. Kodak did not break out its financial results by product line, so it was impossible for analysts to know for sure, but there was a consensus among industry observers that instant photography was a losing proposition for Kodak.[79]

The reasons for this were twofold. First, Polaroid's instant products were clearly beating out Kodak's with consumers, with Polaroid's share of the instant market increasing to seventy-five percent in 1983. Second, while Polaroid had begun to develop new applications for instant photography—most notably Autoprocess, a system for taking instant 35-millimeter color slides, and Palette, a product that produced instant slides from graphic displays on a personal computer—Kodak had lagged behind by at least two years in developing competing products. There was even some speculation that Kodak might give up and abandon the instant market altogether. But a company spokesman quashed those rumors: "We have absolutely no intention to get out of the instant business," announced Henry J. Kaska in September 1983.[80] Nonetheless, Kodak had no ready answers for how it was going to escape the doldrums of declining profitability. "This is a blue-chip company that's dead in the water," declared Barre Little, an analyst for Kidder, Peabody.[81]

The news at Polaroid was also a mixed bag, exacerbated by the uncertainty surrounding the fate of its patent portfolio and thus its dominant position in its core business. Profits for the first three quarters of 1983 had soared by seventy-five percent, but much of this was due to cost savings, since Polaroid revenues had actually continued a downward trend. Even though its market share against Kodak had increased, instant camera and film sales overall continued to decline.[82] The company worked on new products and initiatives to reverse this trend, but a cloud of nervousness permeated the atmosphere while it awaited a decision that would have a profound effect on its future. A loss in which its key patents were invalidated would open the floodgates to competitors. A nominal victory in which a couple of patents were found valid and infringed would be a pyrrhic one, in the sense that the damages it would receive would hardly have an effect on the company's financial situation, and might not even cover the tens of millions of dollars in legal costs it had incurred during the litigation. Key executives at Polaroid were clearly hoping for a much better outcome, and the tension was palpable as they were forced to wait for a decision, the prognostications of a quick result having been proven wrong.

With the case already under consideration for nearly two years and with no decision in sight, the pressure from Polaroid on Schwartz to do something began to mount. The first idea proposed was for Schwartz simply to contact the judge in an effort to focus her on the case and to remind her that there was great anticipation over her decision. On January 24, 1984, Schwartz drafted a simple letter to Judge Zobel, obviously reflecting the message he was getting from his clients, and circulated it for comment.[83] "Polaroid Corporation wishes to advise the Court that the . . . [case] continues to be of substantial importance," he wrote.[84] But Schwartz quickly concluded that sending a letter like that would be a "stupid thing to do."[85] He decided that he couldn't tell her it was important; she already knew that. A week or so later, Schwartz came up with a different tack. This time he drafted a letter informing the judge of some recent decisions that had come down since the posttrial briefs had been submitted.[86] While this was clearly a more conventional and perfectly proper thing to do, after circulating the draft to Polaroid in-house and local counsel, as well as to Land, this letter was also never sent.

In early March, desperate for any kind of information on the status of Judge Zobel's decision-making process, the Polaroid side decided to see whether there was any scuttlebutt circulating in the courthouse that could shed some light on the situation. One of the young Fish & Neave lawyers who had worked on evidentiary and procedural matters during the trial was sent to Boston to see if anything could be learned informally. The lawyer flew to Boston, went to the federal courthouse, and ran into Judge Zobel's clerk, Nina Singer.[87] Unfortunately, the apparently tight-lipped Singer did not provide any useful information, and the lawyer returned to New York empty-handed. Shortly thereafter, Laurence Fordham, a senior partner at Polaroid's local Boston counsel, offered to request a status conference with Judge Zobel. But Fordham's offer was politely declined by Polaroid after consultation with Schwartz, Peck, and Mikulka, who together concluded that "such a request might be counterproductive, particularly in view of the recent visit to Ms. Singer."[88]

The reality was all they really could do was wait. Under the circumstances, this was not a comfortable posture, and so the pressure and the anticipation continued to build with each passing month. By June, Schwartz was having research done on the administrative reporting requirements for federal judges with complex cases pending on their dockets. The Administrative Office of the United States Courts required judges to report all civil cases that had been pending for more than three years and to disclose the reason the case had been pending that long and the prospects for closing it. Since *Polaroid v. Kodak* clearly fell into that category, an effort was made to see whether Judge Zobel had filed any such reports that might provide some insight into when a decision could be expected. Unfortunately, the Administrative Office refused to provide any information about any such report without express instructions from the judge in question.[89] Thus, a request for this report would have to be submitted to Judge Zobel, something that Schwartz was not willing to do.

Polaroid executives, frustrated by the delay, seemed to flail about in an effort to precipitate some movement. Richard DeLima and Gerald Dicker, two of Polaroid's in-house counsel, picked up again on one of DeLima's favorite themes from the early days of the lawsuit—that of threatening or actually bringing an antitrust action against Kodak, accusing the still-reigning and undisputed king of amateur photography of unlawful trade practices.[90] The thrust of DeLima's new theory was that Kodak was engaging in predatory pricing in an effort to devalue the instant photography market. Other Polaroid executives, such as patent counsel Robert Peck, seemed wary of this approach. In particular, there was concern that if so challenged, Kodak could turn around and again attack Polaroid's patent portfolio, as it had done in response to the current patent infringement case. A copy of a draft antitrust complaint was sent to Schwartz for his comments, and he gave Peck research "on the current status of the law with respect to accumulation of patents as providing the basis for a misuse defense or an antitrust counterclaim."[91] Land was kept in this loop, as copies of the correspondence and legal memoranda were sent to Nan Chequer. In the end, Schwartz's advice was that this was all a colossal waste of time and energy and that Polaroid just had to wait for a decision.[92] The matter was never pursued.

While all this was going on, during the period of interminable waiting, Bob Ford, the Polaroid patent counsel who had played such a pivotal role guiding Schwartz and his colleagues through the difficult technological issues in the lawsuit, seemed to be pursuing his own, perhaps independent, strategy for resolving the case. It was, in the end, just a curious sideshow. While attending a legal proceeding in Holland, Ford asked Gerry Battiste, an in-house Kodak patent counsel, whether Kodak wouldn't like to settle the case for the five percent royalty after all.[93] This conversation was dutifully reported back to Cecil Quillen, by now Kodak's general counsel, but was not treated seriously at Kodak, as Quillen was not even sure that Ford had been authorized to make such an offer. Later, Ford called Quillen directly and invited him to Boston to discuss with him and with Richard DeLima whether there might be some way to settle the case before a decision was handed down. Quillen made the trip, but once there, realized that it was a waste of time. Inexplicably, De Lima was not available to meet that day after all, and Ford made no further overtures about a settlement offer. Perhaps he was waiting for Quillen to take the initiative on making a proposal, but Quillen was not there for that purpose. He had come merely to hear what Ford and others at Polaroid had to say, and other than having a "congenial day" with his adversaries, nothing came out of the trip.[94]

By the fall of 1984, there was nothing more Schwartz could do to hold off his client's anxiety about the long-awaited decision. Everyone had been waiting for more than two years, but only silence had come from the judge's chambers. On October 17, at Polaroid's insistence, Schwartz wrote to Judge Zobel requesting a status conference on the case "at the convenience of the Court and counsel."[95] A couple of weeks later, Schwartz received a "pleasant and friendly" call in response from Nina Singer. She told Schwartz that she had been asked to call him and Frank Carr to advise them that Judge Zobel expected to have a decision by the end of the calendar year or, at the worst, early the following year. She said that the judge was "sorry that she had taken so long."[96] Just before receiving this call, Schwartz had learned, for the first time, one of the reasons for the long delay. Judge Zobel had apparently been ill in 1983, and had been out of action for some period of time recuperating. It turned out that this had been a bout with breast cancer.[97] Schwartz told Singer that he had just learned about the judge's illness since having written the letter and told her that had he known he might not have requested the conference at all. Singer thanked him for his understanding and assured him that the judge was now better and back at work.[98]

But Schwartz soon learned, and reported to Polaroid, that a decision would not be handed down before year's end.[99] Nor was a decision forthcoming early in 1985, as Singer had promised. Instead, the big news on the Polaroid front was that Land had decided to sell his remaining 8.3 percent of Polaroid stock, ostensibly to further support the Rowland Institute. In a press release issued by an outside public relations firm (not by Polaroid), Land noted that he and his wife "are both seventy-five years of age and it seems wise and prudent to use our resources for maximum good and effectiveness."[100] He emphasized that they would continue their "lifelong program of channeling a significant portion of . . . [their] income . . . into the fields of pure science and scientific education." Remembering the "hundreds of companions in that unique corporation [Polaroid] . . . with affection," he recalled "happily the joy of our mutual undertaking to create a corporation built around the ideal that an industry is first of all a place to gather together to create together and to grow together." To that end, Land announced, "with all this in mind, my wife and I feel that the time has now come for a final separation from the company we shall never forget."[101]

The months continued to pass, and by spring, Schwartz was pressured to act again. On April 3, Schwartz wrote another letter to Judge Zobel, reiterating the request made the previous October for a status conference.[102] The letter was hand-delivered to the judge's chambers. This time, the response was prompt. That same day, Nina Singer called Schwartz to advise him that the judge would indeed hold a status conference six weeks hence, on May 17. Schwartz advised Polaroid and Land of the development. By this time it had been three years since the last posttrial briefs had been filed, and Polaroid found itself, in effect, still begging for a decision.

In preparing for the status conference, Schwartz roughed out the message he intended to convey to Judge Zobel and shared it with Polaroid and with Land.[103] The crux of the matter was that Polaroid stood before her, *nine* full years after filing suit over Kodak's introduction of its instant cameras and film, still seeking justice. Patents, after all, have a limited life of only seventeen years. Since the trial had ended, two of Polaroid's patents had expired. The other eight had remaining terms in the range of only three to six years. The prejudice to Polaroid from having to wait so long for a decision was clear. Kodak was enjoying nine years in a market that, Polaroid believed, it was not entitled to be in. In sum, Schwartz wrote:

> Patents are in large measure wasting assets. Kodak has now had a nine-year ride. There are but a few years left to run. It is no fault of Polaroid's that this is so. This case should be decided and any appeal expedited as promptly as possible.[104]

In the end, all of the jockeying and prodding, no matter how subtle or blatant it may have been, was unsuccessful in shaking an opinion out of Judge Zobel until she was ready to deliver it. The summer came and went. September 1985 saw the resolution of some other long-pending mysteries and the breaking of long-standing records. It opened with the discovery of the Titanic on the floor of the Atlantic Ocean, seventy-three years after the luxury liner had struck an iceberg and sunk in 12,000 feet of water off the southern end of Newfoundland. On September 11, Pete Rose broke

Ty Cobb's fifty-seven-year-old record for most hits in a baseball career. And finally, two days later, on Friday, September 13, 1985, after three and a half years of deliberation, and nine and a half years after the case was filed, Judge Rya Zobel broke perhaps another record by finally announcing her decision in the landmark case of *Polaroid Corporation v. Eastman Kodak Company*.

CHAPTER 26

VICTORY AT LAST

Shortly after five p.m., with the stock markets closed for the weekend, Herbert Schwartz received a phone call in his office from Bill Cheeseman. He had in his possession, fresh from the courthouse, Judge Zobel's long-awaited opinion in the case of *Polaroid v. Kodak*. As Cheeseman began to read the result to Schwartz, his secretary put out the word to other Fish & Neave lawyers who had worked on the case and, within minutes, those still in the office gathered around Schwartz's desk.[1] It was immediately apparent to all that the result was, from Polaroid's point of view, a good one.

Judge Zobel's decision had not been quick, but it was definitive. Almost nine and a half years after Polaroid had filed suit, it finally had the result it sought. Judge Zobel ruled in Polaroid's favor, finding eight of the ten Polaroid patents valid, seven of which she found Kodak guilty of infringing.[2] Polaroid was victorious on all three of Land's patents: the L-Coat, the Symmetrical Supports and the Rear Pick. Likewise, all three of Howard Rogers' patents were declared valid and infringed—the two covering the "Excedrin" film unit structure and his Negative Dye Developer imaging chemistry patent. Also found valid and infringed was Wareham and Paglia's Light Shield Deflector patent. Lloyd Taylor's Mordant patent was held valid but not infringed. The other two camera patents in suit—the Gear Train and the Detachable Spread Housing—were found to be invalid, but the judge ruled that if these patents were determined to be valid on appeal, they were infringed.

The news of the victory spread quickly to Polaroid, and by 5:42 p.m., the story hit the news wires. United Press International reported, "Judge Rya Zobel upheld the validity of 8 of the 10 Polaroid patents involved, ruling that seven had been infringed by Eastman Kodak."[3] While cautioning that a determination on the amount of damages and other relief remained to be decided, a Polaroid spokesman was quoted as stating, "It is a victory for our company and we are very pleased that we have won this stage of the case. . . . We built our company on inventions of our own making, it's our lifeblood . . . [and] we feel vindicated."[4] According to the United Press International's *Business & Finance Wire Report* that Friday afternoon, "a spokesman for Eastman Kodak . . . declined to comment on the opinion."[5]

When Judge Zobel's opinion in the case finally arrived, it weighed in at a hefty 122 pages. From the outset, its structure echoed clearly that of Polaroid's presentation in court and its posttrial briefs. Before providing a detailed patent-by-patent analysis on the issues of validity and infringement, the opinion began with a discussion of "the big picture"—the history of instant photography, the relationship between the two litigants, and the events surrounding the introduction of their competing instant photography systems. Judge Zobel described the fundamentals of conventional black-and-white photography and how Land adapted those in the first instant systems. It described how this technology was eventually extended to color through the efforts of Howard Rogers and how all of this work over the decades led to Polaroid's introduction of the SX-70 system in 1972. She noted how the SX-70 had been lauded as "an elegant, highly sophisticated camera and film system" and noted that all of the patents in the suit, except for Rogers' Negative Dye Developer process and Excedrin film unit, were incorporated into Polaroid's commercial products.[6]

The history of Kodak's involvement was recounted, including the technical disclosures Polaroid made during that relationship. The opinion reviewed how Land had revealed to Kodak in 1968 the "radically new film" that Polaroid had developed, a film that turned out to be a prototype for SX-70 and how, by early 1969, Kodak had notified Polaroid of its intention to cancel its manufacturing agreement, while simultaneously launching a project to put a competing instant product on the market by 1976.[7]

In chronicling the events that occurred at Kodak in response to the introduction of Polaroid's SX-70 camera and film in 1972, the opinion repeatedly cited the collection of internal Kodak documents produced during discovery. This collection, assembled into a compelling narrative, was apparently as persuasive to Judge Zobel as Schwartz had believed it would be. The opinion described how Kodak's program to develop a peel-apart film that could be used in existing Polaroid cameras was abandoned in 1972. While Judge Zobel acknowledged that

Kodak's stated reason at trial for this decision was its inability to solve certain technical problems, her decision appeared to give more credence to Kodak's contemporaneous memoranda recognizing, as she put it, "that any product created . . . would be obsolete even if it appeared on schedule."[8]

The opinion detailed how Kodak's "expenditures of manpower and money" in its effort to develop an integral instant system that did not require any further manipulation by the user "were enormous." "Despite these efforts," it pointed out, "numerous problems defied solution" as of 1972, when Kodak had as its prototype the Lanyard camera.[9] Quoting Kodak documents, the opinion went on to describe what happened when SX-70 was released. When compared with SX-70, the Kodak prototype was deemed too large and "no longer desirable." After Kodak had purchased and analyzed "large quantities of SX-70 cameras and film," Kodak executives declared its program as "only 'marginally acceptable.'" In Judge Zobel's words: "They recognized that Polaroid had set the standard Kodak would have to meet and that even a 'me-too' program would require more than two and one-half years to produce a less than equal design."[10]

The decision recognized that the facts of this story were not totally dispositive of the case, but Judge Zobel was not reluctant to give them their appropriate weight:

> Although, as Kodak correctly points out, this case involves . . . [ten] different patent disputes and instant photography is not the perpetual domain of Polaroid, the response of Kodak to the SX-70 system has some bearing on the judgment concerning the underlying patents. Also while it is by no means decisive on the question of infringement of any particular patent, so does the suggestion in September 1973 of Kodak's Development Committee that "development should not be constrained by what an individual feels is potential patent infringement."[11]

This excerpt from the court's decision had to be particularly irksome to Kodak and its legal team. In its posttrial brief, Kodak had strenuously protested Polaroid's emphasis on the statement quoted by the court. It believed that Polaroid had taken the phrase from the Kodak memorandum and had used it out of context in an attempt to "darkly suggest a 'damn the patents' attitude at Kodak."[12] Kodak pointed out that the next sentence of the quoted memorandum, which Polaroid had omitted in its brief, advised researchers to consult the Kodak patent office in such cases. Although both Frank Carr and Cecil Quillen later bemoaned the court's unfair interpretation of this document because it had omitted what they considered to be that critical qualifier, there can be little doubt that Judge Zobel was aware of the missing verbiage.[13] Kodak had brought it to her attention repeatedly. Nonetheless, Zobel apparently determined that it neither mitigated nor exonerated Kodak from a policy, or at least an attitude, clearly stated in print. The opinion's reliance on this document, as well as on other aspects of the Polaroid and Kodak development stories, was an example of what happens when, as Stanford professor Paul Goldstein described it, a court "weigh[s] the moral value of creativity in favor of intellectual property owners . . . [and] weigh[s] the opprobrium of piracy against alleged infringers."[14]

In the discussion on the individual patents, another recurring theme emerged that shed some light on the court's decision-making process. It was clear that Judge Zobel found the testimony of Polaroid's inventors, who also served as its experts, more persuasive than that of Kodak's experts. The opinion clearly gave more credence to the technical interpretations and explanations provided by Land and Rogers, men who had worked virtually their entire lives in the field of instant photography.

Land's L-Coat patent was a good example. It covered a process for stabilizing the color image in an instant film unit by using a layer of polymeric acid of a certain composition, and in a certain position, to stop the processing at the correct time and then to purge the image of chemicals that could later harm it. Kodak's expert, Franz Trautweiler, had testified that earlier patents teaching stabilization techniques for previous sepia and black-and-white systems, some of which referred to the use of polymeric acids, had taught Land's L-Coat invention. The court, however, disagreed, concluding that "each type of film, sepia, black-and-white, and color, unexpectedly called for markedly different stabilization techniques."[15]

The opinion specifically gave "credit [to] the testimony of Dr. Land" in dismissing Trautweiler's testimony that Land's patent on sepia techniques taught the color process if one merely dropped the words "color" and "dye" from the L-Coat patent.[16] It recounted, as Land had done on the witness stand, the long history of Land's research into methods for stabilizing the images in each of Polaroid's commercial films and the complications confronted along the way. The court flatly rejected Trautweiler's interpretation:

> [Trautweiler did] not take sufficient account of the essential differences between the processes for sepia and color. . . . The evidence

> is undisputed that the art had been full of experiment for several years, but no one had perceived the particular conformation, position and qualities of the . . . [L-Coat] polymeric acid layer for stabilizing color diffusion transfer film units until Dr. Land did in 1962. He was the first, despite the fact that the parties in this case include on their staffs, without doubt, the world's preeminent experts in the field of instant photography.[17]

The court also failed to accept Kodak's contention that the L-Coat invention as taught in the patent was inoperable because the amounts of polymeric acid disclosed were insufficient to accomplish the assigned tasks. Again, the decision relied on the testimony of Land over that of Trautweiler, noting that Trautweiler's conclusions to that effect were based merely on calculations and not on "experimental data."

Kodak's argument that its PR-10 did not infringe Land's L-Coat patent provided Zobel with an opportunity to display some of the wry wit she had exhibited throughout the trial. The Land patent required the polymeric acid layer to be located in the "photosensitive element" of the film unit. Kodak pointed out that in the manufacture of its PR-10 film units, the polymeric acid layer is coated on one film support, and the photosensitive layers are coated on another support, before they are brought together and sealed into the commercial product. Thus, according to Kodak, its polymeric acid layer was not within PR-10's photosensitive element and thus did not infringe the patent claim. The opinion quickly cut to the fundamental flaw in Kodak's argument, noting that Land's patent covered "a process, not a product." During the forming of an image in the PR-10 film unit, the polymeric acid layer was located precisely as specified in Land's patent, the judge determined. On which support those elements were coated during manufacture was irrelevant.

Moreover, the court forcefully discarded Kodak's citation of other definitions of "photosensitive element" found in other Polaroid patents as supporting its position that PR-10's polymeric acid layer was not within its photosensitive element. Judge Zobel ruled that these "conflicting definitions . . . do not save PR-10 from the claim of infringement," as the term has many possible meanings and Land's patent provided its own "clear definition," as it was required and entitled to do. Judge Zobel couldn't resist quoting Lewis Carroll to support her view:

> "When I use a word," Humpty Dumpty said . . . "it means just what I choose it to mean—neither more nor less."

> "The question is," said Alice, "whether you *can* make words mean so many different things."
>
> "The question is," said Humpty Dumpty, "which is to be master—that's all."[18]

When the opinion discussed Land's Symmetrical Supports patent, it was clear once more that Land's expertise had won the day. Kodak had tried to belittle Land's invention by claiming that the only thing taught in the patent was the use of polyester as film supports, a material that had long been known in the photographic art. The judge plainly rejected this simplistic view, reciting the history of development at both companies in their parallel quests to develop an integral instant film unit that resisted curl or other physical distortion. The decision described how Land had come to a hypothesis that the problem of curl could be solved by using supports that were balanced in terms of their response to the impact of moisture within the film unit, and that "such a balanced response could be attained by using supports that have both good mechanical stability and extremely low permeability to water and water vapor."[19] It described "Dr. Land's insight" that the problem had to do not only with the reaction of the supports to the moist environment but to each other.[20]

Perhaps most importantly, the court's opinion pointed out that prior to Land's breakthrough, the consensus had been that one of the supports needed "to be water permeable so that the film unit could dry fairly quickly. Contrary to that prevailing wisdom," it observed, "Dr. Land believed that with the presence of the [L-Coat] polymeric acid layer in the film unit a wet environment would not harm the image. . . . That Dr. Land's hypothesis is correct has been proven by countless film units produced" by following the teachings of his patent. "Thus," Judge Zobel ruled, "I find that the . . . [Land] patent discloses substantially more than the choice of a known material."[21] To bolster her conclusion, Judge Zobel noted that "the inventiveness of the patent is further underscored by Kodak's unsuccessful efforts to achieve a flat and stable film unit." She pointed out that the Kodak experiments begun in 1969 focused on systems that had at least one permeable support and that more than 600 different support materials had been tried. The facts presented at trial showed that "neither party considered using two polyester supports with integral film units until Dr. Land did so." It was not until "after the introduction of SX-70," she noted, that "Kodak began to use two polyester supports."[22] How then, she inferred, could this invention have been so obvious?

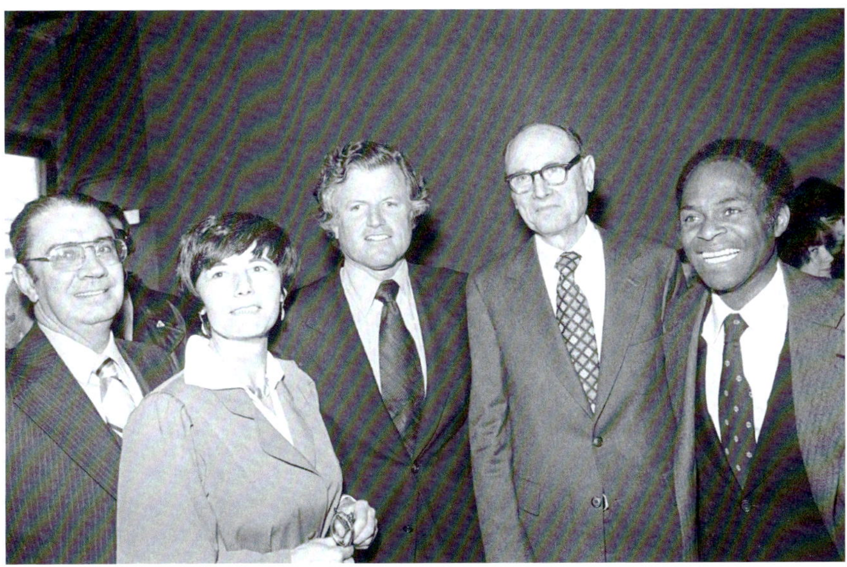

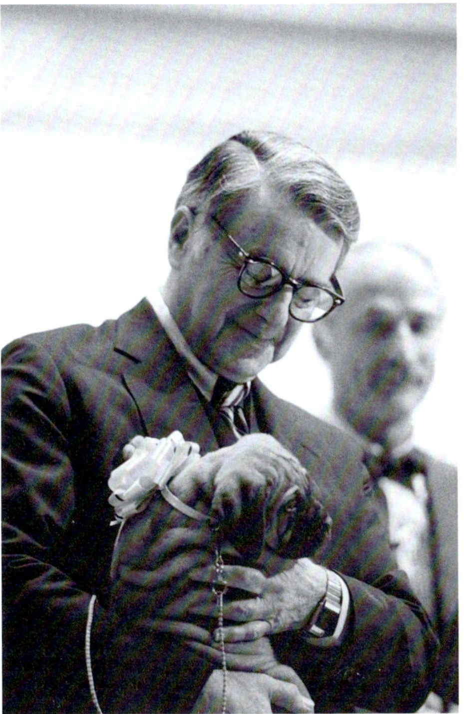

(Top) Figure 17-1: Senator Edward M. Kennedy, D-Mass. (center), poses in 1979 with four judges hc sponsored for seats on the U.S. District Court bench, including Rya W. Zobel, the first woman to sit on the Massachusetts federal bench (second from left) and David S. Nelson, the first black (far right). The other two judges are John J. McNaught (far left) and Robert E. Keeton (second from right).

Courtesy of the Associated Press

(Left) Figure 17-3: Land holding one of two bull mastiff puppies presented to him at the 1980 shareholders meeting in Boston's Symphony Hall. William J. McCune is pictured in the background.

Courtesy Bettman/CORBIS

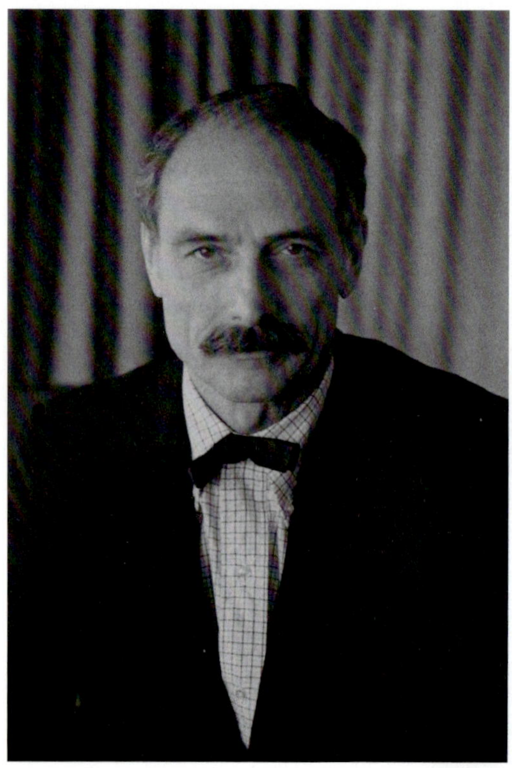

Figure 17-2: William J. McCune in 1975 with his trademark bow tie.

Polaroid Corporation Legal and Patent Records, Box II.230, f. 6, William J. McCune Named President, 1975, Courtesy of Baker Library Historical Collections, Harvard Business School

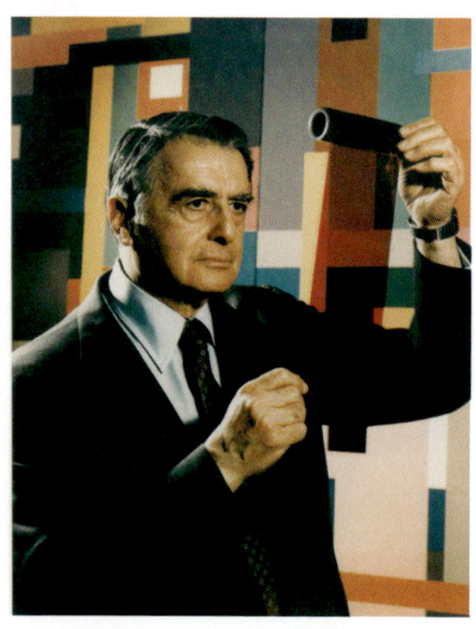

Figure 18-1: Land in his lab in the early 1980s.

Photo by Jay Scarpetti, used with permission of the Scarpetti family and the Rowland Institute

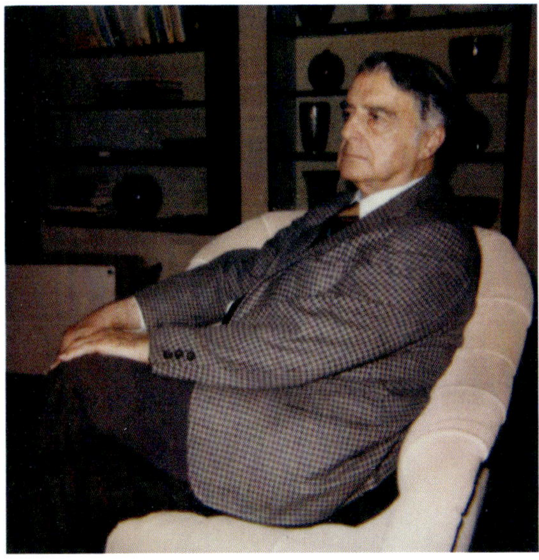

Figure 20-1: Land in his office during a preparation session for his testimony at trial, October 1981 (photo by author).

Polaroid Corporation Legal and Patent Records, Box II.177, f. 2, Edwin H. Land Cross-Examination Notes. Courtesy of Baker Library Historical Collections, Harvard Business School

Figure 20-2: One of the photographs taken by Land from the witness stand on the first day of his trial testimony. Depicted are Polaroid lead counsel William K. Kerr (standing) and, seated at Polaroid counsel table, Herbert F. Schwartz (right), who assumed command of the Polaroid legal team upon Kerr's retirement following Land's opening testimony, and the author (left).

Author's collection

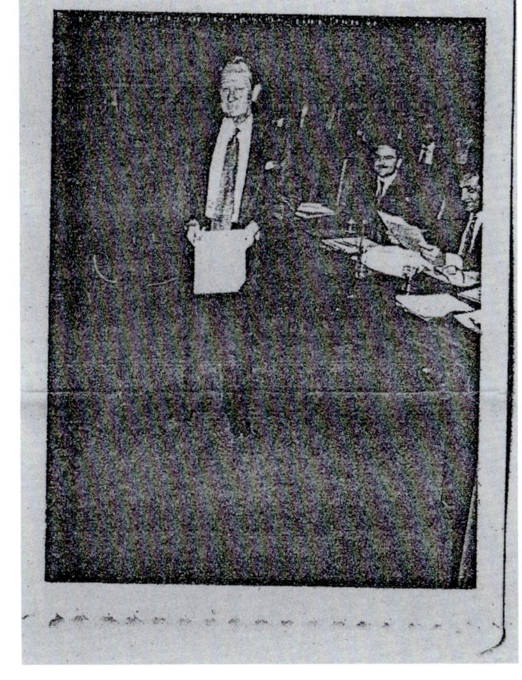

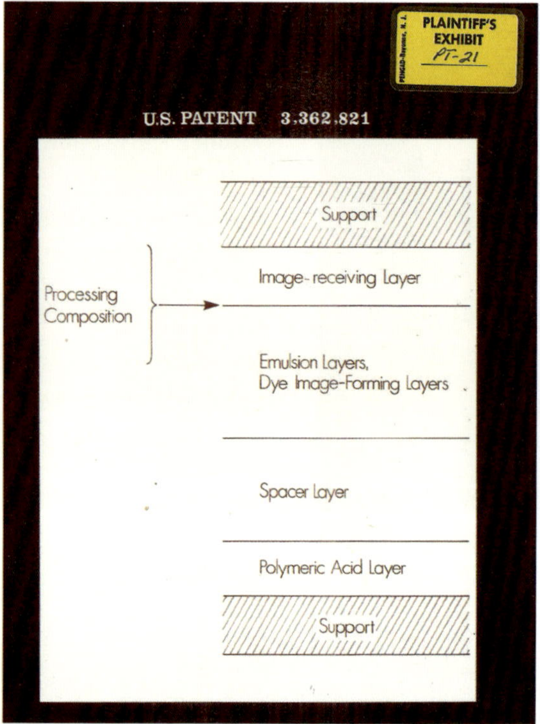

Figure 20-3: A trial exhibit depicting the structure of a film unit during processing that includes Land's L-Coat invention ("Polymeric Acid Layer").

Polaroid Corporation v. Eastman Kodak Company, United States District Court, District of Massachusetts, Civil Action No. 76-1634-Z, Exhibit PT-21

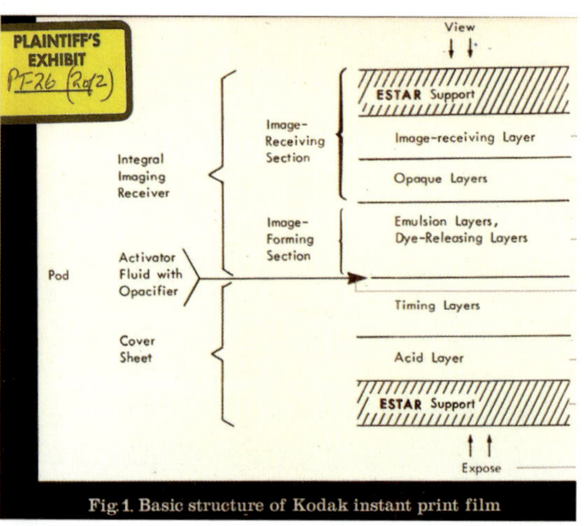

Figure 20-4: Kodak's PR-10 film in a diagram published by its "distinguished scientist" Bunny Hanson. Compare the arrangement of film layers with the Land L-Coat patent diagram above.

Polaroid Corporation v. Eastman Kodak Company, United States District Court, District of Massachusetts, Civil Action No. 76-1634-Z, Exhibit PT-26

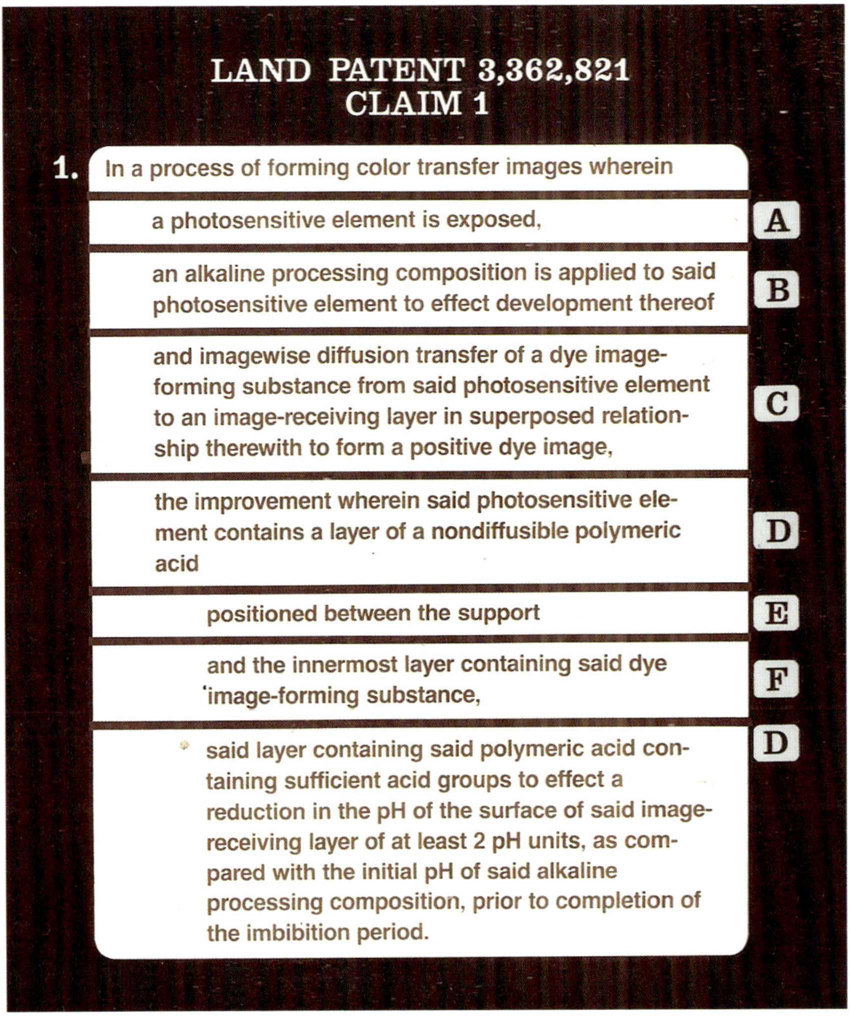

LAND PATENT 3,362,821
CLAIM 1

1.	In a process of forming color transfer images wherein	
	a photosensitive element is exposed,	A
	an alkaline processing composition is applied to said photosensitive element to effect development thereof	B
	and imagewise diffusion transfer of a dye image-forming substance from said photosensitive element to an image-receiving layer in superposed relationship therewith to form a positive dye image,	C
	the improvement wherein said photosensitive element contains a layer of a nondiffusible polymeric acid	D
	positioned between the support	E
	and the innermost layer containing said dye image-forming substance,	F
	said layer containing said polymeric acid containing sufficient acid groups to effect a reduction in the pH of the surface of said image-receiving layer of at least 2 pH units, as compared with the initial pH of said alkaline processing composition, prior to completion of the imbibition period.	D

Figures 20-5 and 20-6: Charts used by Polaroid as trial exhibits to demonstrate how Kodak's PR-10 film unit infringed the Land "L-Coat" patent. The charts show how each element of the patent claim (top) is located in the Kodak film unit (next page), using two diagrams of the PR-10 film unit published by Kodak.

Polaroid Corporation v. Eastman Kodak Company, United States District Court, District of Massachusetts, Civil Action No. 76-1634-Z, Exhibit PT-26

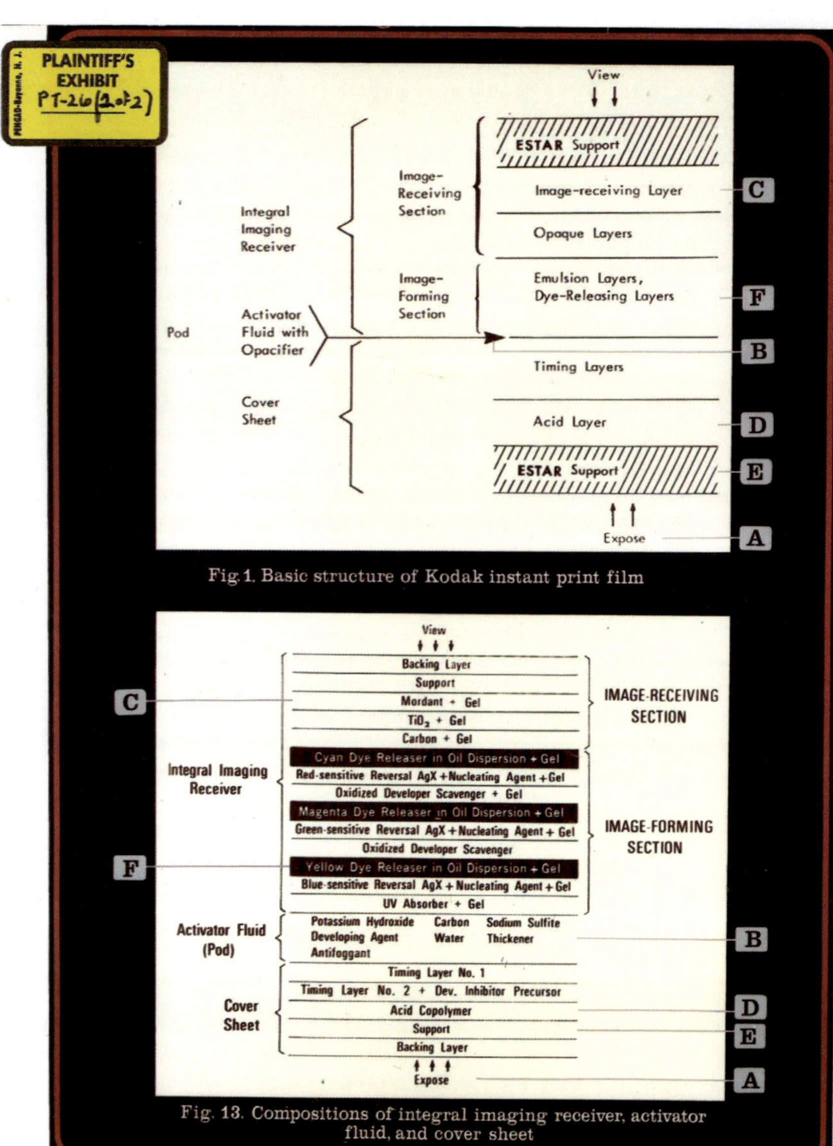

Fig. 1. Basic structure of Kodak instant print film

Fig. 13. Compositions of integral imaging receiver, activator fluid, and cover sheet

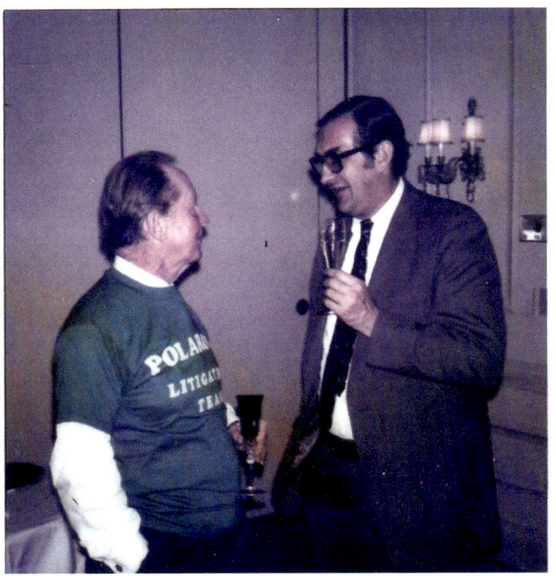

Figure 21-1: William Kerr and Herbert Schwartz (left-right) at an informal celebration held by Polaroid lawyers to mark the conclusion of Polaroid's direct case in December 1981, and to bid Kerr farewell. Kerr is wearing a "Polaroid Litigation Team" T-shirt that had been made for the occasion.

Author's collection

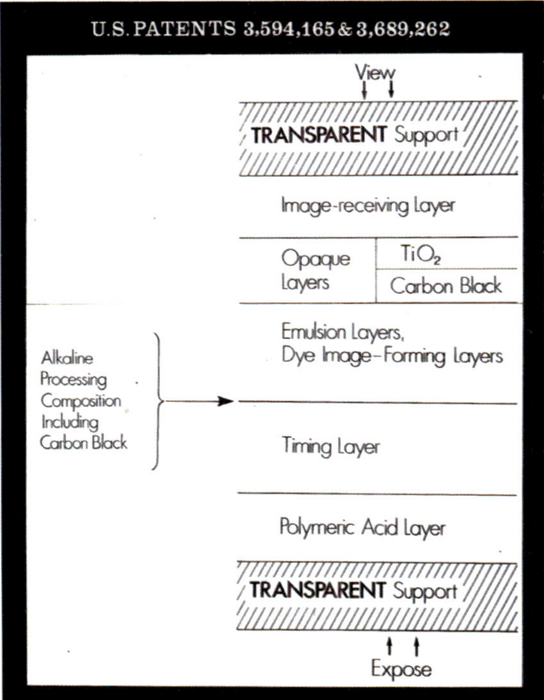

Figure 22-1: Rogers' Excedrin film unit structure as depicted in a chart used at trial. Compare to the arrangement of layers in Kodak's PR-10 film unit (Figure 20-4). This Rogers patent is the one Kodak's chief executive Walter Fallon referred to as a "big block" to Kodak's plans when it was first brought to his attention.

Polaroid Corporation v. Eastman Kodak Company, United States District Court, District of Massachusetts, Civil Action No. 76-1634-Z, Exhibit PT-237

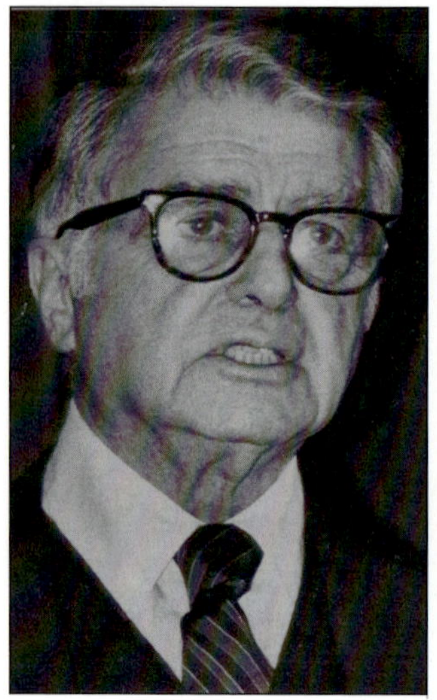

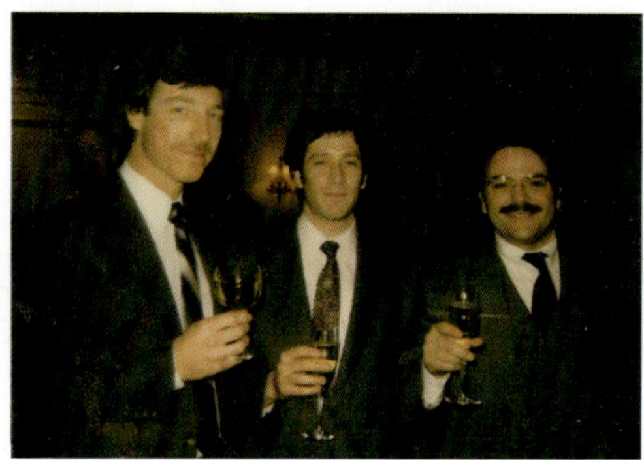

(Top left) Figure 25-1: A beleaguered Land in May 1982, just weeks before he resigned as chairman of Polaroid.

Courtesy of Associated Press

(Top right) Figure 30-1: James M. Williams, Robert J. Goldman, and the author (left-right), celebrating the end of the *Polaroid v. Kodak* trial, February 1982.

Author's collection

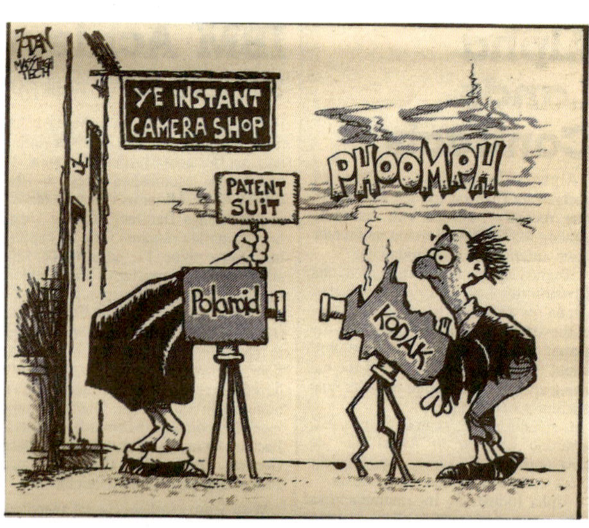

Figure 27-1: The result of the Polaroid v. Kodak decision, as depicted in a cartoon by Glen Foden, *Mass High Tech*, January 20, 1986, p. 10.

Courtesy of Mass High Tech

In the end, the reason Polaroid prevailed on this patent appears to be largely because of the extraordinary effectiveness of Land as a witness, bolstered once again by the big-picture, real-world story Schwartz had worked so hard to present. The irony of this result was inescapable. Schwartz had passed over this patent as a candidate for asserting against Kodak. Now, almost ten years later, Land had clearly succeeded at demonstrating how something that, at first glance, might seem a trivial problem with an easy solution, was actually an obstacle more complex and elusive in the laboratory, both for him and for his rivals at Kodak.

Finally, in terms of Land's patents, the Rear Pick represented another case in which the story of the making of the invention put all of the technical arguments into perspective. Also, it was perhaps another example in which Kodak's expert was proven to have been ill chosen—and no match for Polaroid's inventor in chief. Once again, the court's opinion described the events leading up to Land's invention. Several transport mechanisms had been tried, including friction devices and reciprocating front picks. Film units with tiny holes punched in them to enable front picks to pull them had been tested and discarded.

The opinion noted how Robert Duncan, the SX-70 project manager, had testified that he had believed the introduction date of the system was in jeopardy because of Polaroid's inability to solve this transport problem. It was only then, the court recounted, that Land and some colleagues had "engaged in an intensive effort to find a solution."[23] The effectiveness of Land's solution was clearly persuasive. "The rear pick," the decision observed, "has proven to be very successful and completely reliable. . . . Dr. Land testified, without contradiction, that this pick design has moved billions of pictures and he has never heard of or seen a failure."[24]

In light of the extensive efforts Land and his colleagues had expended in developing the rear pick mechanism, and the special requirements the instant camera process imposed on the transport mechanism for the film units, the court had little trouble disposing of the prior art patents and cameras Kodak had cited. "The differences between the prior art and the invention are important; none of the prior art patents or cameras had to solve the practical problems of a camera that not only stores and exposes film, but also processes it."[25] Then came the *coup de grace*, recalling the dramatic cross-examination by Schwartz of Kodak's expert. "Even Kodak's expert, Mr. Kaprelian," remarked the court, "called the SX-70, which incorporates the . . . [Rear Pick] invention, 'most innovative.'"[26] The conclusion: Land's Rear Pick patent was valid and infringed.

Howard Rogers' patents in suit received similar treatment in the opinion. Once again, the specific technical issues were not considered in isolation, as Kodak had urged the court to do. Instead, the opinion examined the issues through the lens of reality, the actual events that surrounded the efforts of both companies to overcome the developmental obstacles that arose as they worked to produce their integral instant camera and film systems. Kodak's challenges to Rogers' Negative Dye Developer patent mostly turned on fairly narrow and specific technical issues. Time and time again, the opinion "credited" Rogers' testimony in resolving them.

In rejecting Kodak's arguments, it was noted that Kodak's "analysis of the . . . [Negative Dye Developer] process is premised on a strained and narrow reading of the patent," an interpretation the court declined to adopt.[27] Ultimately, Judge Zobel stated her conclusion and the basis for it: "I credit the testimony of Mr. Rogers that the reaction mechanisms . . . [in the Negative Dye Developer patent] and PR-10 are the same." Despite Kodak's attempts to distinguish its imaging chemistry, she ruled that "the sulfonamidophenols are in reality, another species of negative dye developer."[28] Although the opinion did not refer to it, one could not help but recall the Kodak document Schwartz had highlighted during Howard Rogers' direct examination that used the same exact description.

Another technical argument was Kodak's contention that Rogers' patent should be declared invalid for inoperability. It claimed the process taught was so rudimentary that it did not produce quality images. Kodak had introduced into evidence the results of early experiments done by Rogers before applying for his patent. The quality of the images obtained was very poor, and so Kodak believed they demonstrated that the process was inoperable. The court, however, concluded that the poor images Rogers had obtained proved the contrary, that the patent process worked exactly as described. To create better images, the opinion "credit[ed] the testimony of Mr. Rogers that the solution, to add a larger ballast, was an expedient well known in the art."[29] Kodak's attempts to attack the Rogers patent through experimental work conducted by Charles Schallhorn especially for the trial also proved to be ineffective. The court determined that those experiments were not "persuasive" because they employed, as had been demonstrated during the cross-examination, "materials not specified in the patent . . . and which in some instances were not even available" at the time the patent was written.[30]

The court's opinion next turned to Rogers' twin patents on his Excedrin film unit structure, perhaps the most important in the litigation. There was no serious dispute that Kodak's PR-10 film unit used this structure.

Accordingly, Kodak's primary strategy throughout the litigation had been to eliminate these patents by having them declared invalid because the invention described would have been obvious to one skilled in the art of integral instant film unit design.

To set the stage for its legal analysis, the opinion reiterated the story Rogers and Land had presented in their testimony at trial. She described Land's objective for absolute instant photography: "a darkroom mechanism by which the film unit would remain intact before, during and after exposure and would not be physically altered after processing," while allowing "development . . . [to] take place outside the camera and in daylight."[31] It described the film structure invented and patented by Land that was later adopted for use in the SX-70 system and recited the several fundamental complications that this arrangement caused due to the fact that it was exposed and later viewed from the same side. These included the need for a sophisticated temporary opacification system that kept out unwanted light during development after the film unit was ejected from the camera, but then disappeared so the image could be viewed, and a complex camera design that allowed the image to be reversed. The decision recounted how, while Land and some other colleagues worked to overcome those challenges, Rogers had come up with an alternative film unit design that avoided them all together. This design was to become the Excedrin. The court noted, as Rogers had testified, that his design was "not used in the SX-70 because . . . [by that time] the work on the SX-70 was substantially completed and a considerable investment had been made in production machinery."[32]

Given this context, the opinion made short shrift of the prior art patents that Kodak had cited as having taught Rogers' invention. All of them were Polaroid patents, and most were the work of Land. The court found that the similarities between the Excedrin film unit and the prior art were not such that the subject matter of the Excedrin patent as a whole was obvious.[33] Finally, the opinion read: "It is undisputed that Kodak . . . applied for a United States patent on a film unit identical to that of the . . . [Excedrin] patent." The court noted that the Kodak patent attorneys involved in that application had distinguished its structure from the very prior art, including the key Land patent, that they were now arguing rendered the Rogers patent invalid. While the decision made it clear that "Kodak's position at the time of its own application concerning the patentability of the . . . [Excedrin] film structure is not decisive on that issue, it is some evidence that . . . the Rogers patents are valid against the prior art."[34] Once more, it seemed that the judge used a common-sense perspective to help make a determination on what otherwise were highly technical arguments. Kodak's actions spoke louder to her than did its words, at least regarding the Excedrin patent.

News of Judge Zobel's decision was carried in newspapers around the world over the following days. The Saturday headline in the *New York Times* read, "Kodak Infringed on Polaroid Patents."[35] The front page of the hometown *Boston Globe* headlined, "Judge Rules for Polaroid in Patent Suit," and led with the observation that "Polaroid Corp., and especially founder Edwin H. Land, celebrated a landmark legal victory today."[36] Land released a statement from his office at the Rowland Institute declaring how "thrilled" he was with the result.[37] "We worked so hard the first time around on the contents of the film and the camera and the patents on it," he recalled, "and we worked very hard trying to make clear to the court what the essence of every invention was. . . . Even though I no longer have any economic interests," Land pointed out, "I have enormous spiritual and intellectual interests. . . . I shared so much of the adventure with so many, many colleagues. . . . As I sample the . . . [court decision], it's clear that while the technical content is all there, the judge understood the arguments and understood the merits," said Land. "So we feel very happy."

Frank Carr was "crushed" when he learned that he, and his client, had not prevailed.[38] The *Rochester Democrat and Chronicle* ran the initial story of Kodak's defeat on its Saturday front page. "A Kodak Infringement," the headline read. There was still no comment from Kodak; a spokesperson told the local paper that the company had not seen the decision.[39] The Rochester paper pointed out that a determination on the extent of damages that would be assessed against Kodak remained to be made in a future trial. Published speculation on the amount of damages varied widely, from as low as $30–$40 million, according to industry analyst William Relyea, to a potential of $500 million, according to Calvert Crary, the Bear Stearns analyst who had regularly reported on the case over the years.[40] Although both companies denied the possibility, some observers thought that Judge Zobel's decision might lead to a negotiated settlement.[41]

Kodak, however, quickly proved to be defiant and vowed to fight on. On Monday, September 16, a spokesman, Wilbur Prezzano, issued a public letter to Kodak customers, and to the world:

> The recent opinion handed down by the U.S. District Court in Boston concerning the Kodak-Polaroid patent dispute has received widespread media attention. We feel it is important to communicate our position in this matter to you, our customer.

> First, let me assure you that our plans for Kodak instant photographic products have not been altered. Kodak will continue to manufacture and market instant products of traditionally high quality with the same high levels of promotional and service support.
>
> Second, let me say with equal conviction that our position in the litigation is unchanged. When we entered the market, we were confident that Kodak instant photographic products were based on our own distinctive technology and do not infringe patent rights of others. We continue to believe this to be the case, and we will continue to contest this matter vigorously.
>
> In the interim, please remember that our plans for Kodak instant photographic products have not changed.[42]

While Kodak expressed confidence that it would obtain a reversal on appeal of all or most of the trial court's decision, the stakes in this grudge match were about to be raised even higher.

Once a patent holder is successful at having a court declare that its patent is valid and infringed, it is customary for it to ask the court to stop, or in legal parlance, to enjoin, the infringer from further violating its patent rights. Polaroid had sought that remedy against Kodak in its original complaint when the lawsuit began, and there was no doubt that it would continue to seek that relief. There were some who believed that Polaroid would not be granted an injunction in any event.[43] As analyst William Relyea noted, "Seldom is the remedy to push somebody who's been in business for nine years out of the business. One would not expect Kodak to leave," he concluded, "but one probably would expect Polaroid to get some money."[44] That was certainly the prevailing view, given that up to that point, there had never been a major industry shut down by a patent.[45] In any event, even if Polaroid were to succeed in getting Judge Zobel to issue an injunction, it would likely be stayed—that is, its effect delayed—pending Kodak's appeal of Zobel's decision, which could take several more years. Carr, in assessing the situation, took some comfort from a case, involving only one patent rather than seven, in which a federal court of appeals did not render a decision for three years.[46]

Schwartz had a different view. He had been watching decisions coming out of the new federal appellate court that handled patent appeals and believed that there might be a legal basis for urging Judge Zobel not only to enjoin Kodak but also to take the extraordinary step of refusing to stay the implementation of that injunction while Kodak appealed her decision on the merits of the case.[47] Such a ruling would mean that Kodak would immediately have to cease selling its infringing cameras and film. It would be the most drastic of punishments and was virtually unprecedented.[48] Kodak's attorneys themselves were convinced that the judge would not grant Polaroid's request to impose this harsh penalty.[49] Calvert Crary agreed, telling his readers that the "failure to stay the injunction would irreparably destroy . . . [Kodak's] instant photography business, which would be an unjustifiable result if it should happen that Judge Zobel's decision was incorrect."[50] But Schwartz had been developing his strategy for some time and was having extensive legal research and analysis done to provide the arguments he would need. "I thought we ought to take a run at trying to persuade the court not to stay the injunction," and that is exactly what he set out to do.[51]

After a case is decided, the parties are entitled to submit to the court a proposed form of judgment, the legal document that states the conclusions of the decision and the orders of the court that flow from them. On September 23, Schwartz sent to Judge Zobel and to Carr a proposed judgment for the court's consideration. Among its provisions, Schwartz included the proposed injunction Polaroid sought: "Kodak . . . [is] hereby enjoined and restrained from infringing or actively inducing or contributing to the infringement of any one [of the patents Polaroid had prevailed upon]."[52] There was nothing unconventional about this, as every victorious party would at least seek this remedy, regardless of its likelihood of success. It is up to the other side to resist the injunction or at least to insist that it be stayed pending appeal. A few days later, Carr responded on behalf of Kodak.[53] He had his own proposal for the wording of the judgment that reflected Kodak's views on various technicalities. However, his submission focused mostly on its contention that the injunction proposed by Polaroid should not be granted at all or should, at the very least, and pursuant to the usual practice, be stayed pending resolution of Kodak's appeal. According to Kodak, an injunction was unnecessary because even if Polaroid prevailed after its appeal, monetary damages would be sufficient to compensate it for Kodak's infringement.

Kodak's position on this issue seemed compelling. First, Carr argued that "substantial issues" would be presented in its appeal that "could well be ultimately decided . . . in a manner that is not consistent with this Court's decision."[54] This was Carr's diplomatic way of reminding Judge Zobel that her decision might be wrong and might be reversed. Thus, to impose the injunction immediately would be risky. Second, he strenuously argued that any injunction of Kodak's activities in instant photography would be unfair to the company and, more significantly, against the public interest. Kodak's brief pointed out that the company had sold more than sixteen million instant cameras during the pendency of the litigation and that those cameras could only be used with Kodak's PR-10 film. An injunction would prevent those camera owners from using their cameras—and cause great harm to Kodak as well. An injunction "will plainly cause a major disruption in Kodak's business," resulting in the immediate elimination of 800 jobs and the shutting down of "more than $200 million worth of plant and equipment involved in making instant products."[55]

Two days later, on September 30, Judge Zobel held a status conference to consider the form of judgment she would enter. Both Schwartz and Carr stood to argue in favor of the forms they had drafted. During Schwartz's presentation, he pressed his long-shot case for the extraordinary relief he sought. His position was that Polaroid continued to suffer on a daily basis from Kodak's infringement and that to delay providing the relief to which it was entitled for an indeterminate period of time while the appeal was taken, would "only compound the continuing injury to Polaroid."[56] This, he argued, would cause the company unwarranted and irreparable harm, especially as the patents in question had only limited lifetimes and were all heading toward expiration. Much to the surprise of the Kodak contingent present in the courtroom, Judge Zobel immediately seemed to be intrigued by Schwartz's request and asked him if he had any legal authority to back up his position.[57] Schwartz assured her he did, briefly described the cases, and promised to immediately send her a written submission more fully explaining his position. Carr appeared stunned.[58] While the Kodak lawyers continued to believe that there was virtually no likelihood that Zobel would grant this most unusual and drastic request, they were shocked that Schwartz had the audacity to stand up and make such an aggressive push for it. On October 2, Schwartz followed up with a letter laying out the legal authorities he was relying on, and Carr responded immediately, providing Kodak's view that any reliance on the cases Schwartz had cited would be misplaced.[59]

On October 7, Schwartz wrote one additional letter to Zobel, rebutting Carr's arguments. "Kodak's attempts to distinguish the cases cited at oral argument are unavailing," he wrote.[60] The cases "clearly stand for the proposition that irreparable injury to the patentee is presumed once the patent in suit has been held valid, infringed and enforceable," Schwartz contended. He then challenged Kodak's arguments on the equities of the situation, calling its position "unsound." "First," he pointed out, "the public has no right to use cameras which infringe Polaroid's patents. . . . Second, the law places the risk of loss in this situation on Kodak, not on the consumer. . . . Kodak has used the Polaroid patents with impunity for nine years," Schwartz reminded Judge Zobel. "Kodak could have changed its products if it had wanted to. . . . Kodak should not be heard to complain now. . . . In sum, neither law nor logic supports Kodak's position. . . . The injunction in the Judgment should not be stayed pending appeal."[61]

And with that, Judge Zobel once again had to make a decision that had potential consequences of great magnitude—both for Kodak and Polaroid, and their customers.

CHAPTER 27

THE HAMMER FALLS HARDER

In the weeks following the announcement of Judge Zobel's decision, industry analysts hailed it as a great victory for Polaroid, positioning it for a dramatic turnaround, with yet another generation of improved instant products in the pipeline and its finances in excellent condition.[1] Buoyed by the decision, Polaroid's chief executive, William McCune, issued an optimistic assessment of the company's future. "We believe we can change the direction of amateur photography. . . . We can stop its decline and make it grow."[2] The enthusiasm was contagious. When the trading volume of Polaroid's stock rose to abnormally high levels about a month after the decision, it seemed possible that someone was accumulating shares and that the company might even be the target of a takeover bid, an eventuality that never materialized.[3]

The immediate impact of the decision on Kodak was harder to judge. Publicly, Kodak remained defiant and committed to the instant photography market. But sales in the instant photography field appeared to have peaked in 1978 when Polaroid and Kodak sold a combined annual total of 8.2 million instant cameras. By 1984, that number had dropped to just 3.1 million.[4] Moreover, in recent years, both companies had been on paths to diversify from the amateur photography market: Kodak toward electronics and computerized information-retrieval systems and Polaroid into commercial and industrial applications of its instant photography technology. Peter Enderlin, a Smith Barney analyst, noted that the ruling would have been more devastating to Kodak earlier in the game, when the company "had high hopes."[5] In reality, no one had any idea of the ultimate economic consequences for Kodak—the dimension of damages, if any, that would ultimately be awarded in the case or whether the court would enjoin Kodak and thus force it to leave the business.

It had taken Judge Zobel three years, one month, and thirteen days to issue the decision; now she had to rule on Polaroid's request that she immediately enjoin Kodak from further continuing infringement of its patents. This time her judgment was swift. Once again it came down on a late Friday afternoon. It was October 11, 1985, just four days after the final submissions from the parties and less than one month after her original opinion was issued. Even while Kodak publicly protested its innocence and vowed to appeal her decision, Judge Zobel defied convention and, as described by *Business Week*, "didn't show Kodak any mercy."[6] She issued a permanent injunction ordering Kodak to cease using Polaroid's patented technology.[7] More significantly, as Schwartz had so vigorously urged her to do, she refused to stay the injunction pending Kodak's appeal of the liability decision and set the effective date of the injunction for January 9, 1986, just twelve weeks later. This was time enough for Kodak to appeal her refusal to stay the injunction but not nearly enough time for the overall appeal on the merits of the case to run its course.

This was a monumental event. Herb Schwartz was out of the office, and had to be tracked down, probably on his sailboat, to be told.[8] Ironically, Kodak received word of the judge's decision in a most unusual, and awkward, manner. As the news began circulating, Frank Carr was not able to get any information through Kodak's local Boston counsel, so he was forced to call the offices of his opponent, where one of Schwartz's young partners, Patricia Martone, confirmed the decision had been made and told Carr that she had a copy.[9] "Were we enjoined?" Carr asked her. Martone read him the first paragraph: "Polaroid's request for injunctive relief is granted, and Kodak's motion for a stay pending appeal is denied." Martone believed she could tell from Carr's voice, as he thanked her and said goodbye, that he was shaken by the news. After a long war, Carr and Kodak had suffered a devastating defeat.

In making her decision, Zobel soundly rejected Kodak's contention that the equities of the situation tilted the balance in its favor. In support of her brave and proactive ruling, Judge Zobel quoted a New York federal court's observation that "public policy favors the innovator, not the copier."[10] She admitted that Kodak's arguments on the irreparable harm that might result from an immediate injunction "are seductive." She acknowledged the hardship that would befall Kodak's customers who already owned sixteen million Kodak instant cameras, its thousands of employees who would lose their jobs, and the damage to the reputation and goodwill of the company that might result from Kodak being forced from the instant photography business. Yet none of this swayed her from granting to Polaroid the relief to which the law, especially the recent court decisions Schwartz had cited, established that it was entitled. "It is worth noting . . . that the harm Kodak will suffer simply mirrors the success it has enjoyed in the field of instant photography. As one court has observed," she wrote, "the infringer 'should not be heard to complain' when it loses its gamble and reaps predictable results."[11]

Judge Zobel was defiant in dismissing Kodak's arguments that the public interest argued against granting an injunction. "Kodak's characterization of the public interest not only misallocates risk," the judge noted, "it also misconstrues the very concept of public benefit." Citing the words of the Federal Circuit Court of Appeals, the very court that would hear the appeal of this case, she pointed out that "the public policy at issue in patent cases is the 'protection of rights secured by valid patents.'"[12] Judge Zobel explained that, underlying the principle that the owner of a patent ruled valid and infringed is entitled to an injunction against further infringement, the courts had "found money damages deficient against encroachments on rights of exclusivity, and they have judged the public interest better served by a broad policy favoring creativity than by the narrow protection of specific consumers. . . . Kodak's attempts to elude this clear trend in recent patent law are unavailing," she ruled. Of course, unstated in her opinion, but clearly providing the backbone for her trailblazing decision, was Judge Zobel's apparent confidence that her decision on the merits of the case would withstand scrutiny from the appellate courts and not be reversed.

The official judgment, signed by Judge Zobel and issued by the court on October 11, 1985, read:

> Effective January 9, 1986, Kodak, its officers, agents, servants, employees, and attorneys, and those persons in active concert or participation with them who receive actual notice of this judgment by personal service or otherwise, are enjoined and restrained from infringing any one or more of the following . . . [Polaroid patents] including, without limitation, by manufacture, use or sale of PR-10 film and EK-4 and EK-6 cameras.[13]

This was an astounding and history-setting turn of events. As a practical matter, without quick relief from an appellate court in setting aside or delaying Judge Zobel's order, Kodak would indeed be required to get out of the instant photography business altogether, even before its appeal had been heard and decided. The company would literally have to take all of its cameras and film from store shelves across the United States. The story was carried on evening news telecasts that night across the country and in newspapers over the following days. (See Fig. 27-1.)

After having the weekend to absorb and to assess Judge Zobel's decision, Kodak came out on Monday, October 14, as defiant as ever, and totally dismissive of the injunction issued against it. Chairman of the board and CEO Colby Chandler issued a public letter to all Kodak customers that pronounced the company's position. He advised them that Kodak would be appealing the decision and "requesting an immediate order staying the injunction" otherwise set to take effect on January 9, 1986. The rest of his message was baldly contemptuous of the legal process that had led to this predicament:

> Let me assure you that our current plans for Kodak instant photographic products have not changed. Kodak instant photographic products were developed and introduced, based on our own distinctive technology and we continue to believe that our products do not infringe the patent rights of others. We will continue to contest this matter vigorously.
>
> We do not anticipate the injunction will have any effect on current marketing and manufacturing activities. Kodak will continue to manufacture and market instant products of traditionally high quality with the same high levels of promotional and support service.[14]

For the upcoming holiday shopping season, Kodak announced a somewhat reduced but "very aggressive" advertising strategy.[15] Kodak's resolute attitude seemed to be registering. The headline in the *Baltimore News American* read, "Eastman Defies Order."[16] The *Boston Globe*'s headline read: "Kodak Is Undeterred by Polaroid Injunction."[17] The *Boston Herald* ran a story that included a photograph of Judge Zobel under the headline "Kodak to Keep Instant Film on Market Despite Ruling." It pointed out that Kodak intended to continue "despite a sharply worded order from . . . [the] federal judge telling it to stop no later than January 9th."[18]

Kodak's defiance drew sympathetic support within the photographic world.[19] An industry trade paper, *Photo Weekly*, characterized Judge Zobel's action against Kodak as "a precedent-setting decision that has sent shock waves throughout the photographic industry . . . [and] surprised most observers by acceding to the request for an injunction."[20] It conducted a survey of top photography dealers, several of whom appeared skeptical that the judge's harsh penalty would stand. The president of New York City's top photo retailer, Willoughby's, was quoted as declaring: "No way will we stop buying Kodak instant products." The survey concluded that "most observers believe that the appellate court will adjust her decision in some manner to protect the public."[21]

As Kodak began its fight to make these predictions come true, and to avoid its fate as imposed by Judge Zobel, its attorneys and executives threw themselves into the effort with all the passion and persuasiveness they could muster. They were clearly stunned at the predicament they found themselves in and anxious to find a sympathetic ear in the appellate process. The next day, Tuesday, October 15, Kodak filed its notice of appeal of the judgment with the United States Court of Appeals for the Federal Circuit, located in Washington, DC. This court would ultimately hear both pending aspects of this matter—the appeal as to the underlying merits of the liability decision and Kodak's motion to cancel or to at least stay the injunction. Kodak's first order of business was to ask the court to postpone the effective date of the injunction pending its appeal as to the overall case. Its plea on this appeal echoed the arguments that had failed to move Judge Zobel.

The Kodak brief accused Judge Zobel of only "paying lip service to the undeniable harm that an injunction would cause Kodak's business and goodwill, even if such an injunction were ultimately vacated after a decision on appeal."[22] It reiterated the fact that many jobs would be lost and that more than $200 million worth of equipment would be idled. Moreover, "more than sixteen million camera owners in the United States alone would be unable to use their cameras."[23] But then, in a surprise move, Kodak took its argument a step further. It submitted a sworn affidavit from Colby Chandler, Kodak's chief executive, informing the court that if the injunction took effect as scheduled on January 9, the company would terminate forever its instant photography business, whether or not it was finally vindicated on appeal. In his affidavit, Chandler explained, "Kodak has concluded that the injunction if not stayed, will so disrupt Kodak's instant photography business that it will not re-enter the amateur instant photography market even if the injunction is eventually dissolved." He had concluded that "the enormous cost of resuming the business would not be warranted in view of the difficulty that would be encountered in overcoming the disastrous loss of goodwill and confidence by both dealer and customers, which will result if the injunction is not stayed by this Court."[24]

Chandler's announcement surprised even his own attorneys, who tried to caution him about making such an unequivocal statement. But Chandler was adamant.[25] Carr used Chandler's declaration to great dramatic effect. Noting that its instant photography "business will be finally terminated" if the court did not grant Kodak's motion to stay the injunction, he cast Kodak's predicament as the plight "of a condemned prisoner."[26] Referring to Judge Zobel's statement that Kodak had "assumed a 'calculated risk' that it might infringe Polaroid's patents," Carr strenuously argued that "at no time, however, did Kodak assume a risk of corporate capital punishment *even if it was right.*" (emphasis in original)

Carr also tried to cast the public as the "ultimate loser" if the injunction was not stayed. He contended that Judge Zobel's decision would have a chilling effect on industry in general by saying, in effect: "don't risk it; don't commit your resources and energies, don't bring out new competitive technologies, don't even attempt to challenge the entrenched position of a competitor, because if in the unreviewed opinion of a single district judge you fall short, your effort is for naught, even if you were entirely correct. . . . The result will be less innovation, less competition, more licensing of invalid patents, and higher costs and fewer choices for consumers."[27] Carr then went on to detail for the court of appeals, on a patent-by-patent basis, the "substantial" errors Kodak contended Judge Zobel had made in her decision.[28]

In a fascinating, and arguably desperate, move to suggest an improper bias on the part of Judge Zobel, Carr complained that Zobel's decisions on the patents had been "unfairly and unduly influenced by . . . [her] preconceptions of Kodak's bona fides."[29] He cited the fact that Zobel had acknowledged having given "some bearing" to the Kodak developmental story Schwartz had set out at trial. He objected particularly to the judge's reliance on the Kodak document advising its researchers to "not be constrained by what an individual feels is potential patent infringement" when Kodak recast its research program, following the commercial introduction of SX-70. Kodak complained of what Carr called "the sinister connotations the court inferred from this inexcusably incomplete quotation," as well as by the "broad, diffuse and unwarranted implication that Kodak did something underhanded . . . [which] apparently cast Kodak in a role from which it could not emerge unpunished. In allowing these irrelevant perceptions to affect its judgment, the court obviously did not approach the narrow issues of validity and infringement in an even-handed manner."[30]

Finally, Carr reminded the appeals court that an injunction "is permissive, not mandatory."[31] He pointed out how "a determination of infringement does not automatically result in an injunction. Far less," he argued, "should such a determination automatically result in the denial of a stay of an injunction at a transitory stage of a continuing nine-year litigation where, as here, the immediate and direct harm to Kodak and the consuming public far outweighs any monetary [and completely compensable] injury to Polaroid, and where the denial of a stay will ultimately diminish the benefits the public derives from the patent system." When all of these factors were considered together, the Kodak appeal brief maintained that it had made out a "clear case" for a stay of the injunction.

The press quickly picked up on Kodak's desperate plea to the court. One New York newspaper ran a blurb in its business section depicting the Kodak logo with the headline "Death Row."[32] United Press International published its wire report that "Eastman Kodak Co. said yesterday that if it is forced to halt its instant photography business Jan. 9 it will permanently drop that division and cut jobs regardless of future court decisions in its favor."[33] The *Wall Street Journal* also took special note of the Chandler affidavit, observing that "the filing seems designed to put maximum pressure on the appeals court to stay the injunction."[34] When the *Journal*'s reporter asked a Polaroid spokesperson about the hardships Kodak claimed it would suffer should the stay not be granted, he was reminded that it is not Polaroid's "responsibility to save jobs [for Kodak workers] when it's a problem Kodak created."

In its response to Kodak's motion, Polaroid assured the court of appeals that Judge Zobel had neither abused her discretion nor committed egregious error in reaching her decision to deny Kodak's request that the injunction she had issued be stayed pending appeal. Its brief noted how all of Kodak's technical arguments on the specific patents "amount to little more than a rehash of factual issues presented by extensive factual and expert testimony during the 74-day trial. . . . Although instant photography is only a small part of Kodak's business, it is the very heart of Polaroid's. . . . The balance of the equities clearly lies with Polaroid . . . [because of] the actual and irreparable damage" it had suffered due to Kodak's infringement of its patents.[35] "Unpersuasive" is what Schwartz termed Carr's argument that the injunction would discourage competition. "It not only ignores Kodak's arrogant appropriation of Polaroid's patented technology, but also flies in the face of the rationale of the patent system as articulated by the Supreme Court . . . that an infringer 'will not now be heard to say that public policy is in its favor.'"

Schwartz further assailed Carr's public policy arguments, maintaining that Kodak was not "the champion of the public interest" and that its assertions to the contrary were "not only unsound, [but] they come with cynicism and ill grace." Schwartz once again pressed his "Kodak Story," for the most part in Kodak's own words:

> Kodak did not enter the field of instant photography as a protector of the public interest. As the record . . . demonstrates, Kodak's impetus was the threat of competition to its conventional photography business from Polaroid's integral instant photography system. Within weeks after Kodak first learned of Polaroid's experimental work in integral instant photography, Kodak management concluded that if Polaroid's new system were to be "workable, Polaroid will have a tremendously popular, new and perhaps revolutionary system. . . ." As a result, Kodak's "conventional color films run the risk of suffering severely in the market . . ." and "the only certain answer to avoid serious consequences to Kodak's future in the sale of amateur color films is for Kodak itself . . ." to have such a system.
>
> This caused Kodak to embark upon an extensive program to develop an instant photography system to compete with whatever Polaroid might introduce. . . . When Polaroid introduced its revolutionary SX-70 system in April 1972, Kodak was stunned. The SX-70 camera was hailed as "a masterpiece of engineering." Dr. Albert L. Sieg, who was in charge of Kodak's efforts, reported that "[Kodak's] program as presently defined has lost its impact from a marketing point of view."
>
> Kodak's reaction to Polaroid's achievement was swift and direct. In November 1972, Kodak jettisoned a research project for an alternative product [peel-apart film for use in Polaroid cameras] on which it had already spent 94 million dollars, and marshaled its vast resources quickly to commercialize an integral instant photography system.
>
> Kodak rejected the myriad non-infringing camera and film systems which it might have developed had it desired to make a contribution to instant photography in favor of the commercialization of a "me-too" system which appropriated the inventions of the seven patents held to be valid and enforceable here.[36]

Apparently, there was no way that Schwartz was going to stand for Kodak's attempt to don the proverbial white hat at this late moment in this particular story as it was finally being played out.

Moreover, Schwartz also upped the rhetoric considerably in taking on what it termed Chandler's "arrogant threat" that Kodak would abandon instant photography forever if the injunction was not stayed. He characterized Kodak's exhortation as "an irrelevant self-inflicted injury. . . . Kodak urges this Court to put its stamp of approval on this unlicensed and illegal activity because Kodak would be irreparably injured if its infringement is not permitted to continue." Quoting a 1980 decision from a Pennsylvania federal court in a case involving copyright infringement claims, Schwartz observed how "Kodak's position is reminiscent of the apocryphal 'boy who killed both his parents, later to throw himself on the mercy of the court because he was an orphan.'"[37]

By mid-November, the issue of whether to stay the injunction was fully briefed and in the hands of the court of appeals. There were but weeks left before it was to take effect. Both sides were anxious for a ruling. During this period, Kodak's position remained one of defiance. Its president, Kay Whitmore, stated that he was confident the injunction would be reversed and that "Kodak has no contingency plans for shutting down its instant photography business Jan. 9, despite a court order to do so."[38] This was an astonishing performance for the president of a company appealing for relief from "corporate capital punishment."

Whitmore's position perhaps suggested that his company had more faith in the appellate court than in the trial court below. Perhaps it was just a manifestation of brinksmanship by Kodak, based on its assumption that the decision would be reversed. In any event, was Whitmore safe to assume, as he must have, that the appellate court would have no reaction to this kind of public defiance of a valid and pending court order? To do so was, perhaps, yet another example of the isolationist arrogance that Kodak had seemingly displayed throughout the pendency of the litigation. It seemed to believe that it was above the law, announcing publicly that it did not feel compelled to make any plans to comply with a federal court injunction that was to take effect in a matter of weeks. Apparently, that was still the world as Kodak saw it from its insular perch in Rochester, New York.

In fairness, Kodak was not alone in its assessment that Judge Zobel's decision was flawed and susceptible to reversal upon appeal. Around this time, Calvert Crary published his patent-by-patent review of the court's rulings. His conclusion was clear: "Judge Zobel's decision was too generous in Polaroid's favor."[39] The only film patent Crary gave some chance to survive was Rogers' Negative Dye Developer patent, although he expressed his doubts even about that one. In any event, as he pointed out, it had expired during the long wait for Zobel's decision and "therefore could not be a basis for compelling Kodak to withdraw from the market." Although he concluded that the two camera patents on which Polaroid had prevailed might hold up on appeal, he thought that they "can probably be designed around in one way or another and thus would [also] not be a basis for forcing Kodak to withdraw from the market."

Within a few weeks, the court of appeals set an accelerated briefing schedule for Kodak's appeal of the underlying decision on the merits. On December 2, Chief Judge Howard T. Markey issued an order that all of the appeal briefs would be due before the end of the month. Most significantly, he also ordered that Kodak's motion for a stay of the injunction, which had already been fully briefed, would be considered by the same panel of judges as would hear its appeal of the underlying decision.[40] The following week, the court set January 6 as the date for oral argument on Kodak's appeals—just three days before the injunction was to take effect.[41] Chief Judge Markey's handling of the matter actually represented an implicit rejection of Kodak's effort to have the stay issue resolved more expeditiously. It left the defendant on death row. While Kodak's appeal brief was due December 13, Polaroid's was due December 27, and Kodak's reply brief, the last submission, was due on New Year's Eve, December 31. It was not going to be an especially restful holiday season for the hardworking lawyers, who were likely going to be missing in action from their family festivities.

The briefs filed by both sides on Kodak's appeal of the underlying decision shed very little new light on the arguments that had been set forth time and time again through discovery, the trial, and then the posttrial briefings. Kodak explained its theories of invalidity and noninfringement and tried to explain how and why the trial court had, through systematic legal errors, gotten things wrong.[42] Polaroid's brief met Kodak's technical arguments, contending, as one would expect, that in each case the court had "correctly concluded" that a given patent was valid, enforceable, and indeed infringed. It insisted that Kodak had failed to meet its burden under the law of demonstrating why those conclusions were "clearly erroneous" in order for them to be reversed on appeal. Once again Schwartz cited history, noting that "in instant photography . . . Polaroid is the innovator and Kodak the imitator," and accused Kodak's hypertechnical case of "improperly divorc[ing] the patents from the real world which gave rise to them."[43] He reminded the Court how, just the previous year, it had admonished another set of litigants for doing just that when it stated that "lawsuits arise out of the affairs of people, real people facing real problems." In light of that admonition, Schwartz was more than happy in the pages that followed to educate the court on the real people and the real problems that had led to the inventions in suit and to Kodak's appropriation of them.

In Polaroid's brief, Schwartz recounted the development story as he had done in the past, but this time he pointed out how and why Judge Zobel had recognized the import of certain key events in interpreting the patents before her, much as the court of appeals had prescribed the lower courts to do. He also again contrasted the nature of the expert testimony provided by the two sides. Schwartz pointed out how, at trial, the inventors who had "played the key roles in Polaroid's pioneering efforts" had told their stories and explained their patents, while Kodak's principal expert witnesses "had no 'hands-on' experience in instant photography . . . [and] none had designed instant integral film units or instant integral cameras." He summed up for the court of appeals Polaroid's view of the case in these words:

> The "real world" story of these inventions on appeal confirms the soundness of . . . [Judge Zobel's] decision. Kodak's blatant infringement of these seven patents followed from its express belief that it had to have an instant integral photography system similar to Polaroid's instant integral system in order to maintain its position in the general photographic market. Kodak chose to enter instant photography through litigation rather than innovation.[44]

Given the last word in the matter, Carr assailed Schwartz's strategy in Kodak's reply brief. He accused his opponent of "seek[ing] to protect the judgment below . . . by reference to an irrelevant and distorted 'history' of Kodak's entry into instant photography." Describing Polaroid's account as "misleading and replete with false innuendo," Kodak contended that Polaroid's brief was misdirected:

> Polaroid simply does not come to grips with the specific errors of law and fact delineated in Kodak's brief, choosing instead sweepingly and without support to castigate Kodak as a conniving interloper which appropriated Polaroid's SX-70 technology rather than develop its own. In pursuing this diversionary and irrelevant tactic, Polaroid has failed to meet the thrust of Kodak's appeal, which is that . . . [Judge Zobel] fundamentally erred in . . . [her] approach to the construction of the patent[s] . . . and to the questions of . . . [validity]. Polaroid's failure to explain away these demonstrable errors mandates that the judgment be reversed.[45]

The clock was ticking. The court originally allocated twenty minutes for each side to argue its case, but Kodak moved successfully to have that normal period extended to thirty minutes, given the complexity of the issues on appeal.[46] It was the first Monday after the New Year holiday week, and lawyers and executives for both companies traveled to Washington, DC, for the pivotal event. The same two protagonists who had led their teams during trial were again on opposite sides of a tension-packed courtroom to argue their cases. Presiding was Chief Judge Howard T. Markey, the first chief judge in the history of the newly formed court, and the man for whom the very courthouse they sat in was later renamed in 1998 as the Howard T. Markey Federal Courts Building. Circuit Judges Edward Smith and Pauline Newman flanked him on the bench.

Frank Carr was the first to be heard. Surprisingly, he had determined not to address directly the subject of the looming injunction. Instead, he tried a more oblique strategy. He told the court that he intended to deal solely with Kodak's appeal of Judge Zobel's findings of patent validity and infringement. To begin his presentation, Carr borrowed a ploy from Polaroid's playbook—presenting the big "human" picture to impress the court with the gravity of the impact the injunction would have and its unfairness as a remedy for Kodak's actions. He recounted the history of the case and events that led up to Kodak's entry into the instant photography market, asserting that Kodak's work leading to its 1976 entry into the market was "done carefully and well," even though Polaroid was "heavily fortified by layers of patents." "What Kodak did is what the U.S. Patent Office encouraged," Carr insisted, "innovation around existing patents." For dramatic effect, Carr then showed the appellate judges a Kodak EK-6 camera next to an SX-70 model. Mocking a contention Polaroid had made in its brief, Carr pointed to the two cameras and told the court, "If this is a clone of the SX-70, I hope Kodak never gets into biotechnology."[47]

Carr then reiterated for the appellate panel the complaint registered in Kodak's briefs about the unfair innuendos concerning the company's conduct that had been created by Polaroid's recitation of its development story. Judge Newman asked Carr whether he thought Judge Zobel had been "prejudiced" in deciding the factual and legal issues regarding the patents in suit by her perception of Kodak. Carr responded by saying that it appeared that she certainly had been "influenced" by what Kodak believed was an unfair and inaccurate portrayal promulgated by Polaroid. Moving on to the real substance of the appeal, Carr then proceeded to discuss the technical issues with respect to each of the patents, beginning with the "Excedrin" film unit patents. Interestingly, after Carr's recitation of flaws in Judge Zobel's analysis, Chief Judge Markey pointed out that Judge Zobel had apparently been "impressed" by the fact that Kodak had sought patent protection for the same film unit, contending at the time that it was unobvious in view of the same prior art. Carr responded by observing that since it had been a "young Kodak patent attorney" who had come up with the same idea for a film unit as Rogers, this episode was actually further evidence of its obviousness. Without waiting for a response, Carr quickly moved on to the next patent.[48]

When Herb Schwartz rose to plead Polaroid's case, he had a much simpler task. He merely had to convince the appellate panel that there was no basis for them to reverse the detailed decision rendered by Judge Zobel. He pointed out that Kodak was essentially urging the court to reject the trial court's rulings and come to new conclusions, ignoring a five-month trial, weeks of unsuccessful cross-examination of Polaroid witnesses, the concessions at trial of Kodak's own experts, and the fact that Judge Zobel had, after hearing all of the testimony and studying all of the documentary evidence, specifically credited the testimony of Polaroid's experts, casting aside that of Kodak's. Schwartz quoted a recent and apposite Supreme Court case: "The trial on the merits should be 'the main event' . . . rather than a 'tryout on the road.'"[49]

Schwartz then moved on to explain why Judge Zobel had come to the correct conclusions on each of the factual and legal issues relating to each patent and, once again, reviewed for the court the highlights of the real-world stories that had led to the inventions in suit and to Kodak's misappropriation of them. He pressed Polaroid's view that Kodak was startled by the introduction of Polaroid's integral instant system in 1972 as compared with the product it had in its pipeline and, motivated by the perceived challenge to its overall business, "they copied the entire film transport and developing system. . . . It was not a situation where Kodak was trying to do its own independent work. . . . They saw the SX-70, and they copied it," Schwartz asserted.[50]

This last assertion by Schwartz was a tempting one to make, as it provided a tidy conclusion to the real-world story he had woven throughout the case. But it had its dangers in that it was a somewhat simplistic view of the facts. Schwartz had to know that the statement was only correct in a most generalized sense. Yes, it was true that Kodak had not developed a totally new and original approach to instant photography, as its research capabilities and vast resources might have allowed. And yes, even a Kodak executive had termed it a "me too" product. Nonetheless, Schwartz also knew that the original SX-70 system employed only three of the seven patents Polaroid had won on. Putting that in perspective, Polaroid admitted that the SX-70 system incorporated over 150 separately patented features, a fact Kodak pointed out repeatedly. Thus, it did not appear that Kodak's so-called cloning was literally a case of Kodak's having reverse-engineered or slavishly copied the SX-70. Nonetheless, Schwartz obviously concluded that it was worth any risk that the appellate judges might see his argument as disingenuous, thus potentially undermining the rest of his case. Kodak worked hard to point this out in its briefs, and Carr had already emphasized the distinction in his oral argument to the appellate panel, especially with his derisive cloning remark. What impact, if any, this would have on Judge Markey and his colleagues was a big question yet to be answered.

With the oral arguments on Kodak's appeal of the underlying decision now complete, it seemed likely that a decision by the appellate court would take some time. A close observer of the court and the case believed that a decision would not be rendered for "at least several months."[51] Yet Kodak still had the sword of the injunction hanging over its instant photography business with only three days to go. Accordingly, immediately after the oral arguments were done, Kodak acted to address the situation. It filed an emergency motion with the court of appeals asking for a stay of the injunction until one week after the court issued its decision on the appeal. In its motion papers, Kodak reminded the court that it had previously moved for a stay of the injunction pending resolution of its appeal of the underlying case but that the court had not yet ruled on that request either. "If the injunction is not stayed," the Kodak lawyers declared, "Kodak will be required to cease operations precipitously three days from today."[52] It further explained that in the case of an "adverse decision," it sought the additional week "to permit Kodak to close its instant photography business in an orderly way so as to ease the impact on the hundreds of Kodak employees whose jobs will be affected."

Schwartz was delighted when he received his copy of Kodak's latest motion papers. It was just the opening he had been hoping for. Ever since he had read Kodak president Kay Whitmore's published comments that the company had made no contingency plans for shutting down its instant photography business, despite the existence of a court order to do so, Schwartz had been looking for an opportunity to bring these comments to the attention of the judges who had Kodak's fate in their hands.[53] He seized his opportunity the next day, when Polaroid filed its brief in opposition to Kodak's emergency motion. In Polaroid's very short response, he quoted Whitmore's statements verbatim for the judges to read.[54] He also attached a copy of the entire article, hoping that the judges would also see Whitmore's arrogant complaint that Kodak had lost the case only because "we got caught with a judge who doesn't understand the law."[55] Schwartz thought these published statements would have a big impact on the appellate panel. In the vernacular, he thought they "would blow them away."[56]

The court's decision was quick, decisive, and merciless. Within twenty-four hours, on January 7, 1986, Kodak's disastrous foray into instant photography came to a crushing end when the appeals court refused to vacate or even to delay the effect of the injunction.[57] "Having fully considered all of the submissions and having carefully reviewed the well-considered reasons set forth in . . . [Judge Zobel's decision], this court finds no adequate basis for reaching a conclusion different from . . . [hers] with respect to the requested stay of the injunction."[58] The self-described execution of Kodak was going forward as scheduled. Moreover, this prompt decision did not auger well for Kodak on its overall appeal. In discussions immediately following the decision, Kodak counsel surmised that the court must have already concluded that at least one of the patents was valid and infringed.[59] Considering that the court had had the opportunity to review all of the briefs on Kodak's appeal of Judge Zobel's underlying decision and had heard oral argument on the case as well, one would have expected the court at least to stay the injunction if it was tending towards reversing in any meaningful way the judgment that had spawned it. Instead, it chose to allow the dramatic impact of the injunction to stand, and to take effect within forty-eight hours.

The following day, Kodak made one more desperate attempt to avoid its fate by seeking to have the U.S. Supreme Court issue an emergency stay of the injunction. The plea was made to Chief Justice Warren E. Burger, who was the Supreme Court justice responsible for matters originating from the Federal Circuit Court of Appeals. For this task, Kodak enlisted the help of noted Washington, DC, attorney Paul Warnke, an assistant secretary of defense in the Johnson administration, who later served as President Carter's chief arms-control negotiator with the Soviet Union. With this distinguished counsel, Kodak tried a new tack. It now argued that the injunction should be stayed because of antitrust implications. Noting that Kodak would be leaving the instant photography market forever if the injunction was not stayed, it contended that the result would be a "court-ordered monopolization" of the instant photography market. "The elimination of all competition from the billion-dollar domestic instant photograph market, caused by the unreviewed decision of a single federal judge in a complex patent lawsuit, poses an immediate threat to important national interests," Kodak argued. In sum, "the threat . . . is a consequence not of the merits of the underlying patent controversy—of 'fashioning a remedy for an adjudicated harm'—but is a collateral effect of the injunction itself."[60]

Once again, it did not take long for Kodak to get its result. Before the sun had set on that same day, January 8, 1986, the Supreme Court, without comment, ruled on Kodak's request. Chief Justice Burger, without providing a reason, recused himself from the decision, and so it fell to Associate Justice Lewis F. Powell Jr. to render the Court's judgment. It was as simple as it was devastating. "Denied, L.F.P., Jr., 6 PM, Jan. 8, 1986" was stamped upon Kodak's application for a stay.[61] Kodak had lost—again.

"Kodak is out of the instant camera business as of today," announced *USA Today*, as news of the decision, and its practical impact on the average American, quickly spread nationwide.[62] Retailers quickly began announcing that they would immediately pull the Kodak products off their shelves. As noted in the *Washington Post*, "the court ruling left Polaroid in total control of the billion-dollar-a-year U.S. instant photography market."[63] A Kodak spokesman, Hank Kaska, told the Associated Press that "our next steps are uncertain, but as of tomorrow, we will be stopping all production, distribution and marketing of instant products."[64]

At Polaroid, "company officials," it was reported, "were exultant."[65] William McCune issued a simple statement stating that the company was "delighted" with the decision. A Polaroid spokesman, Sam Yanes, confirmed that the company had no plans to manufacture film for the suddenly obsolete Kodak cameras. "We're obviously sympathetic to their customers . . . but we're the biggest victims of this," Yanes explained.[66] "Every time Kodak sold a camera since 1976, we lost sales, profits and jobs because they were infringing our products."[67] Observers from the financial community saw the court's decision as a "sign of better times ahead for Polaroid."[68] Peter Enderlin, a Smith Barney analyst, noted that the industrial sector for instant photography was "a growing business." Given that Polaroid now had the business all to itself, he concluded that it "could do very well there [in the industrial rather than consumer sector] over the long run."[69] The very next day, the Dow Jones Industrial Average "took its biggest one-day drop ever," at least as measured in points on its scale as opposed to a percentage of its overall value. The drop was 2.5 percent, which paled in comparison to the thirteen percent drop in the 1929 market crash, and was apparently prompted by a sudden rise in interest rates. However, the one stock that resisted the downward trend was Polaroid, which was up almost three percent.

Although Edwin Land did not issue a public statement, he had to be immensely gratified that the justice he had sought had finally been achieved. While he had remained silent throughout the appeals process, and did not appear at the oral argument of the appeal in Washington, DC, Land followed every stage of the process carefully. Schwartz had transmitted to him every brief and every significant court document, and his assistant, Nan Chequer, remained actively engaged in the process, as she had throughout discovery and then the trial, shuttling information and Land's insights and advice to Schwartz.

In the next few days, defeat turned into humiliation as Kodak began its retreat from the instant photography business. Kodak employees were frustrated and confounded.[70] "It's awful, how can they do this?" asked one worker.[71] "It's really sad. How can one judge decide the fate of a whole business?" Another Kodak employee glumly said, "I think the judge had an interest in the case. . . . She was in Boston, where Polaroid is, so I don't think it was completely fair."

In reaction to the news, Colby Chandler issued a public statement to the press. In it he expressed Kodak's "deep regret and considerable disappointment" in having to leave the instant photography business. Chandler emphasized, however, that Kodak "continuously believed that our instant products were based on technology which combined the best of our own proprietary research and that which lay in the public domain."[72] That same day, Chandler also issued a letter to Kodak shareowners. In it he recounted the history of the litigation and described the court decisions that had led to this moment. He noted that despite Kodak's continuing appeal of the decision finding it liable for infringing Polaroid's patents, "Kodak and its customers are faced now with a finality: the company cannot continue to supply instant products."[73] The letter addressed the predicaments and the concerns of all those affected—its customers, its dealers, its vendors, and its employees. Chandler concluded with a message that echoed his earlier press statement. It too was drawn clearly in Kodak's defiant and confident corporate tone, without even the slightest hint of remorse:

> It is with regret and disappointment that we leave the instant photography business. The decision to do so is unavoidable in light of the decision of the Federal Courts.
>
> We have served the market for instant photography products for nearly ten years. We believe we have adhered strictly to our policy of respecting the valid patent rights of others.
>
> We will continue to conduct our business in an ethical way, for the benefit of our customers, our employees, our shareowners and the communities in which we operate.[74]

Next on the agenda for Kodak was a public relations campaign that tried to address the problems of its customers, the owners of millions of suddenly obsolete Kodak instant cameras. "Those cameras are worth nothing now," stated a photography industry analyst. "It was a camera yesterday; it's a paperweight today. Or maybe a museum piece."[75] A program was initiated offering a trade-in of the now useless cameras for either a fifty-dollar cash rebate in the form of Kodak coupons, a single share of Kodak common stock, or a Kodak disc camera and two discs of film. The offer left camera owners scrambling as they tried to figure out which was the best deal. Reports surfaced that speculators immediately overwhelmed the toll-free telephone number Kodak had set up and flooded stores to buy cheap Kodak instant cameras for as little as fifteen dollars so that they could be converted into a share of Kodak stock, then trading at $47.50.[76] It was reported that one customer showed up at Lorelski's, a store in Monroeville, Pennsylvania, with cash in hand to buy 1,000 of the Kodak instant cameras. Unfortunately, the store had only a few used models left to sell him.[77]

In response, Kodak quickly limited its offer to three cameras per household and asked retailers to take any remaining cameras off their shelves.[78] The *Wall Street Journal* published the considered opinion of one analyst that the best option was a swap for the fifty dollars in coupons, as he was "bearish" on Kodak stock and "not a fan of the company's disk camera."[79] Ultimately, this entire campaign proved to be ill designed and too hastily implemented. By June, Kodak had to cancel the entire program when more than fourteen class action lawsuits were brought against it by unhappy customers.[80] It also learned, apparently belatedly, that in order to offer the stock swap, it first had to get permission from each of the fifty states.[81] Eventually, the class action lawsuits with the owners of obsolete Kodak instant cameras were reportedly settled for $150 million.[82]

Nothing Kodak did seemed to satisfy an irate and baffled public. An outpouring of increasingly embarrassing ideas abounded on how to dispose of millions of now suddenly useless Kodak instant cameras, comprising more than 16,000 tons of plastic, according to the *Rochester Democrat & Chronicle*.[83] Suggestions included dumping them in Lake Ontario near Rochester to create an artificial reef for fisherman, or melting them down and using them as landfill to build a theme park called "Picture Island." Even Judge Zobel received letters from angry and bewildered Kodak customers, asking her to help them address their camera problems. In one case, Judge Zobel graciously wrote back advising that she was "not personally in a position to help" but promised to forward the correspondence to lawyers for Polaroid and Kodak.[84]

As Kodak wrestled with its public relations problem, the public and the press continued to pile on. A *Fortune* article on recent trends in patent law opened with a dramatic flourish that touched a still-raw nerve for Kodak: "Rushing to introduce an instant camera by 1976, Eastman Kodak's development committee issued a startling directive: 'Development should not be constrained by what an individual feels is potential patent infringement.'"[85] Kodak chairman Colby Chandler wrote a letter to the *Fortune* editor making the same case as its lawyers had repeatedly made to the courts.[86] He complained that the article had "omitted the all-important follow-up sentence . . . [which was] critical to the meaning of the advice given to members of Kodak's developmental committee."[87] The author of the article, in her response, noted that Judge Zobel had "quoted the Kodak . . . directive exactly as *Fortune* does [and] after hearing considerable testimony on the matter . . . apparently concluded that the follow-up sentence was not as important as Mr. Chandler contends."[88]

Whatever faint glimmers of hope may have persisted in Rochester for some moderation of the courts' judgment faded forever within a matter of weeks. On April 25, the Federal Circuit Court of Appeals released its decision on Kodak's appeal of Judge Zobel's underlying decision. Her judgment that Kodak had infringed seven valid Polaroid patents was "in all respects affirmed." Writing for the three-judge panel that heard Kodak's appeal, Chief Judge Markey credited Judge Zobel for her thorough and careful handling of the case:

> With her judgment . . . [Zobel] issued a comprehensive, 122-page Memorandum Decision, in which she exhaustively articulated the parties' contentions, including each of Kodak's myriad defenses.

> Incorporating her fact findings and legal conclusions, she delineated the evidence and testimony introduced at trial and noted her credibility determinations. The . . . decision reflects . . . [Zobel's] grasp of the involved technology and a correct application of the patent law to the facts found in what was at the trial stage a more complex case than even that presented on this appeal.[89]

After noting that it was not the appellate court's job to retry the case "as though no trial had taken place," the appellate court concluded that "though seven patents are involved in this appeal, and we have carefully considered Kodak's arguments on each one, we have been shown no error that might compel us on this record to reverse the judgment or any appealed portion thereof."[90] Before reviewing the arguments on each of the patents, the court tersely yet robustly rejected one of Kodak's overriding contentions, that Polaroid's recitation of the real-world story of Kodak's entry into the market had prejudiced Judge Zobel against it. "The assertion that the trial judge was influenced by preconceptions of Kodak's culpability is unworthy of comment."[91]

There was, however, one final, last-ditch effort for Kodak to try in getting the decision reversed, and that was another shot at the U.S. Supreme Court. Although it had declined to review Kodak's appeal of the court of appeals' refusal to stay the injunction, perhaps Kodak could articulate a rationale that would entice the Supreme Court to review the underlying decision on the merits.[92] Kodak filed its application to have the Supreme Court review the case on July 23. It tried to frame the case in a manner that would pique the interest of the Supreme Court justices by couching the controversy as the kind of legal, as opposed to factual, matter that warranted their review. Kodak argued that the Federal Circuit had disregarded Supreme Court precedents and had gone too far in changing the law on its own; that it had applied new standards of patentability and infringement from those formulated in previous decisions of the Supreme Court.

> If unchecked by this Court, these developments can only chill the use of innovative technology by those seeking to compete with the original entrant into any field. To the costs of research, development and tooling will be added the expense of fighting charges of infringement of dubious patents and the risk of potentially catastrophic loss of investment and damage liability for "infringing" patents on "inventions" that fail to meet the constitutional standard of patentability. The Court should intervene

now to review the marked departure from its decisions that threatens these severe consequences.[93]

In its opposing papers to the Supreme Court, Polaroid characterized Kodak's application to the Court as merely an attempt to get one more "hearing *de novo*" on the facts of the case. It called Judge Zobel's "exhaustive decision," as affirmed in all respects by the Federal Circuit Court of Appeals, "manifestly right."[94] It set out once more the real-life story behind the legal issues, again doing so largely in the words of Kodak's own internal documents. The opening paragraphs of the brief quoted the praise given to Polaroid's SX-70 camera by Kodak's own expert witness, who had called it "remarkable" and "the most innovative and advanced mass-produced amateur camera ever made." When the story of Kodak's reaction to the introduction of SX-70 was described, Polaroid again quoted the now famous Kodak document. Interestingly, this time Polaroid did not omit the balance of the sentence Kodak had so doggedly insisted be included: "Employees were urged not to let development be 'constrained by . . . potential patent infringement' but to leave that concern to Kodak's Patent Department." Dismissing Kodak's pretense for seeking another review of the case, Polaroid contended that "revisitation of the standard of patentability could not alter the outcome of the present case and controversy."[95]

Apparently, the Supreme Court agreed. Just five weeks later, at the opening of its 1986 October term, it announced that Kodak's petition to have the case reviewed was denied. Consistent with Supreme Court practice, no reason was given, the only note to the ruling being that Justice Scalia, who had taken his seat on the Court only days before, "took no part in the consideration or decision of this petition."[96] Polaroid had prevailed, and Land—his work, his company, and his scientific ethic—had been totally vindicated. Kodak had suffered a massive defeat. It endured a huge diminution of its prestige and adverse financial consequences from its foray into the field of instant photography, a technology that never became the threat to Kodak's core business it was originally feared to be and an endeavor that never represented more than five percent of its overall sales. The long battle over instant photography was, once and for all, finally over.

Well, sort of . . .

CHAPTER 28

AFTERMATH

On the first of May, 1989, a full thirteen years after Polaroid filed its original complaint in Boston's federal court accusing Eastman Kodak of infringing its patents, Herbert Schwartz led another team of Fish & Neave lawyers into a different Boston federal courtroom, this time to litigate the amount of damages Kodak would have to pay Polaroid for its transgressions. Between removing all of its cameras and film from store shelves across the country and shuttering its manufacturing operations, its defeat had reportedly already cost Kodak the "staggering" amount of $494 million. It had also cost Kodak an additional $150 million to settle class action lawsuits with unhappy customers. Yet a determination on the amount of compensatory damages Kodak would have to pay to Polaroid was still to be made. From the beginning of the litigation, the case had been bifurcated so that this issue would be addressed after, and if, Kodak's liability was established in the first instance. That time had finally come.

Across the courtroom, the legal team representing Kodak was a new one. Richard E. Carlton and a team of eight lawyers from Sullivan & Cromwell, an old Wall Street firm self-described as the "pre-eminent law firm in America," had replaced Frank Carr and his colleagues from Kenyon & Kenyon. The change was made, at least in part, because Carr, Kenyon's senior partner and lead counsel for the liability phase, was expected to be called as a witness during the damages trial. One of the issues to be considered was whether or not Kodak's infringement had been willful. This could triple the penalty assessed by the judge. The many legal opinions Carr had given to Kodak over the years on the viability of Polaroid's patents would necessarily be of central importance in the willfulness inquiry. So now, Kodak relied on Sullivan & Cromwell, which had successfully represented it in the appeal against the adverse verdict in the

Berkey Camera antitrust case. Its challenge, once again, was to mitigate the impact of an earlier trial court defeat. Kodak could only hope that after the devastating financial losses it had already suffered, Judge Zobel would be merciful and award Polaroid only nominal monetary damages.

Immediately following the issuance of her final judgment in the liability trial, Judge Zobel had commenced the damages phase of the litigation. She presided over two years of discovery and held ten pretrial conferences. Suddenly, on May 9, 1988, she announced at one of those conferences that her mother-in-law, whom she had disclosed as being a Kodak shareholder on the very first day of trial seven years earlier, had died. Since she was a legatee and her husband was executor of the estate, Judge Zobel was prepared to step down. At that conference, counsel for both Polaroid and Kodak "expressed a desire to waive any objection to her participation."[1] Then, on June 15, Kodak's new counsel completely reversed course and filed a motion requesting that Judge Zobel be disqualified from the case and, shockingly, that all of her previous decisions in the case be vacated because of these family stock holdings. In essence, Kodak argued that if the stock ownership was sufficient cause for recusal now, why wasn't it sufficient cause back at the beginning of the trial? It contended that the federal statute did not allow the parties to waive the conflict of interest, and thus the judge had been obligated to step down before the trial had begun.[2] The following week, Judge Zobel went through with her recusal, and the case was assigned to a new judge.

Surprisingly, the judge assigned was one of those who had, more than a decade earlier, bounced the liability trial like a beach ball down the row of federal judges in the District of Massachusetts. Now the Honorable A. David Mazzone found himself presiding over the last chapter of the case. On August 11, he heard oral argument on Kodak's motion to void all of Judge Zobel's decisions. When the lawyers were done, he ruled immediately—Kodak's motion was denied, Judge Mazzone finding that Judge Zobel had been unbiased during the original trial despite the fact that her husband's mother owned some Kodak stock, and especially since neither side had protested after this disclosure.[3] Judge Mazzone went on to praise Judge Zobel's work on the case. "I marveled at the work my distinguished colleague did. . . . I do not believe she was incorrect." He concluded that giving Kodak another trial "would [not] be fair, nor do I think it would be equitable to everyone."[4] Kodak said it would appeal;[5] Polaroid's counsel predictably found the decision "entirely proper and correct and obviously sound."[6]

In February 1989, the Court of Appeals for the Federal Circuit denied Kodak's appeal of Judge Mazzone's decision. Its decision excoriated Kodak for its tactics. In particular, it criticized Kodak for misrepresenting the facts concerning Judge Zobel's recusal. "Exemplifying the adage that 'no good deed ever goes unpunished,'" wrote Chief Judge Markey, "Kodak's brief leads the reader uninformed of the record to believe Judge Zobel disqualified herself only in response to Kodak's motion."[7] He went on to agree completely with Judge Mazzone:

> No basis exists in law, equity or fairness for giving Kodak an additional bite at the apple. It availed itself fully of the judicial process through a fair and complete trial and a full consideration of its arguments on appeal and in its petition . . . [to the Supreme Court]. It failed to establish invalidity of Polaroid's patents and was found to have infringed those patents. In light of the [legal] factors enunciated . . . [by the Supreme Court] and the particular facts of this case listed by Judge Mazzone, we agree with . . . his view that a grant of Kodak's motion would not produce a "sensible and fair outcome of this case."[8]

Persistent as ever, Kodak sought review by the U.S. Supreme Court to overturn the negative decision, but on May 1, the very day on which the damages trial was scheduled to start, the high court announced that it would not review the matter.[9]

Like the liability trial conducted before Judge Zobel, the damages phase was also tried without a jury. It lasted for a total of ninety-six trial days, even longer than the liability trial. Simply put, the object was to determine how Polaroid would have fared financially had Kodak never entered the instant photography market. After hearing bewildering and conflicting opinions from many expert witnesses on both sides, Judge Mazzone clearly felt that the two parties should have reached a settlement and not relied on a court to resolve issues that were more economic and business-related than legal.[10] Prior to trial he had the CEOs of both companies and their counsels meet face-to-face in his chambers in a personal and direct effort to effectuate a resolution.[11] It was to no avail.

The trial, largely a mind-numbing exercise, went on for more than six months, until November 20, 1989. Almost a year later, on October 12, 1990, Judge Mazzone's decision was announced. It included a long and detailed analysis of the amount of damages he deemed adequate to compensate Polaroid for Kodak's infringement. In the end, Kodak was ordered to pay Polaroid almost a billion dollars—$909,457,567, to be exact.[12] Several months later, after a review directed to clerical errors made by the judge in his calculations, the amount of the award to Polaroid was adjusted to $873,158,971.[13] Kodak considered whether it should appeal the amount to a higher court. After Polaroid also threatened to appeal, the two parties, at long last, settled the case for $925 million (more than $1.6 billion in 2014 dollars) on July 16, 1991.[14] It was fifteen years, three months, and twenty days after the lawsuit had begun. The final award was much less than Polaroid had sought and much more than Kodak thought it should pay. But it was a historic and unprecedented penalty. As of this writing in 2014, it still remains the largest satisfied judgment awarded by a court in a patent infringement case in U.S. history. Polaroid's victory was stunning and total.

Beyond the mere numbers, the result in the case also marked a very significant milestone in the ever-developing area of patent law. Polaroid's reliance on its patents to protect the technology it had developed signaled a shift from what had been considered an anti-patent era in the mid-twentieth century to a new period in which patents regained their importance as intellectual property protection for technology companies. During the era that came after the great technological achievements of innovators like Bell, Edison, and Ford, courts were seen as being pro-inventor and pro-patent. However, in the 1930s and 1940s a new concern arose, the danger of monopolies. A fear began to arise that large companies could monopolize fields by buying up patents, which suffered as a result. The government issued fewer patents, and more patents were invalidated when challenged in court.[15]

By the early 1980s, however, "the overemphasis of antitrust feeling prevalent in the 1970s . . . [and earlier began] abating," noted Michael Bloomer, executive director of the American Intellectual Law Association (AILA).[16] Congress recognized and addressed the issue, forming the Federal Circuit Court of Appeals in October 1982 to handle, among other things, all patent appeals, a move designed to provide more certainty and uniformity in patent law and, ultimately, "fair and consistent rulings." Chief Judge Markey denied accusations that the new court's intention was to be "blatantly pro-patent." "We're pro-law, but we're not pro-patent," he insisted.[17]

Nonetheless, it was clear that a new era took hold in the mid-1980s, and many looked on the result in the *Polaroid v. Kodak* case as the turning point. The *New York Times* declared, "The Kodak-Polaroid patent dispute is the most prominent example of an increasingly pro-patent sentiment in American Courts."[18] The effect on America's technology community was clear. A patent attorney at Apple told the *Wall Street Journal* that the result "shows that the patent system is alive, well and very vibrant for those who have inventions."[19] When interviewed about the trend, Schwartz stressed, in words that would surely have elicited a satisfied nod from Edwin Land: "the emphasis [in this new era] is on protecting a climate for innovation."[20]

As a practical matter, the most significant aspect of Polaroid's victory was the court's refusal to stay the injunction, pending appeal, against further infringement of Polaroid's patents by Kodak. In the view of University of Houston Law Center professor Paul Janicke, the "case changed future U.S. patent litigation practice radically in that regard."[21] Patent holders were now free to seek more than just monetary damages—they could reasonably hope to have infringers forced out of the market. In 1986, in response to the decision, *Fortune* interviewed over 100 business executives and lawyers and detected "a growing respect for the power of patents and recognition of the need to manage [companies] differently as a result."[22] AILA Executive Director Bloomer put it succinctly: "A district court put Kodak out of the instant camera business in one day. . . . That's something chief executives understand."[23]

Kodak blamed its loss of the case, among other things, on this shift in judicial attitudes toward patents. Changes in the standards of patentability had occurred between the era of the 1970s, in which Carr originally conducted his evaluation of the Polaroid patents and rendered his opinions to Kodak, and the new post–Federal Circuit environment of the mid-1980s, when the case was finally decided.[24] When the Federal Circuit affirmed Judge Zobel's decision, a Kodak spokesman declared that the ruling "clearly represents" a change in the application of patent law.[25] Stanford Intellectual Property Law professor Paul Goldstein offered the view that "Kodak's intellectual property lawyers could not have anticipated the specific legal changes that tipped the scales in Polaroid's favor over the course of that lawsuit."[26]

Much of this shift was, despite Chief Judge Markey's protestations, attributed to the formation of the Federal Circuit. The contention was that this new court had liberalized the standard for obviousness in the three years from its establishment to the time Judge Zobel issued her opinion in September 1985. The statistics seem to bear this out. When he gave his critical advice to Kodak, Carr was under the impression that only a "relatively small percentage" of patents were upheld, perhaps as low as twenty to thirty percent in his view.[27] According to one study, the percentage of patents found valid by courts rose from around fifty percent prior to the court's formation to a high of seventy-three percent in the years following.[28] Another study found that in 1981 the courts upheld only forty percent of challenged patents, while in 1985 approximately sixty percent were found to be valid.[29] Overall, the general perception among patent litigation practitioners was that the Federal Circuit had moved the percentage of patents being upheld from around thirty-five percent to fifty-five percent.[30] Regardless of the specific figures, the trend was clear. Patents were being accorded renewed respect, and it had become harder for challenges to their validity to succeed.

With regard to the *Polaroid v. Kodak* case, the implication of all this was obvious. While Polaroid's patents may have passed muster with Judge Zobel and the Federal Circuit when considered under what Goldstein termed the "generous . . . standards applied . . . [as] part of an ascent toward a peak that lasted through the 1990s and into the new century," they might not have been upheld as worthy of patent protection under the former, tougher standards Carr had applied when he wrote his opinions.[31] Cecil Quillen, Kodak's in-house attorney who had directed its legal defense against Polaroid, later admitted that, while uncomfortably watching these developments as the case progressed, he dreaded the formation of the Federal Circuit because he knew that thereafter the federal courts were "not going to be a happy place for patent defendants."[32] In his view, "the one thing that had become clear by 1985 was that if a district court judge, ruling in a patent case, wanted to avoid being reversed, they needed to find patents valid and infringed. Those who found otherwise," Quillen contended, "suddenly discovered they got their cases back."[33]

There are, however, two fundamental flaws in both Kodak's and Goldstein's views. First, studies have shown that the success rate of patentees did not change appreciably after the formation of the Federal Circuit. Professor Glynn Lunney Jr. of Tulane University School of Law concluded in 2004:

> Although often perceived as a pro-patent holder court, an empirical examination of appellate patent litigation over the last sixty years suggests that a patent holder is no more likely to succeed on a patent infringement claim under the Federal Circuit than under the circuit courts it replaced.[34]

Second, Judge Zobel did not, in fact, rely on post-shift standards of patentability. She rendered her decision based upon federal appellate law that, for the most part, preexisted the new Federal Circuit. In the entire 122-page decision, the new court is cited on just one single point. Every other citation of legal authority is to precedent established in the various federal courts prior to 1982. Not only that, the overwhelming majority of authority relied on by Judge Zobel dates from the period *prior* to Kodak's introduction of its instant products in 1976—that is, during the period in which Carr wrote his opinion letters. Of the thirty-five cases cited by Judge Zobel, twenty-five are dated 1976 and earlier, and about half of those were issued before Kodak had even started its instant photography efforts in 1969. Thus, in large measure, Carr and Zobel were using the very same legal standards for their opinions.

Whether Kodak's defeat was attributed to the shift in patent law, or on what Quillen, Carr, and others at Kodak have contended was Judge Zobel's flawed analysis,[35] Kodak's decision to pursue a trial rather than come to a negotiated settlement with Polaroid was a puzzling calculation that is, to this day, hard to fathom. Even years after the trial, Carr professed that he "thought we'd be successful [at trial] and I still don't understand why we weren't."[36] Goldstein indicates that Kodak's reliance on finding either a trial judge or, later, an appellate court to concur with Carr's opinions regarding the Polaroid patents was reasonable. "Neither bet was high risk," he writes, "for few patents are bulletproof."[37] Yet for Kodak to believe that it would prevail on ten out of ten close calls is just not realistic. Kodak must have or, at least, should have realized that total victory in a trial was going to be a long shot.

The historical record shows clearly that Kodak decided to proceed with its instant photography program despite these odds. Colby Chandler, who had ascended to Kodak's presidency in 1977, admitted later that "from a probability point of view," he understood it might have been "imprudent for Kodak to enter" the instant photography business because of the large number of Polaroid patents "which might be implicated by the products . . . [Kodak was] going to market."[38] As articulated by J. Allen Jones, Kodak's chief in-house patent counsel, prior to Kodak's entry: "Our optimism [regarding the outcome of a lawsuit] is higher when we look at each patent individually, than it is when we look at the aggregate of all the problems viewed together, since obviously the chance of losing on at least one patent increases as greater numbers of patents are brought into litigation."[39]

But Chandler testified that Kodak elected to go forward nonetheless because management believed that through its patent clearance process—which, incidentally and perhaps conveniently, had found not a single Polaroid patent that Kodak might have infringed to be valid—they had "managed the risk to a low level."[40] Kodak management, in particular Fallon and Chandler, decided to rely on the possibility that not a single asserted Polaroid patent was valid, while disregarding the probability that one or more of those patents might actually hold up in court. This judgment is hard to figure, unless, of course, there was something more going on behind the considerations of Kodak management—either consciously or subconsciously—other than a totally objective evaluation of the business and legal risks.

Apparently nothing could or would divert Kodak from its determined course. Its senior management, isolated and arrogant, had become increasingly frustrated over its inability to wring a license from Polaroid that would have enabled it to enter the instant market without the risk of patent litigation. Moreover, resentment had long brewed in Rochester over what was perceived as Land's inadequate recognition of Kodak's contribution in the early years to his efforts to bring instant photography to market. These were decades-old emotions, deeply felt within the decision-making strata of Kodak upper management. This had either always been, or had at some point become, a grudge match. Kodak's entry into instant photography "was sort of retaliation," one Kodak manager was quoted as saying.[41] As a result, although realizing that it would likely be sued when its instant cameras and film were released, Kodak knowingly accepted the risk and set out to break Polaroid's patent blockade once and for all.[42]

Following the end of the litigation, Kodak seemingly tried to downplay the enormity of its defeat and claim some sort of moral victory from the fact that in the damages trial, Judge Mazzone refused to find Kodak's infringement "willful," a result that, in the judge's discretion, could have tripled the already historically high damages Kodak had to pay.[43] During his opening argument in the damages trial, Schwartz contended that Kodak was aware from the outset that it could not avoid Polaroid's patents in developing an integral system. For example, he recalled how Walter Fallon had written "big block" next to a mention of Rogers' "Excedrin" patent during an early review of Kodak's options for instant photography.[44] "I think that says it all," Schwartz told the court.[45] Ultimately, however, Judge Mazzone found that Polaroid had not adduced at trial sufficient evidence to satisfy the "considerable burden" necessary to support its willfulness assertion.

In opting not to impose this extraordinary punitive remedy, Judge Mazzone relied on two basic and quite reasonable grounds. First, Judge

Mazzone placed great emphasis on the fact that Kodak had done its due diligence with respect to studying Polaroid's patent portfolio. It had engaged Carr to conduct a thorough analysis and to render opinions on the validity of Polaroid's patents and whether Kodak's instant system infringed them. As Judge Mazzone noted, while each and every one of Carr's opinion letters regarding the seven patents adjudged valid and infringed turned out to be wrong, the effort itself undertaken by Kodak was, as the judge described it, "a patent clearance process that could serve as a model for what the law requires."[46] In effect, Kodak's A for effort trumped Carr's F for accuracy.

There was a second, and important, public policy rationale for Judge Mazzone's reluctance to exacerbate Kodak's punishment by tripling its damages. In his opinion, he acknowledged the long-standing judicial philosophy of encouraging parties to challenge patents in court. Citing several legal authorities, he noted the well-established principle that "one who legitimately challenges the validity of a patent should not be overly penalized."[47] Underlying this maxim, he wrote, is "a tension [that] arises between competing interests. . . . On the one hand, the patent system requires a sufficiently severe penalty for infringement to protect the patent owner's exclusive position from pirates, but on the other hand, the public interest requires that there be a real opportunity to test the grants made by the Patent Office, without fear of a ruinous penalty for asserting a position taken in good faith."[48] Clearly, Judge Mazzone was convinced that the penalty imposed on Kodak was already severe enough and that to pile on a finding of willfulness might cross the line, creating a chilling effect on the inclination of future challengers to test the validity of a patent.

In the years following the litigation, Kodak's infringement has been cited as an example of "inadvertent" or "good faith" infringement.[49] The basis for this characterization was Kodak's public proclamation that it had attempted to design around Polaroid's patents, and that some accused features of its instant products differed from those in Polaroid's SX-70 system. For example, Cecil Quillen explained at length how "the Kodak chemistry worked exactly backward from the Polaroid chemistry."[50] He described how Kodak developed "a film unit which you exposed and viewed from opposite sides, just exactly the reverse of the Polaroid camera." Quillen also trumpeted the fact that "we had a rear pick mechanism whose mechanical operation was . . . somewhat different from the mechanical operation of the rear pick in the SX-70 camera."

These observations, of course, completely ignore and perhaps purposefully obfuscate the nature of Polaroid's infringement claim. It did not contend that Kodak's products infringed the SX-70 camera or film *products*. Polaroid alleged, and Kodak was found liable for, infringement of the Polaroid *patents* used in the Kodak instant system. It was completely irrelevant to the litigation whether Polaroid employed those particular patented features in its products or not, or whether the two commercial systems differed in some respects.

Further, these characterizations stretch credulity in the light of the evidence adduced. Kodak opted to move ahead with the development and then the commercial release of its instant camera and film with full knowledge that aspects of its products were covered by Polaroid patents, despite the risk that one or more of those patents might someday be found to be valid and infringed after all. And when it did so, it went further than just registering a defense regarding the specific patents asserted against it but instead attempted to destroy Polaroid's entire patent portfolio by claiming that Polaroid had abused the patent system to keep other companies from entering the instant photography domain. This from a company with thousands of patents of its own—all of which it considers valid—and which "insists [that] . . . others will respect Kodak's patent rights."[51] Although Judge Mazzone ultimately found, essentially for the policy reasons discussed above, that Kodak's conduct was not "willful" under the law controlling punitive damages, history and the reader will have to make their own judgments as to whether the course of action pursued by Kodak with respect to Polaroid's patents was "inadvertent," in "good faith," or something else.

In the end, Kodak's post-decision spin seems like a serious case of denial in the face of the huge defeat it suffered. When offered a settlement by Polaroid just prior to trial that would have granted a license to use the Polaroid patents at a royalty rate that Quillen admitted later they "wish[ed] like hell we'd taken,"[52] Kodak refused even to negotiate and pressed ahead. As one senior Kodak executive admitted in the aftermath of the defeat, "The cost is beyond the money we paid. It was a blow to our prestige."[53] Having been saddled with the worst defeat in the history of patent jurisprudence, Kodak's post-decision rationalizations in this case may be understandable. But they are pure revisionism.

With respect to the trial itself, in the end, Polaroid won in large measure because Judge Zobel adopted the testimony of its expert witnesses over that of Kodak's.[54] Polaroid's key witnesses, of course, were Edwin Land and Howard Rogers. Years later, Judge Zobel recalled that Land "was 'fabulous' as a witness because he was fighting for his intellectual children, 'protecting what was his.'"[55] Kodak believed and argued that Polaroid's experts lacked objectivity because of their financial and emotional investment in the case.[56] Neither Judge Zobel nor the court of appeals agreed. Seemingly of more import to the courts was the fact that Land and Rogers had spent decades, in fact most of their professional lives, immersed in the research and development of instant photographic systems. These men had actually lived with and overcome the obstacles to which the inventions of the patents in suit were directed. Decades earlier, Land had presciently described how experts could be persuasive in a "courtroom only when it is apparent that . . . [they] know their field, know its history, know the great fundamental problems and know, most of all, that to this field they have brought genuine creativity."[57] It appears that this is precisely what he and his colleague, Howard Rogers, had done.

In contrast, Kodak's key experts had none of this hands-on experience in instant photography. Some, like Peter Adelstein, whose overly simplistic views on the Symmetrical Supports patent were obliterated by Land's testimony and courtroom demonstrations during Polaroid's rebuttal case, were Kodak engineers from parts of the company not directly involved with its instant photography program. Others were outside experts like Edward Kaprelian and Franz Trautweiler, who admitted on the witness stand that thcy had no experience working in the nitty-gritty of instant photography research. When matched against the real-world experience of Land and Rogers, their more theoretical or academic interpretations and opinions were simply not as persuasive as the real-life accounts of the men who had actually brought Polaroid's successive instant photography systems to fruition.

Moreover, Kodak made a seemingly major strategic miscalculation in its selection of Kaprelian as an expert, given that, in one of his publications, he had previously described the SX-70 camera as "remarkable" and as "the most innovative and advanced mass-produced amateur camera ever made."[58] This basic blunder was only exacerbated when Kodak's lawyers failed to disarm this statement by bringing it out in court themselves. Instead, they allowed Schwartz the dramatic moment in cross-examination that all trial lawyers dream of, the opportunity to impeach a witness with his or her own words. In the end, Kaprelian's testimony as a trial witness was effectively undermined. Even more significantly, Polaroid was provided with a memorable admission that it used again and again in its court submissions throughout the litigation, ultimately featuring it in the opening paragraphs of its brief to the U.S. Supreme Court.

Kodak's choice of Kaprelian is not the only instance in which it seemed to don blinders to the big picture, while remaining immersed in the minutiae of the case's complicated technical arguments. In assembling his invalidity arguments, Carr had elected not to include any of the real-world factors—what he considered to be "secondary considerations"—in any of the opinions he wrote for his client.[59] Thus, Kodak's management apparently did not consider these factors in its deliberations. Yet, when the trial and appellate courts considered the question of validity, Kodak's obviousness arguments, although technical and detailed in nature, were belied by the big-picture reality.

Judge Zobel apparently recognized that the actual history of the development of one-step photography—at both companies—was the best evidence of whether something was obvious or not. For example, Polaroid had planned in the early 1960s to introduce a print coater with its new Polacolor peel-apart film until Land, working night and day, came up with his L-Coat invention to stabilize the image. This reality was apparently persuasive evidence to Judge Zobel that the techniques of sepia and black-and-white systems Land had known about for decades did not provide an obvious answer, as Kodak urged in hindsight.[60] With his Symmetrical Supports approach to achieving a completely flat instant photograph, Land went against the conventional wisdom that a "forever wet" film unit could remain a flat and stable one. Also persuasive was evidence that Kodak scientists had worked for years in pursuit of its preconceived notion that a film unit needed at least one permeable support and did not adopt Land's approach until after SX-70 had been introduced.[61] Rogers' Negative Dye Developer process, which Kodak had continuously disparaged as useless, had been highlighted in a memo by one of Kodak's leading researchers during its development project as "a basic novel process." Moreover, even though Kodak's lawyers took the position at trial that the sulfonamidophenol imaging compounds used in PR-10 film were not "dye developers," Kodak's researchers used that very term to describe them in their laboratory reports, a "widespread" practice that Carr tried to have stopped for "patent considerations."[62]

Over and over, anecdotal evidence of this sort seemed to sway, in Polaroid's favor, the resolution of tough technical issues with conflicting expert testimony. It was the ultimate validation of the strategy Schwartz had deployed early in the litigation, to develop the story behind the technology so that the making of the inventions could be assessed in a real-life context. In actuality, it was the execution of a fundamental approach articulated by

Charles Neave in 1935, and passed down through generations of Fish & Neave litigators, that the determination of patentable invention depends on

> the history of the invention, or alleged invention and, particularly, an inquiry into whether there had existed a problem, the desirability of the solution of which was recognized, and whether others had unsuccessfully tried to solve it, and whether the final solution had turned out to be a real solution and a real advance in the art.[63]

These factors were the very ones employed by Judge Zobel to resolve the battle of the experts on the myriad technical issues presented in the case. Neave's words of wisdom had indeed passed the test of time, and had brought victory to his professional progeny.

CHAPTER 29

EPILOGUE

Of course, the great irony in this story is the fact that, despite being the technological marvel of its time, instant photography is now, essentially, obsolete and swept away to the vagaries of eBay and local flea markets. And the two companies who fought this epic battle have followed that technology to the business scrap heap. The demise of instant photography came in two sequential waves.

First, in the 1970s came a new generation of 35-millimeter cameras and their high-quality images, used by professionals for decades but made accessible to amateurs thanks to features like autofocusing and autoexposure. Even more importantly, at the same time, the convenience of film processing grew exponentially. The number of local photo "minilabs" grew from 1,000 to 14,000 in the 1980s. Moreover, the speed of the process increased as well. As Judge Mazzone pointed out in his opinion in the damages trial, "Consumers could now get high quality prints in a very short time, cutting into instant's traditional domain."[1] Increasingly affordable 35-millimeter cameras drove instants from peak sales in 1978 to a rapidly declining market in the following years. Despite a brief period of optimism at Polaroid in 1986, caused not only by its victory in the courts but also by the introduction of the very successful Spectra instant camera, the long-term downward trend continued.[2]

The second, and perhaps ultimately more devastating, wave came in the form of digital photography, the first signs of which actually appeared as the *Polaroid v. Kodak* case was coming to trial. In early 1981, *Fortune* got wind of something in the industry pipeline. "Further out over the horizon," it reported, "well beyond any of the camera systems that have Wall Street guessing today, is the electronic camera, a mouthwatering industry prospect."[3] In August 1981, Sony announced in Japan what its chairman and cofounder, Akio Morita, characterized as "the camera of the future" and "another revolution in image recording," the first still camera that uses "electronics and a magnetic disc to capture still pictures."[4] According to photography analyst William Relyea, this "meant an electronic alternative to photography is going to be more competitive earlier than most people had in mind."[5] Even as Land sat on the witness stand in the fall of 1981, Morita came to America to preview Sony's "revolutionary Mavica video still camera."[6] Nonetheless, given that the Sony camera was not scheduled for release until 1983, and even then at a high projected retail cost of $1,000, the consensus in the industry was that this new technology would not "dent consumer markets before 1990."[7]

As a result, both Polaroid and Kodak initially resisted the allure of this new technology and underestimated its long-term impact. Colby Chandler, Kodak's president, spoke in 1981 like the leader of a company built upon conventional silver halide film technology when he reminded observers that "it remains to be seen whether [an all-electronic] camera could be offered at mass-market prices and whether the filmless camera could, or would, offer benefits comparable to those from traditional products."[8] Similarly, Polaroid remained wedded to the technology it had pioneered and reportedly "turned down flat" an offer to partner with Japan's Hitachi Ltd. on its version of a filmless camera.[9] Although it later reversed course and made a belated foray into "electronic photography," Polaroid never became a real player in digital technology, a technology that continues to dominate amateur photography well into the twenty-first century.[10]

While neither 35-millimeter nor the early versions of digital photography offered the essential immediacy of instant photography—the ability to have a print of the subject appear in the photographer's hand seconds after the shutter is snapped—both gradually relegated Land's photo-in-a-minute revelation back to the niche technology that many, including Kodak, first perceived it to be. This was not enough to sustain a major corporation like Polaroid. Instant photography turned out to have a lifespan of decades rather than centuries. Without its visionary founder or a perceptive management able to steer Polaroid on to a new course, the company gradually eroded through the 1990s in tandem with the demise of the technology that had given it life in the first instance. As one business commentator observed, "a lack of imagination and blind faith in an increasingly irrelevant reality produced an inevitable outcome."[11] Similarly, Herbert Keppler, publisher of *Modern Photography*, noted, "There's no longer the wonderful old magic of Polaroid."[12]

In part, Polaroid's demise is also attributable to the fact that it proved unable to compete once its exclusive domain of instant photography began to diminish in importance. As acknowledged by William McCune, Land's successor as Polaroid's chief executive, the company was perhaps not ready for the challenge because of its "unique" early history. "I don't know of any new business today that isn't open to competition," McCune once explained. "Yet, we had a long period without direct competition. Now our competitors are companies like Toshiba and Sony. We are moving closer to what they are doing and they are moving closer to us."[13] Further, Polaroid was no longer the innovator in the fields it found itself competing in. Instead of leading the way, it was joining the parade. Perhaps *Boston Globe* writer Steven Syre put it best in his eulogy to the company: "One day Polaroid stopped dreaming of the future and soon found itself living in the past."[14]

Wracked by a host of other business calamities that only exacerbated the impact of the declining market for instant photography—an unsuccessful but costly 1989 hostile takeover attempt by Roy Disney's Shamrock Holdings, a work force "bloated with hangers on," an overreliance on doing business with developing nations that was undermined by slowing economies and currency issues, and a top management that failed to diversify the company into "cash-flow predictable products"—Polaroid filed for bankruptcy protection in 2001.[15] In 2006, Polaroid quit making instant cameras, and in February 2008, the company announced that it would cease producing instant film. By the end of 2008, it was seeking bankruptcy protection for the second time, largely due to allegations of fraud against the Petters Group Worldwide (PGW), the company that had purchased Polaroid in 2005. Tom Petters, the CEO of PGW, was charged with having concocted a $3 billion Ponzi scheme in order to raise the funds used to, among other things, purchase Polaroid. In 2009, a bankruptcy judge approved the sale of Polaroid's assets to a partnership of venture capital firms. Today, Polaroid exists only as a brand, a still-proud logo that is used for the sale of photography, electronics, and other technology products.

Even Kodak, once the undisputed king of the photographic industry, and one of America's leading companies, is now but a shadow of its former self, its long-term viability questionable. Its decline over the years since the litigation is a more complicated scenario than Polaroid's. Despite its disastrous experience with instant photography, that segment never represented more than five percent of Kodak's business. Although a host of factors seem to have contributed to Kodak's decline, the primary one is that it, too, missed the boat on digital photography. Even though much of the early work on the digital format had been done in Kodak's own facilities, it was so late to the race that it has never been able to catch up.

Kodak, instead, hung on for too long to its traditional business model—rooted in its existing photographic technology—of selling "lots of cameras at low prices to chalk up outsized profits on film, inks, chemicals and papers used for making prints."[16] This is not a unique pitfall for technology companies. In the 1950s, RCA was at the forefront of research on the liquid crystals that would eventually make possible our LCD flat screens of today. Yet in the late 60s, the company dropped its program because of fears that this new technology might adversely affect the revenues it earned from its existing tube-television patents. Around the same time, Sony made a similar decision to eschew the development of LCD technology because it was afraid of losing sales of its Trinitron television tubes.[17]

Despite a belated and desperate attempt through the '90s to transition into digital technology (through consumer products and graphic communications),[18] Kodak's business declined precipitously. By early January 2012, the company's stock price had plummeted from ninety-four dollars a share in 1997 to an all-time low of just $0.37 a share. The final blow occurred on January 19, 2012, when Eastman Kodak filed for bankruptcy protection.[19]

Ironically, its survival strategy was focused on the sale of its vast patent portfolio rather than on the viability of its ongoing business ventures.[20] In August 2012, with no buyer for that portfolio having emerged, Kodak announced that it would be severing its final ties to the technology on which it had been built, selling its "legacy film units" to focus on a future in commercial printing services.[21] Finally, in December 2012, when a consortium of technology companies came forward to purchase Kodak's patents, the sales price was $525 million, far short of its anticipated value of $2.6 billion.[22] The company finally emerged from bankruptcy on September 3, 2013, with little resemblance to its former self. As the *Wall Street Journal* observed: "Gone are the company's units making cameras, film and other consumer photography products that made it famous."[23] Kodak is today a tiny fraction of the company it was in its heyday, with a global work force in late 2013 reduced to 8,500 from a high of 145,300 in 1988, only 700 of whom are in the Rochester area.[24]

The company's fall strikes an emotional chord. "Kodak played a role in pretty much everyone's life in the twentieth century because it was the company we entrusted our most treasured possession to—our memories," explains photography professor, Robert Burley.[25] Tragically, it appears that the company's descent was, in large part, self-inflicted, its management having been slow to react even when the writing was clearly on the wall about the inevitability of the shift from film to digital-based photography.[26] An analysis in *Fortune* observed, "Kodak's lethargy was partly caused by an insular, tradition-bound management."[27] From its view alone atop the amateur photography hill, it espoused a "proprietary view" of the photography business.[28] Kodak's chief technical officer, Edwin Przybylowicz, called it the "Monroe County syndrome."[29] "We're an insular place," he acknowledged. In many ways, these are the same corporate characteristics that seemed to plague Kodak over the years in its decision-making process with respect to instant photography.

The diminution of Rochester's key corporate citizen was a painful one for the community to witness and to endure. For generations, the company had been renowned for its "seemingly endless largess that once allowed . . . George Eastman to provide dental care at little or no cost to every child in town."[30] "Watching Kodak these days is like watching your grandfather die," wrote a local journalist in 2003. "It's sad. You see the life draining out of it; you know that real people are hurting. At the same time, you kind of wish it would just hurry up and die. And then you feel guilty for having such thoughts."[31]

There was also a widespread sense of sadness at Polaroid's demise. When the company announced it was going to cease operations manufacturing film, its hometown *Boston Globe* eulogized the company by reminding its readers just how special an institution it had been:

> Polaroid . . . was the Apple of its day: feisty, ubiquitous, pioneering. The Polaroid camera was like the Mac, with all other consumer cameras PCs. There was the same sense of engineering superiority and cultural cachet. . . . Polaroid uniquely stood at the intersection of science, business and art. . . . It was also a talisman of a lifestyle. So far as the great mass of middle-class Americans were concerned, the embodiments of 60s affluence and liberation . . . weren't bongs or bell-bottoms or even birth control. They were the Ford Mustang and the Polaroid.[32]

Despite Polaroid's disappearance, the SX-70 film unit and camera both seem to have achieved their own status as iconic symbols of another era. For example, New York's Guggenheim Museum sells in its gift shop perhaps the ultimate instant image, a mirror fashioned like an SX-70 film unit, with its reflective surface framed by the familiar thin white border on three sides, with a wider white surface on the bottom. The SX-70 film unit also appears regularly in advertising of all types, as an instantly recognizable graphic design element. For example, if you don't have an image of the item you're selling on eBay, the website will insert an SX-70 graphic to hold the space. These, and many other similar instances, like the Instagram icon based on the OneStep camera design, represent an homage to a photographic phenomenon that had become part of our culture. Like making a "xerox" or using a "kleenex," everyone knew what it meant to take a "polaroid."

Most impressive have been the efforts of some true Polaroid aficionados to bring the system back. After Polaroid announced that it was going to cease manufacturing instant film, a coterie of devotees rallied together to find someone to take over the supply chain. In 2008, Andre Bosman, a Dutch engineer and former Polaroid employee, and Florian Kaps, an Austrian activist in the analog film enthusiast community, joined together in an effort to restart film production. They secured financing from private donors and leased the former Polaroid film manufacturing facility in the Netherlands.[33] Dubbing their effort "Impossible Project—the Re-Invention of Instant Film," they recruited a group of ex-Polaroid engineers and, within months, had a deal in place to have the film distributed under the Polaroid brand name.[34]

Within a few years, the group was marketing film designed for SX-70 and Model 600 cameras, many of which they salvaged and refurbished for resale. Fuji Film of Japan also decided to produce film for this new Polaroid distribution network. In 2010, a new Polaroid camera, the Model PIC 300, was advertised for sale as "a camera for instant gratification."[35] It produces prints that are only 1.8 by 2.4 inches in size, much smaller than either the original peel-apart film or SX-70. In 2012, Impossible announced the release of "large format" eight- by ten-inch instant film, and Polaroid released a new hybrid camera model, the Z2300, that takes digital pictures but can spit out instant prints of selected images.

This "new" Polaroid campaign is being promoted with genuine twenty-first century panache, as a multiyear strategic partnership was announced in early 2010 with the irrepressible and hugely successful popular culture figure Lady Gaga as Polaroid's "creative director." These products are clearly targeted at the "fun" market, people, young and old, who want the immediate experience of sharing photographs at a wide range of gatherings—parties, galas, conventions, and openings of all sorts. The widespread use of photo booths at such events has been described by

Brian Wallis, the chief curator of New York's International Center of Photography, in nostalgic terms: "It's like the return of Polaroid, people are now re-obsessed . . . [with] seeing the image immediately" and walking away with a print.[36] The new Polaroid proudly trumpets the genesis of these new products in long-standing traditions that arose originally from Land's philosophy: the "collaboration between two cultural icons [Polaroid and Lady Gaga] reflects Polaroid's long standing tradition of innovation tracing back to founder Dr. Edwin Land . . . to deliver products that enable creativity for all, celebrate artistry and make sharing instantaneous across the physical and digital."[37]

Notwithstanding this contemporary marketing initiative, the initial impetus for making Polaroid film available once again primarily came from photographic enthusiasts who appreciate instant photography as an artistic medium, as the analog alternative to the digital age, with all of the aesthetics its unique process brings. In some ways, it is the photographic equivalent of the devotees of recorded music played on vinyl records. They have dusted off their turntables and hooked them up to their audiophile amplifiers and speakers, claiming that vinyl still provides better sound quality than a newer technology like the MP3 file. In so doing, they have created their own subculture and in turn have enticed record companies to renew the production of music in that format.

SX-70 film has a similar analog appeal that more modern, sharper films with more sophisticated color characteristics subjectively lack. For instant photos, "the spectrum of color variation and the general unpredictability of the format are a large part of the appeal and the esthetic of the product," explained one report on the attempt to bring the format back.[38] Cambridge photographer Elsa Dorfman, famous for her large-format portraits using Polaroid film, celebrated the return of the "creamy, wonderful, fabulous film."[39] From a commercial photography perspective, one advertising executive admitted that there are clients who still seek that "high-quality, old-fashioned look" that only an instant photo can achieve.[40]

Perhaps, in some ways, instant photography has come around 360 degrees to realize finally Dr. Land's aspirations for it as a creative medium, as inhabiting "the intersection of science and art."[41] As far back as 1947, he pronounced "the purpose of inventing instant photography . . . [to be] essentially aesthetic."[42] Although instant photography had once been a convenient tool—for instance, as a simple and quick way to test composition and lighting in a photo shoot—"now it is the art itself," noted one professional.[43] Land saw his technology as a means of putting an instrument to make art into the hands of the average person:

> The aesthetic purpose is to make available a new medium of expression to the numerous individuals who have an artistic interest in the world around them, but who are not given to drawing, sculpture, or painting. . . . We cannot establish what may be a fact, that artistic ability and need for artistic expression is almost universal, although talent and facility for such expression is obviously not universal. It does seem certain, however, that an enormous number of people who lack talent or inclination for the [other] arts, *do* have the taste for and the need for a simplified medium of artistic expression. (emphasis in original)
>
> This new [instant] photographic program seeks to contribute to satisfying this aesthetic need. By making it possible for the photographer to observe his work and his subject matter simultaneously, and by removing most of the manipulative barriers between the photographer and the photograph, it is hoped that many of the satisfactions of working in the [other] arts can be brought to a new group of photographers.[44]

Throughout his career, no matter the demands of his business life, Land remained focused on the creative implications and applications of his technology through his work with Ansel Adams and others. In the late 1930s, using his polarizer expertise, he created a stereo camera for his longtime friend, art professor Clarence Kennedy, which could create three-dimensional images of famous sculptures.[45] Land used the incredibly high-definition resolution of his black-and-white film to create large format photographic enlargements that grace the walls of museums around the world. He also created a special camera that occupied an entire room in the Boston Museum of Fine Arts. This device, the size of a small Volkswagen, produced seventy-two by forty-inch color prints of paintings and tapestries of extraordinarily accurate color and detail.

To demonstrate the universality of one-step photography as an art form, over the decades, Polaroid cameras and film were given to a host of photographic artists, established and aspiring. This program was originally the suggestion of Land's close collaborator, Meroe Morse. Together with works by Adams and other photographers, a collection of the work that resulted was amassed into what became known as the Polaroid Collections, approximately 23,000 photographs that stand as a testament to the artistic ambitions Land had envisioned.[46] Instant photography as an artistic medium of the highest order was beautifully documented in a

2013 exhibition, *The Polaroid Years*, curated by Mary-Kay Lombino of the Frances Lehman Loeb Art Center of Vassar College.[47]

While Polaroid's products may have achieved an iconic status in our popular culture, their progenitor, Edwin Land, remains largely an unknown and underappreciated figure in our nation's technological history. This is somewhat surprising, as his accomplishments meet or surpass those of many better-known personalities. He died in 1991 with 535 patents to his credit, third in U.S. history to Thomas Edison and Elihu Thomson, the two men whose companies merged in 1892 to form General Electric. His honorary doctorate degrees, too numerous to list, come from the most distinguished academic institutions, including Harvard, Yale, and Columbia. He received virtually every distinction the scientific community has to offer, including the Presidential Medal of Freedom, the National Medal of Science, the National Inventors Hall of Fame, and membership in the prestigious Royal Society of London. Land was included on *Life*'s list of the 100 most important Americans of the twentieth century.

Beyond his contributions to photography, most people use his first invention—the plastic sheet polarizer—just about every day, whether in sunglasses, camera filters, LCD displays, scientific and medical instruments, or car windshields. Land's "Retinex" theory of color vision revolutionized the field and replaced the long-held concept of how we see color, a fact that he dramatically demonstrated at trial. Perhaps most importantly, his contribution to America's defense and intelligence efforts over three decades, and in the service of seven presidents, performed mostly in secret with no public fanfare but to an estimable amount of praise from our country's scientific elite, may be the true measure of Land's stature in the pantheon of great American minds and entrepreneurs.

In business circles, Land is also admired for creating a unique working culture at Polaroid; one that he termed "a noble prototype in industry."[48] As early as 1944, Land had described his dream for the technology corporation of the future. It would feature aspects drawn from institutions of business, academics, and science:

> I believe quite simply that the small company of the future will be as much a research organization as it is a manufacturing company, and that this new kind of company is the frontier for the next generation.
>
> The business of the future will be a scientific, social and economic unit. It will be vigorously creative in pure science, where its

> contributions will compare with those of the universities. Indeed, it will be expected that the career of the pure scientist will be as much in the corporation laboratory as in the university.
>
> We feel that the industrial environment can be as stimulating to the development of pure science as the university has been. We should like to bring into industry the kind of professional ethics that characterize the relationship between pure scientists. We should like to have knowledge of the scientific method permeate our organizations . . . in short, a continuum between pure science in the university and pure science in industry should stimulate and enrich our social system. . . . Internally this business will be a new type of social unit. All will regard themselves as . . . having as their common purpose learning new things and applying that knowledge for public welfare. . . . This new kind of company is the frontier for the next generation.[49]

These strongly held beliefs were principles that Land adhered to and relied on while fashioning his company. He forged strong relationships between Polaroid and Harvard, MIT, Smith College, and other academic institutions, sharing their talent in an endless back-and-forth stream of constructive interaction.

Also, again as early as the 1940s, Land talked about creating a work environment he termed "Semitopia"—that is, a working atmosphere that would approach perfection as closely as it could in an imperfect world.[50] Donald Dery, Polaroid's former communications director, noted that since Land recognized science as "the pursuit of failure," a tolerance for it was a "cornerstone" of his workplace. All that he demanded was that the individual persist in spite of any failure. As Land liked to say, "Every creative act is a sudden cessation of stupidity."[51] According to Dery, part of Land's Semitopia concept was the "sun-satellite theory" of organization. This was a motivational tool that made each worker feel as if he or she were the center of his or her own work effort, with colleagues serving as supporting satellites. Land liked to joke about the sometimes inexact nature of this working environment: "The way we seem to work at Polaroid is to coalesce into little spontaneous groups, each one of which gets itself involved in something highly objectionable to all the other groups."[52] Despite any such tensions, the proof of Land's approach was in the results he achieved. Many of these ideas were completely foreign to U.S. industry, and Land believed that many Japanese corporations were superior because they practiced these tenets.

In so many ways, on so many occasions, Land's life was a manifestation of the indefatigable can-do attitude he embraced and encouraged others to follow. He sought to build an organization in his own image; one that could pursue its dreams instinctively, unshackled by some of the restraints imposed both internally and externally upon other companies. In describing Polaroid, distinguished Harvard Business School professor Joseph L. Bower once noted: "To understand Polaroid, you must understand Land. . . . Land is creative, and he has the well-grounded suspicion that good, careful, systematic planning can kill a creative company."[53] Instead, Land committed Polaroid on a course to pursue the same kind of ambitious challenges as he had set for himself when still a teenager:

> Pick problems that are important and nearly impossible to solve, pick problems that are the result of sensing deep and possibly unarticulated human needs, pick problems that will draw on the diversity of human knowledge for their solution, and where that knowledge is inadequate, fill the gaps with basic scientific exploration—involve all the members of the organization in the sense of adventure and accomplishment, so that a large part of life's rewards would come from this involvement.[54]

Land attracted to this kind of nurturing environment a host of scientists who pursued their interests unfettered by normal corporate constraints. For example, the extensive research done at Polaroid in the field of holography may not have resulted immediately in a commercial product for Polaroid of economic significance, but it helped to carry out, in the words of distinguished MIT professor Stephen Benton, Land's intent that "the huge industrial engine of Polaroid should enable interesting scientific research to be undertaken at will, while providing interesting working lives for all the members of 'the Company.'"[55] Some observers with more of a pure business perspective might point to this kind of activity as wasteful and even antithetical to the primary goal of corporate existence, the maximization of profit. But to Land, it was this amalgam of pragmatic, targeted business considerations, with more creative and free-reining scientific and technical pursuits, that would catalyze a crucible of technological exploration from which useful assets would emerge for his company and, equally as important, from which one might ultimately hope to provide the most benefit to mankind in the long run.

Due to Land's efforts, Polaroid proved to be extremely influential not only in the realm of technological entrepreneurship but also in the domain of corporate citizenry. In many ways, Polaroid tried to operate as a model corporate citizen—internally and externally—during Land's tenure. According to Jeffrey Sonnenfeld, a Harvard Business School professor:

> [Land] fashioned a collegial, knowledge-seeking organization with a highly committed work force. Status distinctions were minimized and career development and job enrichment were emphasized. Long a leader in socially responsible business practices . . . [Polaroid] acted far in advance of legislation regarding employment opportunities for women and minorities and employee benefits.[56]

For example, in the 1970s, urged on by some of its African-American employees, Polaroid was out in front of the effort to combat the white minority government's policy of apartheid in South Africa. It began programs to change the system from within as early as 1971. When those initiatives failed, Polaroid cut off all shipments of its products to the country in 1977, declaring, "We abhor the policy of apartheid."[57] Polaroid's actions in South Africa have been credited as leading other companies and academic institutions to take action.[58] As early as the 1950s, Polaroid had established a comprehensive education program for its employees. A profile of the company in 1982 reported that Polaroid's education department offered more than 100 courses to a work force that included top scientists but also 300 rehabilitating ex-convicts and 100 unskilled laborers.[59] *Business Week* declared: "When it comes to innovative, costly, continuing, and unconventional involvement in social action and community affairs, Polaroid . . . towers . . . high over most U.S. companies."[60] Land believed that, in the end, Polaroid was "one of the strongest companies in the world in terms of those spiritual and intellectual qualities that lead to human strength."[61] In this regard, Land has left a special legacy in the world of business, one that would become a model for companies of the future.

Not surprisingly, Steve Jobs, the chief executive and prime creative force behind Apple, was one of Land's most dedicated fans. In the words of John Sculley, whom Jobs recruited to lead Apple in 1983, "these were two geniuses who totally understood each other from the vantage point that they knew how to take technology and transform it into magic."[62] "Not only was [Land] one of the great inventors of our time," said Jobs in a 1985 interview, "but, more importantly, he saw the intersection of art and science and business and built an organization to reflect that. . . . The man is a national treasure, I don't understand why people like that can't be held up as models. This is the most incredible thing to be—not an astronaut, not a football player—but this."[63] Early in his career, as the founder of a fledgling Silicon Valley company, Jobs had the opportunity to visit with Land, who described to Jobs his vision for the technology company of the future. Jobs confessed to a reporter that getting to meet Land was "like visiting a shrine."[64]

Many years later, Jobs admiringly assured Land that in building Apple, he had tried to emulate the ideals Land had described to him. This occurred in the mid-1980s, when Jobs noticed Land across the room in a Boston restaurant. He hurried over to pay his respects, but Land neither recognized nor remembered him immediately. Jobs, unfazed, introduced himself: "Oh, I'm sorry, I'm Steve Jobs and I came by when we were setting up Apple Computer," he told him.[65] A subsequent meeting in the same period at Land's Rowland Institute lab was organized by Tom Hughes, a former Land colleague and then art director for Macintosh. Hughes described seeing an emotional side of Jobs in Land's presence that he had never previously witnessed. "It was sort of a father and son reunion. Steve was so clearly in admiration of Dr. Land and taken with every word he had to share. It was a very touching moment in time."[66]

The influence that Land had on Jobs is readily apparent to anyone who is familiar with their respective careers. As one journalist noted, they were "a pair of college dropouts with big ideas. Both were driven, demanding, and stubborn—qualities that led them to great things."[67] From the corporate culture Jobs created at Apple, to his widely anticipated product introductions at each Apple shareholders meeting, Jobs arguably became the Edwin Land of his generation. In all but their attire—even in the lab, Land always wore a tie and jacket, while Jobs was known for his more informal look of jeans and a black mock turtleneck—the two men embodied brilliant "troublemakers" (Jobs' characterization) who had the creative vision, as well as the innate intelligence and ability, to build multibillion-dollar companies.

In 2010, when Jobs was previewing Apple's iPad for some journalists prior to its introduction, he was asked what consumer and market research had been conducted to inform Apple's development process. Jobs' reply was pure Land, almost a verbatim reprise of comments Land had made many times throughout his career: "None. It isn't the consumer's job to know what they want."[68] Jobs had earlier told Land how he shared this approach. "I never had to ask customers what they wanted. If it's something truly revolutionary, they won't be able to help you."[69] "Our task is to read things that are not yet on the page," Jobs explained on another occasion.[70] For Land, as well as for Jobs, entrepreneurial scientific investigation and invention was the process of making what the consumer can't even imagine. Ironically, in 1982 Jobs bemoaned to John Sculley the fact that Land had been cast out from Polaroid when all he had done was to "blow a lousy few million [on Polavision]," a statement that foreshadowed his suffering the same fate when he was famously dethroned from Apple by Sculley shortly thereafter.[71]

To a large extent, Land's relative anonymity can perhaps best be explained by his inscrutable personality, his simple shyness, and his blinders-on mentality when it came to his life's work. As related by Mary and John McCann, two of his longtime colleagues, Land "had a true passion for research. He didn't play golf or go to sports events; his idea of fun was to be in his lab. His favorite story was his latest experiment."[72] His capacity for concentration in that pursuit was legendary and inspirational to those around him. He was an outgoing showman when circumstances demanded, whether unveiling his newest surprise at a Polaroid shareholders meeting or adopting a professorial role on the witness stand during the Kodak trial. But day to day, he was known more for being reclusive and exceedingly private. *Time* once described Land as a "paradoxical person, he alternates between lives as laboratory recluse and businessman-philosopher. He can be intensely shy and awkwardly unsure in face-to-face conversation. Yet he is capable of spellbinding audiences with glimpses into new scientific frontiers."[73]

However, Land was not a recluse in the Howard Hughes sense. He was just perpetually engaged, often in something other than the mundane, at least to him, exigencies of the moment. For example, in the second half of 1978, while the Polaroid legal team wondered whether he would ever emerge to participate in the Kodak lawsuit, Land was immersed in a comprehensive investigation into the "exegesis of psychophysics," analyzing and trying to synthesize the work of Jung, Goethe, Helmholtz, and others in an effort to understand, and to be able to measure, the functional relationship between body and mind.[74] It was from cerebral places such as this that Land had to be retrieved occasionally. In a rare interview, his wife, Helen, provided a personal glimpse into life with Edwin Land. "There are occasions when I will be talking with him directly and I know that he's not listening to me," she recalled. "His mind is miles away, but all I have to say to him is 'Polaroid,' and immediately I have his full and complete attention."[75] Later in life, Land seemed to deal with the world in general, and even some colleagues, on a need-to basis. More comfortable operating among a small coterie of close associates, he engaged strangers only when there was a purpose to pushing aside his wizard's curtain.

Yet Land had a warm side that those lucky enough to be in close proximity to the man truly learned to appreciate. Ansel Adams recalled how a mutual friend had characterized him: "Land is not only one of the great minds of the age but he has also one of the great hearts."[76] When Sonnenfeld interviewed Land for a book he was writing on what happens to companies when key CEOs retire, he was struck by the many interruptions they had to endure while they spent time together in various locations. The intrusions "were from enthusiastic well-wishers of all statures who approached with the eagerness of old friends," wrote Sonnenfeld. "Whether custodians, security guards, current research colleagues, middle managers from Polaroid, or top Polaroid executives, Land warmly and promptly greeted them by name, introduced me, and then asked one or two personal questions about their families or their own careers."[77]

Ansel Adams told a wonderful story about Land that demonstrated how much he cared about those friends and family closest to him. Adams and his daughter were visiting with Land in Cambridge and were scheduled to leave on an overnight train to New York. A snowstorm was beginning to blow, so they decided to leave for the station early. Land called a cab and bundled Adams and his daughter into it. When they arrived at the train station, they heard someone calling from behind. "Carry your bag, mister?" Adams said, "[I was] counting my pennies so I declined." But the voice persisted. "Glad to carry your bag, mister." Adams turned around, and sure enough it was Land following behind them. "What in the world are you doing out here in this blizzard?" Adams asked. "Just wanted to make sure you . . . made it safely," Land explained as he said goodbye once again and turned to leave for home.[78]

Of course, despite his prodigious accomplishments and many inspirational and worthy traits, Edwin Land shared with the rest of us all of the frailties of the human experience. To the extent this book may seem like an homage to him, it is not meant to canonize him in total disregard of his shortcomings—notably, his enormous ego and his unrelenting stubbornness. In his heart Land may have wanted to be an ideal leader, but in the end, his impatience for the more mortal abilities of others affected his ability to deal effectively with many colleagues. One wonders whether Land was insensitive to, or just unaware of, his capacity to intimidate others. On the wall behind his desk hung a photograph of a lion perched up in a tree. "That's the chairman," Land told Robert Lenzner of the *Boston Globe*, the journalist who covered Polaroid and its founder more closely than anyone. "That's the image of the way people would like me to be—relaxed and knowing, perspicacious, thoughtful." Awkwardly trying to be self-effacing while referring to at least some of his colleagues, Land added: "When I come down from the tree, I pool my incompetences with their incompetences." This ineffectual stab at humility did not deter Lenzner from describing the Land he knew better after so many years:

> Land's vision of himself does not take into account the possible imbalance between his all-consuming work and his personal life. And it does not include the perspective of some employees who find him difficult, overly demanding and miserly in direct praise of subordinates.
>
> He sees himself as determined, iron-willed and hard driving, a man who will not rest until he has conquered whatever problem is at hand.[79]

Elkan Blout was a Harvard chemist and colleague of Robert Woodward whom Land recruited to join Polaroid in the early 1940s, instead of taking an academic post at Columbia University. He went on to help Howard Rogers in his work to bring color to one-step photography and thereafter joined the faculty at Harvard Medical School while maintaining a career-long consultancy with Polaroid. As someone who worked closely with Land for many years, but then left the company and continued the relationship from a post outside Polaroid, Blout was in a unique position to develop some perspective on the man.

First and foremost, Blout was impressed with Land's competitiveness. Blout liked to tell a story he thought revealed much about his colleague. Around 1950, Land and Blout had side-by-side parking spots in Polaroid's lot. Blout admired Land's Oldsmobile convertible and told him: "I love your car. If you ever want to get rid of it, keep me in mind." Several months later, Land called Blout unexpectedly and asked him if he still wanted to buy the car. Blout said he did, and Land asked if we would pay $800 for it. Blout agreed, and Land said: "Fine. The car is yours. Go ahead and pick it up in Central Square; it has a broken axle."[80] Blout was undeterred. He retrieved the car, had the axle repaired, and enjoyed the car for years. Blout even thought the price was fair, and so never thought that Land was being malicious. Land thought it was all a good joke and, in time, Blout came to agree.

In his oral memoirs, Blout recorded the broader view of his relationship with Land. He describes from a personal perspective the intermingled joys and frustrations, and some of the conflicted feelings that he, and others, experienced:

> I got great satisfaction out of my work at Polaroid, but not complete satisfaction. I think the reason was that Din Land, while being very supportive, was also very secretive. . . . Din was an imaginative, charismatic leader when he wanted to be. He was also a son-of-a-bitch when he wanted to be.
>
> In many ways, he was demanding and imperial. . . . Regardless of our differences . . . I liked him because he was smart enough to let me do what I wanted, and he encouraged my work. He was very self-righteous in what he did, and wanted to get credit for everything he did. In fact, one of the difficulties of being at Polaroid was that he wanted credit for most of the things that everybody did, which created tension in the organization.[81]

When William McCune succeeded Land as Polaroid's chief executive, he endured a turbulent time dealing with his longtime colleague and mentor in what were difficult personal and professional circumstances. Land recognized the difficulties inherent in the kind of long and special relationships he had with some of his key people:

> In a company that grows from nothing, there is always a dual life of friendship and formal relationship. It takes an extraordinary amount of skill to interweave the friendship and the formal relationship in such a way that the individuals retain their individuality at the same time that the corporation gains from the synthesis.[82]

Yet, when all was said and done, McCune fully recognized the magnitude of the personality his career had been shared with. At Land's memorial, he recognized Land's commitment to creating a special kind of company capable of achieving special things:

> One of Din's great creations was the environment he nurtured within the company. He created an environment that attracted and stimulated bright, creative, and interesting people from a wide variety of disciplines, not just scientists. As a result, many extraordinary people who were not scientists found Polaroid an interesting, exciting and rewarding place to work. They contributed greatly to the creative nature of the company and to its extraordinary success.

> Din was one of the most creative persons of our time. He was an inspiring, stimulating, and demanding person to work with. At the same time, he had a great sense of humor, and work could be fun as well as a challenge. He encouraged us and utilized the full capabilities of each of us. For him nothing seemed too difficult, almost nothing impossible. There is a quotation from George Bernard Shaw that reminds me of Din: "I dream things that never were; and I say, Why not?"[83]

After leaving Polaroid following the end of the Kodak litigation, Land spent the final years of his life at the Rowland Institute, which he had founded on the banks of the Charles River in Cambridge, just down the road from his old Polaroid headquarters. It is the lasting legacy to Land's commitment to pure research. He wanted to create an idyllic environment for the pure pursuit of scientific inspiration, and the ongoing productive work conducted there—including the development of a microscopic laser or "tweezer" that is able to manipulate single-cell organisms as small as bacteria—is the best proof that his dream, in fact, has been realized. Although he battled health issues in his last years, Land was content with this work and with this life. In his new surroundings, relieved of the exigencies of business life, he pursued the scientific research that intrigued his ever-questioning mind and published extensively in academic journals.

Land passed away on March 1, 1991, yet his legacy endures. The formula for accomplishment he practiced throughout his life—creative wonderment and intellectual curiosity followed by inexhaustible effort—remains a model that should inform and inspire us all, no matter the particular field of our endeavor. A longtime observer put it another way: "The dominant impressions Land created were artistic sensibility, a sense of drama, delight in experiment, [and] relentless optimism."[84] He was blessed with a potent combination of intellect and energy that enabled him to embrace the challenges of scientific research and discovery with extraordinary zeal, determination, and an almost naïve confidence of success. That confidence proved to be well founded, as repeatedly throughout his career, Land was able to bring to fruition technological miracles of significant practical and artistic benefit to mankind arising directly from his fertile imagination. He once told an interviewer from *Forbes*:

> I'm interested in love and affection and sharing and making beauty part of everyday life. And if I'm lucky enough to be able

> to earn my living by contributing to a warmer and richer world, then I feel that it is awfully good luck. And if I use all of my scientific, professional abilities in doing that, I think that makes for a good life.[85]

That this effort continued for the entirety of his life was in itself one of Land's great lessons. In eulogizing Land, one of his academic colleagues confessed that Land had taught him "the idea that the romantic dream does not belong to romantic youth alone and should not be shelved in one's later years. An individual who has forfeited his dream is an individual who has forfeited the quest to find new things and to contribute in an important way to society and culture."[86]

Land's childhood fascination with light, an all-consuming passion that led him into the great scientific adventures of his life—polarized light, instant photography, and color vision—turned out to be an enduring pursuit. When a memorial service was being planned following his passing, the members of the American Academy of Arts and Sciences had little problem coming up with a title for the event that, to them, defined their colleague: "Light and Life."

When looking back at the meteoric life of instant photography, and especially the historic legal battle fought over it, it is important to keep one's perspective. To dwell on the fact that the technology became largely obsolete over the course of the decade and a half that it took to fight this fight in court is, arguably, to miss the point. When Goldstein wrote that Polaroid and Kodak, "preoccupied with their dazzling technology and their patent lawsuit . . . had been jousting on the decks of the Titanic,"[87] one wonders whether he somehow missed the boat. Seemingly, his observation fails to credit that incredible confluence of scientific accomplishment, creativity, and bold entrepreneurship that created a corpus of such huge significance that it could even engender the colossal battle that Polaroid and Kodak fought. There can be no dispute that when conceived, and even in retrospect, Land's One-Step and then later his Absolute One-Step photography systems were astonishing from a technological perspective. They created a market of inestimable value that lasted for almost half a century. At the time, the technology certainly seemed more than worth fighting over.

Yet, virtually all technological breakthroughs, save some fundamental ones like fire or the wheel, are subject to an arc of utility. They arrive on the scene to great fanfare and wonder, serving the public for some period of time, often incalculable in advance. Ultimately, however, they are destined to be replaced by man's next great discovery and, as a result, inevitably fade into memory, oblivion or a combination thereof. Noted economist Joseph Schumpeter dubbed this phenomenon "Creative Destruction."[88] But this fate should not diminish the significance of the technology in question. For any human achievement, whether in sports, science, or any of the arts, can really only be evaluated within the context of the era in which it was created, and measured against what had come before. Instant photography certainly followed this familiar path over a course of almost fifty years and was fully deserving of the plaudits like "revolutionary" and "miraculous" that it was accorded at the time.

This story, especially the litigation that was its climactic act, was really about a fundamental principle. At its core, it concerned the question of whether an innovative company, and the inspirational leader behind it, could invest its talent, time, and resources in the pioneering of a new field of technology and look to the patent system, enabled by our Founding Fathers to encourage such effort, as protection against predatory competition from another, in this case much larger, firm. Could David's patents serve as the rocks in his slingshot? For decades, Land had emphatically proclaimed the critical importance of the patent system to realizing fully the contribution of the small technology company and the individual inventor to mankind's evolving body of knowledge and understanding. "It should be the role of our patent system to bring encouragement, a sense of reward, and a stimulus to prompt publication to this potential army of great [researchers] in applied science," he proclaimed in 1959.[89] In this regard, Polaroid's resounding and historic victory over Kodak was clearly, for Dr. Land and generations of other innovators to follow, a triumph of genius.

ACKNOWLEDGMENTS

No "last, but not least" for me. First and foremost, I have to acknowledge the help of my wife, Dorsey Regal. She is not only the most supportive (or tolerant, some might say) and perfect life partner, she is the best editor ever. She helped me immeasurably with early drafts of the manuscript, enlightening me to those pesky rules that govern the written word, and helping me find clarity where my thoughts were either obtuse or otherwise undelivered. I owe her big time. And I love her very much.

Our family provides the foundation that makes everything possible. Our sons, Brad and Matt, fill us with the love and pride that make it all worthwhile. My brother, Harvey, has taught me (and lots of others) that it is not only okay but imperative to be who you are and to believe that anything is possible. Our late mother, Jackie, instilled in both of us a love of the arts and passed on some of her librarian genes, which makes, for mc at lcast, thc arduous proccss of research an adventure and a joy. Our late father, Irving, imbued us with a work ethic and commitment to family that is our bedrock.

As anyone who has undertaken a project of this magnitude can attest, one must be inspired to make the commitment necessary to devote a significant portion of one's life to such an endeavor. My motivation came from two disparate sources, the confluence of which pushed me off my personal fence into action.

The first was Edwin H. Land. As a young lawyer, fresh out of law school, I unexpectedly found myself squarely in the middle of an extraordinary event, the epic *Polaroid v. Kodak* lawsuit. More importantly, after a time I somehow found myself working directly with Land. While some people, including many of his own Polaroid colleagues, found Land intimidating, he was never anything less than gracious and supportive to me, a neophyte frantically learning as much as I could about a field of technology that he knew more about than any other person on the planet. My fundamental task—trying to get him to consider the intricacies of this technology in the very specific context of the legal issues presented by the case—was an intellectual exercise I will always cherish. I will forever be grateful to him for indulging my efforts in that regard, and for autographing my copy of his L-Coat patent when we were done. Given his accomplishments, it has long seemed to me that Edwin Land deserves a more prominent position in the pantheon of great Americans than he currently occupies. It was to make some contribution to that end that first moved me to undertake this project.

A second inspiration for writing this book, and for revisiting the world of Edwin Land more than 20 years after I had moved on, was the photographic art of my friend Chip Hooper. Through his work, I rediscovered the magic of light, the motivating force that inspired Land throughout his lifetime. One of the most successful agents in the contemporary music business, Chip has a parallel life as a gifted landscape photographer. His work is influenced by Ansel Adams but is based on a technique he has made his own—extremely long exposures from dusk to sunrise, enabling a limited amount of light to do a very special thing. I am privileged that Chip's work adorns my living and working spaces. His images reignited in me a spark about photography and brought me back to Dr. Land and his special relationship and work with Adams, many examples of which I had the privilege of seeing firsthand decades ago while rummaging around Polaroid back rooms searching for documents relevant to the lawsuit. In this way, Chip's work illuminated the path that led me to undertake this project.

During his trial testimony, Land explained how the work of all scientific researchers stands on the shoulders of those who came before them and those who work beside them. So too does the effort of the researcher and writer of a nonfiction work like this. I am indebted to Victor McElheny, whose comprehensive research provided a great resource for me, particularly with respect to the early chapters of this book. Rita Rosenkranz, my literary agent, was supportive and pragmatic throughout the process, and I appreciated her guidance and perseverance. When my original manuscript came in at an unwieldy length, I was privileged to have James Wade work with me for the better part of a year as a sage and experienced, if ruthless, editor. His encouragement and positive reinforcement were invaluable. University of Houston Law Center Professor Paul Janicke reviewed the manuscript and offered important suggestions, as did Vivian Walworth, one of Land's closest collaborators, from a technical perspective. Melissa Bailey made a huge contribution to this effort as a careful proofreader who helped me polish the manuscript at the end of my process.

Every aspiring author needs someone who steps forward to embrace his or her work. I will always be grateful to my editor, Jonathan Malysiak of ABA Publishing, for his enthusiastic support. His colleagues did a wonderful job in turning my manuscript into a book, especially Amanda Fry, who designed the cover, and Jason Stauter, who supervised the editing and production process. Thanks to everyone at ABA Publishing who contributed. Our project is privileged to have Sandi Mendelson and David Kass directing our publicity efforts.

I am especially indebted to all of the people who gave so generously of their time in interviews and correspondence during my research, in particular, Herbert F. Schwartz, Cecil Quillen, Nan Chequer Schwartz, Robert Goldman, James Galbraith, Patricia Martone, Richard Barnes, and Mark Kleiman. Thomas Sciattara was invaluable with the logistics of accessing the archives my former colleagues at Fish & Neave, now part of Ropes & Gray, graciously made available. I am also grateful to everyone else who helped me in one way or another, including Deborah Douglas of the MIT Museum, Tim Mahoney and his colleagues in the Historical Collections Department at Harvard Business School's Baker Library, Harry McCracken, Christina Folz of Optics & Photonics News, Heather Lacey, Suzanne Grinnan of the Society for Imaging Science and Technology, David Caruso of the Chemical Heritage Foundation, Angela Healy, Michael Burns and Frans Spaepen of the Rowland Institute at Harvard, Christopher Bonanos, Dr. Michael Pritchard of the Royal Photographic Society, Denise Scarpetti, Jesi Barron, Matthew Lutts of the Associated Press, Michael Duncan of the Royal Courts of Justice in London, Fred Bremer and Louis Argenziano of Merrill Lynch, Betsy Becker, Carolyn French, and Sharon O'Neill. Alan Earls was most helpful with his advice and with the fantastic images he collected with Nasrin Rohani for the volume on Polaroid produced for the Images of America series. Mary-Kay Lombino of the Frances Lehman Loeb Art Center at Vassar College was a generous and enthusiastic guide to the exhibition of Polaroid art she has curated—her work is so very much at the heart of Land's creative aspirations for his technology. It was Brad Kullberg and Barbara Hitchcock, two of Polaroid's last and most dedicated employees, who helped me set sail on this journey. Thank you both. I appreciate the cooperation of April Lunde and her colleagues at the "new" Polaroid, PLR Holdings, Inc.

DEDICATION

This book is dedicated to the memory of James M. Williams. Jim, our friend Robert J. Goldman, and I comprised what I've always thought of as the Three Musketeers of the *Polaroid v. Kodak* lawsuit. We were young associates who worked incredibly hard in the trenches of that litigation during the liability phase—through years of discovery and then a trial that was, at least for me, and I bet for them, the most intense experience of my life. If I can be immodest here, we were, in so many ways, the nucleus of the huge team of lawyers—under the leadership of Herbert F. Schwartz—that successfully brought Edwin Land and Polaroid the justice it sought. As the junior member of the trio, I relied on Bob and Jim for guidance and direction. They were wonderfully supportive teachers. Bob has continued his career as a distinguished patent litigator. I segued into a career in the entertainment business. Jim—a good man, a good lawyer, a good friend, and a loving husband to Sheree and father to Brandon—passed at an obscenely young age. I thought of him often in reliving so much of this story while researching and writing this book. (See Fig. 30-1.)

ENDNOTES

Note concerning endnotes:

The endnotes used in this text are meant to direct the interested reader to the source of the fact(s) or quotation(s) appearing in that section. In order to keep the number of endnotes as small as possible, when a single source is used repeatedly in succession in a given paragraph, not every fact or quotation is notated. In such cases, where an endnote is omitted, the source can readily be located by using neighboring citations.

Abbreviations used in endnotes (after first appearance):

HBS = Baker Library Historical Collections, Harvard Business School.
PvK = Polaroid Corporation vs. Eastman Kodak Company, United States District Court, District of Massachusetts, Civil Action No. 76-1634-Z (liability trial).
PvKdam = Polaroid Corporation v. Eastman Kodak Company, United States District Court, District of Massachusetts, Civil Action 76-1634-MA (damages trial).
F&N = Fish & Neave, Litigation Files, author's collection.
PvB = Polaroid Corporation v. Berkey Photo, Inc., United States District Court, District of Delaware, Civil Action No. 75-179.

Cover photo: Courtesy of Polaroid Corporation Archives

Preface

1. Steve Huntley, "Polaroid Landmark Comes into View," *Chicago Sun-Times*, September 17, 1987.
2. Edwin H. Land, "On Some Conditions for Scientific Profundity in Industrial Research," Charles F. Kettering Award Address, June 17, 1965, in *Edwin H. Land's Essays*, Vol. II, 58, ed. Mary McCann (Society for Imaging Science and Technology 1993); F.W. Campbell, "Edwin Herbert Land," *Biographical Memoirs of Fellows of the Royal Society*, Vol. 40 (Nov. 1994), 202.
3. Elkan Blout, "Polaroid: Dreams to Reality," *Daedalus*, Vol. 125, No. 2 (American Academy of Arts and Sciences, March 22, 1996).

Chapter 1

1. Victor K. McElheny, *Insisting on the Impossible, the Life of Edwin Land* (Perseus Books, 1998), 13–14.
2. Kurt Grunwald, "Turkenhirsch: Study of Baron Maurice de Hirsch" (Israel Program for Scientific Translations, 1966).

3. McElheny, *Insisting on the Impossible*, 6.

4. Barry Meisel, interview by author, September 2008, Chappaqua, NY.

5. McElheny, *Insisting on the Impossible*, 14.

6. Ibid., 13.

7. Alex Beam, "Edwin Land's Place in History," *Boston Globe*, August 11, 1987.

8. McElheny, *Insisting on the Impossible*, 13.

9. Victor K. McElheny, "Edwin Herbert Land," *Biographical Memoirs* (National Academy of Sciences, 1999), 5.

10. Robert Lenzner, "Land: The Man behind the Camera," *Boston Globe*, October 17, 1976, B7.

11. Robert Lenzner, "The Promised Land," *Boston Globe*, August 3, 1982, 43.

12. McElheny, "Edwin Herbert Land," 6.

13. Edwin H. Land, "Some Aspects of the Development of Sheet Polarizers," *Journal of the Optical Society of America*, December 1951, 957.

14. Stephen A. Benton, "Edwin Land, 3-D, and Holography," *Optics & Photonics News*, October 1994, 41.

15. Edwin H. Land, "Our 'Polar Partnership' with the World around Us," *Harvard Magazine* 80 (1978).

16. Ibid.

17. F.W. Campbell, "Edwin Herbert Land," *Biographical Memoirs of Fellows of the Royal Society* 40 (Nov. 1994), 198.

18. Edwin H. Land, "Pointillism and Laser Scintillation" (lecture, Johns Hopkins University, November 21, 1975), in *Edwin H. Land's Essays*, vol. 3, ed. Mary McCann (Society for Imaging Science and Technology, 1993), 119; McElheny, *Insisting on the Impossible*, 20.

19. McElheny, *Insisting on the Impossible*, 20.

20. McElheny, "Edwin Herbert Land," 6.

21. Robert Wood, *Physical Optics* (1914), 288, quoted in McElheny, *Insisting on the Impossible*, 20.

22. McElheny, "Edwin Herbert Land," 6.

23. "Polaroid's Big Gamble on Small Cameras," *Time*, June 26, 1972.

24. Campbell, "Edwin Herbert Land," 198.

25. Howard Rogers, "Biography, Edwin H. Land 1909–1991," in *Edwin H. Land's Essays*, vol. 1, 5.

26. Edwin H. Land, "On Some Conditions for Scientific Profundity in Industrial Research" (Charles F. Kettering Award address, Washington, DC, June 17, 1965), in *Edwin H. Land's Essays*, vol. 2, 57–58.

27. Edwin H. Land, "Polaroid," *Society of Automotive Engineers Journal* (1937), 20.

28. Land, "Some Aspects of the Development of Sheet Polarizers."

29. Lenzner, "Land: The Man behind the Camera," B7.

30. Campbell, "Edwin Herbert Land," 198.

31. McElheny, *Insisting on the Impossible*, 27.

32. McElheny, "Edwin Herbert Land," 4.

33. Ibid.

34. Edwin H. Land, "Addiction as a Necessity and Opportunity," *Science*, January 15, 1971, reprinted in *Edwin H. Land's Essays*, vol. 2, 71; McElheny, *Insisting on the Impossible*, 27.

35. Lenzner, "Land: The Man behind the Camera," B1.

36. McElheny, *Insisting on the Impossible*, 31.

37. Liz Roman Gallese, "I Am a Camera," *Boston Business*, Fall 1966, 51; McElheny, *Insisting on the Impossible*, 28.

38. Campbell, "Edwin Herbert Land," 199.

39. Mitchell Lynch, "Polaroid Tries to Get Itself in Focus," *New York Times*, May 15, 1983, p. 54; McElheny, *Insisting on the Impossible*, 75.

40. Campbell, "Edwin Herbert Land," 200.

41. Donald L. Brown, "Protection through Patents: The Polaroid Story," *Journal of the Patent Office Society* (1960), vol. 42, 439, 441.

42. Edwin H. Land, "Polaroid and the Headlight Problem," *Journal of the Franklin Institute*, September 1937, reprinted in *Edwin H. Land's Essays*, vol. 1, 5.

43. Ibid.

44. McElheny, *Insisting on the Impossible*, 34.

45. Ibid., 36.

46. Ibid., 37.

47. Land, "Some Aspects of the Development of Sheet Polarizers," 99.

48. McElheny, *Insisting on the Impossible*, 38.

49. Karel Svoboda, "Light and Life, Optical Matter, a Symposium in Honor of Edwin Land," *Bulletin of the American Academy of Arts and Sciences*, April 1992, vol. 45, 23.

50. Harland Manchester, "Pictures in 60 Seconds," *Scientific American*, April 1947, 168.

51. McElheny, *Insisting on the Impossible*, 38.

52. Land, "Polaroid and the Headlight Problem," 5.

53. McElheny, *Insisting on the Impossible*, 39.

54. Brown, "Protection through Patents: The Polaroid Story," 442–43.

55. McElheny, *Insisting on the Impossible*, 41.

56. Ibid., 40.

57. Ibid., 41.

58. "In the Light of Polaroid," *Fortune*, September 1938, 76, Polaroid Corporation Administrative Records, Box I225, f. 4, Baker Library Historical Collections, Harvard Business School.

59. Claudia Deutsch, "G.W. Wheelwright III, 97, Dies; Co-founder of Polaroid," *New York Times*, March 3, 2001.

60. McElheny, "Edwin Herbert Land," 7.

61. McElheny, *Insisting on the Impossible*, 45.

62. J.D. Ratcliff, "The $50,000,000 Quinine Sandwich," *True*, August 1948, 82.

63. McElheny, *Insisting on the Impossible*, 45.

64. Land, "Some Aspects of the Development of Sheet Polarizers."

65. Campbell, "Edwin Herbert Land," 199.

66. McElheny, *Insisting on the Impossible*, 44.

67. McElheny, "Edwin Herbert Land," 7.

68. McElheny, *Insisting on the Impossible*, 47.

69. Ibid., 87.

70. Ibid., 88.

71. Ibid., 90.

72. Ibid., 91.

73. Ibid., 49.

74. "Auto Glare Patent Won after 17 Years," *New York Times*, July 25, 1937.
75. "In the Light of Polaroid," 77.
76. McElheny, *Insisting on the Impossible*, 51.
77. Ibid., 62–63.
78. Ibid., 49–50.
79. Ibid., 53–67.
80. Ibid., 68.
81. Ibid., 68.
82. Land, "Polaroid."

Chapter 2

1. Campbell, "Edwin Herbert Land," 199.
2. McElheny, *Insisting on the Impossible*, 63.
3. Jane Poss, "Edwin Land Dead at 81, Polaroid Chief Left Image on Industry," *Boston Globe*, March 2, 1981.
4. McElheny, *Insisting on the Impossible*, 64.
5. Ratcliff, "The $50,000,000 Quinine Sandwich," 82; John Case, "The Problems at Polaroid," *Boston Magazine*, April 1982, 109.
6. McElheny, *Insisting on the Impossible*, 65.
7. Advertisement, *New York Times*, December 11, 1937, Polaroid Corporation Administrative Records, Box I.86, f. 23, Press Clippings—Sunglasses, 1937, HBS.
8. Land, "Pointillism and Laser Scintillation," 119.
9. McElheny, *Insisting on the Impossible*, 50–56.
10. Watson Davis, "Strange Light of Service in Work of Peace and War," *Science News Letter*, November 9, 1940, 294; James O. Spearing, "At the Wheel, Killing Headlight Glare," *New York Times*, January 26, 1936.
11. McElheny, *Insisting on the Impossible*, 53–55.
12. Spearing, "At the Wheel, Killing Headlight Glare."
13. McElheny, *Insisting on the Impossible*, 70.
14. Land-Wheelwright Laboratories, "Invention of Polaroid Gives Science a New Tool," Press Release, January 30, 1936, Polaroid Corporation Legal and Patent Records, Box II.3, f. 1, Land, Edwin H., HBS.
15. "New 'Glass' Cuts Glare of Light; Aid to Movies and Science Seen," *New York Times*, January 31, 1936.
16. Spearing, "At the Wheel, No More Glare?"
17. "New 'Glass.'"
18. "Invention of Polaroid Gives Science a New Tool."
19. Spearing, "At the Wheel, No More Glare?"
20. "New 'Glass.'"
21. "3-D Dimension Movies with Polarized Glass," *St. Louis Post-Dispatch*, December 11, 1936, Polaroid Corporation Administrative Records, Box I.86, f. 19, Press Clippings—3-D Movies, 1936, HBS.
22. McElheny, *Insisting on the Impossible*, 70–71.
23. Ibid., 59.
24. "Three Living Leaders, Edwin Herbert Land," *Fortune*, March 23, 1981, 106.
25. Edwin H. Land, "If You Are Able to State a Problem, It Can Be Solved," *Life*, October 27, 1972, 48.

26. McElheny, *Insisting on the Impossible*, 59–60.
27. Poss, "Edwin Land Dead at 81."
28. McElheny, *Insisting on the Impossible*, 60.
29. Howard Gardner Rogers (Polaroid Corporation Biography), author's collection.
30. "3d Dimension Given to Films by Parlor Trick," *New York Herald Tribune*, May 14, 1936, Polaroid Corporation Administrative Records, Box I.86, f. 1, Press Clippings—3-D Movies, 1935–1938, HBS.
31. Land, "Polaroid."
32. McElheny, *Insisting on the Impossible*, 75.
33. Ibid.
34. Ibid., 62, 66.
35. Ibid., 77.
36. Daniel D. Nossiter, "No Instant Success," *Barron's*, July 5, 1982, 11.
37. McElheny, *Insisting on the Impossible*, 78.
38. Ibid., 77–79.
39. "Auto Glare Patent Won after 17 Years," *New York Times*, July 25, 1937.
40. McElheny, *Insisting on the Impossible*, 79–81.
41. Ibid., 81–82.
42. "In the Light of Polaroid," *Fortune*, September 1938, 118, Polaroid Corporation Administrative Records, Box I.225, f. 4, HBS.
43. Land, "Some Aspects of the Development of Sheet Polarizers."
44. McElheny, *Insisting on the Impossible*, 101.
45. "In the Light of Polaroid," 74.
46. Arthur A. Riley, "Polaroid Grew from Non-glare Headlight Search," *Boston Globe*, June 21, 1959; McElheny, "Edwin Herbert Land," 8.
47. McElheny, *Insisting on the Impossible*, 84.
48. McElheny, *Insisting on the Impossible*, 72.
49. "'Modern Pioneers' Will Get Plaques," *New York Times*, February 27, 1940.
50. "Trade Is Defended in 'Labor Surplus,'" *New York Times*, February 28, 1940.
51. Associated Press, "'Modern Pioneer' Plaques Awarded," February 28, 1940, Polaroid Corporation Administrative Records, Box I.86, f. 1, Press Clippings—General, 1940, HBS.
52. McElheny, *Insisting on the Impossible*, 82–83.
53. "G.W. Wheelwright III, 97, Dies; Co-founder of Polaroid," *New York Times*, March 3, 2001.
54. "George Wheelwright III; Co-founded Polaroid Corp.," *Los Angeles Times*, March 9, 2001.
55. Victor K. McElheny, "Polaroid's Coup: Instant Movies," *New York Times*, April 24, 1977.
56. William McCune, Concord Oral History Program, July 11, 1996, 3.
57. Ibid.
58. Robert Reinhold, "Dr. Vannevar Bush Is Dead at 84," *New York Times*, June 30, 1974, 1.
59. Ibid., 36.
60. Davis, "Strange Light of Service."
61. McElheny, *Insisting on the Impossible*, 128.
62. McCune Oral History, 3.
63. Ibid.

64. William McCune, "Light and Life, a Symposium in Honor of Edwin Land," *Bulletin of the American Academy of Arts and Sciences* 45, no. 7 (April 1992), 42.

65. McElheny, *Insisting on the Impossible*, 127.

66. Ibid.

67. "Edwin Land on Polaroid—and the Kodak Patent Suit," *Boston Globe*, October 18, 1976.

68. "Light Control," *Life*, February 7, 1944, Polaroid Corporation Administrative Records, Box I226, f. 7, HBS.

69. McElheny, *Insisting on the Impossible*, 130.

70. *Newsweek*, July 26, 1943.

71. N.R. French, "Memo to Mr. Brown," January 23, 1943, Polaroid Corporation Legal and Patent Records, Box II.104, f. 15, DLB Correspondence, 1942–43, HBS.

72. Polaroid Corporation, Rogers Biography, Fish & Neave, Litigation Files.

73. Ratcliff, "The $50,000,000 Quinine Sandwich"; McElheny, *Insisting on the Impossible*, 128–31.

74. Ibid., 130–31.

75. Edward Sheehan, "The Rise and Fall of a Soviet Agent," *Saturday Evening Post*, February 15, 1964; Handwritten Letter, Tracy to Donald Brown, February 15, 1964, Polaroid Corporation Legal and Patent Records, Box II.103, f. 4, Publicity, Polaroid Net Rises, 1964, HBS.

76. Louis Rosenblum, "Quick Turnaround on Night Goggles," *Optics & Photonics News*, October 1994, 14–16.

77. Ibid.

78. "Brookline Man Predicts Three-Dimension Movies," *Boston Globe*, March 13, 1944, Polaroid Corporation Administrative Records, Box I.88, f. 3, Press Clippings—Vectograph, HBS.

79. Stephan A. Benton, "Edwin Land, 3-D, and Holography," *Optics & Photonics News*, October 1994, 41.

80. Edwin H. Land, "Vectographs: Images in Terms of Vectorial Inequality and Their Application in Three-Dimensional Representation," *Journal of the Optical Society of America*, June 1940, 230.

81. Waldemar Kaempffert, "Vectographs," *Science in the News*, June 30, 1940.

82. McElheny, *Insisting on the Impossible*, 116–17.

83. Benton, "Edwin Land, 3-D," 41.

84. Catherine Coyne, "Stereopticon Has Gone to War to Make More Deadly Fire Aim," *Boston Herald*, April 23, 1944, Polaroid Corporation Administrative Records, Box I.88, f. 3, Press Clippings—Vectograph, 1944, HBS.

85. Alden P. Armagnac, "Polarized Light for the Eyes of the Army," *Popular Science*, March 1943, 58; Martin Sheridan, "Three Dimensional Photos Big Help in War Effort," *Boston Globe*, May 25, 1943, Polaroid Corporation Administrative Records, Box I.87, f. 17, Press Clippings—General, 1943, HBS.

86. Mary A. McCann and John J. McCann, "Land's Chemical, Physical, and Psychological Images," *Optics & Photonics News*, October 1994, 34; McElheny, "Edwin Herbert Land," 8.

87. McElheny, *Insisting on the Impossible*, 117–18.

88. Ibid., 142–44.

89. "Quinine Synthesized," *Science News Letter*, May 13, 1944, Polaroid Corporation Administrative Records, Box I.88, f. 2, Press Clippings—Quinine, 1944, HBS.

90. Brown, "Protection through Patents: The Polaroid Story," 447.

91. Ibid.

92. McElheny, *Insisting on the Impossible*, 153.

93. McCune Oral History, 3–4.

94. McElheny, *Insisting on the Impossible*, 135–36.

95. David S. Grey, "Digging for the Bomb," *Optics & Photonics News*, October 1994, 20.

96. Ibid.

97. Ibid.

98. Ibid.; McElheny, *Insisting on the Impossible*, 139.

99. Grey, "Digging for the Bomb," 20.

100. McCune Oral History, 3–4.

101. McElheny, *Insisting on the Impossible*, 139.

102. McCune Oral History, 4.

103. Robert Schwartz, "Atomic Bomb Away, Tank," *Army Weekly*, September 7, 1945, Polaroid Corporation Administrative Records, Box I.88, f. 29, Press Clippings—General, 1945, HBS.

104. Benton, "Edwin Land, 3-D," 41.

Chapter 3

1. "Banner Year for Polaroid Corp.," *Boston Mass. News Bureau*, February 26, 1945, Polaroid Corporation Administrative Records, Box I.88, f. 21, Press Clippings Profits and Company Growth, 1945, HBS.

2. Thomas B. Adams, "Light and Life, a Symposium in Honor of Edwin Land," *Bulletin of the American Academy of Arts and Sciences* 45, no. 7 (April 1992).

3. *Polaroid 3-D News*, July 1953, Polaroid Corporation Legal and Patent Records, Box II.3, f. 1, Land, Edwin H., HBS.

4. Arthur A. Riley, "Polaroid Grew from Non-glare Headlight Search," *Boston Sunday Globe*, June 21, 1959, A-61; "Photography Innovator," *New York Times*, April 26, 1972.

5. "60-Second Film," *Time*, November 28, 1955, Polaroid Corporation Administrative/Organizational Records, Non Polaroid Publications, General Magazines, Box 3, f. 17, HBS.

6. "Three Living Leaders."

7. Bert Pierce, "Automobiles: Lights," *New York Times*, November 23, 1947.

8. McElheny, *Insisting on the Impossible*, 101–2.

9. Riley, "Polaroid Grew from Non-glare Headlight Search," A-61.

10. Bert Pierce, "'49 Traffic Relief Predicted for City," *New York Times*, December 4, 1947.

11. *Engineering News Record*, December 25, 1947, 43, Polaroid Corporation Administrative Records, Box I.89, f. 22, Press Clippings—Headlight Glare, 1947, HBS.

12. Edwin H. Land, "The Polarized Headlight System," *Highway Research Board Bulletin* (1948), in *Edwin H. Land's Essays*, vol. 1, ed. Mary McCann, 78.

13. *Engineering News Record.*

14. McElheny, *Insisting on the Impossible*, 102–3.

15. Ibid., 107.

16. "Polaroid—Turning Away from Land's One-Product Strategy," *Business Week*, March 2, 1981.

17. Blout, "Polaroid: Dreams to Reality."

18. Transcript of SPSE Session (San Francisco, May 10, 1972), D.A. Delwiche, Kodak memorandum (August 11, 1972), Polaroid Corporation v. Eastman Kodak Company, District of Massachusetts, Civil Action 76-1634-MA, Trial Exhibits, PT 269(b), tab 27.

19. Land, "If You Are Able to State a Problem."

20. Ibid.

21. "Three Living Leaders," 106.

22. Charles Mikulka, Handwritten Memo, "Prepared from Disclosure Made by E.H. Land on December 16, 1943," Polaroid Corporation Legal and Patent Records, Box II.26, f. 6, Project SX70 Disclosures, December 1943, HBS.

23. Frederick J. Binda, Untitled Memo, December 23, 1943, Polaroid Corporation Legal and Patent Records, Box II.26, f. 6, Project SX70 Disclosures, December 1943, HBS.

24. Richard Kriebel, Handwritten Memo, December 27, 1943, Polaroid Corporation Legal and Patent Records, Box II.26, f. 6, Project SX70 Disclosures, December 1943, HBS.

25. Land, "If You Are Able to State a Problem," 48.

26. Transcript of SPSE Session.

27. Edwin H. Land, "The Universe of One-Step Photography, Pioneers of Photography," ed. Eugene Ostroff, in *Edwin H. Land's Essays*, vol. 1, 265, 275.

28. McCune, "Light and Life, a Symposium in Honor of Edwin Land."

29. McElheny, "Edwin Herbert Land," 9.

30. McElheny, "Polaroid's Coup: Instant Movies."

31. Stanley Mervis, Handwritten note, June 14, 1993, Polaroid Corporation Legal and Patent Records, Box II.29, f. 4, Stanley Mervis Correspondence, 1971–93, HBS.

32. Richard Wareham, "Dr. Land's SX-70 Camera: A Brief," *Optics & Photonic News*, October 1994, 46.

33. Transcript of SPSE Session.

34. McElheny, *Insisting on the Impossible*, 171.

35. Manchester, "Pictures in 60 Seconds."

36. Edwin H. Land, "A New One-Step Photographic Process," *Journal of the Optical Society of America* 37, no. 2, February 1947, in *Edwin H. Land's Essays*.

37. Manchester, "Pictures in 60 Seconds."

38. Testimony of Edwin H. Land, Trial Transcript, PvK, October 5, 1981, 46.

39. Ibid., 32–33.

40. Land, "A New One-Step Photographic Process," 123.

41. Edith Weyde, "Pioneering the Diffusion-Transfer-Reversal Process," Pioneers of Photography, *The Society for Imaging Science and Technology* (1987), 216–17.

42. Land, "A New One-Step Photographic Process."

43. Land, "The Universe of One-Step Photography," 265.

44. Trial Testimony of Edwin H. Land, PvK, October 9, 1981, 421–22.

45. O.W. Hayes, Memo to File, January 18, 1946, Polaroid Corporation Legal and Patent Records, Box II.23, f. 19, Project SX70 Disclosures to File, HBS; O.W. Hayes, Memo to Mr. E.H. Land, May 2, 1946, ibid.

46. Transcript of SPSE Session.
47. Land, "If You Are Able to State a Problem," 48.
48. McElheny, "Edwin Herbert Land," 10.
49. Charles Mikulka, Memorandum to Donald Brown, October 17, 1944, Polaroid Corporation Legal and Patent Records, Box II.29, f. 4, Stanley Mervis Correspondence 1971–1993, HBS.
50. Ibid.; Transcript of Charles Mikulka deposition, PvK, 6–7, Defendant Eastman Kodak Trial Exhibit DF-500.
51. Edwin H. Land, Photographic product comprising a rupturable container carrying a photographic processing liquid, U.S. Patent 2,543,181, issued February 27, 1951, as continuation in part of application Serial No. 539,550, filed June 9, 1944.
52. Mikulka, memorandum.
53. Ibid.
54. Ibid.
55. Ibid.
56. Ibid.
57. "Banner Year for Polaroid Corp."
58. O.W. Hayes, Memorandum to File, January 11, 1946, Polaroid Corporation Legal and Patent Records, Box II.23, f. 19, Project SX-70 Disclosures to File, HBS.
59. Philip Boone, Memorandum to Donald L. Brown, "SX-70 Cases to Be Filed," October 31, 1945; O.W. Hayes, Memorandum to Donald L. Brown, "SX70 Applications to Be Filed," October 31, 1945; O.W. Hayes, Memorandum to Donald L. Brown, "SX70 Applications to Be Filed," January 15, 1946; O.W. Hayes, Memorandum to Donald L. Brown, "SX70 Applications to Be Filed," January 16, 1946, all located in Polaroid Corporation Legal and Patent Records, Box II.23, f. 18, Applications to be Prepared SX-70, HBS.
60. Trial Testimony of Edwin H. Land, PvK, October 9, 1981, 422.
61. Douglas Collins, *The Story of Kodak* (Harry N. Abrams Inc., 1990), 305.
62. O.W. Hayes, Memorandum to Mr. E.H. Land, May 2, 1946, Polaroid Corporation Legal and Patent Records, Box II.23, f. 18, Applications to be Prepared SX-70, HBS.
63. Transcript of SPSE Session.
64. O.W. Hayes, Memorandum to Mr. Edwin H. Land, "Applications of SX-70 Process," October 3, 1946, Polaroid Corporation Legal and Patent Records, Box II.23, f. 18, Applications to Be Prepared SX-70, HBS.
65. McElheny, *Insisting on the Impossible*, 183.
66. Ibid., 184.
67. Ibid., 183.
68. McCune Oral History, 4.
69. Ibid.
70. William I. Laurence, "One-Step Camera Is Demonstrated," *New York Times*, February 22, 1947.
71. McCune Oral History, 4.
72. Herbert Nichols, "It's a Snap—with a New Twist: Camera Prints Own Pictures," *Christian Science Monitor*, February 25, 1947.
73. "Picture of the Week," *Life*, March 3, 1947, 33; "Camera Coughs Out Finished Prints," *Popular Science*, May 1947.
74. "The Camera Does the Rest," *New York Times*, February 22, 1947.
75. Nichols, "It's a Snap."

76. Manchester, "Pictures in 60 Seconds."

77. "One-Step Camera," *Business Week*, March 1, 1947, Polaroid Corporation Administrative Records, Box I.233, f. 10, HBS.

78. Manchester, "Pictures in 60 Seconds"; Laurence, "One-Step Camera Is Demonstrated"; "Science—Snap and See," *Newsweek*, March 3, 1947, Polaroid Corporation Administrative Records, Box I.234, f. 18, HBS.

79. Land, "A New One-Step Photographic Process."

80. "The Camera Does the Rest"; Collins, *The Story of Kodak*, 46, 54–59.

81. "Science—Snap and See."

82. "One-Step Camera."

83. Robert Lenzner, "The Promised Land," *Boston Globe*, August 3, 1982.

84. "Three Living Leaders," 106.

Chapter 4

1. McCune Oral History, 4.

2. Trial Testimony of Edwin H. Land, PvK, October 9, 1981, 423.

3. Cecil Quillen, interview by author, December 9, 2009, Washington, DC.

4. Trial Testimony of Edwin H. Land, PvK, October 9, 1981, 423.

5. Quillen, interview.

6. McCune Oral History, 5.

7. Trial Testimony of Edwin H. Land, PvK, October 6, 1981, 97–100.

8. McCune Oral History, 4.

9. Trial Testimony of Edwin H. Land, PvK, October 6, 1981, 54–55, 97.

10. Jacob Deschin, "Tax Law Benefits," *New York Times*, November 4, 1951; McCune Oral History, 5.

11. "Dr. Land Redesigns His Camera Company," *Business Week*, April 15, 1972; McCune Oral History, 4.

12. Testimony of William McCune, Trial Transcript, Polaroid Corporation v. Eastman Kodak Company, District of Massachusetts, Civil Action 76-1634-MA, May 2, 1989, 2–34; McElheny, *Insisting on the Impossible,* 189.

13. Edwin H. Land, Developing camera utilizing a film, U.S. Patent 2,435,717, filed Oct. 6, 1945, and issued Feb. 10, 1948; Edwin H. Land, Photographic process and apparatus, U.S. Patent 2,435,718, filed Jan. 11, 1946, and issued Feb. 10, 1948; Edwin H. Land, Photographic apparatus for subjecting a photographic film to a processing fluid, U.S. Patent 2,435,719, filed Feb. 12, 1946, and issued Feb. 10, 1948; Edwin H. Land, Apparatus for exposing and processing photographic film, U.S. Patent 2,435,720, filed Aug. 29, 1946, and issued Feb. 10, 1948.

14. "Instant Developers—Grant Patents for New-Type Cameras," *Atlanta Constitution*, February 15, 1948, Polaroid Corporation Administrative Records, Box I.92, f. 5, Press Clippings—Land Camera, 1948, HBS.

15. Campbell, "Edwin Herbert Land," 210.

16. Francis Cassidy, Memorandum to Miss Natalie Fultz, "Patents Issued to Dr. Land," May 17, 1960, Polaroid Corporation Legal and Patent Records, Box II.3, f. 1, Land, Edwin H., HBS.

17. Brown, "Protection through Patents: The Polaroid Story," 450.

18. "Land Camera This Year," *New York Times*, February 29, 1948.

19. Edwin H. Land, "Summary of Communication Presented before the Academy," May 12, 1948, Polaroid Corporation Legal and Patent Records, Box II.3, f. 1, Land, Edwin H., HBS.

20. Jack Dudley, "Camera Prints Photo in Minute!," *Cincinnati Enquirer*, November 6, 1948, Polaroid Corporation Administrative Records, Box C7, f. I.92, Press Clippings—Land Camera, 1948, HBS.

21. Ibid.

22. Jacob Deschin, "One-Step Camera," *New York Times*, November 14, 1948.

23. George Green, "Editor Previews Polaroid Pic-in-a-Minute Camera," *Boston Globe*, November 7, 1948, Polaroid Corporation Administrative Records, Box I.92, f. 3, Press Clippings—Land Camera, 1948, HBS.

24. Deschin, "One-Step Camera."

25. Bernard Brown, "Land's Camera Is Here!," *National Photo Dealer*, November 1948, Polaroid Corporation Administrative Records, Box I.92, f. 7, Press Clippings—Land Camera, 1948, HBS.

26. McCune Oral History, 5.

27. Ibid.

28. "Three Living Leaders" at 106; Trial Testimony of William McCune, PvKdam, May 2, 1989, 2–25.

29. Alan Earls and Nasrin Rohani, *Polaroid, Images of America* (Arcadia, 2005), 76.

30. Steven Syre, "Polaroid Slowly Fades Away but at One Time Edwin Land and his Audacious Firm Could Do No Wrong," *Boston Globe*, October 14, 2001.

31. McCune Oral History, 5.

32. *Dow Jones Ticker*, Boston, December 14, 1948, Polaroid Corporation Administrative Records, Box I.92, f. 6, Press Clippings—Land Camera, 1948, HBS.

33. McCune Oral History, 5.

34. "New Products—Pictures in a Minute," *Time*, May 30, 1949, Polaroid Corporation Administrative Records, Box I.235, f. 17, HBS.

35. John Case, "The Problems at Polaroid," *Boston Magazine*, April 1982, 160.

36. Fred Bayles, "Latter-Day Edison Leaves Company He Built," *Associated Press*, August 1, 1982.

37. Jimmy Powers, "The Powerhouse," *New York Daily News*, December 8, 1948, Polaroid Corporation Administrative Records, Box I.92, f. 7, Press Clippings—Land Camera, 1948, HBS.

38. "Camera Gives Print in a Minute," *Popular Science*, February 1949.

39. "Truman Takes a One-Minute Photo," *San Francisco Chronicle*, December 9, 1948, Polaroid Corporation Administrative Records, Box I.92, f. 10, Press Clippings—One Minute Photos, 1948, HBS.

40. "Three Living Leaders"; Subrata Chakravarty and Ruth Simon, "Has the World Passed Kodak By?," *Forbes*, November 5, 1984, 84; "The Camera Boom," *Newsweek*, May 3, 1976, 69; Victor K. McElheny, "In the Beginning Was Dr. Land," *New York Times*, April 18, 1976.

41. Walter Clark, "The Photographic Year: A Review of Progress in 1947," *Popular Photography*, February 1948, Polaroid Corporation Administrative Records, Box I.92, f. 2, Press Clippings—Land Camera, 1948, HBS.

42. Subrata Chakravarty, "An Interview with Dr. Edwin Land," *Forbes*, June 1, 1975.

43. Riley, "Polaroid Grew from Non-glare Headlight Search," A-61.

44. Collins, *The Story of Kodak*, 111.

45. Testimony of Howard Rogers, trial transcript, PvK, November 30, 1981, 3092–93.

46. Blout, "Polaroid: Dreams to Reality," 2–3.

47. Trial Testimony of Edwin H. Land, PvK, October 9, 1981, 426.

48. William McCune, memorandum to M. Wren Gabel, February 8, 1957, PvK, Trial Exhibit DF-500; Trial Testimony of Land, PvK, October 9, 1981, 426; McCune Oral History, 6.

49. Trial Testimony of William McCune, PvKdam, May 2, 1989, 2–41.

50. Trial Testimony of Howard Rogers, PvK, November 30, 1981, 3093–98.

51. Ibid., 3117–30.

52. Jane Poss, "Edwin H. Land, The Soul of an Inventor," *Boston Globe*, March 3, 1991.

53. McElheny, *Insisting on the Impossible*, 203.

54. Ibid., 213.

55. Mary A. McCann and John J. McCann, "Land's Chemical, Physical, and Psychological Images," *Optics & Photonics News*, October 1994, 34.

56. Edwin H. Land, "The Universe of One-Step Photography," *Pioneers of Photography*, ed. Eugene Ostroff, in *Edwin H. Land's Essays*, vol. 1, 265, 269.

57. Trial Testimony of Land, PvK, October 6, 1981, 216.

58. Ibid.

59. McCann and McCann, "Land's Chemical, Physical, and Psychological Images," 34.

60. Seymour Spector, "Along Camera Row," *New York Journal-American*, May 7, 1950, Polaroid Corporation Administrative Records, Box I.96, f. 20, Press Clippings—Black and White Film, 1950, HBS.

61. Arthur Juntunen, "Land Camera Gets Black-White Film," Detroit Free Press, May 11, 1950, Polaroid Corporation Administrative Records, Box I.96, f. 20, Press Clippings—Black and White Film, 1950, HBS.

62. McElheny, *Insisting on the Impossible*, 206.

63. Trial Testimony of Land, PvK, October 6, 1981, 212–21; McElheny, *Insisting on the Impossible*, 209.

64. Trial Testimony of Land, PvK, October 6, 1981, 212.

65. McElheny, *Insisting on the Impossible*, 209.

66. "Background Notes—Polaroid Cameras" (Polaroid, 1958), Polaroid Corporation Legal and Patent Records, Box II.27, f. 25, Misc. Administrative Papers, HBS.

67. Blout, "Polaroid: Dreams to Reality," 3.

68. "Background Notes"; Brown, "Protection through Patents: The Polaroid Story," 440.

69. McElheny, *Insisting on the Impossible*, 198.

70. "Background Notes."

71. "Edwin H. Land, a Scrapbook" (video tribute, Polaroid Corporation, 1991).

72. Land, "One-Step Photography."

73. Barbara Hitchcock, "When Land Met Adams," *The Polaroid Book* (Taschen, 2004), 24; Ansel Adams, *An Autobiography* (Little Brown & Company, 1985), 293.

74. Adams, *An Autobiography*, 295.

75. Ibid., 297.

76. Ibid., 300.
77. Ibid., 302.
78. McElheny, *Insisting on the Impossible*, 205.
79. Ibid., 208.
80. Ibid., 213, 219.
81. "Land Heads Arts Academy," *New York Times*, May 10, 1951.
82. Donald Brown, Memorandum to Dr. Edwin H. Land, "Professor Melman's Report re the Impact of Patent Systems on Research," September 3, 1958, Polaroid Corporation Legal and Patent Records, Box II.3, f. 1, Land, Edwin H., HBS.
83. Donald Brown, Memorandum to Dr. Edwin H. Land, "1959 BPLA Dinner Speaker," June 30, 1958, Polaroid Corporation Legal and Patent Records, Box II.3, f. 1, Land, Edwin H., HBS.
84. Brown, "Professor Melman's Report."
85. Donald Brown, Memorandum to Mrs. L.G. Billings, "Dr. Land's 1959 BPLA Talk," August 18, 1958, Polaroid Corporation Legal and Patent Records, Box II.104, f. 13, Correspondence between Donald Brown and others re: Dr. Land's speech @ BPLA Dinner, April 2, 1959, HBS.
86. Donald Brown, Memorandum to Dr. Edwin H. Land, "April 2nd Talk to the BPLA," February 27, 1959, Polaroid Corporation Legal and Patent Records, Box II.104, f. 13, Correspondence between Donald Brown and others re: Dr. Land's speech @ BPLA Dinner, April 2, 1959, HBS.
87. Edwin H. Land, "Thinking Ahead: Patents & New Enterprises," *Harvard Business Review*, September–October 1959, in *Edwin H. Land's Essays*, vol. 2, 43–44.
88. H.R. Mayers, Letter to Mr. R.E. Hosley, June 4, 1959, Polaroid Corporation Legal and Patent Records, Box II.104, f. 13, Correspondence between Donald Brown and others re: Dr. Land's speech @ BPLA Dinner, April 2, 1959, HBS.
89. Ibid.
90. Donald Brown, Memorandum to Dr. Edwin H. Land, "BPLA Talk, June 5, 1959," Polaroid Corporation Legal and Patent Records, Box II.104, f. 13, Correspondence between Donald Brown and others re: Dr. Land's speech @ BPLA Dinner, April 2, 1959, HBS.
91. Fred Warshofsky, *The Patent Wars* (John Wiley & Sons, 1994), 77.
92. Stanley Mervis, "Edwin Land—Champion of Patents," *Optics & Photonics News*, October 1994, 50.

Chapter 5

1. Land, "On Some Conditions for Scientific Profundity"; Benton, "Edwin Land, 3-D," 41.
2. Trial Testimony of Howard Rogers, PvK, November 30, 1981, 3140–44.
3. "Now It's 60-Second Pictures in Color," *Life*, January 25, 1963, Polaroid Corporation Administrative Records, Box I.227, f. 2, HBS.
4. PvK, Plaintiff Polaroid Corporation's Brief after Trial, Rogers U.S. Patent 3,245,789, 5.
5. Trial Testimony of Rogers, PvK, November 30, 1981, 3144–51.
6. Polaroid Brief after Trial, Rogers U.S. Patent 3,245,789, 5.

7. Trial Testimony of Rogers, PvK, November 30, 1981, 3190–91.

8. Trial Testimony of Edwin H. Land, PvK, October 29, 1981, 1279; ibid., October 6, 1981, 427.

9. Deposition Testimony of M. Wren Gabel, PvK, December 14, 1980, Trial Exhibit DF-500, tab 2, 24–25; ibid., tab 10, 9.

10. William McCune, memorandum to M. Wren Gabel, February 8, 1957, PvK, Trial Exhibit DF-500, tab 2.

11. *Polaroid Corporation Annual Report for 1974* (March 25, 1975), PvK, Trial Exhibit PT-79.

12. Trial Testimony of Land, PvK, October 29, 1981, 1279–80; ibid., October 6, 1981, 427.

13. McCune, memorandum, February 8, 1957.

14. M. Wren Gabel, letter to William McCune, March 25, 1957, and attached unsigned memorandum dated March 22, 1957, PvK, Trial Exhibit DF-500, tab 2.

15. McCune, memorandum, February 8, 1957.

16. Gabel, letter, March 25, 1957.

17. "81-Year Highs Set by R.J. Reynolds," *New York Times*, February 7, 1957.

18. "Eastman Kodak Sets 2 Records," *New York Times*, February 25, 1957.

19. Burton Crane, "Market Climbs on Broad Front," *New York Times*, May 21, 1958.

20. Trial Testimony of William McCune, PvKdam, May 3, 1989, 291–93.

21. Memorandum, Summary of Views Expressed at Meeting at Rochester October 2, 1963, between Representatives of Eastman Kodak Company and Polaroid Corporation (undated), 5, PvK, Trial Exhibit DF-500, tab 4.

22. Gabel, letter, March 25, 1957.

23. "The Blurry Picture at Eastman Kodak," *Forbes*, September 15, 1974, Polaroid Corporation Legal and Patent Records, Box II.203, f. 8, Articles of Interest—Kodak, 1976–1981, HBS.

24. Collins, *The Story of Kodak*, 148–50, 158–59.

25. Gabel, letter, March 25, 1957.

26. McCune Oral History, 6.

27. Trial Testimony of Land, PvK, October 29, 1981, 1280–81.

28. William McCune, letter to M. Wren Gabel, May 10, 1957, with memorandum from M. Wren Gabel to William J. McCune of same date attached, PvK, Trial Exhibit DF-500, tab 2.

29. McElheny, *Insisting on the Impossible*, 281.

30. David Welzenbach, "Din Land: Patriot from Polaroid," *Optics & Photonics News*, October 1994, 22.

31. James R. Killian Jr., *Sputnik, Scientists, and Eisenhower* (MIT Press, 1977), 68.

32. Ibid., 67–68; Welzenbach, "Din Land," 22.

33. Welzenbach, "Din Land," 22–23.

34. Ibid., 23.

35. Ibid.

36. Killian Jr., *Sputnik, Scientists, and Eisenhower*, 79.

37. McCune Oral History, 2.

38. McElheny, *Insisting on the Impossible*, 283.

39. Welzenbach, "Din Land," 23–24.
40. Killian Jr., *Sputnik, Scientists, and Eisenhower*, 82.
41. Welzenbach, "Din Land," 25–28.
42. McElheny, *Insisting on the Impossible*, 298.
43. Ibid., 299.
44. Welzenbach, "Din Land," 26; Killian Jr., *Sputnik, Scientists, and Eisenhower*, 82.
45. Campbell, "Edwin Herbert Land," 202.
46. Richard Garwin, "Impressions of Edwin H. Land," *Optics & Photonics News*, October 1994, 24.
47. Welzenbach, "Din Land," 28.
48. McElheny, *Insisting on the Impossible*, 308.
49. Killian Jr., *Sputnik, Scientists, and Eisenhower*, 122–42.
50. "Round the World in 96 Minutes," *New York Times*, October 6, 1957.
51. Harry Schwartz, "Soviet Exploits Its New 'Sputnik' Diplomacy," *New York Times*, October 20, 1957.
52. Killian Jr., *Sputnik, Scientists, and Eisenhower*, 7–8.
53. Ibid.
54. Arthur Krock, "The Effects of the Sputnik So Far," *New York Times*, October 10, 1957.
55. Killian Jr., *Sputnik, Scientists, and Eisenhower*, 15.
56. McElheny, *Insisting on the Impossible*, 311.
57. Killian Jr., *Sputnik, Scientists, and Eisenhower*, 16.
58. McElheny, *Insisting on the Impossible*, 306.
59. Robert Divine, *The Sputnik Challenge* (Oxford University Press, 1993), 15.
60. John W. Finney, "Sputnik Acts as a Spur to U.S. Science and Research," *New York Times*, November 3, 1957.
61. "Text of the Address by President Eisenhower on Science in National Security," *New York Times*, November 8, 1957.
62. "Text of President's Talk in Oklahoma City Citing Need for Rise in Funds for Science," *New York Times*, November 14, 1957.
63. Killian Jr., *Sputnik, Scientists, and Eisenhower*, 16.
64. Ibid., 88.
65. Ibid., 87.
66. Welzenbach, "Din Land," 28.
67. Lenzner, "Land: The Man behind the Camera."
68. Robert Lenzner, "Edwin Land on Polaroid—and the Kodak Patent Suit," *Boston Globe*, October 18, 1976.
69. Ibid.
70. Ibid.
71. Killian Jr., *Sputnik, Scientists, and Eisenhower*, 43.
72. Marjorie Hunter, "President Names 31 for Freedom Medal," *New York Times*, July 5, 1963; "President Awards National Science Medals to 12," *New York Times*, February 14, 1968; Victor K. McElheny, "White House Pomp Attends Honors for 3 Hub Scientists," *Boston Globe*, February 14, 1968.
73. Campbell, "Edwin Herbert Land," 209.

74. Trial Testimony of Land, PvK, October 7, 1981, 268.

75. Deposition Testimony of Charles Mikulka, PvK, August 12, 1980, 194–95, Trial Exhibit DF-500, tab 2.

76. Agreement between Polaroid Corporation and Eastman Kodak Company as of December 12, 1957, PvK, 17–20, 23, Trial Exhibit DF-500, tab 2; Memorandum, "Historical Context," attached to handwritten note, October 28, 1977, Polaroid Corporation Legal and Patent Records, Box II.216, f. 8, Eastman Kodak 1980, HBS.

77. Stanley Mervis, memorandum to William McCune and H.C. Yutzy, February 7, 1958.

78. Ibid.

79. McCune Oral History, 6.

80. Mervis, "Edwin Land—Champion of Patents," 51.

81. Blout, "Light and Life," 3.

82. "Official Says Kodak Has No Plans to Enter Picture-a-Minute Field," *Wall Street Journal*, April 29, 1959.

83. Jacob Deschin, "Polaroid in Color," *New York Times*, April 17, 1960.

84. Ibid.

85. Frederick McCarthy, "Polaroid Sets 1962 Date for Color Film," *Boston Globe*, April 12, 1961.

86. Trial Testimony of Land, PvK, October 6, 1981, 222–23.

87. Ibid., 224–25.

88. "Now It's 60-Second Pictures in Color."

89. Edwin H. Land, Letter from the Chairman of the Board, *Polaroid Corporation Annual Report for 1976*, in Edwin H. Land, *Selected Papers on Industry* (Polaroid Corporation, 1983).

90. Trial Testimony of Land, PvK, October 6, 1981, 226.

91. Mervis, "Edwin Land—Champion of Patents," 52.

92. "Now It's 60-Second Pictures in Color."

93. Stanley Mervis, Memo to Nasrin Rohani, June 16, 1993, Polaroid Corporation Legal and Patent Records, Box II.29, f. 4, Stanley Mervis Correspondence, 1971–93, HBS.

94. Polaroid Corporation, press release, January 24, 1963, PvK, Trial Exhibit PT-80, 2–3.

95. Eastman Kodak Company, "Statement about Polaroid Color," press release, January 24, 1963, PvK, Trial Exhibit DF-500.

96. Ibid.

97. "Land Reports on Production of Polacolor," *Boston Globe*, April 10, 1963; Jacob Deschin, "New Items on Market," *New York Times*, April 7, 1963.

98. William Siegrist, "Instant Color Photography," *Science News Letter* (Society for Science & the Public, April 20, 1963).

99. "Now It's 60-Second Pictures in Color."

100. Polaroid Corporation, "Background Information about Polaroid Land Color Film," press release, 7, PvK, Trial Exhibit PT-16.

101. Trial Testimony of William McCune, PvKdam, May 2, 1989, 2–42.

102. Quillen Interview.

103. George Peterson Deposition Testimony, 156–57, PvK, Trial Exhibit DF-500, tab 4.

104. *Polaroid Corporation Annual Report for 1974*, March 25, 1975, PvK, Trial Exhibit PT-79.

105. Quillen, interview.

Chapter 6

1. McCune Oral History, 3; Trial Testimony of William McCune, PvKdam, May 2, 1989, 2–7.

2. "Polaroid Corp. Plans Low-Priced Camera," *Boston Globe*, April 15, 1964; *Dow Jones Report*, attached to letter from Daniel G.W. del Rio to Donald Brown, April 17, 1964, Polaroid Corporation Legal and Patent Records, Box II.103, f. 12, Personal Files, Stock 1960–1966, HBS.

3. Daniel G.W. del Rio, *Institutional Research Bulletin* (Delafield & Delafield, June 10, 1965), Polaroid Corporation Legal and Patent Records, Box II.103, f. 12, Personal Files, Stock 1960–1966, HBS.

4. "Now It's 60-Second Pictures in Color."

5. Edwin H. Land, letter to shareholders, March 1972, quoted in Polaroid Corporation press release, May 10, 1972.

6. "Dr. Land's Latest Bit of Magic," *Life*, October 27, 1972, 44.

7. Edwin H. Land, "Absolute One-Step Photography," *Photographic Science and Engineering*, July–August 1972, in *Edwin H. Land's Essays*, vol. 1, 179.

8. "Polaroid's Big Gamble on Small Cameras," *Time*, June 26, 1972.

9. Kodak Research Laboratories, *Apparatus for Experimental Processes—In-Camera Processing of Image Transfer Materials (Initial Report)*, November 1, 1961, PvK, Trial Exhibit DF-500, tab 3.

10. Kodak's General Post-Trial Brief, PvK, 2.

11. Richard Martin, "Polaroid's McCune Tackles Big Problems as Its New President," *Wall Street Journal*, April 25, 1975, Polaroid Corporation Legal and Patent Records, Box II.217, f. 3, Wall Street Journal Interview, January 1975, HBS.

12. Memorandum, Historical Context, attached to handwritten note, October 28, 1977, 10, Polaroid Corporation Legal and Patent Records, Box II.216, f. 8, Eastman Kodak 1980, HBS.

13. McCune Oral History, 6.

14. Charles Mikulka, Letter to George Petersen, December 20, 1962, Polaroid Corporation Legal and Patent Records, Box II.252, f. 1, Kodak Agreement, 1962–1966, HBS.

15. Cyril J. Staud, letter to M. Wren Gabel, October 23, 1962, PvK, Trial Exhibit DF-500, tab 3.

16. Memorandum, Meeting on In-Camera Process Methods, February 11, 1963, PvK, Trial Exhibit PT-269(B), tab 2.

17. H.C. Yutzy, letter to Cyril J. Staud et al., March 19, 1963, PvK, Trial Exhibit DF-500, tab 3.

18. George Petersen, Letter to Charles Mikulka, February 26, 1963, Polaroid Corporation Legal and Patent Records, Box II.252, f. 1, Kodak Agreement 1962–1966, HBS; Memorandum, "Historical Context."

19. Charles Mikulka, "Memo to the Files," July 2, 1963, Polaroid Corporation Legal and Patent Records, Box II.250, f. 5, Directors Meetings, 1959–1969, HBS.

20. Ibid.

21. Deposition Testimony of Charles Mikulka, August 13, 1980, 367, PvK, Trial Exhibit DF-500, tab 4.

22. Trial Testimony of William McCune, PvKdam, May 2, 1989, 2–43.

23. Memorandum, Summary of Views Expressed at Meeting at Rochester October 2, 1963, between Representatives of Eastman Kodak Company and Polaroid Corporation, 1–3, PvK, Trial Exhibit DF-500, tab 4.

24. Ibid., 4.

25. Petersen Deposition Testimony, January 21, 1981, 148–49, 155.

26. Ibid., 148–55.

27. Ibid., 151.

28. Mikulka Deposition Testimony, March 12, 1981, 502.

29. Memorandum, Summary of Views, 6.

30. Petersen Deposition Testimony, January 21, 1981, 152–53.

31. Kodak Summary of Discovery Exhibits, PvK, Trial Exhibit DF-500, tab 4; Petersen Deposition Testimony, January 21, 1981, 152.

32. Petersen Deposition Testimony, January 21, 1981, 152.

33. Alexander Auerbach, "Dr. Land's Topic: Polaroid's Profit Up 34%," *Boston Globe*, April 12, 1967.

34. "Here Come Those Great New Cameras," *Time*, June 26, 1972.

35. Dan Cordtz, "How Polaroid Bet Its Future on the SX-70," *Fortune*, January 1974, 85–87.

36. Ibid., 87.

37. Ibid.; *Neblette's Handbook of Photography and Reprography*, ed. John Sturge (Van Nostrand Reinhold Company, 1977, 7th ed.), 322.

38. Mervis, "Edwin Land—Champion of Patents," 51; Peter Gosselin, "How Polaroid Beat Kodak," *Boston Globe*, January 14, 1986.

39. *Neblette's Handbook*, 262.

40. Trial Testimony of Edwin H. Land, PvK, October 6, 1981, 284–85, 293–94.

41. Nan Chequer, notes attached to Ronald Fierstein, memorandum to James Williams and Robert Goldman, May 29, 1980, F&N.

42. Richard Wareham, "Dr. Land's SX-70 Camera: A Brief," *Optics & Photonics News*, October 1994, 46.

43. Cordtz, "How Polaroid Bet Its Future," 144.

44. Gallese, "I Am a Camera," 49.

45. Robert Goldman, telephone interview by author, January 11, 2011.

46. Cordtz, "How Polaroid Bet Its Future," 87.

47. Land, "Absolute One-Step Photography."

48. Wareham, "Dr. Land's SX-70 Camera," 46.

49. Kodak's Post-Trial Brief with Respect to U.S. Patent 3,810,211, PvK, 14–15.

50. Goldman, telephone interview.

51. Chequer, notes.

52. Plaintiff Polaroid Corporation's Brief after Trial, Blinow and Leduc U.S. Patent 3,709,122, PvK, 2.

53. Cordtz, "How Polaroid Bet Its Future," 87.

54. Ibid.

55. Trial Testimony of Howard Rogers, PvK, December 1, 1981, 3290–91; Trial Testimony of Howard Rogers, PvK, December 4, 1981, 3616–19.

56. Howard Rogers, sketch, January 1, 1968, PvK, Trial Exhibit DF-449.

57. Gosselin, "How Polaroid Beat Kodak."

58. Trial Testimony of Howard Rogers, PvK, December 1, 1981, 3293–94.

59. Edwin H. Land, "Absolute One-Step Photography," *The Photographic Journal*, July 1974, in *Edwin H. Land's Essays*, vol. 1, 185.

60. McElheny, *Insisting on the Impossible*, 369.

61. Cordtz, "How Polaroid Bet Its Future," 87.

62. McElheny, *Insisting on the Impossible*, 370.

63. "Polaroid's Big Gamble on Small Cameras"; Cordtz, "How Polaroid Bet Its Future."

64. Robert Reinhold, "Land Achieves His Dream with New Polaroid SX-70," *New York Times*, October 30, 1972.

Chapter 7

1. Memorandum, "E.K. Meeting—A Morning of Philosophy—An Afternoon of Accomplishment," December 26, 1967, Polaroid Corporation Legal and Patent Records, Box II.227, f. 3, Kodak 1967–1971, HBS; Memorandum, "List of Polaroid/Kodak Meetings, 1963–1969," undated, Polaroid Corporation Legal and Patent Records, Box II.250, f. 3, Kodak Meetings 1967–1969, HBS.

2. M. Wren Gabel, memorandum, February 12, 1968, PvK, Trial Exhibit DF-500.

3. Philip Siekman, "Kodak and Polaroid: An End to Peaceful Coexistence," *Fortune*, November 1970, 85; Alexander Auerbach, "Dr. Land's Topic: Polaroid's Profit Up 34%," *Boston Globe*, April 12, 1967; "Polaroid's Big Gamble on Small Cameras."

4. Siekman, "Kodak and Polaroid," 82.

5. Victor K. McElheny, "In the Beginning Was Dr. Land," *New York Times*, April 18, 1976.

6. Siekman, "Kodak and Polaroid," 85.

7. G. Bramhall, Memorandum to Mr. W.J. McCune, August 12, 1969, Polaroid Corporation Legal and Patent Records, Box II.227, f. 3, Kodak, 1967–1971, HBS; Bro Uttal, "Eastman Kodak's Orderly Two-Front War," *Fortune*, September 1976, 125.

8. "Polaroid's Big Gamble on Small Cameras."

9. Siekman, "Kodak and Polaroid," 85.

10. "Polaroid's Big Gamble on Small Cameras."

11. Siekman, "Kodak and Polaroid," 85.

12. William Carley, "Polaroid Seen Wary, Worried as It Girds for Kodak Arrival in Instant-Photo Field," *Wall Street Journal*, April 16, 1976.

13. Victor K. McElheny, "Polaroid Is Suing Kodak, Charges Patent Violation," *New York Times*, April 28, 1976.

14. Quillen interview; Fred Warshofsky, *The Patent Wars* (John Wiley and Sons, 1994), 74.

15. Robert Wright, "Living Up to Kodak Past Is Part of Picture," *New York Times*, February 9, 1969.

16. Quillen, interview.

17. Siekman, "Kodak and Polaroid," 118.

18. "Dr. Land Redesigns His Camera Company," *Business Week*, April 15, 1972; Siekman, "Kodak and Polaroid," 83.

19. Quillen, interview.

20. Petersen Deposition Testimony, 27–28, 131–35, quoted in M.P. Brook, memorandum to Herbert Schwartz and Edward Mullowney, November 6, 1986, F&N.

21. Henry Yutzy, memorandum, Image Color a la Polaroid, March 1965, PvK, Trial Exhibit PT-269(B), tab 3; Plaintiff Polaroid Corporation's General Brief after Trial, PvK, 3.

22. Gabel, memorandum.

23. Edwin H. Land, letter to M. Wren Gabel, April 9, 1968, PvK, Trial Exhibit PT-82.

24. M. Wren Gabel, draft letter to Edwin H. Land, May 8, 1968, PvK, Trial Exhibit DF-500.

25. Paul Goldstein, *Intellectual Property* (Portfolio, 2007), 22.

26. Louis Eilers, note to M. Wren Gabel, May 8, 1968, PvK, Trial Exhibit DF-500.

27. M. Wren Gabel, letter to Edwin H. Land, May 8, 1968, PvK, Trial Exhibit PT-83.

28. "Kodak's No. 3 Man, M.W. Gabel, Resigns," *Wall Street Journal*, August 23, 1968, Polaroid Corporation Legal and Patent Records, Box II.227, f. 3, Kodak 1967–1971, HBS.

29. Louis Eilers, memorandum for file, September 12, 1968, PvK, Trial Exhibit DF-500.

30. George Petersen, letter to Charles Mikulka, October 9, 1968, Polaroid Corporation Legal and Patent Records, Box II.251, f. 4, Kodak Agreement 1966–1969, HBS.

31. Siekman, "Kodak and Polaroid," 82.

32. Ibid., 83.

33. McCune Oral History, 6; Trial Testimony of William McCune, PvKdam, May 2, 1989, 246–47.

34. George Petersen, memorandum for file, November 6, 1968, PvK, Trial Exhibit DF-500.

35. Ibid.

36. Trial Testimony of Edwin H. Land, PvK, October 9, 1981, 438–40.

37. Polaroid Corporation v. Eastman Kodak Company, 641 F. Supp. 828, 831 (D. Mass 1986); PvK, Trial Exhibits PT-84 and PT-85; Petersen Deposition Testimony, 195–97, PvK, Trial Exhibit PT-269 (B), tab 6.

38. Trial Testimony of Edwin H. Land, PvK, October 9, 1981, 440; Trial Testimony of William McCune, PvKdam, May 2, 1989, 244.

39. Siekman, "Kodak and Polaroid," 87.

40. Trial Testimony of Edwin H. Land, PvK, October 9, 1981, 438–39.

41. Memorandum, "Historical Context."

42. George Petersen, Status Report—Polaroid Negotiations, November 18, 1968, PvK, Trial Exhibit DF-500.

43. Ira Werle, memorandum for file of meeting, November 27, 1968, PvK, Trial Exhibit DF-500.

44. "Instant Battle: Kodak v. Polaroid," *Time*, April 26, 1976, 61.

45. Petersen, Status Report.

46. Ibid.

47. Plaintiff Polaroid Corporation's General Brief after Trial, PvK, 5; Eastman Kodak, Petitioner v. Polaroid Corporation, Respondent, United States Supreme Court, Petition for Writ of Certiorari, Respondent's Brief in Opposition, 4, F&N.
48. Petersen, Status Report, 5.
49. Ibid.
50. Werle, memorandum for file of meeting.
51. Kodak's General Post-Trial Brief, PvK, 2.
52. Werle, memorandum for file of meeting.
53. Trial Testimony of William McCune, PvKdam, May 2, 1989, 246; Paul Goldstein, *Intellectual Property* (Portfolio, 2007), 22.
54. McCune Oral History, 7.
55. Ibid.; Trial Testimony of William McCune, PvKdam, May 2, 1989, 245–48.
56. Paul Sullivan, Memorandum to Dr. Edwin H. Land, April 10, 1969, 3, Polaroid Corporation Legal and Patent Records, Box II.250, f. 3, Kodak Meetings 1967–1969, HBS.
57. Louis Eilers, letter to Edwin H. Land, April 22, 1969, PvK, Trial Exhibit PT-86; Trial Testimony of William McCune, May 2, 1989, 294–95; Envelope (front and back), addressed to Dr. Edwin H. Land, postmarked April 23, 1969, Polaroid Corporation Legal and Patent Records, Box II.226, f. 3, Eastman Kodak Correspondence, 1957–1969, HBS.
58. William McCune, Handwritten notes on Letter, Louis Eilers to Dr. Edwin H. Land, April 22, 1969, Polaroid Corporation Legal and Patent Records, Box II.226, f. 3, Eastman Kodak Correspondence, 1957–1969, HBS.
59. Trial Testimony of William McCune, PvKdam, May 3, 1989, 299–300.
60. John Rumsey, "The Times of Instant's Life," *Rochester Times-Union*, January 9, 1986, Polaroid Corporation Administrative Records, Box I.122, f. 2, Polaroid vs. Kodak Clippings 1986, HBS.
61. Trial Testimony of William McCune, PvKdam, May 3, 1989, 299–300.
62. Edwin H. Land, letter to Louis Eilers, May 13, 1969, PvK, Trial Exhibit PT-87.
63. W.J. McCune, Memorandum to Dr. Edwin H. Land, May 23, 1969, Polaroid Corporation Legal and Patent Records, Box II.226, f. 2, Eastman Kodak Correspondence, 1957–1969, HBS.
64. Ibid.
65. "Resume of Meeting June 27, 1969, with Eastman Kodak at the Plaza," July 1, 1969, Polaroid Corporation Legal and Patent Records, Box II.250, f. 3, Kodak Meetings, 1967–1969, HBS.
66. Trial Testimony of William McCune, PvKdam, May 3, 1989, 300–302, 307.
67. "Resume of Meeting June 27, 1969."
68. Ibid.
69. Louis Eilers, Letter to Dr. Edwin H. Land, June 30, 1969, Polaroid Corporation Legal and Patent Records, Box II.251, f. 4, Kodak Agreement, 1966–1969, HBS.
70. Memorandum, "Polaroid-Kodak Meeting, September 11, 1969," 1–2, Polaroid Corporation Legal and Patent Records, Box II.250, f. 3, Kodak Meetings 1967–1969, HBS.
71. Edwin H. Land, Draft letter to L.K. Eilers, September 11, 1969, Polaroid Corporation Legal and Patent Records, Box II.251, f. 4, Kodak Agreement, 1966–1969, HBS.
72. Memorandum, Polaroid-Kodak Meeting, September 11, 1969, 3.

73. Memorandum, "Historical Context."

74. Siekman, "Kodak and Polaroid," 118.

75. Polaroid Corporation, Press Release, December 10, 1969, Polaroid Corporation Legal and Patent Records, Box II.250, f. 6, Kodak Agreement, 1968–1970, HBS.

76. McCune Oral History, 6.

77. Petersen Deposition Testimony, 271–76.

78. Agreement between Polaroid Corporation and Eastman Kodak Company, as of December 10, 1969, Article IV(D), PvK, Trial Exhibit PT-89, 9.

79. Polaroid Press Release, December 10, 1969.

80. "Comment," *British Journal of Photography*, January 30, 1970, Polaroid Corporation Legal and Patent Records, Box II.227, f. 3, Kodak Meetings, 1967–1969, HBS.

81. Polaroid Press Release, December 10, 1969.

82. Testimony of Robert Duncan, Trial Transcript, PvK, October 30, 1981, 1340–41.

83. Ibid., 1344.

84. C.R. Kalina, memorandum to R.C. Duncan, Notes on the Staff Meeting of August 26, August 29, 1969, PvK, Trial Exhibit PT-141; Trial Testimony of Robert Duncan, October 30, 1981, 1344–47.

85. Polaroid Corporation, Manufacturing Drawings, PvK, Trial Exhibit PT-142; Trial Testimony of Robert Duncan, October 30, 1981, 1344–47, 1354–56.

86. Robert C. Duncan, letter to Edwin H. Land, September 12, 1969, PvK, Trial Exhibit PT-143.

87. Plaintiff Polaroid Corporation's Brief after Trial, Land U.S. Patent 3,753,392, PvK, 2–3; Trial Testimony of Robert Duncan, PvK, October 30, 1981, 1356–57.

88. Trial Testimony of Robert Duncan, PvK, October 30, 1981, 1357–58.

89. Duncan, letter, September 12, 1969.

90. Cordtz, "How Polaroid Bet Its Future," 144.

91. McElheny, *Insisting on the Impossible*, 362.

92. Trial Testimony of Robert Duncan, PvK, October 30, 1981, 1365.

93. Plaintiff Polaroid Corporation's Brief after Trial, Land U.S. Patent 3,753,392, PvK, 3–4; Trial Testimony of Robert Duncan, October 30, 1981, 1363.

94. William K. Kerr, memorandum for file re: Conference with Dr. Land on Wednesday, June 27, 1979, June 28, 1979, F&N.

95. Plaintiff Polaroid Corporation's Brief after Trial, Land U.S. Patent 3,753,392, PvK, 3–4.

96. Carley, "Polaroid Seen Wary."

97. McElheny, *Insisting on the Impossible*, 366, 375.

98. Land, "Absolute One-Step Photography," *The Photographic Journal*, 2.

99. William Plummer, "The SX-70 Camera: The Optics," *Optics & Photonics News*, October 1994, 44.

100. "Transcript of SPSE Session."

101. Plummer, "The SX-70 Camera: The Optics," 44.

102. Cordtz, "How Polaroid Bet Its Future," 144.

103. Ibid.

104. J.J. Mohr and R. Wareham, memorandum to R.C. Duncan, Camera Description, March 16, 1970, PvK, Trial Exhibit PT-146; Trial Testimony of Robert Duncan, PvK, October 30, 1981, 1370-71.

105. Plaintiff Polaroid Corporation's Brief after Trial, Blinow and Leduc U.S. Patent 3,709,122, PvK, 4–5; *Polaroid*, 641 F. Supp. at 680–81.
106. Trial Testimony of Robert Duncan, PvK, 1376–79.
107. Polaroid Corporation, press release, May 10, 1972, F&N.
108. Land Video Tribute.
109. Plummer, "The SX-70 Camera: The Optics," 48.
110. Edwin H. Land, Letter from the Chairman of the Board, Polaroid Corporation Annual Report for 1970, in *Selected Papers on Industry*.

Chapter 8

1. Cordtz, "How Polaroid Bet Its Future," 83.
2. Blout, "Light and Life."
3. Plaintiff Polaroid Corporation's Brief after Trial, Wareham and Paglia U.S. Patent 3,810,211, PvK, 3–4.
4. Peter C. Wensberg, *Land's Polaroid* (Houghton Mifflin Company, 1987), 169–70.
5. Cordtz, "How Polaroid Bet Its Future," 87.
6. "Dr. Land Redesigns His Camera Company."
7. Cordtz, "How Polaroid Bet Its Future," 142.
8. Subrata Chakravarty, "An Interview with Dr. Edwin Land," *Forbes*, June 1, 1975.
9. McElheny, *Insisting on the Impossible*, 387.
10. David Brand and Liz Roman Gallese, "Company Proves Adept at Ballyhoo in Putting New Camera on Market," *The Wall Street Journal*, October 23, 1972.
11. McElheny, *Insisting on the Impossible*, 375.
12. Liz Roman Gallese, "Polaroid Holders Get Instant Spectaculars at Annual Meetings," *Wall Street Journal*, April 25, 1978.
13. Robert Reinhold, "Dr. Land Introduces 2 Camera Models," *New York Times*, April 28, 1971.
14. Brand and Gallese, "Company Proves Adept."
15. H.C. Barrett, memorandum to Herbert Schwartz, December 8, 1975, reporting recollection of Gerald Dicker, F&N.
16. Donald White, "It's Annual Meeting Time. . .," *Boston Globe*, April 28, 1971.
17. Reinhold, "Dr. Land Introduces 2 Camera Models."
18. White, "It's Annual Meeting Time."
19. Reinhold, "Dr. Land Introduces 2 Camera Models."
20. Polaroid Corporation, *The Polaroid Newsletter*, May 17, 1971, F&N.
21. Polaroid Corporation, Transcript of Polaroid Shareholders Meeting of April 27, 1971, F&N.
22. Nan Chequer, interview by author, January 30, 2009, Riverside, Conn.; Wensberg, *Land's Polaroid*, 175–76.
23. Brand and Gallese, "Company Proves Adept."
24. Frank Corrigan, "Kodak vs. Polaroid: Two Winners?," *Newsday*, July 4, 1971.
25. Ibid.
26. "Dr. Land Redesigns His Camera Company."
27. Plaintiff Polaroid Corporation's General Brief after Trial, PvK, 5; Eastman Kodak, Petitioner v. Polaroid Corporation, Respondent, United States Supreme Court, Petition for Writ of Certiorari, Respondent's Brief in Opposition, 4, F&N.

28. McElheny, "In the Beginning Was Dr. Land."

29. W.M. Bush, memorandum to L.J. Thomas, July 8, 1970, PvK, Trial Exhibit PT-269(B), tab 12, F&N.

30. Corrigan, "Kodak vs. Polaroid: Two Winners?"

31. Ibid.

32. Gene Smith, "Eastman Shows Low-Light Color Film; Instant Pictures Promised in Future," *New York Times*, April 28, 1971.

33. Testimony of Walter Fallon, Trial Transcript, PvKdam, vol. 61, August 28, 1989, 7669.

34. Ibid., 7640–41, 7644–47.

35. Ibid., 7724–25.

36. Trial Testimony of Walter Fallon, PvKdam, vol. 60, August 25, 1989, 7650–52; Trial Testimony of Walter Fallon, PvKdam, vol. 61, August 28, 1989, 7725–26.

37. Smith, "Eastman Shows Low-Light Color Film."

38. Trial Testimony of Walter Fallon, PvKdam, vol. 60, August 25, 1989, 7601–2, 7667–69.

39. Smith, "Eastman Shows Low-Light Color Film."

40. Polaroid Corporation, Transcript of Polaroid Shareholders Meeting of April 27, 1971, F&N, Litigation Files.

41. Trial Testimony of Walter Fallon, PvKdam, vol. 61, August 28, 1989, 7677.

42. Plaintiff Polaroid Corporation's Brief after Trial, Wareham and Paglia U.S. Patent 3,810,211, PvK, 3–4.

43. *Polaroid*, 641 F.Supp. 868–69.

44. Goldman interview.

45. W.J. McCune, Memorandum to Sheldon Buckler, March 9, 1972, Polaroid Corporation Legal and Patent Records, Box II.227, f. 6, SX-70, 1972–1978, HBS.

46. McCune Oral History, 7; Cordtz, "How Polaroid Bet Its Future," 142.

47. "Breast Pocket Polaroid," *Time*, May 8, 1972.

48. "Photography Innovator, Edwin Herbert Land," *New York Times*, April 26, 1972.

49. Donald White, "Polaroid Allows Peek at Its 'Mini,'" *Boston Globe*, April 26, 1972; Land Video Tribute.

50. "Polaroid-Kodak," Case 376-266 (rev. March 1984), Harvard Business School, 5.

51. McElheny, *Insisting on the Impossible*, 383.

52. "Breast Pocket Polaroid."

53. White, "Polaroid Allows Peek."

54. "Breast Pocket Polaroid."

55. McElheny, *Insisting on the Impossible*, 383.

56. Robert Reinhold, "Land Achieves His Dream with New Polaroid SX-70," *New York Times*, October 30, 1972.

57. "Polaroid's Big Gamble on Small Cameras."

58. "Dr. Land's Latest Bit of Magic," *Life*, October 27, 1927.

59. "Dr. Land's Latest Fantasy," Businessmen in the News, *Fortune*, June 1972, 31.

60. Ibid.

61. Gallese, "I Am a Camera," 48.

62. Herbert Koshetz, "Shareholders Watch 5 Pictures Taken in 10 Seconds," *New York Times*, April 26, 1972.

63. Brand and Gallese, "Company Proves Adept."
64. White, "Polaroid Allows Peek."
65. Wensberg, *Land's Polaroid*, 179.
66. "Breast Pocket Polaroid."
67. Rob Walker, "Photo Finish," *New York Times Magazine*, March 16, 2008.
68. "Dr. Land's Latest Fantasy."
69. "Transcript of SPSE Session."
70. Ibid.
71. Ibid.
72. Ibid.
73. Ibid.
74. Ibid.
75. D.A. Delwiche, memorandum, Eastman Kodak, August 11, 1972, PvK, Trial Exhibit PT 269(b), tab 27.
76. Memorandum, Eastman Kodak, Early P-130 History as Recalled by D.A. Delwiche, PvK, Trial Exhibit PT 275(B), tabs 1–3.
77. G.E. Kindig, P-130 General Review, September 20, 1971, PvK, Trial Exhibit PT 275(B), tab 30.
78. S.H. Smith Jr., P-130 Committee Meeting Notes, Covering April 7, 16, 23 and 30, 1971, PvK, Trial Exhibit PT 275(B), tab 27; memorandum, Eastman Kodak, P-130 Committee Meeting Notes, May 21, 1971, PvK, Trial Exhibit PT 275(B), tab 28; D.A. Delwiche, P-130 Mini-Task Force Progress Report, May 14, 1971, PvK, Trial Exhibit PT 275(B), tab 29.
79. Kindig, P-130 General Review.
80. Plaintiff Polaroid Corporation's General Brief after Trial, PvK, 6, F&N.
81. Early P-130 History as Recalled by D.A. Delwiche.
82. Deposition Testimony of Albert Sieg, Plaintiff Polaroid Corporation's General Brief after Trial, PvK, 6, F&N.
83. G.E. Kindig, KAD Status Report for P-130, February 11, 1972, PvK, Trial Exhibit PT 275(B), tab 32.
84. Ibid.
85. Albert Sieg, P-130 Committee Meeting Notes, March 17, 1972, PvK, Trial Exhibit PT 274(B), tab 9.
86. Ibid.
87. G.E. Kindig, KAD Status Report for P-130, April 14, 1972, PvK, Trial Exhibit PT 275(B), tab 34.
88. G.E. Kindig Deposition Testimony, Trial Exhibit, PvK, PT 275(B), tab 35; Testimony of Albert Sieg, Trial Transcript, PvK, December 10, 1981, 4090–91.
89. Deposition Testimony of Albert Sieg, Plaintiff Polaroid Corporation's General Brief after Trial, PvK, 6.
90. G.E. Kindig, Aladdin Camera—as Described in Polaroid Annual Meeting April 25, 1972, May 4, 1972, PvK, Trial Exhibit PT 275(B), tab 36.
91. G.E. Kindig Deposition Testimony.
92. Testimony of Albert Sieg, PvK, December 10, 1981, 4095–107.
93. D.A. Delwiche, memorandum, August 11, 1972, PvK, Trial Exhibit PT 269(b), tab 27.

94. Robert Mahone, Observations about P-130, May 8, 1972, PvK, Trial Exhibit PT 269(B), tab 28.

95. Ibid.

96. Albert Sieg, P-130 Committee Meeting, May 11, 1972, PvK, Trial Exhibit PT 274(B), tab 13.

97. Albert Sieg, P-130 Committee Meeting, May 19, 1972, PvK, Trial Exhibit PT 269(B), tab 29.

98. Howard Rudnitsky, "Why Kodak Will Buck the Recession," *Forbes*, September 3, 1979; Trial Testimony of Walter Fallon, PvKdam, vol. 60, 7594–95.

99. William McCune, Letter to Walter Fallon, May 30, 1972, Polaroid Corporation Legal and Patent Records, Box II.200, f. 3, Kodak, 1972, HBS.

100. David Brand, "Kodak Dominates Field, Rolls Up Huge Profits, but Antagonizes Many," *Wall Street Journal*, November 15, 1972, Polaroid Corporation Legal and Patent Records, Box II.203, f. 8, Articles of Interest—Kodak, 1976–1981, HBS.

101. Rudnitsky, "Why Kodak Will Buck the Recession."

102. Brand, "Kodak Dominates Field."

103. Ibid.

104. Ibid.

105. Bro Uttal, "Eastman Kodak's Orderly Two-Front War," *Fortune*, September 1976.

106. Quillen interview; Alecia Swasy, *Changing Focus* (Random House, 1997), 20–21.

107. Ibid.

108. Warshofsky, *The Patent Wars*, 69.

109. Trial Testimony of Walter Fallon, PvKdam, vol. 60, 7603.

110. Albert Sieg, P-130 Committee Meeting, May 31, 1972, PvK, Trial Exhibit, tab 26.

111. Brand, "Kodak Dominates Field."

112. Leonard Sloane, "Kodak Goal: Beat Polaroid," *New York Times*, June 11, 1972.

113. Ibid.

114. Ibid.

115. Douglas W. Cray, "Polaroid Pocket Camera to Miss Christmas Sales," *New York Times*, September 4, 1972.

116. Brand and Gallese, "Company Proves Adept."

117. Reinhold, "Land Achieves His Dream with New Polaroid SX-70."

118. McElheny, *Insisting on the Impossible*, 341.

119. Brand and Gallese, "Company Proves Adept."

120. "Here Come Those Great New Cameras," *Time*, June 26, 1972.

121. "Dr. Land's Latest Bit of Magic."

122. "Polaroid's Big Gamble on Small Cameras"; "A Genius and His Magic Camera," *Life*, October 27, 1972.

123. A.R. Andrew, Project 129, November 6, 1972, PvK, Trial Exhibit PT 269(B), tab 30.

124. Uttal, "Eastman Kodak's Orderly Two-Front War."

125. Trial Testimony of Leo Thomas, PvK, December 10, 1981, 3997–4001; Trial Testimony of Colby Chandler, PvKdam, vol. 53, August 16, 1989, 6753–57.

126. Andrew, Project 129.

127. William D. Smith, "Kodak Drops Plans to Market Films for Polaroid Use," *New York Times*, November 18, 1972.

128. David Brand, "Kodak Won't Make Polaroid Instant Film, Plans to Sell Rival Film, Camera Instead," *Wall Street Journal*, November 20, 1972, Polaroid Corporation Legal and Patent Records, Box II.203, f. 8, Articles of Interest—Kodak, 1976–1981, HBS.

129. Ibid.

130. Trial Testimony of Colby Chandler, PvKdam, vol. 51, August 14, 1989, 6495–97; Quillen, interview.

131. Uttal, "Eastman Kodak's Orderly Two-Front War."

132. Brand, "Kodak Won't Make Polaroid Instant Film."

133. Ibid.; Uttal, "Eastman Kodak's Orderly Two-Front War."

134. J.M. Streb, Marketing Comments on P-130, January 18, 1973, PvK, Trial Exhibit PT 269(B), tab 32; Plaintiff Polaroid Corporation's General Brief after Trial, PvK, 8; Eastman Kodak, Petitioner v. Polaroid Corporation, Respondent, Supreme Court, Petition for Writ of Certiorari, Respondent's Brief in Opposition, 3, 5.

135. Memorandum, P-130 Photographic Committee Meeting, October 7, 1974, PvK, Trial Exhibit PT 269(B), tab 31.

136. Streb, Marketing Comments on P-130.

137. Ibid.

138. Ibid.

139. Isadore Barmash, "Goals at Kodak and at Polaroid," *New York Times*, April 25, 1973.

140. *Polaroid*, 641 F. Supp. 831.

141. Robert A. Mahone, "Kodak Park Development Committee Meeting," September 21, 1973, PvK, Trial Exhibit PT 269(B), tab 33.

142. Uttal, "Eastman Kodak's Orderly Two-Front War"; Subrata Chakravarty and Amy Feldman, "How Kodak Paid the Price for Losing Its Focus," *BRW International Edition* 3, no. 10, November 1993, 42.

143. Mahone, "Kodak Park Development Committee Meeting."

Chapter 9

1. Trial Testimony of William McCune, PvKdam, vol. 2, May 2, 1989, 2–23.
2. "Polaroid's Latest Gamble," *Newsweek*, November 5, 1973, Polaroid Corporation Administrative Records, Box I.234, f. 20, HBS.
3. Wensberg, *Land's Polaroid*, 201.
4. Robert Metz, "Polaroid Plummets Despite Explanation," *New York Times*, October 20, 1973.
5. Charles J. Elia, "Heard on the Street," *Wall Street Journal*, December 3, 1973.
6. Cordtz, "How Polaroid Bet Its Future," 83.
7. Chakravarty, "An Interview with Dr. Edwin Land."
8. "Dr. Land Redesigns His Camera Company," *Business Week*, April 15, 1972.
9. Ibid.
10. Cordtz, "How Polaroid Bet Its Future," 85.
11. Ibid., 83.
12. "Polaroid Says Customers Love SX-70 Cameras, Demand Tops Supply," *Wall Street Journal*, January 8, 1974.
13. Trial Testimony of McCune, PvKdam, vol. 2, 276–77; Victor K. McElheny, *Insisting on the Impossible*, 353, 405.

14. Polaroid Corporation v. Eastman Kodak Company, 16 U.S.P.Q.2d 1481, 1505 (D. Mass. 1990); Nan Chequer, interview by author, January 30, 2009, Riverside, Conn.; Trial Testimony of McCune, PvKdam, vol. 3, May 3, 1989, 319–20.

15. Richard Martin, "Polaroid Announces Low-Priced Camera, Details SX-70 Woes to Securities Group," *Wall Street Journal*, July 22, 1974.

16. Ernest Holsendolph, "Polaroid Will Focus on a New Inexpensive Camera," *New York Times*, July 20, 1974.

17. "Polaroid's Latest Gamble."

18. James J. Nagle, "Berkey Unveils Instant Camera," *New York Times*, October 13, 1972; "Polaroid, Berkey Enter Camera Licensing Pact," *Wall Street Journal*, December 17, 1974.

19. Robert Lenzner, "The Instant World of Polaroid: A Preview," *Boston Globe*, April 22, 1975.

20. Goldstein, *Intellectual Property*, 9.

21. Trial Testimony of Colby Chandler, PvKdam, vol. 54, August 17, 1989, 6919–20; Plaintiff Polaroid Corporation's Post-Trial Submissions, PvK, facts IX–4.

22. Chequer, interview.

23. Donald L. Brown, Letter to William Rymer, January 31, 1963, Polaroid Corporation Legal and Patent Records, Box II.103, f. 5, Q-R, 1960–1965, HBS.

24. *History of the First Century of the Firm of Fish & Neave, 1878–1978* (Fish & Neave publication), author's collection; John Nathan, "Fish & Neave, Leaders in the Laws of Ideas," *The Newcomen Society*, April 1997, 16.

25. Fred Howard, *Wilbur and Orville* (Dover Publications, 1987), 380–81.

26. Tom Crouch, *The Bishop's Boys* (Norton, 1989), 447–48.

27. "Fish & Neave," Funding Universe, www.fundinguniverse.com; Nathan, "Fish & Neave," 20.

28. "Fish & Neave" (promotional booklet, Fish & Neave), author's collection.

29. Jill Abramson, "Princes of the Patent Bar," *American Lawyer*, April 1983.

30. Learned Hand, "In Memory of Charles Neave," *Spirit of Liberty, Papers and Addresses of Learned Hand* (Alfred A. Knopf, 3rd ed. 1974), 128.

31. Donald L. Brown, Letter to William Rymer, September 25, 1963, Polaroid Corporation Legal and Patent Records, Series II, Donald L. Brown Papers, Box 103, f. 5, Q-R, 1960–1965, HBS.

32. Fish & Neave, commitment memorandum, December 4, 1973, F&N.

33. "Kodak's Instant-Film Progress Report Fells Polaroid Stock," *Boston Globe*, March 22, 1974.

34. David Brand, "Eastman Kodak Says It Has 'Feasible' Film for Instant Photos," *Wall Street Journal*, March 21, 1974.

35. Vartanig Vartan, "Polaroid Falls on Wall St.," *New York Times*, March 22, 1974.

36. *Eastman Kodak Company 1975 Annual Report*, F&N.

37. Don Langer, "'Instant' Pictures: Polaroid vs. Kodak," *New York Post*, March 28, 1974.

38. Rynn Berry, letter to William Rymer, April 2, 1974, F&N.

39. "Kodak's Latest Hint," *Business Week*, March 30, 1974.

40. Vartanig Vartan, "Analysts Differ over Polaroid," *New York Times*, April 18, 1974.

41. Susan Trausch, "Polaroid's Mini-Giant," *Boston Globe*, April 24, 1974.

42. Ibid.
43. Vartan, "Analysts Differ over Polaroid."
44. Trial Testimony of McCune, PvKdam, vol. 2, 256–57.
45. Robert Metz, "Pictures Awry for Polaroid," *New York Times*, July 3, 1974.
46. Martin, "Polaroid Announces Low-Priced Camera."
47. Ernest Holsendolph, "Polaroid's Profit Down," *Wall Street Journal*, July 17, 1974.
48. Martin, "Polaroid Announces Low-Priced Camera."
49. T.H. Wyman, "Memorandum to Management Executive Committee," May 13, 1974, Polaroid Corporation Legal and Patent Records, Box II.216, f. 1, Management Executive Committee Meetings 1974, HBS.
50. Dan Dorfman, "Is Polaroid Through on Wall Street?," *New York Magazine*, August 26, 1974.
51. "Up and Down Wall Street," *Barron's*, January 14, 1974.
52. Ibid.
53. Rynn Berry, letter to William Rymer, January 15, 1974, F&N.
54. Herbert Schwartz, interview with author, December 8, 2008, Riverside, Conn.
55. Rynn Berry, handwritten memorandum, July 24, 1974, F&N.
56. Obituary, "William K. Kerr," *New York Times*, November 28, 2007.
57. William Rymer, letter to William Kerr, July 25, 1974, F&N.
58. William K. Kerr, letter to Robert Peck, November 19, 1974, F&N.
59. Herbert Schwartz, interview with author, January 23, 2009, Riverside, Conn.; Schwartz, interview, December 8, 2008.
60. Cameron Foote, Memorandum to Peter Wensberg, November 18, 1974, Polaroid Corporation Legal and Patent Records, Box II.200, f. 3, Kodak, 1972, HBS.
61. Trial Testimony of McCune, PvKdam, vol. 3, 312–14; Ronald K. Fierstein, memorandum for file, 1973–74 Polaroid-Kodak Discussions, February 7, 1980, F&N.
62. Trial Testimony of Walter Fallon, PvKdam, vol. 61, August 28, 1989, 7743–44.
63. Ibid., 7744; Fierstein, 1973–74 Polaroid-Kodak Discussions; McCune Oral History, 7.
64. Trial Testimony of Walter Fallon, PvKdam, vol. 61, 7744; Quillen interview; McCune Oral History, 7.
65. Fierstein, 1973–74 Polaroid-Kodak Discussions.
66. Trial Testimony of McCune, PvKdam, vol. 3, 313–14.
67. Quillen, interview.
68. Swasy, *Changing Focus*, 17.
69. Fierstein, 1973–74 Polaroid-Kodak Discussions.
70. Quillen, interview.
71. William K. Kerr, memorandum to file, Agenda, November 15, 1974, F&N.
72. Schwartz, interview, January 23, 2009.
73. William K. Kerr, letter to Robert Peck, November 19, 1974, F&N.
74. William K. Kerr, letter to Robert Peck, December 5, 1974, F&N.
75. "Mr. Ford's Economics," *New York Times*, August 20, 1974.
76. Peter N. Carroll, *It Seemed Like Nothing Happened, America in the 1970s* (Rutgers University Press, 1982).
77. Charles J. Elia, "Polaroid's Loyal Following Is Seen Facing Further Testing," *Wall Street Journal*, January 27, 1975.

78. Cordtz, "How Polaroid Bet Its Future," 147.

79. Siekman, "Kodak and Polaroid," 124.

80. Richard Martin, "Polaroid's McCune Tackles Big Problems as Its New President," *Wall Street Journal*, April 25, 1975, Polaroid Corporation Legal and Patent Records, Box II.217, f. 3, Wall Street Journal Interview, January 1975, HBS.

81. Polaroid Corporation, Press Release, January 22, 1975, Polaroid Corporation Legal and Patent Records, Box II.230, f. 6, William J. McCune Named President, 1975, HBS.

82. Martin, "Polaroid's McCune Tackles Big Problems."

83. Peter T. Kilborn, "Polaroid President Appointed by Land," *New York Times*, January 23, 1975.

84. Polaroid Corporation, Press Release, January 24, 1975, Polaroid Corporation Legal and Patent Records, Box II.230, f. 6, William J. McCune Named President, 1975, HBS.

85. Douglas Martin, "Thomas Wyman, Ex-CBS Chief, Dies at 73," *New York Times*, January 10, 2003; Martin, "Polaroid's McCune Tackles Big Problems."

86. Ronald Rosenberg, "The Realist behind Genius Edwin Land Retires Today," *Boston Globe*, May 7, 1991; Lenzner, "The Instant World of Polaroid."

87. Rosenberg, "The Realist behind Genius Edwin Land Retires Today."

88. Susan Trausch, "McCune In, Wyman Out in Polaroid Corp. Shift," *Boston Globe*, January 23, 1975.

89. Polaroid, Press Release, January 22, 1975.

90. Ansel Adams, Letter to William McCune, January 24, 1975, Polaroid Corporation Legal and Patent Records, Box II.29, f. 3, William J. McCune Congratulatory Letters for Polaroid president 1975, HBS.

91. Chakravarty, "An Interview with Dr. Edwin Land."

92. Lenzner, "The Instant World of Polaroid."

93. Ibid.

94. Martin, "Polaroid's McCune Tackles Big Problems."

95. Polaroid Corporation, Weekly Company Comments, January 28, 1975, author's collection.

96. Martin, "Polaroid's McCune Tackles Big Problems."

97. Mitchell Lynch, "New Polaroid Chief Is Tough Fellow with Tough Task," *Wall Street Journal*, March 10, 1980.

98. Martin, "Polaroid's McCune Tackles Big Problems."

99. W.J. McCune, "Memorandum to Company Members," March 28, 1975, Polaroid Corporation Legal and Patent Records, Box II.212, f. 12, Management Executive Committee Meetings Minutes, 1975, HBS.

100. "Kodak Begins Work on 'Major Program' for a New Camera," *Wall Street Journal*, March 19, 1975.

101. Robert Metz, "Market Place," *New York Times*, March 19, 1975.

102. "Kodak Unveils Its New Ektaprint Copier, Instant-Film Pictures at Annual Meeting," *Wall Street Journal*, April 30, 1975.

103. "Berkey Photo Develops Camera That Will Use Polaroid SX-70 Film," *Wall Street Journal*, March 20, 1975.

104. "Minutes, Polaroid Management Executive Committee, February 13 and August 23, 1973," Polaroid Corporation Legal and Patent Records, Box II.215, f. 14, Management Executive Committee Meetings Minutes, 1973, HBS; "Minutes, Polaroid Management

Executive Committee, April 16, 1974," Polaroid Corporation Legal and Patent Records, Box II.214, f. 1, Management Executive Committee Meetings Minutes, 1974, HBS.

105. "Polaroid, Berkey Enter Camera Licensing Pact," *Wall Street Journal*, December 17, 1974.

106. "Message for Bill McCune," March 17, 1975, Polaroid Corporation Legal and Patent Records, Box II.212, f. 11, Berkey Photo, HBS.

107. "Message for Bill McCune, March 20, 1975," Polaroid Corporation Legal and Patent Records, Box II.212, f. 11, Berkey Photo, HBS.

108. "Berkey Photo Develops Camera."

109. Ibid.

110. Schwartz, interview, December 8, 2008.

111. "The 100 Most Influential Lawyers in America," *National Law Journal*, April 28, 1997.

112. Mark Kleiman, memorandum to Edwin H. Land, January 19, 1976, Polaroid Corporation Legal and Patent Records, circa 1905–1995, Polaroid v. Kodak, circa 1950–1991, D. Administrative, 1957–1985, Box 218, f. 1, Polaroid v. Kodak 1975–1976, HBS; Mark Kleiman, telephone interview by author, March 22, 2011.

113. Schwartz, interview, January 22, 2009.

114. Schwartz, interview, December 8, 2008.

115. Schwartz, interview, January 22, 2009.

116. Complaint, Polaroid Corporation v. Berkey Photo, Inc., Civil Action 75-179 (United States District Court, District of Delaware), par. 16.

117. Herbert Schwartz, memorandum to William Kerr, Considerations re: Which Polaroid Patents Should Be Asserted against Berkey, May 9, 1975, F&N.

118. William K. Kerr, letter and enclosure to Robert Peck, May 12, 1975, F&N.

119. William K. Kerr, memorandum to Herbert Schwartz, May 22, 1975, F&N.

120. Ibid.

121. William K. Kerr, letter and enclosure to Robert Peck, June 10, 1975, F&N.

122. Herbert Schwartz, memorandum for file, June 27, 1975, F&N.

123. "Polaroid Sues Berkey on Patents," *New York Times*, July 4, 1975.

124. "'No Valid Claims' Found by Berkey in Polaroid Suit," *New York Times*, July 8, 1975.

Chapter 10

1. "Polaroid Says Earnings in 2nd Quarter Were Better Than Expected," *Wall Street Journal*, July 9, 1975.

2. "New Polaroid Camera Due," *New York Times*, December 27, 1975.

3. Norman A. Berg, "Polaroid-Kodak," Harvard Business School Case 376-266 (rev. March 1984), 7.

4. "Polaroid to Market Fourth SX-70 Camera, Carrying Lowest Price," *Wall Street Journal*, December 29, 1975.

5. "Pronto to the Rescue," *Time*, February 2, 1976.

6. Ibid.

7. Bob Donath, "Polaroid Trots Out Lower-Price Pronto," *Advertising Age*, January 19, 1976.

8. William D. Smith, "Polaroid's Economy SX-70 Is Unveiled," *New York Times*, January 14, 1976.

9. "Instant Battle: Kodak v. Polaroid," *Time*, April 26, 1976, 61.

10. "Polaroid Profit Jumped by 166% in 4th Quarter," *Wall Street Journal*, February 20, 1976.

11. "Kodak Product Development," *Wall Street Journal*, October 23, 1975.

12. Donath, "Polaroid Trots Out Lower-Price Pronto."

13. Bob Donath, "New Polaroid SX-70 Ready; Kodak Entry Near," *Advertising Age*, January 5, 1976.

14. Anthony Broy, "Kodak's Challenge to Polaroid," *Financial World*, February 15, 1976.

15. Ibid.

16. William M. Carley, "Polaroid Seen Wary, Worried as It Girds for Kodak Arrival in Instant-Photo Field," *Wall Street Journal*, April 16, 1976.

17. Donath, "New Polaroid SX-70 Ready."

18. Ulysses A. Yannas, *A Report on the Photography Industry, the Instant Photography Market* (William D. Witter, Inc., March 24, 1975), Polaroid Corporation Legal and Patent Records, Box II.203, f. 9, Eastman Kodak vs. Polaroid, circa 1980, HBS.

19. Memorandum from Herbert F. Schwartz to W.K. Kerr, February 13, 1976, F&N.

20. "Kodak Plans to Make Small Electronic Parts Used in Its Products," *Wall Street Journal*, February 24, 1976; "Last Minute News," *Advertising Age*, March 8, 1976.

21. "Kodak Expects Good Year in 1976, Cites New Instant Camera and Office Copier," *Wall Street Journal*, March 17, 1976.

22. "Kodak Says Its Instant Color Photography Ready for Market," *Boston Herald American*, March 19, 1976.

23. Broy, "Kodak's Challenge to Polaroid."

24. Charles J. Elia, "Kodak Report Titillates Analysts with Photo from New Instant Camera; Stock Price Leaps," *Wall Street Journal*, March 24, 1976.

25. "Kodak Says Its Instant Color Photography Ready for Market."

26. Memorandum from Herbert F. Schwartz to W.K. Kerr, March 22, 1976, F&N.

27. Memorandum from Herbert F. Schwartz to W.K. Kerr, March 23, 1976, F&N.

28. Ibid.

29. Memorandum from William Kerr to Henry Zafian, October 9, 1979, F&N.

30. "Actions Speak Louder," *Fortune*, April 1976, Polaroid Corporation Legal and Patent Records, Box II.200, f. 3, Kodak 1972, HBS.

31. Ray Gniewek, "Kodak Will Unveil Instant System April 27," *Rochester Democrat and Chronicle*, March 30, 1976.

32. Letter from Robert F. Peck to William K. Kerr, March 26, 1976, F&N.

33. Herbert F. Schwartz, interview, January 23, 2009.

34. Memorandum from Kleiman to Edwin H. Land.

35. Uttal, "Eastman Kodak's Orderly Two-Front War."

36. Fierstein, "1973–74 Polaroid-Kodak Discussions."

37. Eastman Kodak Company, *1975 Annual Report*, 26–27.

38. John Rumsey, "Kodak Announces 'Instant' Camera, Film on Apr. 20," *Rochester Times-Union*, April 12, 1976, Polaroid Corporation Legal and Patent Records, Box II.230, f. 1, Kodak Instant Products Introduction, April 1976, HBS; "Kodak's Instant Camera to

Be Shown Next Week," *Wall Street Journal*, April 13, 1976; Victor K. McElheny, "Kodak to Unveil Instant Camera," *New York Times*, April 13, 1976.

39. Memorandum by Thomas G. Quinn to File, November 17, 1976, F&N.

40. "The Camera Boom," *Newsweek*, May 3, 1976.

41. Eastman Kodak Company, "Kodak Announces Instant Cameras, Instant Film for Self-Developing Prints," press release, April 20, 1976, F&N.

42. John Dancer, "It's Here! What Now?," *Photo Marketing*, May 1976, 3.

43. Robert Lenzner, "New Cameras Available by Late June," *Boston Globe*, April 21, 1976.

44. Victor K. McElheny, "Eastman Kodak Demonstrates System for Instant Pictures," *New York Times*, April 21, 1976.

45. Uttal, "Eastman Kodak's Orderly Two-Front War," 132.

46. Eastman Kodak Company, "Kodak President Cites Extensive Testing of New Instant Cameras and Instant Film," press release, April 20, 1976, F&N.

47. Eastman Kodak Company, "Kodak Announces Instant Cameras."

48. Ibid.

49. Polaroid Corporation Memorandum, "Kodak Report, 6-1, 12:45 pm, 4/20," Polaroid Corporation Legal and Patent Records, Box II.230, f. 1, Kodak Instant Products Introduction, April 1976, HBS.

50. Memorandum by Quinn to File.

51. Trial Testimony of William McCune, PvKdam, vol. 2, May 2, 1989, 296–97.

52. Polaroid Corporation, Management Executive Committee Meeting Agenda, April 20, 1976, Polaroid Corporation Legal and Patent Records, Box II.212, f. 1, Management Executive Committee, 1976, HBS; Herbert F. Schwartz, interview, December 8, 2008.

53. Schwartz, interview, December 8, 2008.

54. Trial Testimony of William McCune, PvKdam, vol. 2, May 2, 1989, 296–97.

55. Polaroid Corporation, press release, April 20, 1976, Polaroid Corporation Legal and Patent Records, Box II.216, f. 2, Press Releases, 1976, HBS; McElheny, "Eastman Kodak Demonstrates System."

56. Daniel Corcoran, "Kodak enters the instant photography market with its EKs, Polaroid officials breathe easier," *Boston Globe*, April 21, 1976, Polaroid Corporation Legal and Patent Records, Box II.230, f. 1, Kodak Instant Products Introduction, April 1976, HBS.

57. Uttal, "Eastman Kodak's Orderly Two-Front War," 132; Trial Testimony of Peter Wensberg, PvKdam, vol. 7, May 9, 1989, 739.

58. Mark Kleiman, telephone interview by author, March 22, 2011.

59. Polaroid Corporation, "Kodak Report 8-1, Impressions," Polaroid Corporation Legal and Patent Records, Box II.230, f. 1, Kodak Instant Products Introduction, April 1976, HBS.

60. "Kodak Report, 6-1."

61. Polaroid Corporation, "Kodak Report, 9-1," Polaroid Corporation Legal and Patent Records, Box II.230, f. 1, Kodak Instant Products Introduction, April 1976, HBS.

62. William Carley, "Kodak Unveils Instant-Picture Cameras, Strategy for Challenging Polaroid's Line," *Wall Street Journal*, April 21, 1976.

63. Victor K. McElheny, "Kodak and Polaroid: Color Systems Differ," *New York Times*, April 23, 1975; Schwartz, interview, December 8, 2008.

64. "A Hard Tussle between Friends," *Time*, May 3, 1976; Linda Snyder Hayes, "What's Kodak Developing Now?," *Fortune*, March 23, 1981; Robert Metz, "Kodak Stock a Worry for Investors," *New York Times*, March 28, 1977.

65. *Photo Marketing Newsletter*, April 28, 1976, Polaroid Corporation Legal and Patent Records, Box II.230, f. 2, Polaroid/Kodak Lawsuit, 1976, HBS.

66. Charles Elia, "Bid by Kodak for a Share of Polaroid's Market Gains Respect as Analysts Focus on Its Entry," *Wall Street Journal*, April 22, 1976.

67. Lenzner, "New Cameras Available by Late June."

68. Gosselin, "How Polaroid Beat Kodak."

69. McElheny, *Insisting on the Impossible*, 448.

70. "Edwin Land on Polaroid—and the Kodak Patent Suit," *Boston Globe*, October 18, 1976.

71. "The Camera Boom."

72. "Edwin Land on Polaroid."

73. "Polaroid—Turning Away from Land's One Product Strategy," *Business Week*, March 2, 1981, 111.

74. "Edwin Land on Polaroid."

75. "Polaroid Sues Kodak," *Time*, May 10, 1976, 64.

76. Quillen, interview.

77. Quinn, memorandum to file.

78. Sanitary Refrigerator Co. v. Winters, 280 U.S. 30, 42 (1929), quoting Union Paper-Bag Machinery Company v. Murphy, 97 U.S. 120, 125 (1878).

79. Herbert F. Schwartz, *Polaroid Co. v. Eastman Kodak Co.: A Retrospective (1991)*, 4–5; Schwartz, interview, December 8, 2008.

80. Schwartz, interview, January 23, 2009.

81. Schwartz, interview, December 8, 2008.

82. Schwartz, interview, January 23, 2009; William K. Kerr, letter to Robert Peck, June 30, 1975, F&N.

83. Alfred E. Corrigan, letter with attachment to William K. Kerr, July 9, 1975, F&N.

84. "Polaroid Sues Kodak."

85. Complaint, PvK, F&N.

86. "Polaroid Sues Kodak."

87. "Photography: Dueling Cameras," *Newsweek*, May 10, 1976, 86.

88. "Polaroid Sues Kodak."

89. "Polaroid Suit Charges Eastman Kodak's Instant Camera, Film Infringe on Patents," *Wall Street Journal*, April 28, 1976.

90. "Polaroid-Kodak," HBS Case, 1–2.

91. "Polaroid Suit Charges Eastman Kodak's Instant Camera."

92. "Photography: Dueling Cameras."

93. "Polaroid Suit Charges Eastman Kodak's Instant Camera."

94. Ibid.

95. Victor K. McElheny, "Polaroid Is Suing Kodak, Charges Patent Violation," *New York Times*, April 28, 1976.

96. "Polaroid-Kodak," HBS Case, 8.

97. Gosselin, "How Polaroid Beat Kodak," 46.

Chapter 11

1. "Polaroid Suit Charges Eastman Kodak's Instant Camera."
2. Victor K. McElheny, "Polaroid Is Suing Kodak."
3. "Polaroid Suit Charges Eastman Kodak's Instant Camera."
4. Trial Testimony of Walter Fallon, PvKdam, vol. 60, August 25, 1989, 7654, 7758–59; ibid., vol. 61, August 28, 1989, 7741–42.
5. Trial Testimony of Francis T. Carr, PvKdam, vol. 55, August 18, 1989, 7017–18.
6. Jim Galbraith, memorandum to author, October 28, 2009.
7. Houston W. Kenyon Jr., *An Informal History, the First Half Century of the Kenyon Firm 1879–1933.*
8. Trial Testimony of Francis T. Carr, 6991.
9. "Dueling Cameras," *Newsweek*, May 10, 1976.
10. Trial Testimony of Colby Chandler, PvKdam, vol. 54, August 17, 1989, 6952.
11. "Up and Down Wall Street," *Barron's*, January 14, 1974.
12. "Polaroid, Kodak Feud over Patent Rights Girdles the Globe," *Wall Street Journal*, April 29, 1976.
13. "Polaroid Sues Kodak."
14. "Talking Point, Eastman Kodak and Polaroid," *Reuters*, May 12, 1976.
15. "Polaroid, Kodak Feud over Patent Rights."
16. Ibid.
17. Schwartz, interview, January 23, 2009.
18. George W. Petersen, letter to Edwin Land, April 4, 1975, Polaroid Corporation Legal and Patent Records, Box II.29, f. 4, Stanley Mervis Correspondence, 1971–93, HBS.
19. M. Wren Gabel, letter to William McCune, April 17, 1975, Polaroid Corporation Legal and Patent Records, Box II.29, f. 3, William J. McCune Congratulatory Letters for Polaroid President 1975, HBS.
20. William McCune, letter to M. Wren Gabel, April 30, 1977, Polaroid Corporation Legal and Patent Records, Box II.29, f. 3, William J. McCune Congratulatory Letters for Polaroid President 1975, HBS.
21. "Mr. Polaroid's Portrait," *London Sunday Times*, May 2, 1976, Polaroid Corporation Legal and Patent Records, Box II.230, f. 2, Polaroid/Kodak Lawsuit, 1976, HBS.
22. "Polaroid, Kodak Feud over Patent Rights."
23. R.D. Lorbach, letter to Edwin Land, April 30, 1976, F&N.
24. William McCune, letter to William K. Kerr, May 5, 1976, F&N; Lorbach, letter.
25. McCune Oral History, 7.
26. Answer and Counterclaims, PvK, par. 30, F&N.
27. Ibid., par. 10–12.
28. Ibid., Third Counterclaim, par. 35–42.
29. Quillen, interview.
30. Defendant's Memorandum in Opposition to Plaintiff's Motions to Dismiss Defendant's Amended Counterclaims and Affirmative Defense, PvK, August 31, 1976, 3–4.
31. "Kodak Claims Patents of Polaroid Are Invalid and Unenforceable," *Wall Street Journal*, May 19, 1976, 12.

32. "Kodak Asks Court to Void Patents Held by Polaroid," *New York Law Journal*, May 19, 1976, 1, F&N.

33. Brand, "Kodak Dominates Field."

34. Plaintiff's Motions to Dismiss Amended Counterclaims with Respect to Added Patents and Patent Misuse and to Strike the Affirmative Defense of Patent Misuse, PvK, June 28, 1976, F&N.

35. Reply to Portions of Defendant's Amended Counterclaims and Plaintiff's Counterclaim, PvK, June 28, 1976, F&N.

36. Plaintiff's Reply Memorandum in Support of Its Motions to Dismiss Amended Counterclaims with Respect to Added Patents and Patent Misuse and to Strike the Affirmative Defense of Patent Misuse, PvK, October 15, 1976, 4, F&N.

37. Ibid.

38. Schwartz, interview, January 23, 2009.

39. Robert Peck, letter with attachment to William K. Kerr, May 10, 1976, F&N.

40. "Polaroid May License SX-70 Film-Type Camera," *Wall Street Journal*, May 17, 1976.

41. William D. Smith, "I.B.M. Challenges Xerox in Copier Field," *New York Times*, April 22, 1970, 67, 72.

42. Ibid., 42.

43. Ibid.

44. Peter T. Kilborn, "Xerox Complaint Settled by F.T.C.," *New York Times*, November 16, 1974, 41.

45. Schwartz, interview, December 8, 2008.

46. Ibid.

47. Schwartz, interview, January 23, 2009.

48. Clipping, *Wall Street Journal*, August 6, 1976, F&N.

49. Neil Ulman, "Polaroid Patent Win in UK Court Arose from Technicalities," *Wall Street Journal*, October 7, 1976.

50. Victor K. McElheny, "Polaroid-Kodak Battle," *New York Times*, October 13, 1976.

51. Richard Lannamann, *Polaroid's Patent Infringement Suits against Eastman Kodak*, (Smith Barney, Harris Upham & Co., August 9, 1976), Polaroid Corporation Legal and Patent Records, Box II.218, f. 1, Polaroid vs. Kodak, 1975–1976, HBS.

52. Calvin D. Crary, Polaroid v. Eastman Kodak*, the British Injunction Proceeding: Window on the U.S. Case* (Bache Halsey Stuart, Litigation Service, September 8, 1976).

53. "Kodak Wins Appeal against British Ban in Polaroid Battle," *Wall Street Journal*, November 11, 1976.

54. Schwartz, interview, January 23, 2009.

55. Ibid.

Chapter 12

1. "Instant Kodak vs. Polaroid," *Modern Photography*, August 1976.

2. James Grant, "Instant Rivalry," *Barron's*, September 20, 1976, 23; "Fuji Instant Camera Is Claimed as Rival to Polaroid, Kodak," *Wall Street Journal*, September 30, 1976; "Fuji Film Instant Photo," *New York Times*, September 30, 1976.

3. "Kodak vs. Polaroid in Market and Court," *Business Week*, September 6, 1976.

4. "Kodak's New Model of 'Instant' Camera Delayed until 1977," *Wall Street Journal*, September 10, 1976.

5. David Gumpert, "Kodak Instant Film Has Fading Problem, Publication Asserts," *Wall Street Journal*, November 5, 1976, 18.

6. "Instant Picture Cameras," *Consumer Reports*, November 1976, 622.

7. Arthur M. Louis, "Polaroid's OneStep Is Stopping Kodak Cold," *Fortune*, February 13, 1978.

8. W.S. Allen, Eastman Kodak Company, "Letter to Dealers in Kodak Products," November 9, 1976, Polaroid Corporation Legal and Patent Records, Box II.203, f. 8, Articles of Interest—Kodak, 1976–1981, HBS.

9. John Stewart, "Kodak Finds Its Instant Photos Fade More," *Rochester Democrat and Chronicle*, June 3, 1977, Polaroid Corporation Legal and Patent Records, Box II.203, f. 8, Articles of Interest—Kodak, 1976–1981, HBS.

10. Eastman Kodak Company, "Kodak President Cites Extensive Testing of New Instant Cameras and Instant Film," press release, April 20, 1976.

11. Grant, "Instant Rivalry."

12. "Photography: Kodak's Blitz," *Newsweek*, April 4, 1977, Polaroid Corporation Administrative Records, Box I.234, f. 21, HBS.

13. Robert Metz, "Marketplace: Camera Problems Depress Kodak's Stock," *New York Times*, September 15, 1976.

14. Ed Bedrosian, note to Edwin Land, December 24, 1976, transmitting Ulysses Yannas, *Polaroid Corporation* (Drexel Burnham Lambert Research, December 15, 1976), Polaroid Corporation Legal and Patent Records, Box II.218, f. 1, Polaroid vs. Kodak, 1975–1976, HBS.

15. "Edwin Land on Polaroid."

16. Grant, "Instant Rivalry."

17. "Kodak vs. Polaroid in Market and Court."

18. "At the Company That Never Rests," *Fortune*, January 1977, 19.

19. Walter Fallon, "Planning a Profitable Future" (presentation, Security Analysts of San Francisco, October 19, 1976), F&N.

20. Polaroid-Kodak, Harvard Business School Case 378-173 (February 1983), 2.

21. "Kodak Says Shipments of Its Instant Cameras Will Be at Record Pace," *Wall Street Journal*, October 20, 1976.

22. E.R. Bedrosian, memorandum to Polaroid Management Executive Committee, October 21, 1976, F&N.

23. Robert Peck, letter to William K. Kerr, October 27, 1976, F&N.

24. Robert Peck, letter with attachments to William K. Kerr, November 5, 1976, F&N.

25. Richard DeLima, letter to Hammond Chaffetz (attachment to Peck letter of November 5, 1976 to Kerr), October 28, 1976, F&N.

26. Schwartz, interview, January 23, 2009.

27. Francis Carr, letter to Herbert Schwartz, December 16, 1987, F&N.

28. Plaintiff's Motion for a Protective Order, PvK, December 23, 1976, F&N.

29. Protective Order, PvK, January 5, 1977, F&N.

30. William K. Kerr, letter to Francis Carr, January 6, 1977, F&N; Francis Carr, letter to William K. Kerr, January 10, 1977, F&N.

31. Edward Handler, letter to William K. Kerr, January 31, 1977, with handwritten notation, F&N.

32. William K. Kerr, letter to Edward Handler, February 1, 1977, F&N.

33. Francis Carr, letter to William Kerr, February 16, 1977, F&N.

34. Answer and Counterclaim, Polaroid Corp. v. Berkey Photo, Inc., Civil Action No. 75-179 (D. Del. October 6, 1975), 12, F&N.

35. Opinion, PvB, December 30, 1976, F&N.

36. Sidney Neuman, letter to William Kerr, August 2, 1976, F&N.

37. William Kerr, memorandum to Schwartz, December 8, 1976, F&N.

38. William Kerr, letter to Sidney Neuman, February 16, 1977, F&N.

39. Plaintiff's Motion under Rule 42(b), FRCP, for Separate Trial of Patent Invalidity and Infringement and for a Stay of Proceedings with Respect to Defendant's Unclean Hands Defense and Antitrust Counterclaim, PvB, February 22, 1977, F&N.

40. William Kerr, letter to Richard Corroon, March 31, 1977, F&N.

41. Sidney Neuman, letter to William Kerr, June 27, 1977, F&N.

42. "Kodak Will Offer Rebates of $5 on Its Instant Cameras and Film," *New York Times*, May 10, 1977.

43. Robert Metz, "Market Place: Kodak Stock a Worry for Investors," *New York Times*, March 28, 1977.

44. Ibid.

45. "Kodak Unveils Instant Camera, Other Products," *Wall Street Journal*, March 24, 1977.

46. Victor McElheny, "17 Photo Products Unveiled by Kodak," *New York Times*, March 24, 1977, D1.

47. "Eastman Kodak to Cut Work Force by Using Rotating Furloughs," *Wall Street Journal*, April 22, 1977.

48. "Kodak Sets Rebates for Buyers, Dealers on Instant Products," *Wall Street Journal*, May 10, 1977, 10.

49. "Kodak Will Offer Rebates of $5."

50. "Eastman Kodak Company, Survey of Pending Litigation," *Wood Walker Litigation Service*, February 3, 1976, F&N.

51. U.S. Department of Commerce News, "Five Inventors to be Inducted into the National Inventors Hall of Fame," press release, January 28, 1977, F&N.

52. Campbell, "Edwin Herbert Land," 203.

53. Stanley Mervis, letter to Arthur Seidel, January 28, 1977, F&N; Fish & Neave, memorandum to William Kerr, February 2, 1977, F&N.

54. Michael Jensen, "Kodak Makes Its Response to Polaroid—at Last," *New York Times*, April 18, 1976, 12.

55. Remarks of Wesley Hanson on the occasion of the induction of George Eastman into the Inventors Hall of Fame, February 6, 1977, F&N.

56. Ibid.

57. Remarks of Edwin Land on the Occasion of His induction into the Inventors Hall of Fame, February 6, 1977, F&N.

58. Mervis, "Edwin Land—Champion of Patents," 53.

59. "Land's New Wonder," *Newsweek*, May 9, 1977, 75.

60. Polaroid-Kodak (B-6), Harvard Business School Case 378-178 (February 1983), 1.

61. Neil Ulman, "Polaroid Introduces Color Instant Movie Camera at Meeting," *Wall Street Journal*, April 27, 1977.

62. Polaroid-Kodak (B-6), Harvard Business School Case 378-178 (February 1983), 1.

63. McElheny, "Polaroid's Coup: Instant Movies."

64. Ulman, "Polaroid Introduces Color Instant Movie Camera."

65. Polaroid-Kodak (B-5), Harvard Business School Case 378-177 (February 1983), 1.

66. Polaroid-Kodak (B-3), Harvard Business School Case 378-175 (February 1983), 1.

67. Herbert Keppler, *Photo Weekly*, April 4, 1977; Polaroid-Kodak (B-3), Harvard Business School Case 378-175 (February 1983), 1–2.

68. Polaroid-Kodak (B-3), Harvard Business School Case 378-175 (February 1983), 2.

69. "For Land, a Cinematic Triumph," *New York Times*, May 1, 1977; Polaroid-Kodak (B-5), Harvard Business School Case 378-177 (February 1983), 2.

70. "Kodak Says It, Polaroid Have Discussed a Possible Settlement of Patent Dispute," *Wall Street Journal*, April 27, 1977.

71. "Instant Movies System Introduced by Polaroid," *New York Times*, April 27, 1977, Polaroid Corporation Legal and Patent Records, Box II.198, f. 3, Annual Meeting 1977, HBS.

72. William Kerr, letter to Robert Peck, April 29, 1977, F&N.

73. Quillen, interview.

74. Ulman, "Polaroid Introduces Color Instant Movie Camera"; Polaroid-Kodak (B-5), HBS, 2.

75. Schwartz, interview, January 23, 2009.

76. Gerald H. Rubin, letter to Richard DeLima, April 27, 1977, F&N.

77. Calvert Crary, "Polaroid v Eastman Kodak, Settlement Prospects," *Litigation Review* (Bache Halsey Stuart, May 11, 1977), Polaroid Corporation Legal and Patent Records, Box II.203, f. 8, Articles of Interest—Kodak, 1976–1981, HBS.

78. Ibid.

79. Quillen, interview.

80. William Kerr, memorandum for Herbert Schwartz, April 21, 1977, F&N.

81. Mario Milletti, "Kodak Net Off 3.7% in Latest Drop while Rival Polaroid Rises 10.8%," *New York Times*, July 21, 1977.

82. Louis, "Polaroid's OneStep Is Stopping Kodak Cold."

83. Charles Elia, "Polaroid's Record Results Fail to Quell Doubts as Some Shade Estimates for 1977," *Wall Street Journal*, August 3, 1977.

84. Herbert Schwartz, memorandum to William Kerr, July 1, 1977, F&N.

85. Herbert Schwartz, memorandum to William Kerr, "Status of Discovery from Eastman Kodak," July 26, 1977, F&N.

86. Notice of Taking Deposition, PvK, September 20, 1977, F&N.

87. Sidney Neuman, letter to William Kerr, July 28, 1977, F&N.

88. William Kerr, letter to Sidney Neuman, August 2, 1977, F&N.

89. Order, PvB, October 4, 1977, F&N.

Chapter 13

1. Deposition Transcript of Albert Louis Sieg, PvK, November 16, 1977, 17–19, F&N.
2. Ibid., 30–31.

3. Ibid., 47.
4. Ibid., 253.
5. Ibid., 256.
6. Ibid., 257.
7. Ibid., 257–58.
8. Ibid., 259.
9. Ibid.
10. Hugh C. Barrett, Summary of Discovery Disputes Discussed at 11/29/77 Meeting, December 3, 1977, 27–28, F&N.
11. Herbert Schwartz, letter to Francis T. Carr, December 8, 1977, F&N.
12. Herbert Schwartz, letter to Francis T. Carr, December 19, 1977, F&N.
13. Robert Metz, "Surprising Gains for Berkey," *New York Times*, January 11, 1978, D2.
14. John McQuiston, "U.S. Jury Finds a Kodak Monopoly in Amateur-Photography Business," *New York Times*, January 22, 1978.
15. "Kodak Is Found to Have Broken Antitrust Laws," *Wall Street Journal*, January 23, 1978.
16. Robert Peck, memorandum to Charles Mikulka, February 28, 1978, F&N.
17. Arnold Lubasch, "Kodak Wins Trust Suit Appeal," *New York Times*, June 26, 1979.
18. Consent Judgment, PvB, par. 3, F&N.
19. "Berkey Drops Instant Cameras," *International Herald Tribune*, April 14, 1978.
20. Hugh C. Barrett, Summary of January 12, 1978 Meeting, F&N.
21. Herbert Schwartz, letter to Robert Peck, January 23, 1978, F&N.
22. Herbert F. Schwartz, memorandum to William Kerr, January 24, 1978, F&N.
23. Herbert Schwartz, Topics to Discuss at Polaroid, February 1, 1978, F&N.
24. William Kerr, letter to Francis T. Carr, April 18, 1978, F&N.
25. Francis T. Carr, letter to William Kerr, April 27, 1978, F&N.
26. William Kerr, Estimates of Time Commitments for Dr. Land, July 25, 1977, F&N.
27. William Kerr, letter to Stanley Mervis, April 14, 1978, F&N.
28. William Kerr, letter to Francis T. Carr, May 1, 1978, F&N.
29. William Kerr, letter to Francis T. Carr, May 3, 1978, F&N.
30. Francis T. Carr, letters (3) to William Kerr, May 4, 1978, F&N.
31. William Kerr, letter to Robert Peck, May 5, 1978, F&N.
32. Herbert Schwartz, memorandum to William Kerr, April 3, 1978, F&N.
33. Robert Peck, letter to Herbert Schwartz, November 15, 1977, F&N.
34. Robert Frank, letter to Honorable A. David Mazzone, June 2, 1978, F&N; William Cheeseman, letter to Honorable A. David Mazzone, June 2, 1978, F&N.
35. Memorandum and Order, PvK, June 12, 1978, F&N.
36. "Kodak Patent Talks with Polaroid Fail," *New York Times*, June 13, 1978.
37. John White, "Polaroid Steps into Echo-Focusing," *Boston Globe*, April 26, 1978.
38. "Polaroid Corp. to Push Suits against Kodak, Says Talks Have Failed," *Wall Street Journal*, June 14, 1978.
39. Herbert F. Schwartz, memorandum to William Kerr, June 12, 1978, F&N.
40. Calvert Crary, *Litigation Review* 3, no. 21 (Bache, June 28, 1978), F&N.
41. Victor K. McElheny, "Kodak Cuts Prices and Introduces Instant and Conventional Cameras," *New York Times*, February 9, 1978.

42. "Cameras That See by Sound," *Time*, May 8, 1978; Arthur M. Louis, "Polaroid's OneStep Is Stopping Kodak Cold," *Fortune*, February 13, 1978.

43. Louis, "Polaroid's OneStep Is Stopping Kodak Cold."

44. Clare Reckert, "Kodak's Earnings for First Quarter Up 50%, a Record," *New York Times*, April 24, 1978.

45. "Polaroid Foresees Record Earnings," *New York Times*, February 1, 1978.

46. Walter McLaughlin, letter to John Hall and William Cheeseman, June 20, 1978, F&N.

47. Defendant's Motion under Rule 37(a) F.R.Civ.P. to Compel Answers to Interrogatories and to Compel the Production of Documents and Things, PvK, June 22, 1978, F&N.

48. Fish & Neave, hearing notes, June 21, 1978, F&N.

49. Order, PvK, July 6, 1978, F&N.

50. Francis T. Carr, letter to William Kerr, July 31, 1978, F&N.

51. Robert Peck and Robert Ford, litigation notes, January 3, 1978, F&N.

52. Robert Peck, letter to William Kerr, July 31, 1978, F&N.

Chapter 14

1. Walter McLaughlin, letter to William Cheeseman and Robert Frank, August 14, 1978.

2. Gosselin, "How Polaroid Beat Kodak."

3. Schwartz, interview, Riverside, CT, July 22, 2009.

4. Defendant's Motion under Rules 37(a) and 26(c), Fed. R. Civ. P., to Compel Witnesses to Answer Deposition Questions, and to Regulate Conduct at Depositions, PvK, 4, F&N.

5. Memorandum and Order, PvK, October 4, 1978, 5–6, F&N.

6. Ibid., 2.

7. Ibid., 3.

8. Ibid., 6–7.

9. Plaintiff's Memorandum in Opposition to Defendant's Motion to Compel Answers to Deposition Questions, PvK, October 16, 1978, 5–6, F&N.

10. Order, PvK, November 7, 1978, F&N.

11. Herbert Schwartz, memorandum to William Kerr, Topics to Be Discussed at Polaroid on 10/18/78, October 15, 1978, F&N.

12. Schwartz, interview, January 30, 2009.

13. Schwartz, Topics to Be Discussed at Polaroid on 10/18/78 (see Kerr handwritten notes), F&N.

14. William Kerr, letter to Francis T. Carr, December 14, 1978, F&N.

15. Schwartz, interview, January 30, 2009.

16. Chequer, interview, January 23, 2009.

17. Herbert Schwartz, letter to Robert Peck, November 29, 1978, F&N.

18. Special Master's Advisory Report to the Court in re: Questions Propounded to Doctor Rogers (an Inventor of One or More of the Patents in Suit) Relative to Prior Art, PvK, January 19, 1979, 4, F&N.

19. Ibid.

20. Order, PvK, January 19, 1979, F&N.
21. Ibid., 2.
22. Francis T. Carr, letter to Hugh C. Barrett, January 26, 1979, F&N.
23. Application for Review of the Special Master's Order and Advisory Report of January 19, 1979, PvK, February 1, 1979, F&N.
24. Ibid., 16; Reply Brief in Support of Polaroid's Application for Review of the Special Master's Order and Advisory Report of January 19, 1979, PvK, February 23, 1979, 2, F&N.
25. Francis T. Carr, letter to Walter McLaughlin, March 1, 1979, F&N.
26. Ibid, handwritten notation by William Kerr.
27. Walter McLaughlin, letter to Francis T. Carr, March 2, 1979, F&N.
28. William Kerr, letter to Walter McLaughlin, March 5, 1979, F&N.
29. Order, PvK, March 6, 1979, F&N.
30. William Kerr, memorandum to file, March 19, 1979, F&N.
31. Tom Goldstein, "The Pretrial Stage of a Suit," *New York Times*, March 16, 1979.
32. John Fogarty, letter to Herbert Schwartz, February 6, 1979, F&N.
33. William Kerr, letter to Francis T. Carr, February 8, 1979, F&N.
34. Robert J. Goldman, letter to John Fogarty, February 13, 1979, F&N.
35. Edward J. Handler, letter to Herbert Schwartz, February 16, 1979, F&N.

Chapter 15

1. Herbert Schwartz, interview by author, January 23, 2009.
2. Herbert Schwartz, memorandum to William Kerr, January 23, 1979, F&N.
3. William Kerr, letter to Robert F. Peck, January 30, 1979, F&N.
4. Ibid.
5. Schwartz, interview, January 30, 2009.
6. Chequer, interview by author, January 23, 2009.
7. Ibid.
8. Schwartz, memorandum, January 23, 1979.
9. "60-Second Film," *Time*, November 28, 1955, Polaroid Corporation Administrative Records, Box I.235, f. 18, HBS.
10. Stan Luxenberg, "Instant Camera War Developments," *New York Times*, April 1, 1979.
11. Richard A. Shaffer, "Polaroid's Recent Sharp Price Drop Is Tied by Analyst to Company's Policy of Secrecy," *Wall Street Journal*, March 1, 1979.
12. "Investment Firm Files Suit against Polaroid and Two Top Officers," *Wall Street Journal*, March 5, 1979.
13. "Up and Down Wall Street," *Barron's*, March 5, 1979.
14. Luxenberg, "Instant Camera War Developments."
15. William K. Kerr, Letter to Robert Peck, April 2, 1979, F&N.
16. Herbert F. Schwartz, memorandum to William Kerr, "Status of Polaroid v. Kodak Litigation," June 13, 1979, F&N.
17. Robert S. Frank, letter to clerk, U.S. District Court, Boston, MA, April 12, 1979, F&N.

18. Francis T. Carr, letter to William Kerr, April 10, 1979, F&N.

19. Robert Peck, memorandum to Edwin Land, April 19, 1979, F&N.

20. Ibid.

21. Peter Schuyten, "Polaroid Unveils Faster Film," *New York Times*, December 4, 1979.

22. Defendant's Motion under Rule 56, F.R.C.P., for Partial Summary Judgment, PvK, April 25, 1979, F&N.

23. Francis T. Carr, "Polaroid v. Kodak, a Reminiscence," 15, author's collection.

24. Plaintiff Polaroid Corporation's Post-Trial Submissions, PvK, facts IX-8, F&N.

25. Trial Testimony of Walter Fallon, PvKdam, vol. 60, August 25, 1989, 7653–54, 7638–39.

26. Plaintiff Polaroid Corporation's Post-Trial Submissions, PvK, Facts IX-8, F&N.

27. Ibid.; Trial Testimony of Walter Fallon, PvKdam, vol. 60, August 25, 1989, 7640–41.

28. Trial Testimony of Colby Chandler, PvKdam, vol. 54, August 17, 1989, 6928–30.

29. Ibid., 6927–28; Plaintiff Polaroid Corporation's Post-Trial Submissions, PvK, facts IX-11, F&N.

30. Herbert Schwartz, letter to Francis T. Carr, May 1, 1979, F&N; Francis T. Carr, letter to Herbert Schwartz, May 11, 1979, F&N.

31. U.S. Patent Law, 35 U.S.C. § 103.

32. Defendant's Motion under Rule 56, F.R.C.P., for Partial Summary Judgment, 45.

33. Ibid., 34.

34. Richard M. Barnes, interview with author in Pelham, NY, July 22, 2010.

35. Clare M. Reckert, "Kodak Profit Rose 40.1% in Quarter," *New York Times*, April 26, 1979.

36. "Kodak Sets Rebates as Latest Weapon in Polaroid Battle," *Wall Street Journal*, June 5, 1979.

37. "New Kodak Quarterback in Berkey Litigation," *New York Times*, May 1, 1979.

38. "2d Circuit Hears Appeals of Kodak and Berkey Photo," *New York Law Journal*, April 19, 1979.

39. Plaintiff's Motion for a Protective Order, PvK, December 13, 1976, F&N.

40. Defendant's Motion to Declare Certain Materials Not Confidential under the Protective Order of January 5, 1977, PvK, April 25, 1979, F&N.

41. William Kerr, letter to Francis T. Carr, May 4, 1979, F&N; Francis T. Carr, letter to William Kerr, May 10, 1979, F&N; William Kerr, letter to Francis T. Carr, May 14, 1979, F&N; Francis T. Carr, letter to William K. Kerr, May 15, 1979, F&N.

42. Special Master's Advisory Opinion and Order on Motion of Defendant Eastman Kodak to Declare Certain Materials "to Be Not Confidential," PvK, July 23, 1979, 4, F&N.

43. "Polaroid, Citing Slowing in Market, to Lay Off 800," *New York Times*, May 16, 1979.

44. "Layoffs at Polaroid: What Went Wrong?," *New York Times*, May 20, 1979.

45. Isadore Barmash, "Polaroid's Future out of Focus," *New York Times*, May 21, 1979.

46. Herbert Schwartz, memorandum to William Kerr, May 4, 1979, F&N.

47. William Kerr, letter to Francis T. Carr, May 17, 1979, F&N.

48. Frank Carr, letter to William Kerr, May 17, 1979, F&N.

49. Ibid. (handwritten notation).

50. Francis T. Carr, letter to William Kerr, June 7, 1979, F&N.

51. Herbert Schwartz, memorandum to William Kerr, May 4, 1979, F&N.

52. Richard Barnes, memorandum, Land and Bachelder United States Patent No. 3,732,101, May 16, 1979, F&N.

53. Barnes, interview.

54. Schwartz, interview, January 30, 2009.

55. William Kerr, draft of letter to Edwin Land, May 15, 1979, F&N.

56. Richard Barnes, memorandum to William Kerr, May 15, 1979, handwritten notation, F&N; Richard Barnes, letter to Robert Ford, May 16, 1979, F&N; Robert M. Ford, letter to Richard Barnes, May 29, 1979, F&N.

57. Herbert Schwartz, Topics to Be Discussed at Meeting with Polaroid on Monday, June 18, 1979, F&N.

58. Richard Barnes, letter to Robert M. Ford, June 4, 1979, F&N.

59. Barnes, interview.

60. Wolfgang Berg, "Polaroid One-Step Color Photography," *Die Naturwissenschaften*, January 1977.

61. Herbert Schwartz, Topics to Be Discussed at Meeting with Polaroid on Monday, June 14, 1979, F&N.

62. Herbert Schwartz, memorandum to William Kerr, Status of *Polaroid v. Kodak* Litigation, June 13, 1979, 4, F&N.

63. Ibid.

64. Ibid., 5.

Chapter 16

1. "Justices Bar Review in Berkey Award Cut," *New York Times*, February 20, 1980.

2. Arnold Lubasch, "Kodak Wins Trust Suit Appeal," *New York Times*, June 26, 1979.

3. "Justices Bar Review."

4. "Berkey to Appeal Cut in Kodak Award," *New York Times*, July 6, 1979.

5. "Kodak, Berkey Settle Suit for $6.8 Million," *New York Times*, September 24, 1981.

6. William K. Kerr, memorandum for file, Conference with Dr. Land on Wednesday, June 27, 1979, June 28, 1979, F&N.

7. Nan Chequer, handwritten notes, June 27, 1979, Polaroid Corporation Legal and Patent Records, Box II.189, f. 6, Nan B. Chequer Research Notebook, September 19, 1979, 43 (back), HBS.

8. Kerr, memorandum, Conference with Dr. Land.

9. Robert Peck, letter to William Kerr, July 30, 1979, F&N.

10. Herbert Schwartz, letter to Francis T. Carr, July 13, 1979, F&N.

11. Howard Rudnitsky, "Why Kodak Will Buck the Recession," *Forbes*, September 3, 1979.

12. Herbert Schwartz, letter to Francis T. Carr, August 7, 1979, F&N.

13. Polaroid's Memorandum in Opposition to Kodak's Motion under Rule 56, F.R.Civ.P., for Partial Summary Judgment, PvK, 4, F&N.

14. Ibid., 19.

15. Ibid., 19, 36–38.

16. Ibid., 38.
17. Ibid., 41.
18. Polaroid's Rejoinder Memorandum in Opposition to Kodak's Motion under Rule 56, F.R.Civ.P., for Partial Summary Judgment, PvK, 20, F&N.
19. Polaroid's Supplement to Its Rejoinder Memorandum in Opposition to Partial Summary Judgment, PvK, 2–3, F&N.
20. William Kerr, letter to Francis T. Carr, August 13, 1979, F&N.
21. Francis T. Carr, letter to William Kerr, August 17, 1979, F&N.
22. Francis T. Carr, letter to William Kerr, September 12, 1979.
23. William Kerr, letter to Francis T. Carr, September 13, 1979.
24. William Kerr, Notes Taken during Land Deposition, September 25–28, 1979, F&N.
25. Ibid.
26. Galbraith, memorandum.
27. Schwartz, interview, June 16, 2010.
28. Warshofsky, *The Patent Wars*, 69; Quillen, interview.
29. Edwin H. Land, Deposition Transcript, vol. 1, Polaroid Corporation Legal and Patent Records, Box II.173, f. 7, September 25, 1979, HBS.
30. Quillen, interview.
31. Robert Peck, letter to William Kerr, November 8, 1979, F&N.
32. Rosenberg, "The Realist behind Genius Edwin Land"; McCune Oral History, 8.
33. Mitchell Lynch, "Instant Movies Falter; Is Polaroid Chairman Wrong for Change?," *Wall Street Journal*, August 9, 1979.
34. Jeffrey Sonnenfeld, *The Hero's Farewell* (Oxford University Press, 1988), 120.
35. Lynch, "Instant Movies Falter."
36. Ibid.
37. Steven Solomon, "Polaroid: Instant Turnaround?," *Financial World*, December 1979.
38. Ibid.
39. Robert Cole, "Polaroid Writes Off $68 Million," *New York Times*, September 13, 1979.
40. Ralph Kaplan, *Polaroid: Acquisition Target?* (Brean Murray, Foster Securities Inc., November 5, 1979), Polaroid Corporation Legal and Patent Records, Box II.198, f. 2, Financial News Statements, 1979, HBS; "Inside Wall Street, a Polaroid Takeover?," *Business Week*, November 19, 1979; Mitchell Lynch, "Blurry Picture at Polaroid Shows Net Surging but Difficulties Still Developing, Analysts Say," *Wall Street Journal*, February 27, 1980.
41. Francis T. Carr, letter to William Kerr, October 16, 1979, F&N.
42. Francis T. Carr, letter to William Kerr, November 5, 1979, F&N.
43. Ibid.
44. Chequer, interview.
45. Francis T. Carr, letter to William Kerr, November 20, 1979, F&N.
46. Herbert F. Schwartz, memorandum to William Kerr, November 6, 1979, F&N; Chequer, interview.
47. Hugh C. Barrett, memorandum for file, October 1, 1979, F&N.
48. William Cheeseman, letter to Walter McLaughlin, October 30, 1979, F&N; Walter McLaughlin, letter to William Cheeseman, October 30, 1979, F&N.

49. William Kerr, handwritten note on Robert Frank letter to Walter McLaughlin, December 11, 1979, F&N.

50. Robert J. Goldman, 1979 Christmas card note, author's collection.

Chapter 17

1. Special Master's Advisory Opinion to the Court and Order on the Defendant Eastman Kodak's Fifth Motion under Rule 37(a), F.R.Civ.P., to Compel Answers to Interrogatories and to Compel the Production of Documents and Things, PvK, February 8, 1980, 2–3, F&N.

2. Herbert F. Schwartz, letter to Francis T. Carr, January 9, 1980, F&N.

3. Ronald K. Fierstein, memorandum for file, Rochester Notebook Inspection, February 13, 1980, F&N.

4. Ronald K. Fierstein, memorandum for file, Rochester Notebook Inspection, March 12, 1980, F&N; Ronald K. Fierstein, memorandum for file, Eastman Kodak Research Laboratory Notebook Inspection, March 25, 1980, F&N.

5. Herbert F. Schwartz, memorandum to William Kerr, November 8, 1979, F&N.

6. William Kerr, Outline for Appearance before the Court, November 16, 1979, F&N.

7. Ibid.

8. Frank Dello Russo, letter to William Kerr, December 11, 1979, F&N.

9. Ibid.

10. William Cheeseman, letter to Robert Ford, December 18, 1979, F&N.

11. Laurence Fordham, letter to Frank Dello Russo, December 18, 1979, F&N.

12. Robert Frank, letter to Frank Dello Russo, December 20, 1979, F&N.

13. Memorandum of Decision on Question of Disqualification, January 29, 1980, PvK, filed March 17, 1980, 2, F&N.

14. Ibid., 6.

15. Ibid., 10.

16. Judge David Nelson, letter to William Kerr, March 17, 1980, F&N.

17. Office of the Clerk, U.S. District Court, Boston, Massachusetts, Memorandum to Counsel, March 17, 1980, F&N.

18. Schwartz, interview, January 30, 2009.

19. William Kerr, letter to Julius Silver, April 10, 1980, F&N.

20. Lawrence Fordham, letter to Honorable Rya Zobel, March 21, 1980, F&N.

21. Robert Frank, letter to Nina Singer, March 28, 1980, F&N.

22. Samuel Hoar, letter to Laurence Fordham, April 10, 1980, F&N; Samuel Hoar, letter to William Cheeseman, April 3, 1980, F&N.

23. Nina Singer, letter to Robert Frank et al., April 9, 1980, F&N.

24. Polaroid Corporation's Application for a Pre-Trial Conference, PvK, filed April 11, 1980, F&N.

25. Eastman Kodak's Response to Polaroid Corporation's Application for Pre-Trial Conference, PvK, filed April 16, 1980, F&N.

26. Laurence Fordham, letter to Honorable Rya Zobel, May 6, 1980, F&N.

27. Herbert F. Schwartz, memorandum to William Kerr, May 9, 1980, F&N.

28. Peter Schuyten, "Polaroid's Land to Quit Chief Executive Position," *New York Times*, March 7, 1980.

29. Polaroid Corporation, press release, February 21, 1980, Polaroid Corporation Legal and Patent Records, Box II.216, f. 7, Press Releases 1980, HBS.

30. Peter Bernstein, "Polaroid Struggles to Get Back into Focus," *Fortune*, April 7, 1980.

31. Schuyten, "Polaroid's Land to Quit."

32. "Land Will Become Polaroid Consultant," *New York Times*, April 15, 1980.

33. Chequer, interview.

34. "Polaroid—Turning Away from Land's One-Product Strategy," *Business Week*, March 2, 1981.

35. Chequer, interview.

36. Herbert F. Schwartz, letter to Nan Chequer, April 11, 1980, F&N.

37. Mitchell Lynch, "Polaroid Applauds Parting Chief Land but Doesn't Introduce Startling Products," *Wall Street Journal*, April 23, 1980.

38. Polaroid Corporation, Music Selections, Empire Brass Quintet, Polaroid Corporation Legal and Patent Records, Box II.230, f. 7, Annual Meeting, 1981, HBS.

39. Lynch, "Polaroid Applauds Parting Chief."

40. Edwin Land, address to shareholders, April 22, 1980, Polaroid Corporation Legal and Patent Records, Box II.218, f. 2, Annual Meeting . . . Transcript of Edwin H. Land Speech, 1980, HBS.

41. "Land Will Become Polaroid Consultant."

42. Lynch, "Blurry Picture at Polaroid."

43. Bernstein, "Polaroid Struggles."

44. Ibid.

45. Robert Metz, "Polaroid's Outlook," *New York Times*, April 29, 1980.

46. Schwartz, interview, January 30, 2009.

47. Ibid.

48. Galbraith, memorandum.

49. Trial Testimony of Francis T. Carr, PvKdam, vol. 58, August 23, 1989, 7394–97.

50. Schwartz, interview, January 30, 2009.

51. Goldstein, *Intellectual Property*, 44.

52. Ibid., 45.

53. Ibid.; P.J. Federico, "Adjudicated Patents, 1948–54," *Journal of the Patent Office Society* 38 (1956), 233.

54. Goldstein, *Intellectual Property*, 45; Jungersen v. Ostby & Barton Co., 335 U.S. 560, 572 (1949).

55. Edwin H. Land, "The Role of Patents in the Growth of New Companies," *Journal of the Patent Office Society* 41 (1952), 502.

56. Steven Z. Szczepanski, "Licensing or Settlement: Deferring the Fight to Another Day," *AIPLA Quarterly Journal* 15 (1987), 298, 301; Gloria K. Koenig, "Patent Invalidity: A Statistical and Substantive Analysis," in *Patent Law Perspectives* (1980), 4–23; Paul Janicke, memorandum to author, November 18, 2013.

57. Herbert F. Schwartz, letter to Robert Ford, June 9, 1980, F&N.

58. Herbert F. Schwartz, letter to Milton Solomon, June 9, 1980, F&N.

59. Goldstein, *Intellectual Property*, 10–11; Polaroid Corp. v. Eastman Kodak Co., 16 U.S.P.Q.2d 1481, 1536–38 (D. Mass. 1990).

60. Schwartz, interview, January 30, 2009.

61. Ibid.

62. Herbert F. Schwartz, letter to Francis T. Carr, July 11, 1980, F&N.

63. William Kerr, Notes for July 14 Conference with Judge Zobel, F&N.

64. Ibid.

65. Francis T. Carr, memorandum, Polaroid Corporation v. Eastman Kodak Company, Pending Matters before the Court, July 14, 1980, F&N.
66. William Kerr, letter to Julius Silver, July 16, 1980, F&N.
67. Schwartz, interview, January 30, 2009.
68. Francis T. Carr, letter to Herbert Schwartz, July 15, 1980, F&N.
69. Kenneth Herman, memorandum for the file, August 15, 1980, F&N.
70. William Kerr, letter to Francis T. Carr, August 15, 1980, F&N.
71. Walter McLaughlin, letter to Francis T. Carr, August 18, 1980, F&N.

Chapter 18

1. William Kerr, memorandum to Schwartz, Nathan, August 21, 1980, F&N.
2. William Kerr, letter to Julius Silver, July 16, 1980, F&N.
3. William Kerr, letter to Julius Silver, October 3, 1980, F&N.
4. Defendant Eastman Kodak's Memorandum in Support of Its Second Motion under Rule 56, F.R.C.P. for Partial Summary Judgment, PvK, filed September 19, 1980.
5. William Kerr, letter to Julius Silver, October 3, 1980, F&N.
6. Francis T. Carr, letter to Herbert F. Schwartz, November 13, 1980, F&N; Herbert F. Schwartz, letter to Frank Carr, November 14, 1980, F&N.
7. Francis T. Carr, letter to Nina Singer, November 7, 1980, F&N.
8. William Kerr, letter to Nina Singer, November 12, 1980, F&N.
9. Herbert Schwartz, memorandum to Polaroid lawyers, November 26, 1980, F&N.
10. William Kerr, letter to Julius Silver, November 26, 1980, F&N.
11. Herbert F. Schwartz, memorandum to William Kerr, December 10, 1980, F&N.
12. William Kerr, letter to Rya Zobel, December 24, 1980, F&N.
13. Francis T. Carr, letter to Rya Zobel, December 29, 1980, F&N.
14. Schwartz, interview, January 30, 2009.
15. William Kerr, letter to Julius Silver, February 17, 1981, F&N.
16. Ibid.
17. Frank Carr, letter to Walter McLaughlin, December 5, 1980, F&N.
18. Walter McLaughlin, letter to all counsel, December 12, 1980, F&N.
19. Defendant Eastman Kodak's Eleventh Motion under Rule 37(a), F.R.C.P. to Compel the Production of Documents, PvK, filed December 19, 1980, F&N.
20. Herbert F. Schwartz, letter to Walter McLaughlin, December 22, 1980, F&N.
21. Francis T. Carr, letter to Walter McLaughlin, December 23, 1980, F&N.
22. Herbert F. Schwartz, letter to Walter McLaughlin, December 28, 1980, F&N.
23. Walter McLaughlin, letter to Herbert F. Schwartz, December 30, 1980, F&N.
24. "Polaroid Corp. Enlarges Office of the President with Four Promotions," *Wall Street Journal*, October 16, 1980.
25. Donald White, "It's Annual Meeting Time," *Boston Globe*, April 28, 1971.
26. "Polaroid—Turning Away from Land's One-Product Strategy," *Business Week*, March 2, 1981.
27. Ibid.
28. Ibid.
29. Dan Dorfman, "Report Polaroid Boss Land Exploring 160M Stock Sale," *New York Daily News*, February 18, 1981.
30. "McCune and Land," *Polaroid Newsletter*, February 1981, F&N.

31. Ibid.
32. Chequer, interview.
33. Edwin Land, Final, President's Script, Annual Meeting of Shareholders, April 21, 1981, Polaroid Corporation Legal and Patent Records, Box II.230, f. 7, Annual Meeting, 1981, HBS.
34. Herbert F. Schwartz, letter to Francis T. Carr, February 2, 1981, F&N.
35. John Fogarty, letter to Walter McLaughlin, February 19, 1981, F&N.
36. William Kerr, letter to Julius Silver, February 17, 1981, F&N.
37. Francis T. Carr, letter to Rya Zobel, February 6, 1981, F&N.
38. William Kerr, letter to Rya Zobel, February 9, 1981, F&N.
39. Notice, PvK, February 17, 1981, F&N.
40. Memorandum of Decision, PvK, February 25, 1981, F&N.
41. Ibid.
42. William Kerr, letter to Julius Silver, March 3, 1981, F&N.
43. Schwartz, interview, January 30, 2009.
44. Calvert Crary, *Bear Stearns Litigation Review*, March 18, 1981, 4, F&N.
45. Quillen, interview.
46. Herbert F. Schwartz, memorandum to Fish & Neave legal team, March 11, 1981, F&N.
47. William Cheeseman, letter and attachment to Clerk, March 20, 1981, F&N.
48. Order, PvK, March 24, 1981, F&N.
49. William Kerr, letter to Julius Silver, March 23, 1981, F&N.
50. Herbert F. Schwartz, memorandum to William Kerr, March 26, 1981, F&N.
51. Defendant Eastman Kodak's Memorandum in Support of Its Second Motion under Rule 56, F.R.C.P. for Partial Summary Judgment, PvK, filed September 19, 1980, F&N.
52. Francis T. Carr, letter to Rya Zobel, April 14, 1981, F&N.
53. William Kerr, letter to Rya Zobel, April 15, 1981, F&N.
54. Francis T. Carr, letter to Rya Zobel, April 20, 1981, F&N.
55. William Kerr, letter to Rya Zobel, April 23, 1981, F&N.
56. Francis T. Carr, letter to Rya Zobel, April 28, 1981, F&N.
57. First Pretrial Order, PvK, filed May 11, 1981, F&N.
58. Ibid.
59. Ibid.; Patricia Martone, Memorandum Concerning Pretrial Conference of May 8, 1981, May 11, 1981, F&N.
60. Calvert Crary, *Bear Stearns Litigation Review*, June 1, 1981, 7, F&N.
61. Schwartz, interview, January 30, 2009.
62. William Kerr, letter to Edwin Land, May 11, 1981, F&N.
63. Schwartz, interview, January 30, 2009.

Chapter 19

1. William Kerr, letter to Nan Chequer, May 18, 1981, F&N.
2. William Kerr, memorandum, Polaroid Patent Policy, attached to William Kerr letter to Nan Chequer, May 19, 1981, F&N.
3. Ibid.
4. Schwartz, interview, January 30, 2009.
5. Kerr, letter, May 19, 1981, F&N.

6. William Kerr, letter to Edwin Land, May 27, 1981, F&N.

7. Memorandum of Decision, PvK, filed July 10, 1981, 13, F&N.

8. Ibid., 10.

9. Ibid., 11.

10. Ibid.

11. Quillen, interview, December 9, 2009.

12. William Kerr, letter to Julius Silver, July 14, 1981, F&N.

13. Robert Ford, memorandum to Edwin Land, July 13, 1981, Polaroid Corporation Legal and Patent Records, circa 1905–1995, Polaroid v. Kodak, circa 1950–1991, C. Discovery and Trial Preparation, circa 1950–1991, Box 185, f. 3, Patents & Claims, HBS.

14. William Kerr, letter to Rya Zobel, August 10, 1981, F&N.

15. Ronald K. Fierstein, memorandum to Schwartz, August 10, 1981, F&N.

16. William Kerr, letter to Rya Zobel, August 10, 1981, F&N.

17. Francis T. Carr, letter to Rya Zobel, August 12, 1981, F&N.

18. Defendant's Third Motion under Rule 56, F.R.Civ.P., for Partial Summary Judgment, PvK, filed August 14, 1981, F&N.

19. Kodak's Response to: Polaroid Corporation's Motion to Strike Eastman Kodak Company's Third Motion for Partial Summary Judgment and Polaroid Corporation's Opposition to Eastman Kodak Company's Third Motion for Partial Summary Judgment and Polaroid Corporation's Motion to Enlarge Its Time to File Papers in Opposition to Eastman Kodak Company's Third Motion for Partial Summary Judgment, PvK, filed August 27, 1981, F&N.

20. William Kerr, letter to Rya Zobel, August 18, 1981, F&N.

21. Ibid.

22. William McCune, letter to Walter Fallon, August 18, 1981, F&N.

23. Polaroid Corporation's Memorandum in Support of Its Motion to Strike Eastman Kodak Company's Third Motion under Rule 56, F.R.Civ.P. for Partial Summary Judgment, PvK, filed August 21, 1981, 2, F&N.

24. Quillen, interview, December 9, 2009.

25. Francis T. Carr, letter to Rya Zobel, August 27, 1981, F&N.

26. Schwartz, interview, January 30, 2009; William Kerr, letter to Rya Zobel, August 28, 1981, F&N.

27. Patricia Martone, memorandum to Kerr and Schwartz, August 28, 1981, F&N.

28. Polaroid Corporation, memorandum, Ritz, August 28, 1981, Polaroid Corporation Legal and Patent Records, Box II.200, f. 3, Kodak 1972, HBS; William Kerr, letter to Rya Zobel, September 24, 1981, HBS.

29. Quillen, interview, December 9, 2009.

30. Ibid.

31. Goldstein, *Intellectual Property*, 23.

32. William Kerr, letter to Rya Zobel, September 24, 1981, F&N.

33. Agreement between Polaroid Corporation and Eastman Kodak Company as of December 12, 1957, PvK, Trial Exhibit DF-500, 17–20, 23, F&N.

34. William McCune, conversation with Fallon, September 3, 1981 (handwritten notes), Polaroid Corporation Legal and Patent Records, Box II.200, f. 3, Kodak 1972, HBS.

35. Ibid.

36. Ibid.

37. William Kerr, letter to Rya Zobel, September 18, 1981, F&N.
38. Francis T. Carr, letter to Rya Zobel, September 17, 1981, F&N.
39. Kerr, letter, September 18, 1981.
40. William McCune, phone call with Fallon, September 15, 1981 (handwritten notes), Polaroid Corporation Legal and Patent Records, Box II.200, f. 3, Kodak 1972, HBS.
41. Francis T. Carr, letter to Rya Zobel, September 17, 1981, F&N.
42. William Kerr, letter to Rya Zobel, September 18, 1981, F&N.
43. Ibid.
44. Francis T. Carr, letter to Rya Zobel, September 23, 1981, F&N.
45. Ibid.
46. William Kerr, letter to Rya Zobel, September 24, 1981, F&N.
47. Francis T. Carr, letter to Rya Zobel, September 24, 1981, F&N.
48. Robert Lenzner, "Polaroid to Trade Shots with Kodak," *Boston Globe*, October 4, 1981.
49. Alecia Swasy, *Changing Focus* (Times Business Random House, 1997), 25.
50. Quillen, interview.
51. Swasy, *Changing Focus*, 17–27.
52. Schwartz, interviews, March 15, 2007, and January 30, 2009.

Chapter 20

1. Robert Lenzner, "Land Gives Polaroid's Side," *Boston Globe*, October 6, 1981.
2. Robert Lenzner, "Polaroid to Trade Shots with Kodak," *Boston Globe*, October 4, 1981, 1.
3. Chequer, interview; John Rumsey, "The Kodak-Polaroid Trial Left Some Strong Images," *Rochester Times-Union*, January 9, 1986, Polaroid Corporation Administrative Records, Box I.122, f. 2, Polaroid vs. Kodak Clippings 1986, HBS.
4. Patricia Martone, memorandum to Kerr and Schwartz, Proposed Agenda for September 28, 1981 Pretrial Conference, F&N; Stipulation and Order, PvK, October 2, 1981, F&N.
5. Robert J. Goldman, memorandum to Schwartz, A Brief Overview of the Polaroid v. Kodak United States Patent Litigation, June 7, 1985, 4; Schwartz, interview, January 23, 2009.
6. William Kerr, Notes for Polaroid Witnesses and Others in Attendance at Trial, September 16, 1981, F&N.
7. Robert Lenzner, Meeting Polaroid's Land, *Boston Globe*, October 11, 1981.
8. Barnaby J. Feder, Kodak's Fight with Polaroid, *New York Times*, October 8, 1981.
9. Schwartz, interview, June 16, 2010.
10. Carr, "Polaroid v. Kodak, a Reminiscence," 3–4.
11. Quillen, interview.
12. Lenzner, "Land Gives Polaroid's Side."
13. Trial Transcript, PvK, October 5, 1981, 3–4, F&N.
14. Ibid., 4–6.
15. Lenzner, "Land Gives Polaroid's Side."
16. Trial Transcript, PvK, 8–10.
17. Ibid., 10.

18. Ibid., 16–17; "Polaroid Says Kodak's Entry into Instant Photos Injured It," *New York Times*, October 6, 1981.
19. Trial Transcript, PvK, 19–20.
20. Lenzner, "Land Gives Polaroid's Side."
21. Trial Transcript, PvK, 20.
22. Lenzner, "Land Gives Polaroid's Side."
23. Trial Transcript, PvK, 21–22.
24. Ibid., 23.
25. Ibid., 28.
26. Ibid., 29.
27. John Case, "The Problems at Polaroid," *Boston Magazine*, April 1982, 108.
28. William Kerr, letter to Nan Chequer, September 29, 1981, F&N.
29. Trial Testimony of Edwin H. Land, PvK, October 5, 1981, 32.
30. Lenzner, "Land Gives Polaroid's Side."
31. Trial Testimony of Edwin H. Land, PvK, 32.
32. Ibid., 33–35.
33. Steve Wilson, "Land Takes the Stand in Patent Case," *Philadelphia Inquirer*, October 7, 1981.
34. Trial Testimony of Edwin H. Land, PvK, October 5, 1981, 54–55.
35. Lenzner, "Land Gives Polaroid's Side."
36. Trial Testimony of Edwin H. Land, PvK, October 5, 1981, 56–57.
37. Ibid., 88–94, 97, 115–16.
38. Ibid., 88–89.
39. Ibid., 83.
40. Ibid., 86, 110–11.
41. Ibid., 102–04.
42. Ibid., 121.
43. Ibid., 123.
44. Ibid., 122–34
45. Ibid., 129–31.
46. John Rumsey, "The Kodak-Polaroid Trial Left Some Strong Images."
47. Trial Testimony of Edwin H. Land, PvK, October 6, 1981, 16.
48. Ibid., 17–21.
49. Lenzner, "Meeting Polaroid's Land."
50. Trial Testimony of Edwin H. Land, PvK, October 6, 1981, 26.
51. Ibid., 27–31.
52. Ibid., 44–46.
53. Wilson, "Land Takes the Stand in Patent Case"; "Land Takes Stand Again," *Boston Globe*, October 7, 1981.
54. Trial Testimony of Edwin H. Land, PvK, October 6, 1981, 80–81.
55. Ibid., 84–85.
56. Ibid., 93–94.
57. Ibid., 95–98.
58. Trial Testimony of Edwin H. Land, PvK, October 7, 1981, 245–47, 256–61.
59. Ibid., 268.
60. Ibid., 268–71.
61. Ibid.

62. "Land Knew of Kodak's Film," *New York Times*, October 8, 1981.
63. George Croft, "Land: No 'Conflict' on Kodak Film," October 8, 1981.
64. Trial Testimony of Edwin H. Land, PvK, October 7, 1981, 297.
65. Ibid., 297–302.
66. Ibid., 307–08.
67. Ibid., 326–27.
68. Ibid.
69. Ibid., 329.
70. Trial Testimony of Edwin H. Land, PvK, October 9, 1981, 393–409.
71. Ibid., 421–22.
72. Ibid., 426.
73. Ibid., 427.
74. Ibid., 431–33.
75. Ibid., 435.
76. Ibid., 438–40.
77. Ibid., 438, 442.
78. Ibid., 443–44.
79. "Land Told Kodak about Color Film," *Boston Herald American*, October 10, 1981.
80. Robert Lenzner, "Land Tells of Polaroid's Falling Out with Eastman Kodak," *Boston Globe*, October 10, 1981.
81. Goldstein, *Intellectual Property*, 14.

Chapter 21

1. Trial Testimony of Edwin H. Land, PvK, October 9, 1981, 445–56.
2. Ibid., 456–57.
3. Ibid., 460–64, 473–80.
4. Robert Lenzner, "The Kodak, Polaroid Trial a Picture of Splitting Hairs," *Boston Globe*, October 14, 1981.
5. Ibid.
6. Phil Ebersole, "Polaroid's Land Is a Tough Trial Witness," *Rochester Democrat and Chronicle*, October 17, 1981.
7. Trial Testimony of Edwin H. Land, PvK, October 13, 1981, 509.
8. Ibid., 534–42.
9. Ebersole, "Polaroid's Land Is a Tough Trial Witness."
10. Trial Testimony of Edwin H. Land, PvK, October 13, 1981, 544.
11. Trial Testimony of Edwin H. Land, PvK, October 14, 1981, 626–27.
12. Ebersole, "Polaroid's Land Is a Tough Trial Witness."
13. Ibid.
14. Phil Ebersole, "Land Asked about Kodak 'Help' on Film," *Rochester Democrat and Chronicle*, October 15, 1981.
15. Phil Ebersole, "Land Questioned about Patent Similarities," *Rochester Democrat and Chronicle*, October 16, 1981.
16. Trial Testimony of Edwin H. Land, PvK, October 15, 1981, 725.
17. Ibid., 772.
18. Ibid., 764.
19. Ibid., 788–90.

20. Ebersole, "Polaroid's Land Is a Tough Trial Witness."

21. Trial Testimony of Edwin H. Land, PvK, October 16, 1981, 824.

22. Ibid., 829–31.

23. Ebersole, "Polaroid's Land Is a Tough Trial Witness"; Robert Lenzner, "Polaroid Case Focuses on Layer of Polymeric Acid," *Boston Globe*, October 17, 1981.

24. Trial Testimony of Edwin H. Land, PvK, October 16, 1981, 851.

25. Ibid., 852–55.

26. Ibid., 858.

27. Ibid., 880.

28. Ibid., 902–06.

29. Ibid., 919–20.

30. Trial Testimony of Edwin H. Land, PvK, October 20, 1981, 937–39.

31. Ibid., 938–39.

32. Ibid., 961.

33. Ibid., 979–82.

34. Polaroid Corporation's Memorandum in Support of Its Motion Pursuant to F.R.Evid. 403 for an Order Limiting the Time Allowed for Cross-Examination, PvK, October 23, 1981, F&N.

35. Patricia Martone, interview with author in New York City, June 30, 2010.

36. William Kerr, letter to Francis T. Carr, October 22, 1981, F&N.

37. Francis T. Carr, letter to William Kerr, October 22, 1981, F&N.

38. Trial Testimony of Edwin H. Land, PvK, October 21, 1981, 1011–47.

39. Polaroid Corporation's Motion Pursuant to F.R.Evid. 403.

40. Ibid., 2.

41. Ibid., 2–3.

42. Martone, interview.

43. Trial Testimony of Edwin H. Land, PvK, October 28, 1981, 1079–80.

44. Trial Testimony of Edwin H. Land, PvK, October 29, 1981, 1184.

45. Ibid., 1186.

46. Trial Testimony of Edwin H. Land, PvK, October 30, 1981, 1287.

47. Ibid., 1298.

48. Ibid., 1316.

Chapter 22

1. Trial Testimony of Robert Duncan, PvK, October 30, 1981, 1339–40.

2. Ibid., 1363.

3. Trial Transcript, PvK, November 12, 1981, 1929–30.

4. Ibid.

5. Barnaby Feder, "Fuji Ready to Sell Its Instant Cameras in Japanese Market," *New York Times*, October 13, 1981.

6. Mitchell Lynch, "Fuji Photo Film's New Instant Camera Appears on Par with Polaroid and Kodak," *Wall Street Journal*, October 21, 1981.

7. Quillen, interview.

8. Feder, "Fuji Ready to Sell Its Instant Cameras."

9. William K. Kerr, letter to Rya Zobel, November 23, 1981, F&N.
10. Ibid., 2579–619.
11. Trial Testimony of Howard Rogers, PvK, November 30, 1981, 3092–93.
12. Ibid., 3094–142.
13. Ibid., 3149–53.
14. Ibid., 3171–72; Trial Testimony of Howard Rogers, PvK, December 1, 1981, 3246–55.
15. Ibid., 3258.
16. Ibid., 3261.
17. Ibid., 3262–74.
18. Ibid., 3276.
19. Ibid., 3278.
20. Ibid., 3279–80.
21. Ibid., 3290–91.
22. Ibid., 3292–93.
23. Ibid., 3294.
24. Kerr, letter to Zobel.
25. Ibid.
26. Trial Transcript, PvK, December 2, 1981, 3322.
27. Ibid., 3323.
28. Ibid., 3324.
29. Ibid., 3379–85.
30. Ibid., 3394–95.
31. Ibid., 3644–45.
32. Ibid., 3886.
33. Ibid., 3890.
34. Ibid., 3891.

Chapter 23

1. Trial Transcript, PvK, December 9, 1981, 3897.
2. Memorandum in Support of Defendant's Motion under Rule 41(b), F.R.Civ.P., for Dismissal with Respect to U.S. Patent 3,362,821, PvK, December 9, 1981, 1–2, F&N.
3. Memorandum in Support of Defendant's Motion under Rule 41(b), F.R.Civ.P., for Dismissal with Respect to U.S. Patent 3,578,540, PvK, December 9, 1981, 1–2, F&N.
4. Memorandum in Support of Defendant's Motion under Rule 41(b), F.R.Civ.P., for Dismissal with Respect to U.S. Patent 3,753,392, PvK, December 9, 1981, 1–2, F&N.
5. Defendant's Motion under Rule 41(b), F.R.Civ.P., for Dismissal with Respect to U.S. Patent 3,810,220, PvK, December 9, 1981, F&N; Defendant's Motion under Rule 41(b), F.R.Civ.P., for Dismissal with Respect to U.S. Patent 3,709,122, PvK, December 9, 1981, F&N.
6. Polaroid Corporation's Memorandum in Opposition to Eastman Kodak Company's Motions to Dismiss under F.R.Civ.P. Rule 41(b), PvK, December 18, 1981, F&N.
7. Trial Testimony of Leo Thomas, PvK, December 9, 1981, 3913–14.
8. Ibid., 3915–33.

9. Ibid., 3963–65.
10. Ibid., 3996–98.
11. Ibid., 3998–4001.
12. Ibid., 4001.
13. Trial Testimony of Albert Sieg, PvK, December 10, 1981, 4042–56.
14. Ibid., 4049–50.
15. Ibid., 4090–99.
16. Ibid., 4135–44.
17. Ibid., 4142–43.
18. Ibid., 4144–47.
19. Ibid., 4147–49.
20. Trial Testimony of Charles Schallhorn, PvK, December 14, 1981, 4284–4306.
21. Ibid., 4315–20.
22. Ibid., 4318.
23. Trial Testimony of R. Frederick Porter, PvK, December 14, 1981, 4330–36.
24. Ibid., 4867.
25. Ibid., 4867–69.
26. "Court Denies Kodak's Bid to Water Down Polaroid Suit," *Boston Globe*, January 5, 1981, 52.
27. Calvert Crary, "Polaroid v. Eastman Kodak," *Bear Stearns Litigation Review* 2, no. 9, February 1982, F&N.
28. Trial Transcript, PvK, January 4, 1982, 4869.
29. Trial Testimony of Donald Smith, PvK, January 4, 1982, 4917–18.
30. Ibid., 5146–51.
31. Ibid., 5173.
32. Ibid., 5177–80.
33. Ibid., 5180.
34. Trial Testimony of Peter Adelstein, PvK, January 7, 1982, 5306.
35. Ibid., 5307.
36. Trial Transcript, PvK, January 8, 1982, 5328.
37. Ibid., 5328–29.
38. Ibid., 5335–36.
39. Trial Testimony of Adelstein, January 8, 1982, 5349–50.
40. Trial Testimony of Edwin Land, October 28, 1981, 1100–1101.
41. Trial Testimony of Adelstein, January 8, 1982, 5350.
42. Ibid., 5358–411.
43. Ibid., 5412.
44. Ibid., 5546–48.
45. Ibid., 5646–52.
46. Ibid., 5653.

Chapter 24

1. Statement of Eastman Kodak Company with Respect to Expert Witnesses, PvK, August 17, 1981, F&N; Trial Testimony of Franz Trautweiler, PvK, January 15, 1982, 6074–75.

2. Trial Testimony of Trautweiler, PvK, January 12, 1982, 5682–95.
3. Patents, 35 U.S.C. § 112.
4. Trial Testimony of Trautweiler, PvK, January 13, 1982, 5705–9, 5851–53.
5. Ibid., 5754.
6. Ibid., 5755.
7. Trial Testimony of Trautweiler, PvK, January 14, 1982, 5906.
8. Ibid., 5906–9.
9. Ibid., 5910–14.
10. Frank Carr, letter to Herbert Schwartz, January 20, 1982, F&N.
11. Herbert Schwartz, letter to Frank Carr, January 22, 1982, F&N.
12. Trial Testimony of Trautweiler, PvK, January 14, 1982, 5949.
13. Ibid., 5955.
14. Trial Testimony of Trautweiler, PvK, January 18, 1982, 6039–42.
15. Ibid., 6052–55.
16. Ibid., 6018–20.
17. Ibid., 6069–81.
18. Ibid., 6087–89.
19. Trial Testimony of Trautweiler, PvK, January 19, 1982, 6091–95.
20. Ibid., 6166–72, 6183–86.
21. Trial Testimony of Trautweiler, PvK, January 20, 1982, 6214–33.
22. Ibid., 6346–47.
23. Statement of Eastman Kodak Company with Respect to Expert Witnesses.
24. Trial Testimony of Edward Kaprelian, PvK, February 11, 1982, 8302–4.
25. Edward K. Kaprelian, "135 Years of American Cameras," *Photomethods*, August 1976; Trial Exhibit PT-448, PvK.
26. Ibid., 8304–10.
27. Martone, interview; Schwartz, interview, July 22, 2009.
28. Ibid.
29. Schwartz, interview, June 16, 2010.
30. Frank Carr, letter to Herbert Schwartz, February 5, 1982, F&N.
31. Frank Carr, letter to Herbert Schwartz, February 8, 1982, F&N.
32. Herbert Schwartz, letter to Frank Carr, February 9, 1982, F&N.
33. Herbert Schwartz, letter to Frank Carr, February 19, 1982, F&N.
34. Trial Testimony of Edwin H. Land, PvK, February 25, 1982, 8732.
35. Ibid., 8733–40.
36. Ibid., 8748–54.
37. Ibid., 8754–55.
38. Ibid., 8756–58.
39. Ibid., 8761–62.
40. Ibid., 8765.
41. Ibid., 8765–66.
42. Schwartz, interview, July 22, 2009.
43. Schwartz, interview, June 16, 2010.
44. Schwartz, interview, July 22, 2009.
45. Schwartz, interview, June 16, 2010.
46. Trial Testimony of Land, PvK, February 25, 1982, 8767.

47. Schwartz, interview, July 22, 2009.
48. Trial Testimony of Land, PvK, February 25, 1982, 8770.
49. Ibid., 8770–72.
50. Schwartz, interview, June 16, 2010.
51. Trial Testimony of Land, PvK, February 25, 1982, 8772–73.
52. Ibid., 8772–74.
53. Ibid., 8778–79.
54. Ibid., 8809–10.

Chapter 25

1. *1982 Annual Report of the Director of the Administrative Office of the United States Courts*, quoted in Robert Goldman, memorandum to Herbert Schwartz, June 7, 1985, F&N.
2. Herbert Schwartz, letter to Edwin H. Land, March 10, 1982, F&N.
3. Robert Lenzner, "Land: The Man behind the Camera," *Boston Globe*, October 17, 1976.
4. Schwartz, interviews, June 16, 2010, and September 30, 2010.
5. Plaintiff Polaroid Corporation's General Brief after Trial, PvK, 1–5, F&N.
6. Ibid., 6–8.
7. Ibid., 10–12.
8. Plaintiff Polaroid Corporation's Brief after Trial, Land U.S. Patent 3,362,821, PvK, 1–5, 20–21, F&N.
9. Kodak's General Post-Trial Brief, PvK, 1–3, F&N.
10. Pre-Trial Brief of Eastman Kodak Company, PvK, October 2, 1981, 3–4, F&N.
11. Marconi Wireless Co. v. United States, 320 U.S. 1, 37–38 (1943).
12. Kodak's General Post-Trial Brief, 2.
13. Kodak's Post-Trial Brief with Respect to U.S. Patent 3,362,821, PvK, 1, F&N.
14. Kodak's Post-Trial Brief with Respect to U.S. Patent 3,753,392, PvK, 21.
15. Ibid., 11–13.
16. Kodak's Post-Trial Brief with Respect to U.S. Patent 3,578,540, PvK, 2, F&N.
17. Ibid., 5–9.
18. Ibid., 14–15.
19. Kodak's Post-Trial Brief with Respect to U.S. Patent 3,245,789, PvK, 1, 4–6, F&N.
20. Ibid., 13–16.
21. Kodak's Post-Trial Brief with Respect to U.S. Patents 3,594,165 and 3,689,262, PvK, 1, F&N.
22. Ibid., 4–14.
23. Ibid., 15; James Avtges, memorandum to Stanley Mervis, New Aspirin Systems, January 30, 1968, DF-450, PvK, F&N.
24. Kodak's Post-Trial Brief with Respect to U.S. Patents 3,594,165 and 3,689,262, 15–17.
25. Ibid., 25.
26. Plaintiff Polaroid Corporation's Reply Briefs after Trial, PvK, July 12, 1982, 1–2, F&N.
27. Ibid., 2–4.
28. Ibid., 9.

29. Kodak's Post-Trial Reply Brief, PvK, July 12, 1982, 1, F&N.
30. Ibid., 2.
31. Herbert Schwartz, memorandum, July 12, 1982, F&N.
32. Fred Bayles, "Final Briefs Are Due in Kodak-Polaroid Suit," *Boston Globe*, July 27, 1982.
33. Lenzner, "The Promised Land," 46.
34. Herbert Schwartz, letter to William Kerr, July 12, 1982, F&N.
35. Trial Testimony of Frank Carr, PvKdam, vol. 58, August 23, 1989, 7406.
36. Quillen, interview.
37. Calvert Crary, "Polaroid v. Eastman Kodak," *Bear Stearns Litigation Review*, Vol. III, No. 3, September 1982, F&N.
38. Bayles, "Final Briefs Are Due."
39. Ibid.
40. "Polaroid Profit Plunged 95% in 4th Quarter," *Wall Street Journal*, February 19, 1982.
41. "A 94.7% Drop for Polaroid," *New York Times*, February 19, 1982.
42. Joan Fitzgerald, "Land, at 73, Leaves Polaroid for Pure Scientific Research," *Boston Globe*, July 28, 1982.
43. Michael Blumstein, "Era Ends as Land Leaves Polaroid," *New York Times*, July 28, 1982.
44. "Polaroid Profit Drops; Founder Ends Active Role," *Wall Street Journal*, July 28, 1982.
45. Liz Roman Gallese, "I Am a Camera," *Boston Business*, Fall 1986, 79.
46. William McCune, "Letter to Stockholders, 9/1 R. DeLima—for your comments," Polaroid Corporation Legal and Patent Records, Box II.229, f. 8, Annual Report Letter 1982, HBS.
47. Ibid.
48. William McCune, "Letter to Stockholders, Final," Polaroid Corporation Legal and Patent Records, Box II.229, f. 8, Annual Report Letter 1982, HBS.
49. Blumstein, "Era Ends as Land Leaves Polaroid."
50. Fitzgerald, "Land, at 73, Leaves Polaroid"; Daniel Nossiter, "No Instant Success," *Barron's*, July 5, 1982.
51. Ibid.
52. Lenzner, "The Promised Land."
53. Blumstein, "Era Ends as Land Leaves Polaroid."
54. Lenzner, "The Promised Land."
55. Subrata Chakravarty, "An Interview with Dr. Edwin Land," *Forbes*, June 1, 1975.
56. Dan Baum, "Polaroid's Founder to Sell Remaining 8.3% Stake in Firm," *Wall Street Journal*, April 30, 1985, Polaroid Corporation Administrative Records, Box I.122, f. 1, 1985 News Clippings, HBS; Fitzgerald, "Land, at 73, Leaves Polaroid."
57. Lenzner, "The Promised Land."
58. Wendy Fox, "Land Bids Farewell to Polaroid," *Boston Globe*, April 30, 1985.
59. Lenzner, "The Promised Land."
60. Lynn Kettleson, "Camera Pioneer Severs Last Ties with Polaroid," *Boston Herald*, April 30, 1985, Polaroid Corporation Administrative Records, Box I.122, f. 1, 1985—News Clippings, HBS.
61. Baum, "Polaroid's Founder to Sell Remaining 8.3% Stake in Firm."

62. Jeffrey Sonnenfeld, *The Hero's Farewell* (Oxford University Press, 1988), 121.

63. John Case, "The Problems at Polaroid," *Boston Magazine*, April 1982, 108.

64. Ibid., 162.

65. Nossiter, "No Instant Success."

66. Ibid.

67. Blumstein, "Era Ends as Land Leaves Polaroid."

68. Nossiter, "No Instant Success."

69. Gay Jervey, "Fuji Seen Bringing Instant Camera to U.S.," *Advertising Age*, March 1, 1982.

70. Sam Jameson, "Japan: Selling There a Snap, Polaroid Says," *Los Angeles Times*, April 20, 1982.

71. Joan Fitzgerald, "Kodak Displays Four New Instant-Photography Cameras," *Boston Globe*, January 14, 1982.

72. Barnaby Feder, "Kodak Introduces an Improved Line of Instant Cameras," *New York Times*, January 14, 1982.

73. William Cheeseman, letter to Nina Singer, November 1, 1982, F&N.

74. Patricia Martone, memorandum to file, November 4, 1982, F&N.

75. Patricia Martone, memorandum to file, November 9, 1982, F&N.

76. Subrata Chakravarty and Ruth Simon, "Has the World Passed Kodak By?," *Forbes*, November 5, 1984.

77. Ibid.

78. Ibid.

79. Gary Putka, "Growing Belief Kodak to Drop Instant Lines Helps to Strengthen Its Stock and Polaroid's," *Wall Street Journal*, September 28, 1983.

80. Thomas Lueck, "Kodak Upgrading Disk Film," *New York Times*, September 29, 1983.

81. Chakravarty and Simon, "Has the World Passed Kodak By?"

82. Mitchell Lynch, "Polaroid Tries to Get Itself in Focus," *New York Times*, May 15, 1983.

83. Herbert Schwartz, letter to Robert Ford, January 24, 1984, PvK.

84. Herbert Schwartz, draft letter to Honorable Rya Zobel, January 24, 1984, F&N.

85. Schwartz, interview, June 16, 2010.

86. Herbert Schwartz, draft letter to Honorable Rya Zobel, February 2, 1984, F&N.

87. Martone, interview.

88. Robert Peck, letter to Laurence Fordham, March 30, 1984, F&N.

89. Robert Goldman, memorandum to Herbert Schwartz, June 27, 1984, F&N.

90. Robert Peck, letter to Herbert Schwartz, September 19, 1984, F&N.

91. Herbert Schwartz, letter to Robert Peck, September 13, 1984, F&N.

92. Schwartz, interview, June 16, 2010.

93. Quillen, interview.

94. Ibid.

95. Herbert Schwartz, letter to Honorable Rya Zobel, October 17, 1984, F&N.

96. Herbert Schwartz, letter to Robert Peck, November 5, 1984, F&N.

97. Warshofsky, *The Patent Wars*, 81.

98. Schwartz, letter to Peck, November 5, 1984.

99. Nealda Corallo, memorandum to Herbert Schwartz, December 12, 1984, F&N.

100. Albert Frank-Guenther Law, News Release, April 29, 1985, Polaroid Corporation Legal and Patent Records, Box II.229, f. 3, Polaroid Corporation Annual Report 1984, HBS.

101. Ibid.

102. Herbert Schwartz, letter to Honorable Rya Zobel, April 3, 1985, F&N.

103. Herbert Schwartz, letter to Robert Ford, April 26, 1985, F&N.

104. Herbert Schwartz, Outline of Thoughts re: Conference on May 16, 1985, April 26, 1985, F&N.

Chapter 26

1. Martone, interview; Barnes, interview.
2. *Polaroid*, 641 F. Supp. at 828.
3. United Press International, Business & Finance Wire, New York, NY, September 13, 1985, F&N.
4. Polaroid Corporation, Press Release, September 12, 1985, Polaroid Corporation Legal and Patent Records, Box II.240, f. 20, Press Releases and Publicity Relating to the Trial, 1963–1988, HBS; "Polaroid Wins in Patent Suit against Kodak," *Wall Street Journal*, September 16, 1985.
5. United Press International, Business & Finance Wire.
6. *Polaroid*, 641 F. Supp. at 828, 829–31.
7. Ibid., 831.
8. Ibid.
9. Ibid.
10. Ibid., 831–32.
11. Ibid., 832.
12. Kodak's General Post-Trial Reply Brief, PvK, 4, F&N.
13. Goldstein, *Intellectual Property*, 14; Warshofsky, *The Patent Wars*, 84–85.
14. Goldstein, *Intellectual Property*, 14.
15. *Polaroid*, 641 F. Supp. at 833.
16. Ibid., 837.
17. Ibid.
18. Ibid., 838.
19. Ibid., 855–56.
20. Ibid., 857.
21. Ibid.
22. Ibid., 858.
23. Ibid., 859–60.
24. Ibid., 860.
25. Ibid., 867.
26. Ibid.
27. Ibid., 843.
28. Ibid.
29. Ibid., 840–41.
30. Ibid., 844.
31. Ibid., 844–45.

32. Ibid., 845.

33. Ibid., 846–47.

34. Ibid., 848.

35. David Sanger, "Kodak Infringed on Polaroid Patents," *New York Times*, September 14, 1985.

36. Wendy Fox, "Judge Rules for Polaroid in Patent Suit," *Boston Globe*, September 14, 1985.

37. United Press International, Business & Finance Wire, New York, NY, September 16, 1985, F&N.

38. Trial Testimony of Frank Carr, PvKdam, 7405.

39. Phil Ebersole, "A Kodak Infringement," *Rochester Democrat and Chronicle*, September 14, 1985.

40. Fox, "Judge Rules for Polaroid"; Calvert Crary, "Polaroid v. Eastman Kodak," *Bear Stearns Litigation Review*, October 1985, F&N.

41. John Crudele, "A New View of Polaroid," *New York Times*, October 10, 1985; Geoffrey Smith, "Polaroid Lands Win vs. Kodak," *Boston Herald*, September 14, 1985; John Rumsey, "Polaroid, Kodak May Try for Deal," *Rochester Times Union*, September 16, 1985.

42. Wilbur Prezzano, "Letter: Dear Kodak Customer," Eastman Kodak Company, September 16, 1985, F&N.

43. Ebersole, "A Kodak Infringement"; Crary, "Polaroid v. Eastman Kodak," October 1985; John Rumsey, "Polaroid, Kodak May Try for Deal," *Rochester Times Union*, September 16, 1985.

44. Fox, "Judge Rules for Polaroid."

45. Schwartz, interviews, January 30, 2009, and June 16, 2010.

46. Carr, "Polaroid v. Kodak, a Reminiscence," 19–20.

47. Schwartz, interview, June 16, 2010.

48. Eric Schmitt, "Judicial Shift in Patent Cases," *New York Times*, January 21, 1986; "Polaroid vs. Kodak: The Decisive Round," *Business Week*, January 13, 1986.

49. Quillen, interview.

50. Crary, "Polaroid v. Eastman Kodak," October 1985.

51. Goldstein, *Intellectual Property*, 14.

52. Herbert Schwartz, letter to Honorable Rya Zobel, September 23, 1985, F&N.

53. Frank Carr, letter to Honorable Rya Zobel, September 28, 1985, F&N.

54. Response of Eastman Kodak Company to Proposed Judgment of Polaroid Corporation, PvK, 3–4, F&N.

55. Ibid., 4–7.

56. Herbert Schwartz, letter to Honorable Rya Zobel, October 2, 1985, F&N.

57. Schwartz, interview, June 16, 2010.

58. Ibid.

59. Schwartz, letter, October 2, 1985; Frank Carr, letter to Honorable Rya Zobel, October 3, 1985, F&N; Frank Carr, letter to Honorable Rya Zobel, October 4, 1985, F&N.

60. Herbert Schwartz, letter to Honorable Rya Zobel, October 7, 1985, F&N.

61. Ibid.

Chapter 27

1. Bob Davis, "New Line of Cameras Developed by Polaroid Could Brighten Concern's Financial Picture," *Wall Street Journal*, June 4, 1985; John Hillkirk, "Polaroid Stock Develops," *USA Today*, September 17, 1985; John Crudele, "A New View of Polaroid," *New York Times*, October 10, 1985.

2. Bob Davis, "Polaroid Hopes to Snap Out of a Sales Slump," *Wall Street Journal*, November 11, 1985.

3. Crudele, "A New View of Polaroid"; Davis, "New Line of Cameras Developed by Polaroid."

4. Phil Ebersole, "A Kodak Infringement," *Rochester Democrat and Chronicle*, September 14, 1985.

5. "Polaroid Wins in Patent Suit against Kodak."

6. Alex Beam and William Glaberson, "Polaroid vs. Kodak: The Decisive Round," *Business Week*, January 13, 1986.

7. Memorandum of Decision and Order, PvK, October 11, 1985, F&N.

8. Schwartz, interview, June 16, 2010.

9. Martone, interview.

10. Memorandum of Decision and Order, 7.

11. Ibid., 4–5.

12. Ibid., 6–7.

13. Judgment, PvK, October 11, 1985, par. 15, F&N.

14. Colby Chandler, letter to Kodak customers, October 14, 1985, F&N.

15. Phil Ebersole, "Instant-Photography Advertising Strategies Diverge for This Christmas," *Rochester Democrat and Chronicle*, October 22, 1985.

16. "Eastman Defies Order," *Baltimore News American*, October 15, 1985.

17. "Kodak Is Undeterred by Polaroid Injunction," *Boston Globe*, October 15, 1985.

18. Geoffrey Smith, "Kodak to Keep Instant Film on Market Despite Ruling," *Boston Herald*, October 15, 1985.

19. Phil Ebersole, "Questions Still Remain in Kodak Photo Suit," *Rochester Democrat and Chronicle*, October 15, 1985; John Rumsey, "Will Instant Camera Owners Ask for Money Back?," *Rochester Times Union*, October 14, 1985.

20. Bill Clark, "Judge Bars EK from Future Sales of Instant Products," *Photo Weekly*, October 21, 1985.

21. Ibid.

22. Memorandum of Appellant Eastman Kodak Company in Support of Its Motion under Rule 8(a), F.R.A.P., for a Stay of Injunction Pending Appeal, Polaroid Corporation v. Eastman Kodak Company, Appeal No. 86–604 (Fed. Cir. November 4, 1985), 4, F&N.

23. Ibid., 4–5.

24. Affidavit of Colby H. Chandler, Polaroid Corporation v. Eastman Kodak Company, Appeal No. 86-604, 2–3, F&N.

25. Quillen, interview; Goldstein, *Intellectual Property*, 13.

26. Memorandum of Appellant Eastman Kodak, 5, 10.

27. Ibid., 8–9.

28. Ibid., 53.

29. Ibid., 13–14

30. Ibid., 14–15.

31. Ibid., 53.

32. "Death Row," *Newsday*, November 6, 1985; "Photo Finish," *New York Daily News*, November 7, 1985, 59.

33. United Press International, "Kodak Threatens Job Cuts if Injunction Not Halted," November 7, 1985.

34. J. Ernest Beazley, "Kodak to Drop Its Instant Camera Line if Ban on Sales after Jan. 9 Isn't Lifted," *Wall Street Journal*, November 7, 1985.

35. Polaroid's Memorandum in Opposition to Kodak's Motion for a Stay of Injunction Pending Appeal, Polaroid Corporation v. Eastman Kodak Company, Appeal No. 86-604, 3–5, F&N.

36. Ibid., 9–11.

37. Ibid., 15–16.

38. John Rumsey, "Kodak Not Planning to Close Its 'Instant' Operations," *Rochester Times Union*, November 21, 1985.

39. Calvert Crary, "Polaroid v. Eastman Kodak," *Bear Stearns Litigation Review*, November/December 1985, 3, F&N.

40. Order, Polaroid Corporation v. Eastman Kodak Company, Appeal No. 86-604, December 2, 1985, F&N.

41. Francis X. Gindhart, clerk, letter to Frank Carr and Herbert Schwartz, December 9, 1985, F&N.

42. Brief for Appellant, Polaroid Corporation v. Eastman Kodak Company, Appeal No. 86-604, 43–44, F&N.

43. Brief for Appellee, Polaroid Corporation v. Eastman Kodak Company, Appeal No. 86-604, 3, F&N.

44. Ibid., 28–29.

45. Reply Brief for Appellant, Polaroid Corporation v. Eastman Kodak Company, Appeal No. 86-604, 1.

46. Francis X. Gindhart, clerk, letter to Frank Carr and Herbert Schwartz, December 23, 1985, F&N.

47. Lee Byrd, "Kodak Seeks Reversal of Instant-Photo Ban," *Washington Post*, January 7, 1986; Kevin Maney, "Kodak Fights to Save Its Instant Film," *USA Today*, January 7, 1986.

48. Ronald Fierstein, contemporaneous notes taken during oral argument, January 6, 1986, author's collection.

49. Herbert Schwartz, notes for oral argument, January 6, 1986, F&N, quoting Anderson v. City of Bessemer City, NC, 105 S. Ct. 1504, 1511–13 (1985).

50. Byrd, "Kodak Seeks Reversal of Instant-Photo Ban."

51. Calvert Crary, "Polaroid v. Eastman Kodak," *Bear Stearns Litigation Review*, January 1986, 6, F&N.

52. Emergency Motion of Appellant Eastman Kodak Company for a Stay of Injunction Pending a Decision on Its Motion under Rule 8(a), F.R.A.P., Polaroid Corporation v. Eastman Kodak Company, Appeal No. 86-604, January 6, 1986, 2, F&N.

53. Schwartz, interview, June 16, 2010.

54. Polaroid Corporation's Opposition to Emergency Motion of Eastman Kodak Company for a Stay of Injunction, Polaroid Corporation v. Eastman Kodak Company, Appeal No. 86-604, January 7, 1986.

55. Rumsey, "Kodak Not Planning to Close Its 'Instant' Operations"; Schwartz, interview, June 16, 2010.

56. Ibid.

57. Ernest Beazley, Carol Hymowitz, and Bob Davis, "Kodak Loses a Decisive Round in Fight with Polaroid over Instant Photo Lines," *Wall Street Journal*, January 9, 1986.

58. Polaroid Corporation v. Eastman Kodak Company, 833 F.2d 930–31 (Fed. Cir. 1986).

59. Quillen, interview.

60. Application for Stay of District Court Injunction Pending Review by the United States Court of Appeals for the Federal Circuit, Polaroid Corporation v. Eastman Kodak Company, Pending Federal Appeal No. 86-604, January 8, 1986, 2, 9, F&N.

61. Joseph F. Spaniol Jr., clerk, Supreme Court of the United States, letter to Paul C. Warnke and Frank Carr, January 8, 1986, F&N.

62. John Hillkirk, "Kodak to Replace Instants," *USA Today*, January 9, 1986; Daniel Cuff, "Patent Case Plea Lost by Kodak," *New York Times*, January 9, 1986; "Kodak Gets the Picture Instant-Photo Line Shuttered," *Chicago Sun-Times*, January 9, 1986.

63. Michael Isikoff, "Kodak to End Its Instant Camera Line," *Washington Post*, January 9, 1986, Polaroid Corporation Administrative Records, Box I.122, f. 2, Polaroid vs. Kodak Clippings 1986, HBS.

64. "Instant Exit by Kodak," *Chicago Sun-Times*, January 8, 1986.

65. Beazley, Hymowitz, and Davis, "Kodak Loses a Decisive Round in Fight with Polaroid."

66. Isikoff, "Kodak to End Its Instant Camera Line."

67. Ibid.

68. John Jones, "Court Ruling, New Products Improve Outlook for Polaroid," *Investor's Daily*, January 14, 1986, Polaroid Corporation Administrative Records, Box I.122, f. 2, Polaroid vs. Kodak Clippings 1986, HBS.

69. "Instant Exit from Instant Cameras," *Fortune*, February 3, 1986, Polaroid Corporation Administrative Records, Box I.122, f. 2, Polaroid vs. Kodak Clippings 1986, HBS.

70. John Rumsey, "EK Costs Mounting Quickly," *Rochester Times Union*, January 9, 1986, Polaroid Corporation Administrative Records, Box I.122, f. 2, Polaroid vs. Kodak Clippings 1986, HBS.

71. Ellen Rosen and Victoria Shannon, "Kodak Workers Say: How Can They Do This?," *Rochester Times Union*, January 9, 1986, Polaroid Corporation Administrative Records, Box I.122, f. 2, Polaroid vs. Kodak Clippings 1986, HBS.

72. "Kodak Gets the Picture Instant-Photo Line Shuttered."

73. Colby Chandler, letter to Kodak shareowners, January 9, 1986, F&N.

74. Ibid.

75. Michael Isikoff, "A Market Develops Instantly," *Washington Post*, January 10, 1986, Polaroid Corporation Administrative Records, Box I.122, f. 2, Polaroid vs. Kodak Clippings 1986, HBS.

76. Jerri Stroud, "Kodak Offering a Trade," *St. Louis Post-Dispatch*, January 10, 1986; "Speculators on Kodak Flood Stores," *Greenwich Time* (Knight-Ridder Newspapers), January 10, 1986; Robin Schatz, "Kodak Delays Camera Trade-In Promotion," *Newsday*, January 16, 1986.

77. Carol Hymowitz, "Kodak Owners Advised: 'Take Rebate and Run,'" *Wall Street Journal*, January 10, 1986.

78. Daniel Cuff, "Callers Besiege Kodak on Trade-Ins," *New York Times*, January 10, 1986.

79. Hymowitz, "Kodak Owners Advised: 'Take Rebate and Run.'"

80. "Kodak Is Sued on Plan Offering Swap to Users of its Instant Cameras," *Wall Street Journal*, February 4, 1986, Polaroid Corporation Administrative Records, Box I.122, f. 2, Polaroid vs. Kodak Clippings 1986, HBS; "Suit Puts Kodak's Refund Program on Hold," *Investor's Daily*, February 16, 1986, Polaroid Corporation Administrative Records, Box I.122, f. 2, Polaroid vs. Kodak Clippings 1986, HBS.

81. "Kodak Postpones Refunds to Instant-Camera Owners," *Wall Street Journal*, June 9, 1986; Mark Lewyn, "Kodak Plan Is Sent Back to Darkroom," *USA Today*, June 6, 1986; "3.4M Eligible in Kodak Suit," *New York Newsday*, July 9, 1986.

82. Desiree French, "U.S. Judge Upholds Polaroid's Victory," *Boston Globe*, August 12, 1988.

83. Clare Ansberry, "No Doubt, Kodak Has a Few Ideas about What to Do with These," *Wall Street Journal*, January 17, 1986.

84. Lois Jenness, secretary to Judge Rya Zobel, letter to Herbert Schwartz, Frank Carr, and Richard Carlton, with attachments, March 18, 1986, F&N.

85. Nancy Perry, "The Surprising New Power of Patents," *Fortune*, June 23, 1986, 57.

86. Carr, "Polaroid v. Kodak, a Reminiscence," 9.

87. Colby Chandler, "Patent Power," *Fortune*, July 21, 1986, 17.

88. Nancy Perry, "Author Responds," *Fortune*, July 21, 1986, 17.

89. Polaroid Corporation v. Eastman Kodak Company, 229 U.S.P.Q. 561, 562 (Fed. Cir. 1986).

90. Ibid., 562–63.

91. Ibid.

92. "Kodak Asks Court Review on Polaroid," *New York Times*, July 25, 1986.

93. Petition for a Writ of Certiorari to the United States Court of Appeals for the Federal Circuit, Eastman Kodak Company, Petitioner v. Polaroid Corporation, Respondent (July 23, 1986), 13–14, F&N.

94. Respondent's Brief in Opposition to Petition for a Writ of Certiorari to the United States Court of Appeals for the Federal Circuit, Eastman Kodak Company v. Polaroid Corporation (August 22, 1986), 2, 27, F&N.

95. Ibid., 2–5.

96. Joseph F. Spaniol Jr., clerk, letter to Herbert F. Schwartz, October 6, 1986, F&N.

Chapter 28

1. Polaroid Corporation v. Eastman Kodak Company, 867 F.2d 1415, 1416 (Fed. Cir. 1989).

2. Carr, "Polaroid v. Kodak, a Reminiscence," 2–3.

3. "Court Rules for Polaroid," *New York Times*, August 12, 1988.
4. *Polaroid*, 867 F.2d 1415, 1417.
5. French, "U.S. Judge Upholds Polaroid's Victory."
6. "Court Rules for Polaroid."
7. *Polaroid*, 867 F.2d 1415, 1417.
8. Ibid., 1421; Stephen Labaton, "Polaroid v. Kodak: Future on the Line," *Business and the Law*, May 1, 1989.
9. Polaroid Corporation v. Eastman Kodak Company, 490 U.S. 1047 (1989).
10. Polaroid Corporation v. Eastman Kodak Company, 16 U.S.P.Q.2d 1481, 1507 (D. Mass. 1990).
11. Schwartz, interview, September 30, 2010.
12. *Polaroid*, 16 U.S.P.Q.2d 1481, 1541; John Holusha, "Kodak Told It Must Pay $909 Million," *New York Times*, October 13, 1989.
13. Polaroid Corporation v. Eastman Kodak Company, 17 U.S.P.Q.2d 1711 (D. Mass. 1991); Kurt Eichenwald, "Company News; Kodak's Payment to Polaroid Reduced," *New York Times*, January 12, 1991.
14. "Kodak to Pay Polaroid $925 Million to Settle Suit," *Wall Street Journal*, July 16, 1991; Lawrence Edelman, "Polaroid, Kodak Both Ready to Appeal $873M Decision," *Boston Globe*, February 8, 1991.
15. Jill Andresky, "A Weapon at Last," *Forbes*, March 10, 1986; Perry, "The Surprising New Power of Patents."
16. Eric Schmitt, "Judicial Shift in Patent Cases," *New York Times*, January 21, 1986.
17. Perry, "The Surprising New Power of Patents," 60.
18. Schmitt, "Judicial Shift in Patent Cases."
19. Bob Davis, "Computer Firms Turn to Patents, Once Viewed as Weak Protection," *Wall Street Journal*, January, 1986.
20. Gosselin, "How Polaroid Beat Kodak."
21. Paul Janicke, memorandum to author, November 18, 2013.
22. Perry, "The Surprising New Power of Patents," 57.
23. Schmitt, "Judicial Shift in Patent Cases."
24. Quillen, interview; Goldstein, *Intellectual Property*, 4–5, 11–12; Warshofsky, *The Patent Wars*, 77, 80.
25. "Polaroid Wins Another Round in Kodak Suit," Boston Globe, April 26, 1986, Polaroid Corporation Administrative Records, Box I.122, f. 5, Press Clippings April 1986, HBS.
26. Goldstein, *Intellectual Property*, 5.
27. Trial Testimony of Francis T. Carr, PvKdam, vol. 59, August 24, 1989, 7507.
28. James Bessen and Michael J. Meurer, *Patent Failure* (Princeton University Press, 2008), 70.
29. Andresky, "A Weapon at Last."
30. Janicke, memorandum.
31. Goldstein, *Intellectual Property*, 5, 11.
32. Quillen, interview.
33. Warshofsky, *The Patent Wars*, 80–81.
34. Glynn S. Lunney Jr., "Patent Law, the Federal Circuit, and the Supreme Court: A Quiet Revolution," *Supreme Court Economic Review* (University of Chicago, 2004), 1, 7–8.

35. Carr, "Polaroid v. Kodak, a Reminiscence," 10–13; Warshofsky, *The Patent Wars*, 81–85.

36. Trial Testimony of Francis T. Carr, PvKdam, vol. 58, August 23, 1989, 7397.

37. Goldstein, *Intellectual Property*, 11.

38. Trial Testimony of Colby Chandler, PvKdam, vol. 51, August 14, 1989, 6511–12.

39. J. Allen Jones, handwritten notes, May 1975, PvK, Trial Exhibit PT-2684; Trial Testimony of Colby Chandler, vol. 54, August 17, 1989, 6955–56.

40. Trial Testimony of Colby Chandler, vol. 51, 6512.

41. Swasy, "Changing Focus," 24.

42. Trial Testimony of Colby Chandler, vol. 51, 6511–13; "Kodak Defense in Camera Trial," *New York Times*, August 15, 1989.

43. Warren Brown, "Judge Orders Kodak to Pay $909 Million," *Washington Post*, October 13, 1990; "What Every Kodak Customer and Investor Should Know," Eastman Kodak Full Page Advertisement, October 1990, F&N; Bessen and Meurer, *Patent Failure*, 50–51.

44. Trial Testimony of Walter Fallon, PvKdam, vol. 60, August 25, 1989, 7640–41; Alan Cooperman, "Polaroid Case May Turn on Scribble," *Washington Post*, May 2, 1989.

45. Anthony Flint, "Polaroid, Kodak Damage Suit Opens," *Boston Globe*, May 2, 1989.

46. *Polaroid*, 16 U.S.P.Q.2d 1481, 1538.

47. Ibid., 1539.

48. Ibid.

49. Bessen and Meurer, *Patent Failure*, 47–48, 249.

50. Ibid., 48; Warshofsky, *The Patent Wars*, 78–79.

51. Plaintiff Polaroid Corporation's Post-Trial Submissions, facts IX–4, PvK, F&N.

52. Goldstein, *Intellectual Property*, 23.

53. Swasy, "Changing Focus," 24.

54. Carr, "Polaroid v. Kodak, a Reminiscence," 9–13.

55. Goldstein, *Intellectual Property*, 14.

56. Schwartz, "*Polaroid Co. v. Eastman Kodak Co.*: A Retrospective," 8.

57. Gosselin, "How Polaroid Beat Kodak."

58. Respondent's Brief in Opposition, Eastman Kodak v. Polaroid Corporation Petition for Writ of Certiorari, 2, F&N.

59. Trial Testimony of Francis T. Carr, PvKdam, vol. 55, August 18, 1989, 7010–11; ibid., vol. 59, August 24, 1989, 7477, 7481–83, 7510–11, 7518–19, 7570–71; Plaintiff Polaroid Corporation's Post-Trial Submissions, facts IX–4.

60. Carr, "Polaroid v. Kodak, a Reminiscence," 13.

61. Ibid., 15.

62. Plaintiff Polaroid Corporation's Post-Trial Submissions, facts IX–13; Trial Testimony of Francis T. Carr, PvKdam, vol. 56, August 21, 1989, 7140–41, 7147–77; ibid., vol. 58, August 23, 1989, 7447–51, 7459–61; ibid., vol. 59, August 24, 1989, 7541–43.

63. Charles Neave, letter to Judge Evan E. Evans, Seventh Circuit Court of Appeals (1935), as recounted in *History of the First Century of the Firm of Fish & Neave* (1978), 21.

Chapter 29

1. *Polaroid*, 16 U.S.P.Q.2d 1481, 1507.

2. Alex Beam, "Spectra's Instant Success Gives Polaroid a Shot in the Arm," *Business Week*, November 3, 1986; Brian Dumaine, "How Polaroid Flashed Back," *Fortune*, February 16, 1987; Nicholas Kristof, "Polaroid Bets on New Camera," *New York Times*, April 3, 1986.

3. Linda Snyder Hayes, "What's Kodak Developing Now?," *Fortune*, March 23, 1981.

4. David Warsh, "Instant Photos and Polaroid," *Boston Globe*, October 27, 1981; John Marcom Jr., "Sony Corp. Introduces Video Still Camera That Uses Magnetic Discs to Replace Film," *Wall Street Journal*, August 25, 1981.

5. Steve Lohr, "Filmless Video Camera Is Introduced by Sony," *New York Times*, August 25, 1981; Barnaby Feder, "Kodak's Fight with Polaroid," *New York Times*, October 8, 1981.

6. "Press Release," Sony Corporation, October 10, 1981, Polaroid Corporation Legal and Patent Records, Box II.177, f. 2, Edwin H. Land Cross-Examination Pre-Trial, HBS; Memorandum, Polaroid Corporation, "Sony Participants in the Mavica Press Conference," Polaroid Corporation Legal and Patent Records, Box II.177, f. 2, Edwin H. Land Cross-Examination Pre-Trial, HBS.

7. Alex Beam and Otis Port, "The Filmless Camera Is Here, but Will It Sell?," *Business Week*, April 15, 1985; Warsh, "Instant Photos and Polaroid."

8. "Kodak President Says Concern Could Make an Electronic Camera," *Wall Street Journal*, October 29, 1981.

9. Beam and Port, "The Filmless Camera Is Here."

10. Alex Beam, "Is Polaroid Playing to a Market That Just Isn't There?," *Business Week*, April 7, 1986; Claudia Deutsch, "Deep in Debt since 1988, Polaroid Files for Bankruptcy," *New York Times*, October 13, 2001.

11. Gal Borenstein, *What Really Counts for CEOs* (Borenstein Executive University Press, 2009).

12. Leslie Wayne, "Polaroid Gropes for Another Winner," *New York Times*, February 3, 1985.

13. Ibid.

14. Syre, "Polaroid Slowly Fades Away but at One Time Edwin Land and His Audacious Firm Could Do No Wrong."

15. Deutsch, "Deep in Debt since 1988, Polaroid Files for Bankruptcy"; Claudia Deutsch, "For Polaroid, the Bad News Seems to Be the Only News," *New York Times*, October 4, 2001.

16. Claudia Deutsch, "Kodak Gets the (Digital) Picture," *International Herald Tribune*, January 26, 2005; Matthew Daneman, "Kodak Bets on Film Lovers," *Rochester Democrat and Chronicle*, January 4, 2009.

17. Nicholson Baker, "A Fourth State of Matter," *New Yorker*, July 8 and 15, 2013, 65, 69.

18. "Former Trailblazer Kodak Files for Chapter 11," *Wall Street Journal*, January 19, 2012; "Eastman Kodak," *Standard & Poor's Stock Report*, September 25, 2010; "Kodak Retires Kodachrome Film," *Leisure & Travel Business*, July 12, 2009.

19. Michael De La Merced, "Eastman Kodak Files for Bankruptcy," *New York Times*, January 19, 2012.

20. Andrew Martin and Michael De La Merced, "Bankrupt, Kodak Vows to Rebound," *New York Times*, January 19, 2012.

21. Michael De La Merced, "Kodak to Sell Legacy Film Units as Part of Bankruptcy Plan," August 23, 2012.

22. Andrew Martin, "Kodak to Sell Digital Imaging Patents for $525 Million," *New York Times*, December 19, 2012.

23. Rachel Feintzeig and Tom Gara, "Kodak Is Back, but Not the Kodak You Remember," *Wall Street Journal*, September 3, 2013.

24. Mike Dickinson, "Kodak Closes on $848 Million Financing," *Rochester Business Journal*, March 22, 2013; "Kodak, Smaller and Redirected, Leaves Bankruptcy," *New York Times*, September 3, 2013; Michael Daneman, "Kodak Bankruptcy Officially Ends," *USA Today*, September 3, 2013.

25. "Former Trailblazer Kodak."

26. Damon Darlin, "Always Pushing beyond the Envelope," *New York Times*, August 7, 2010; Swasy, "Changing Focus."

27. Alex Taylor, "Kodak Scrambles to Refocus," *Fortune*, March 3, 1986.

28. Uttal, "Eastman Kodak's Orderly Two-Front War," 124.

29. Swasy, "Changing Focus," 25.

30. Peter Applebome, "Despite Long Slide by Kodak, Company Town Avoids Decay," *New York Times*, January 16, 2012.

31. Mark Hare, "Kodak's Future and Rochester's Are No Longer Linked," *Rochester Democrat and Chronicle*, September 28, 2003.

32. Mark Feeney, "Instant Karma; Before Polaroid Fades into History, Let's Remember How Influential—and Cool—the Art of the Snapshot, and the Cameras Themselves, Could Be," *Boston Globe*, March 16, 2008.

33. Jenna Wortham, "Polaroid Fans Try Making New Film for Old Cameras," *New York Times*, January 22, 2009; Adam McCauley, "Reinventing Instant Film in an Age of Instant Imagery," *New York Times*, August 22, 2012.

34. "Polaroid Instant Film to Return in 2010," *Rochester Democrat & Chronicle*, November 8, 2009.

35. Advertisement, Bloomingdales, *New York Times*, December 5, 2010, 42.

36. Christopher Barnard, "Photo Booths, the Party's Hot Spot," *New York Times*, September 27, 2013.

37. "Polaroid and Lady Gaga Announce Grey Label," press release, Polaroid Corporation, January 6, 2011.

38. Wortham, "Polaroid Fans Try Making New Film."

39. Feeney, "Instant Karma."

40. Carter Dougherty, "Dutch Group Looks to Revive Polaroid Instant Film," *New York Times*, May 26, 2009.

41. Gallese, "I Am a Camera."

42. Feeney, "Instant Karma."

43. Dougherty, "Dutch Group Looks to Revive Polaroid Instant Film."

44. Edwin H. Land, "One-Step Photography," *Photographic Journal*, January 1950, in *Edwin H. Land's Essays*, vol. 1, 139.

45. Adams, *An Autobiography*, 296–304.

46. Hitchcock, "When Land Met Adams," 24–29.

47. *The Polaroid Years, Instant Photography and Experimentation*, ed. Mary-Kay Lombino (Frances Lehman Loeb Art Center, Vassar College, Delmonico Books, 2013), Exhibition Catalog.

48. Lenzner, "The Promised Land."

49. Edwin H. Land, "Research by the Business Itself" (presented at the Standard Oil Development Company forum in New York, NY, October 5, 1944), in *Edwin H. Land, Selected Papers on Industry* (Polaroid Corporation, 1983), author's collection; R.T. Kriebel, Memorandum to All Employees, Polaroid Corporation, March 2, 1956, Polaroid Corporation Legal and Patent Records, Box II.3, f. 1, Land, HBS.

50. Gallese, "I Am a Camera," 50.

51. Land, "A Scrapbook."

52. Ibid.

53. "Dr. Land Redesigns His Camera Company," *Business Week*, April 15, 1972.

54. Land, "On Some Conditions for Scientific Profundity in Industrial Research," 57.

55. Stephen A. Benton, "Edwin Land, 3-D, and Holography," *Optics & Photonics News*, October 1994, 41–43.

56. Jeffrey Sonnenfeld, *The Hero's Farewell* (Oxford University Press, 1988), 119.

57. Paul Hofmann, "Aide Says Polaroid Move to Help South African Blacks Is Success," *New York Times*, October 31, 1971; Michael C. Jensen, "Polaroid Severs Business Links to South Africa," *New York Times*, November 23, 1977; "S. Africa's Loss," *New York Times*, November 27, 1977.

58. Eric Morgan, "The World Is Watching: Polaroid and South Africa," *Enterprise & Society* 7, no. 3 (Oxford University Press, 2006).

59. Case, "Problems at Polaroid," 158.

60. "Polaroid: An Award for Developing Human Resources," *Business Week*, June 30, 1973.

61. "Edwin H. Land, a Scrapbook" (video tribute).

62. Steven Syre, "A Magical Meeting," *Boston Globe*, October 7, 2011.

63. "The Playboy Interview: Steven Jobs," *Playboy*, February 1985; Walter Isaacson, *Steve Jobs* (Simon & Schuster, 2011), xix, 567.

64. Lenzner, "The Promised Land."

65. Sonnenfeld, *The Hero's Farewell*, 119–20.

66. Syre, "A Magical Meeting."

67. Ibid.

68. Steve Lohr, "The Missing Tastemaker," *New York Times*, January 19, 2011, B-1.

69. John Sculley, "No Bozos. Ever.," *Bloomsberg Businessweek*, October 10–16, 2011, 27.

70. Isaacson, *Steve Jobs*, 567.

71. Ibid., 207; Sculley, "No Bozos. Ever."

72. Mary A. McCann and John J. McCann, "Land's Chemical, Physical, and Psychophysical Images," *Optics and Photonics News*, October 1994, 34.

73. "Polaroid's Big Gamble on Small Cameras"; "Dr. Land Redesigns His Camera Company."

74. Nan Chequer, Handwritten notes, *Nan B. Chequer Notebook*, September 19–November 5, 1978, Polaroid Corporation Legal and Patent Records, Box II.189, f. 6, Nan B. Chequer Research Notebook, HBS.

75. Lenzner, "Land: The Man behind the Camera."
76. Adams, *An Autobiography*, 306.
77. Sonnenfeld, *The Hero's Farewell*, 119–20.
78. Adams, *An Autobiography*, 298–99.
79. Lenzner, "Land: The Man behind the Camera."
80. Blout, "Polaroid: Dreams to Reality."
81. Elkan R. Blout, interviews by James J. Bohning and Arnold Thackray at Harvard Medical School, Harvard School of Public Health, Cambridge, Massachusetts, 30 May 1991, 13 September 2002, and 22 November 2002 (Philadelphia: Chemical Heritage Foundation, Oral History Transcript # 0263), 43–44.
82. "Edwin H. Land, a Scrapbook" (video tribute).
83. McCune, "Light and Life," 43.
84. McElheny, "Edwin Herbert Land."
85. Chakravarty, "An Interview with Dr. Edwin Land."
86. Semir Zeki, "Light and Life, a Symposium in Honor of Edwin Land," *Bulletin of the American Academy of Arts and Sciences* 45, no. 7 (April 1992), 52.
87. Goldstein, *Intellectual Property*, 15.
88. Joseph A. Schumpeter, *Capitalism, Socialism and Democracy* (Harper & Brothers, 1942).
89. Land, "The Role of Patents in the Growth of New Companies," 509.

INDEX

A

Abel, Rudolf, 91
Academy of Sciences, 39
Adams, Ansel, 71–73, 142, 157, 158, 189, 522, 529
Adams, Harkness & Hill, 237
Adelstein, Peter, 415–419, 435–436, 443, 511
Administrative Office of the United States Courts, 460–461
Aerial photographs, 34
Agfa-Gevaert, 47, 235
AILA. *See* American Intellectual Law Association (AILA)
Alda, Alan, 198
Allied Chemical, 177
American Academy of Arts and Sciences, 62, 74, 533
American Intellectual Law Association (AILA), 504, 505
American Optical, 19–20
Anaglyph, 33
Apple, 526
Aspirin film format, 113, 180, 203
Associated Press, 28, 451
AT&T, 23, 176, 177
Atabrine, 35
Atomic bomb, 30–31, 36–37
Automobile Manufacturers Association, 27, 40

B

Bache Halsey Stuart, 233
Baker, James, 142–143
Baker, William, 90, 94
Baltimore News American, 482
Barnes, Richard, 287–288
Barrett, Terry, 300
Barron's, 237
Bartholin, Erasmus, 4
Bartlett's Familiar Quotations, 297
Baruch, Bernard, 26
Batchelder, Al, 115
Bates, Joseph, 3
Battiste, Gerry, 461
Beacon Hill, 88
Bell & Howell, 183
Bell, Alexander Graham, 57, 176, 243
Bellows, Alfred, 117
Benton, Stephen, 34, 37, 525
Berg, Wolfgang, 287–288, 292
Bergen, Candace, 198
Berkey, Ben, 173, 192
Berkey Keystone Wizard, 191, 196
Berkey Photo
- antitrust lawsuit against Kodak, 242, 249–250, 255–257, 283, 289
- development of instant camera, 173, 182
- Polaroid licensing policy and, 228
- Polaroid infringement suit against, 196, 249–250, 256–257
- Polaroid patent issues, 191–196

Berry, Rynn, 178, 180, 181
Bethlehem Steel, 177
B.F. Goodrich, 177
Binda, Frederick, 43, 48
Bissell, Richard, 91
Black, Susan, 241
"Black Thursday" stock market crash, 12
Blinow, Igor, 215

Bloom, Stanley, 117–119
Bloomer, Michael, 504, 505
Blout, Elkan, 67, 96–97, 295, 530–531
Bosman, Andre, 520
Boston Globe, 62, 189, 208, 210, 237, 344, 356, 363, 368, 371, 453, 474, 482, 517, 519
Boston Herald, 368, 482
Boston Herald American, 200
Boston Patent Law Association (BPLA), 74–75
Bower, Joseph L., 525
BPLA. *See* Boston Patent Law Association (BPLA)
Brewster, David, 3
Bristol-Myers, 221
British Journal of Photography, 139
Brown, Donald, 11–12, 15, 27, 42, 49, 61, 74–75, 111, 175–177
Browne & Witter, 220
Brush, Charles, 220
Buckler, Sheldon, 113, 116
Buckley, Denys Burton, 233
Burger, Warren E., 493, 494
Burley, Robert, 519
Bush, Vannevar, 30, 35
Business Week, 148, 172, 210, 323, 324, 480, 526

C

Camp Moween, 4
Campbell, F.W., 8
Campbell, John E., 215, 320, 328
Capstaff, John, 67
Carlton, Richard E., 501
Carr, Francis T. "Frank"
  Adelstein examination, 415–417
  background, 221
  discovery process, 239–240, 257, 285
  due diligence of, 509
  files briefs with court of appeals, 488–489
  Land cross-examination, 369–388
  Land deposition, 294–297, 299–300, 311–312
  Land rebuttal presentation, 436–439
  on last day of trial, 452
  on motion to limit cross-examination, 390–391
  on news of permanent injunction, 480
  notice of appeal of judgment, 484–485
  opening statement, 352–354
  oral arguments on appeals, 490–491
  plea for extension of the discovery deadline, 321, 322
  Polaroid rebuttal witnesses, 432
  Porter examination, 411
  proposed agenda for pre-trial conference, 329
  proposed form of judgment, 476
  reaction to court decision, 474
  Rogers deposition, 266–273, 400
  Rule 41 motions, 398–399, 403
  seeks to extend length of Land's deposition, 291
  settlement conference, 337–338
  settlement discussions, 338–344
  Sieg deposition, 252–255
  status conference to consider form of judgment, 477
  Thomas examination, 404–405
  Trautweiler examination, 421–422
  witness in damages trial, 501
Carter, Jimmy, 309, 493
Case, Raymond, 5
CBS, 188, 199
Chaffetz, Hammond, 201
Chandler, Colby
  background, 237–238
  chairman of Kodak, 458
  dismissive of injunction issued, 482–483
  letter to *Fortune*, 497
  president of Kodak, 241
  on problems of cameras and film, 235–236
  remarks on electronic cameras, 516
  remarks on entry in instant photography business, 507–508
  report on Kodak instant photography business, 245

statement on leaving instant photography business, 495
Chapman, Albert, 82, 97
Chase Manhattan Bank, 151
Cheeseman, William, 248, 260, 457, 465
Chequer, Nan, 276, 290, 293, 299, 300, 325, 334, 354–355, 386, 417, 441–442, 495
Choate Hall & Stewart LLP, 299
Christian Science Monitor, 13, 25, 41, 55
Chrysler, 20, 28
Chubb, Lewis Warrington, 16–17, 20–21, 26–27, 32
CIA, 89, 90, 91, 94
Ciba-Geigy AG, 421
Cincinnati Enquirer, 62
City of Los Angeles, 27–28
Clark, Joan, 275, 276–277, 290
Coca-Cola, 221
Cole, Harold, 293
Colgate University, 181
"Color coupler" chemistry, 68
Color diffusion transfer process, 204
Color photography, 426
Color vision, 421, 426–428, 434–435
Columbia University, 7–8, 10, 88
Committee on Investigation of New Devices, 15
Compton, Karl T., 28, 30
Conant, James Bryant, 30, 35
Consumer Reports, 236
Copyrapid, 47
Crank patent, 227, 347–348
Crary, Calvert, 232, 234, 246, 261, 327, 412, 452, 474, 476, 487–488
"Creative Destruction," 533–534
Curtiss, Glenn, 176

D

Damschroder, Rudolph E., 129–130
Dann, C. Marshall, 244
Dean, John, 95
Declaratory judgment action, 180
DeForrest, Lee, 28
DeLima, Richard, 183, 196, 201, 238, 452, 454, 461–462
Delwiche, Donald, 161
Democrat and Chronicle, 371, 373, 474
Dery, Don, 206, 208, 524
Detachable Spread Housing patent, 215, 390, 465
Dewey, Thomas, 62
Dicker, Gerald, 461
Diesel, Rudolf, 243
Dietz, Milton, 390
Diffusion transfer imaging process, 46–47, 51, 426
Disney, Roy, 517
Doctrine of Equivalents, 212
Doering, William, 35
Donaldson, Lufkin & Jenrette, 241
Donovan, Leisure, Newton & Irvine, 283
Doolittle, James H., 89, 91
Dorfman, Elsa, 521
Dove Project, 35–36
Drexel Burnham Lambert, 237
Driscoll, John, 390
Dulles, Alan, 90
Duncan, Robert, 140, 190, 389–390, 471
DuPont, 177, 324

E

Eastman, George, 51, 55, 85, 175, 197, 243–244, 519
Eastman Kodak Company
answer and counterclaims to complaint, 225–229
antitrust lawsuits against, 85, 183, 242, 249
appeals court decision, 497–498
appointment of Special Master, 259–260
Berkey Photo lawsuit, 242, 249, 255–257, 283, 289, 502
black-and-white film, 67, 84
British appeals court overturns injunction, 233
business relationship with Land-Wheelwright Laboratories, 17–18

Eastman Kodak Company (*continued*)
business relationship with Polaroid, 58, 67, 82–87, 366–368, 384
campaign to challenge Polaroid foreign patents, 174–175, 203–204, 221–222
Chandler becomes president, 458
change in attitude toward Polaroid, 122–125
choice of experts, 511–512
claim of inadvertence, 510
class action lawsuits, 496, 501
color film, 66–67, 82–87, 95–103
court decision on damages, 503–504
defects in instant film, 236
demise of, 517–519
determination on amount of compensatory damages, 501
discovery process, 239–240, 248–249, 257–260, 263, 279–280, 285
dismissive of injunction issued against, 482–484
Dove Project, 36
earnings, 84, 241, 247–248
entering instant photography field, 151–154, 184–185, 198–201, 443, 507–508
Fallon appointed president, 163–165
Fallon retirement, 458
files application for Supreme Court review, 498–499, 503
files briefs with court of appeals, 488–489
files emergency motion asking for stay of the injunction, 492
files for bankruptcy protection, 518
files for summary judgment, 280–284, 320, 328, 336–340
files new motion to compel compliance with discovery, 322
files reply brief on the discovery motion, 300–301
film disc technology, 456
Fuji Photo Film Co. and, 391–392
High Court of London issues temporary injunction against, 232–234
issue of damages, 452, 474
issues public letter to customers, 474–475
joint press release marking supply and license agreements, 139
on Judge Nelson recusal, 305–306
on Judge Zobel recusal, 306–307
Land's polarizer patents and, 16–17
lanyard camera, 160–163, 407–410
leaves instant photography business, 494–497
legal strategy, 251–255
license issues with Polaroid, 107–111, 123–133, 184–185, 367–368
license to manufacture a peel-apart film, 138–139
Mees role at, 51–52
misuse allegations, 227–229, 231–232
motion to void all of Zobel's decisions, 502–503
negative manufacturing facility, 148
notice of appeal of judgment, 483
observation of Land's SX-70 introductions, 161–162
offers license on PR-10 film to Polaroid, 224
oral arguments on appeals, 490–491
P-129 project, 153, 167–169, 405, 407
P-130 project, 152, 160, 161–162, 169, 208
patent on Kodak imaging chemistry, 184, 205
permanent injunction against, 480–482
PL-974 project, 405
PL-976 project, 404–405
polarizers, 16–21
Polaroid announces end of exploratory talks with, 260–261
Polaroid files infringement suit against, 214–217
post-decision spin, 509–510
post-trial brief, 444–450
problems with cameras and film, 235–236, 242
products for instant photography, 108–111, 178
proposed form of judgment, 476–477
proposed takeover of Land-Wheelwright Laboratories, 25–26

reaction to Polaroid negative manufacturing facility, 134
reaction to release of SX-70, 167–169, 408
regime change in 1968 at, 124
reply brief, 450–451
reviews Polaroid patents, 281–283, 508–509
Rule 41 motions, 392, 397–398, 403, 412–413
seeks emergency stay of the injunction from Supreme Court, 493
settlement discussions, 338–344, 461–462
sponsorship of Summer Olympic Games, 458
summary judgment motions, 280–281, 287, 292–293, 304, 317, 319, 326–329
Supreme Court decision, 494, 499
survival strategy, 518
three-dimensional movies, 23
United Kingdom patent revocation action by, 180
unveiling of instant photography system, 205–207
U.S. Department of Justice investigation, 242–243
work on integral instant photography system, 160, 165–166
Ebersole, Phil, 373
Edison, Thomas, 6, 176, 243, 523
Eilers, Louis, 124–129, 134–138, 163, 223, 281, 367
Einstein, Albert, 30
Eisendrath, David, 131
Eisenhower, Dwight D., 62, 88–90, 91–93, 94, 95, 362
Electric Boat, 2
Electronic cameras, 515–516
Elia, Charles, 248
Elliott Cresson Medal, 28
Enderlin, Peter, 479, 494
Enola Gay, 36
Excedrin film format, 118, 180
Excedrin film unit patents, 203–204, 221, 234, 280–284, 393, 395–397, 421, 422–426, 428, 448–449, 465, 472–473
Excess profits tax, 57
Exhausted Developer Process, 47, 68

F

Fairchild Semiconductor, 140
Fallon, Walter
announces increase in profits and sales, 283
announces retirement, 458
appointment as president of Kodak, 163–165
becomes chairman of board of directors, 241
considerations of, 508
deposition of, 291–292
on Kodak's interest in instant photography, 237–238
informal talks with Polaroid, 245–246
meeting to discuss negotiated resolution, 321
negotiations with Polaroid, 131
P-129 project, 406
Polaroid patents, 152, 281
on problems of cameras and film, 235–236
remarks after filing of lawsuit, 219
settlement discussions, 340–344
unveiling of instant photography system, 205–207
on validity of Polaroid patents, 315
work on integral instant photography system, 169, 178, 184–185, 200
Family Circle, 206
Fermi, Enrico, 62
Fier, Robert, 252, 407
Fish & Neave, 177, 178, 181, 203, 207, 220, 230, 239–240, 252, 276, 315, 318, 322, 333, 383, 390
Fish, Frederick P., 176
Fish, Richardson & Neave, 175–176, 192–193

Fleckenstein, Lee, 184, 205, 410
Fogarty, John, 326, 390, 429, 432
Foley, Hoag & Eliot, 260
Foote, Cameron, 183
Forbes, 454, 532
Ford, Gerald, 95, 186–187, 363
Ford, Henry, 28
Ford Motor Company, 20
Ford, Robert M., 194, 201–203, 207, 252, 287, 290, 293, 295, 314, 461
Fordham, Laurence, 460
Fortune, 27, 28, 122, 124, 138, 148, 172, 177, 187, 311, 497, 515, 519
Frankel, Marvin, 256
Fraser, William W., 11
Friedman, Joseph, 9
FTC. *See* U.S. Federal Trade Commission (FTC)
Fuji Photo Film Company, 235, 391, 456, 458, 520

G

Gabel, Wren, 82–87, 91, 95, 108, 124–127, 223, 367
GAF, 242
Galbraith, James, 419
Gear Train patent, 215, 390, 465
General Electric, 20, 27, 41, 77–78, 140, 176, 177, 523
General Motors, 15, 20, 28, 29, 40, 177, 188
Girden, Barney "Cap," 4–5, 11
Godowsky, Leopold, Jr., 68
Goggles, 32–33, 36–37
Goldfaden, Mattie. *See* Land, Mattie Goldfaden
Goldman, Robert J., 300
Goldstein, Paul, 468, 505–507, 533
Goodwin Film & Camera Co., 85
Goodwin, Proctor & Hoar, 306
Goodyear, Charles, 243
Govatos, Ty, 452
Graham, John Patrick, 232
Graham v. John Deere Co., 447
Great Depression, 12
Green, George, 62–63
Green Giant Company, 188
Greenlaw, David, 246, 247, 321, 340–341
Grey, David, 35
Guggenheim Museum, 519

H

Hand, Learned, 177
Handler, Edward J., 273
Hanson, Wesley "Bunny," 184, 243–244
Hardy, Arthur C., 55
Harvard Business Review, 77
Harvard University, 5, 12–13, 30, 61, 117
Harvey, Douglass, 242
Headlight glare, 5, 9, 15–16, 20–24, 27, 40
Herapath, William, 8–9
Herapathite, 8–10
Highway Research Board, 41
Hiroshima, 36–37
Hirsch, Maurice "Baron," 1
Hitachi Ltd., 516
Hoar, Samuel, 307
Holmes, Jon, 206
Holmes, Oliver Wendell, 3, 177
Hopkins, Harry, 30
Hunt, J.H., 15, 41

I

IBM, 229–231, 242
Iceland spar, 4
The Impossible Project, 520–521
In Tune with Tomorrow, 28
Ingraham, Christopher, 112
Instagram, 520
International Center of Photography, 521
Invalidity, 193, 219
Invention, 513

J

J. Walter Thompson, 198
Jackson, Robert, 313
James, Ted, 238
Janicke, Paul, 505

Japanese "kamikaze" suicide attacks, 32
Jefferson, Thomas, 447
Jennings, Eugene C., 187
Jewish Colonization Association, 1
Jobs, Steve, 41, 148, 155, 526–528
Johnson, Lyndon B., 91, 94, 363
Jordan Marsh, 64

K

Kaleidoscopes, 3
Kaprelian, Edward R., 429–432, 443, 446, 471, 511–512
Kaps, Florian, 520
Kaska, Hank, 494
Kaufman, Irving, 289
Kennedy, Clarence, 18, 68, 71, 522
Kennedy, John F., 87, 91, 94, 363
Kenway, Herbert, 75
Kenyon & Kenyon, 220, 240, 280, 315, 333
Kenyon, Houston, 220
Kenyon, Theodore, 221
Keppler, Herbert, 516
Kerr, Andy, 181
Kerr, William K.
- application for pre-trial conference, 307–308
- background, 181–182
- Berkey Photo case, 240
- on Carr's proposed agenda for pre-trial conference, 329–330
- delivery of lawsuit, 216
- discovery process, 239–240, 248–249, 257–260, 263, 285
- Fallon's informal talks with Polaroid, 246
- final dinner with Land, 386–387
- files motion to limit cross-examination, 382–383
- initial pre-trial conference with Zobel, 316–317
- Kodak summary judgment motions, 319–320, 326–329
- Land deposition, 295–297, 299–300, 311–312
- Land deposition preparation, 285–287, 290–291, 293
- Land direct examination at trial, 354–368
- legal strategy discussions, 184, 185–186, 192–196, 199, 238
- letter to Land, 335–336
- meeting with Land, 275–276
- opening statement at trial, 348–352
- Polaroid Patent Policy memorandum, 334–335
- review of litigation with Polaroid senior management, 288
- review of Polaroid patents, 214
- Rogers deposition, 269–270
- sent post-trial submissions by Schwartz, 451–452
- settlement discussions, 338–344
- Sieg deposition, 254–255
- Wareham deposition, 325–326
- Young deposition, 266

Kettering, Charles, 15, 28
Khrushchev, Nikita, 92
Killian, James, 87–88, 90, 91, 93
Kirkland & Ellis, 201
Kodachrome, 67, 68, 85, 426
Kodacolor, 68, 426
Kodak. *See* Eastman Kodak Company
Kodak Bimat system, 362–363, 413–415, 447
Kodak Carousel slide projector, 221
Kodak EK-4 camera, 208, 236
Kodak EK-6 camera, 208, 214, 236, 365, 408–409
Kodak EK-8 camera, 208, 236, 242
Kodak Handle camera, 245, 262
Kodak Instamatic Camera, 122, 221
Kodak Lanyard Camera, 160–161, 407, 408–409, 467
Kodak Park Development Committee (KPDC), 169–170
Kodak Pocket Instamatic, 256
Kodak PR-10 instant film, 208, 236, 262, 358–359, 363, 373–375, 393, 395, 396, 403, 410, 413, 417, 419, 426, 447–448, 469

KPDC. *See* Kodak Park Development Committee (KPDC)
Kriebel, Richard, 43, 64
Kuhn, Loeb & Company, 26

L

L-Coat, 99, 114, 367, 393, 512
L-Coat patent, 213, 295, 357–361, 362, 371–376, 403, 421, 425–426, 428–429, 439, 444, 446, 465, 468
Lady Gaga, 520–521
Land, Abraham, 1
Land, Edwin Herbert "Din"
  acquisition of nickname, 2
  announcement of low-cost camera, 106
  awards and honors, 28–29, 61, 243–244, 523
  as benefactor, 15
  birth of, 2
  boyhood, 3
  Christmas Eve presentation to employees, 53
  Cold War activities, 87–94
  commits to participating in litigation, 275
  cross-examination by Carr, 369–388
  death of, 532
  deposition, 294–297, 299–300, 311–312
  deposition preparation, 285–287, 290–291
  on development of color film, 97–99
  dinner at Ritz Carlton Hotel, Boston, 386
  direct examination at trial of, 354–368
  dream for technology corporation of future, 522–523
  dream of "absolute one-step photography," 106–107, 112–119
  early experiments with polarized light, 5–6
  early one-step process demonstration for Kodak, 51–52
  elected president of American Academy of Arts and Sciences, 74
  first experiments on polarizers, 7
  formation of Polaroid Corporation, 26
  full technical demonstration of Aladdin, 157–160
  goal for instant photography, 46
  on a good life, 532–533
  at Harvard University, 5, 12–13
  high school years, 5
  Highway Research Board presentation, 41–42
  instant photography demonstrations, 51–55, 62, 69, 166
  interest in optical science, 4–5
  on intersection of art and science, 522
  interview in *Boston Globe*, 237
  interviews with *Business Week*, 323–324
  on joint development program with Kodak, 102–103
  Kodak's purchase of Land's polarizer material, 17–18
  leaves presidency of Polaroid, 187–190
  letter of introduction to Mees, 12
  memorial service, 531–532, 533
  move to New York City, 6–7
  on necessity for suing Kodak, 211
  in negotiations with Kodak, 135–138
  on Nixon's enemies list, 94–95
  organizational skills, 37, 524–525
  on patent system, 76–78
  patents, first application, 11, 15–16
  patents, instant photography, 49, 60–61, 212–214
  patents, light polarization, 61
  patents, optics, 61
  patents, three-dimensional technology, 33
  patents, Vectograph, 33
  personality, 529–532
  philosophy of intellectual investigation, 6–7
  philosophy of product development, 147–148
  polarizer, invention of, 9–11
  Polaroid Patent Policy memorandum, 334–335
  reaction to court decision, 474
  reaction to Kodak instant photography system, 207–210

rebuttal presentation, 433–439
refusal to give up Polavision, 297–299
relationship with Ansel Adams, 71–73, 529
relationships with academic institutions, 524
relationships with U.S. presidents, 94–95
religious faith of, 2
remarks at shareholders meeting after filing of lawsuit, 216–217
on research process, 48–49
on role in war effort, 31
Rowland Institute, 454–455, 532
on scientific method, 43–44
sells remaining Polaroid stock, 462–463
shows Kodak representatives integral film unit, 130, 367–368
Society of Automotive Engineers meeting, 25
steps down as chairman of the board of directors, 453
steps down as chief executive officer, 309–311
strategy for identifying employment prospects, 24
team research experience under, 144–145
unveiling of Aladdin, 149–150, 154–160
vision of himself, 529–530
Waldorf-Astoria demonstration of polarizer, 22–23
as witness, 510–511
on work environment, 524, 525
working relationship with Rogers, 295
Land, Ella, 1
Land, Harry, 1, 2, 6
Land, Helen (Terre), 8, 12, 13, 26, 528
Land, Jennifer, 12
Land, Louis, 1
Land, Mattie Goldfaden, 2
"Land Panel," 91
Land, Samuel, 1
Land, Valerie, 12
Land-Wheelwright Laboratories, 14, 17, 20–21, 24
Landry, Brenda Lee, 454
Latham, Jack, 29
LCD, 518
Leduc, Robert, 215
Lefevre, Belfield, 46
Lenzner, Robert, 56, 371, 454, 529
Libby-Owens-Ford, 17
Life, 55, 69, 101, 167, 523
Light-Shield/Deflector patent, 215, 390, 465
Linowitz, Sol, 231
Lockheed, 90
Lombino, Mary-Kay, 523
Lorbach, R.D., 224
Lum, Kin Kwong, 411
Lunney, Glynn, Jr., 506
Lyman, Theodore, 13

M

Madsen, Kenneth, 300, 400, 413
Mahler, Joseph, 33
Maislen, Helen (Terre). *See* Land, Helen (Terre)
Mannes, Leopold, 68
Marconi, Guglielmo, 243, 445
Markey, Howard T., 488, 490, 497, 504, 505
Martone, Patricia, 451, 480
Mason, Lowell, 65
Mayer, Harry, 77
Mazzone, A. David, 259, 261, 272, 502, 508–510, 515
McCann, John, 528
McCann, Mary, 528
McCloud, John, 13
McCormack, Cyrus, 243
McCune, Elizabeth, 54
McCune, William
ability of overcoming problems, 70
background, 190
becomes president of Polaroid, 187–190
as chief executive officer, 309–310
confronts Land on stock sale, 324
demise of Polaroid, 517
elected chairman of the board of directors, 453
faces long-anticipated challenges, 191–192

McCune, William (*continued*)
first public demonstration of instant photography, 54–55
Gabel's letter to, 223
introduction of Land at shareholders meeting, 325
issues optimistic assessment of company's future, 479
joins Polaroid, 29, 190
on Land's dream of instant photography, 44, 105
on Land's tunnel vision, 113
legal strategy discussions, 185
on marketing approach of Polaroid products, 64
meeting on patents with Kodak, 184–185
meeting to discuss negotiated resolution, 321–322
meetings with Kodak, 367
on motion picture technology, 295
negative manufacturing facility, 133–134, 139, 149, 154
negotiations with Kodak, 82–87, 91, 95–97, 101–102, 133–134
Polaroid licensing policy, 229
on progress of agreement with Kodak, 261
on public release of SX-70, 171
recollection of Land's talk on war effort, 31
remarks at Land's memorial service, 531–532
settlement discussions, 340–344
Supreme Court decision, 494
McElheny, Victor, 2, 31
McGovern, George, 147
McLaughlin, Walter, 260, 261, 262, 267–268, 271, 318, 322, 330
Mees, C.E. Kenneth, 12, 16, 51, 58, 62, 67, 68, 286, 366
Melman, Seymour, 74–76
Mervis, Stanley, 96, 99
Mikulka, Charles, 42–43, 44, 49, 95, 96, 108–112, 176, 181, 273, 290, 321, 340
Miller, Robert, 134, 135
MIT, 24, 28, 30, 34, 55, 411
alumnus working at Polaroid, 29, 117, 140
Junior Science Symposium, 6
Modern Photography, 131, 235, 516
Modern Pioneers on the Frontiers of Industry, 28
Monomoy Point, 36
"Monroe County syndrome," 519
Monthly Newsletter on Corporate Litigation, 412
Mordant patent, 215, 392–393, 413, 465
Morgan, J.P., 26
Morgan Stanley, 454
Morita, Akio, 516
Morse, Meroe, 68–69, 70, 71, 73–74, 295, 522
Morse, Samuel, 243
Muller, Eudoxia, 45, 58
Mullowney, Edward, 410–411, 413
Murray, Frank J., 214, 239, 248, 259, 284
Museum of Fine Arts, Boston, 522
Museum of Science, 75, 171

N

NASA, 91, 140
Nathan, John, 287–288
National Association of Manufacturers, 28
National Defense Research Committee (NDRC), 30–31
National Inventors Hall of Fame, 243, 523
National Law Journal, 193
National Medal of Science, 523
National Photo Dealer, 63
National Security Council, 90
NDRC. *See* National Defense Research Committee (NDRC)
Neave, Alexander, 193
Neave, Charles, 176–178, 193, 513
Negative Dye Developer patent, 205, 394–395, 410, 421–422, 465, 472, 487–488
Negative Dye Developer process, 204–205, 394–395, 410–411, 447, 466, 512

Nelson, David S., 279, 305
Neuman, Sidney, 240–241
New York City Museum of Science and Industry, 25
New York Daily News, 65
New York Herald Tribune, 24
New York Public Library, 6
New York Times, 23, 30, 55, 61, 62, 92, 147, 166, 178, 187, 230, 232, 241, 242, 245, 246, 247, 273, 279, 283, 348, 363, 474, 504
New York University, 6
New York World's Fair, 1939–1940, 28
Newman, Pauline, 490
Newsday, 152
Newsweek, 55, 221, 236, 244
Nichols, Herbert, 25, 41
Nimslo camera, 456
Nixon administration, 95
Nixon, Richard, 91, 94, 147, 186, 363
Noninfringement, 193, 219
Norwich Free Academy, 5

O

O'Connor, Sandra Day, 347
Office of Defense Mobilization, 88
Office of Scientific Research and Development (OSRD), 30–31
Olivier, Laurence, 166
One-step photography, 46
Optical Society of America, 33, 53, 55
OSRD. *See* Office of Scientific Research and Development (OSRD)
Otto, John, 413–415
Oxidized Developer Process, 47

P

P-108 process, 122
P-109 process, 122
P-110 process, 121–122, 125, 131–133
Paglia, Richard, 154, 215, 216, 465
Patent infringement
- Berkey Photo lawsuit for, 249–250, 256–257
- Kodak found guilty of, 465–478
- Kodak Park Development Committee memo, 170
- lawsuits filed in foreign countries, 222
- legal defenses to, 193, 219–220
- Polaroid files suit against Kodak for, 214–217

Patent system, 74–78, 353
Patents
- of Aladdin components, 151
- asserted against Kodak, 214–216, 227
- Crank patent, 347–348
- Detachable Spread Housing patent, 215, 390
- dye developer patents, 357, 377, 393, 472
- Excedrin film unit patent, 203–204, 221, 234, 280–284, 393, 395–397, 421, 422–426, 428, 448–449, 465, 472–473
- Fish & Neave review of Polaroid, 201–205
- Gear Train patent, 215, 390
- for instant photography, 49, 60–61, 212–214
- for integral negative-positive system, 130
- issued on Kodak imaging chemistry, 184, 205
- L-Coat patent, 213, 295, 357–361, 362, 371–376, 403, 421, 425–426, 428–429, 439, 444, 446, 465, 468
- for Land's sheet polarizer, 11, 15–17
- Light-Shield/Deflector patent, 215, 390, 465
- Mordant patent, 215, 392–393, 413, 465
- Negative Dye Developer patent, 394–395, 410, 421–422, 465, 472, 487–488
- Polaroid patent practice, 174–175, 224–226
- problems with Berkey Photo, 191–196
- Rear Pick patent, 213, 295, 364–365, 384, 403, 433, 446, 471

Patents (*continued*)
requirements for granting, 193–194
Symmetrical Supports patent, 213, 361–363, 377–385, 403, 413, 415, 419–420, 436–438, 447, 465, 470–471, 511, 512
Trap patent, 215, 320, 328, 345, 353
for Vectograph, 33
Whitmore-Mader patent, 410–411
Patton, George S., 32
Pavelle Corporation, 242
Pearl Harbor attack, 31, 32
Peck, Robert
defense of Excedrin patents in foreign jurisdictions, 287
determination of damages, 314
Kodak misuse allegations, 228, 231
Land deposition, 295
Land deposition preparation, 290, 293
large-scale litigation story, 273
legal strategy discussions, 199, 201–202
meeting with Land, 275–276
on new discovery initiative, 263
possible settlement considerations, 328
report on Kodak's position, 238
retaining outside legal counsel, 175–176
settlement discussions, 340
Pentagon Papers, 147
Perkin-Elmer Co., 90
Perkin, Richard, 90
Perkins, Mahlon, 283
Perry, Holly, 276, 290
Petersen, George, 108–111, 130–132
Petters Group Worldwide (PGW), 517
Petters, Tom, 517
Philby, Kim, 32
Phillip Morris, 58
Photo Weekly, 482
Photographic Society of America, 62, 69
Photography, conventional, 66–67, 79–80
Photography, digital, 515–516
Photography, instant (also referred to as one-step photography)
black-and-white film, 68–70
black-and-white images, 47
color diffusion transfer process, 204
color film, 67–68, 79–87, 91, 95–103
competition in worldwide, 456
consumer version, 59
demise of, 515
development of, 42–43
first public demonstration of, 53–55
patent applications, 49, 51–52, 60
press reaction, 55–56
process of conventional photography, 45–46
processing solution/solvent pod for, 48
sepia film, 58–59, 69
sepia-toned images, 47
use of standard-issue Kodak negatives, 47
Physical Optics (Wood), 4
Piel, William, Jr., 283
Pittsburgh Plate Glass, 20
Plummer, William, 142–143
Polarized Lights, 16, 20, 27
Polarizers
history of, 4–5, 8
issuance of patent, 15–16
K Sheet, 27
making plastic sheets of, 9–11
patent rivalry, 16–17, 20–21, 27
for photographic lenses, 16
publicity for, 23–24, 27–28
sunglasses, 19–20
use in war effort, 31
Polaroid Collections, 522
Polaroid Corporation
acquisition of Polarized Lights, 27
announces end of exploratory talks with Kodak, 260–261
Berkey Photo lawsuit, 196, 249–250, 256–257
Berkey Photo patent problems, 191–196
British appeals court overturns ruling, 233
business decline post–WWII, 57
business relationship with Kodak, 17–18, 58, 67, 82–87, 366–368, 384
ceases camera and film production, 517
collaboration with Ansel Adams, 72–73

color film, 95–103
as corporate citizen, 525–526
counteroffensive in Kodak litigation, 285–288
court decision on damages, 503–504
decision in Kodak case, 465–474
decline of instant camera and film sales, 515–516
delay in release of Aladdin, 166
demise of, 516–517, 519
determination on amount of compensatory damages, 501
development agreement with Kodak, 107–108
discovery process, 239–240, 248–249, 257–260, 263, 279–280, 285
disposable polarizing eyeglasses, 39
Dove Project, 35–36
earnings, 70, 84, 106, 171, 197, 236, 244, 247–248, 262, 277–278, 308, 452–453
education department, 526
entry into consumer photography business, 52
files brief in opposition to emergency motion, 492–493
files briefs with court of appeals, 488–489
files for bankruptcy protection, 517
files infringement suit against Kodak, 214–217
foray into electronic photography, 516
formation of, 26
High Court of London issues temporary injunction against Kodak, 232–234
hostile takeover attempt, 517
instant home-movie camera system, 245 (*see also* Polaroid Polavision)
introduction of non-folding instant camera, 197–198
issue of damages, 452, 474
joint press release marking supply and license agreements, 139
on Judge Nelson recusal, 305–306
Kodak files motion to compel discovery from, 262–263
Kodak summary judgment motions, 280–281, 287, 292–293, 304, 317, 319, 326–329
Land leaves presidency, 187–190
Land steps down as chairman of the board of directors, 453
Land steps down as chief executive officer, 309–311
layoffs at camera production facility, 284–285
legal strategy discussions, 181–186, 201
license issues with Kodak, 107–111, 123–133, 184–185, 367–368
license policy, 228–229
license to manufacture a peel-apart film, 138–139
McCune becomes president, 187–190
McCune joins, 29, 190
misuse allegations, 227–229, 231–232
negative manufacturing facility, 133–134, 148–149
new challenges for, 455–456
New York investment firm lawsuit, 278–279
number of employees, 37, 71
opposing papers to Supreme Court, 499
oral arguments on appeals, 491–492
patent department, 42, 175
patent infringement lawsuits filed in foreign countries, 222
patent policy, 370, 444–445
patent practices, 174–175, 224–225
policy of apartheid in South Africa and, 526
post-trial brief, 442–444
print coater for black-and-white film, 69–70
on progress of settlement with Kodak, 261
proposed form of judgment, 476
public release of first consumer products, 64–66
reaction to Kodak instant photography system, 207–209
rebuttal case, 432–438
reply brief, 449–451

Polaroid Corporation (*continued*)
response to Kodak's answer and counterclaims, 227–229
response to filing of reply brief on discovery motion, 300–301
retaining outside legal counsel, 175–178
Rogers employment at, 25
role in war effort, 31–37, 50–51
sales growth, 122, 261–262
seeks status of Zobel's decision-making process, 460
settlement discussions, 461–462
Polaroid Corporation v. Eastman Kodak Company, 248, 349, 440, 464, 465, 504, 506, 515
Polaroid Day Glasses, 20
Polaroid Highlander, 71
Polaroid, genesis of name, 18
Polaroid Land Camera Model 95, 59, 64–65, 356
Polaroid Land Model 95A, 71
Polaroid Land Type 40 film, 58–59
Polaroid Land Type 41 film, 69
Polaroid Model PIC 300, 520
Polaroid on Parade, 25
Polaroid OneStep, 245, 247, 262, 520
Polaroid Polacolor Film, 99–100, 108, 114, 122
Polaroid PolaPan 200, 71
Polaroid PolaPan black-and-white films, 71
Polaroid Polavision, 245, 278–279, 297–299
Polaroid Pronto, 197–198, 206, 236
Polaroid Spectra Instant Camera, 515
Polaroid Sun 600 Instant Film, 457
Polaroid Sun Glasses, 20
Polaroid Sun Instant Camera, 457
Polaroid Swinger, 106, 122, 123
Polaroid Time-Zero Supercolor Film, 280
The Polaroid Years, 523
Polaroid Z2300, 520
Polascreen, 19
Popular Photography, 66, 71
Porter, Michael E., 456
Porter, R. Frederick, 411
Positive dye developer process, 204, 394–395
Powell, Lewis F., Jr., 494
Powers, Francis Gary, 91
Powers, Jimmy, 65
Presidential Medal of Freedom, 523
Prezzano, Wilbur, 474
Prime, William G., 230
Progress Medal of the Photographic Society of America, 62
Project Charles, 87–88
Przybylowicz, Edwin, 519
Purcell, Edward, 91

Q

Quillen, Cecil, 247, 321, 328, 340–341, 431, 452, 461–462, 506–507, 509
Quinine, synthesis of, 34–35

R

Radio Corporation of America (RCA), 177, 199, 518
"Rainbow" hologram, 34
Ray-O-Vac, 140
RCA. *See* Radio Corporation of America (RCA)
Reagan, Ronald, 247
Rear Pick patent, 214, 295, 364–365, 384, 403, 433, 446, 471
Reinhold, Robert, 166
Relyea, William, 474, 475, 516
R.J. Reynolds, 58
Rogers, Howard
crafting plastic lenses, 32
cross-examination of, 400
deposition of, 263, 265–268, 400
development of color diffusion transfer process, 204
development of K Sheet, 27
direct examination at trial of, 393–397, 399–400
dye developer patents, 357, 377, 393, 472
eulogy at service for Morse, 73–74
hiring of, 24
as Land's primary scientific collaborator, 29

negative dye developer process patent, 205, 394–395, 410, 421–422, 465, 472, 487–488
rebuttal witness, 432
as witness, 510–511
work on instant color films, 67, 79, 80–82, 96, 101, 113, 366
work on film unit for SX-70 camera, 118, 180, 203, 280–281, 448–449, 472–473
working relationship with Land, 295
Rogers, Nick, 24
Roosevelt, Franklin Delano, 30–31, 57, 313
Roosevelt-Willkie presidential election campaign, 33
Rosenblum, Louis, 33
Rothschild, L.F., 198
Rott, Andre, 47
Rowland Foundation, 278
Rowland Institute, 310, 442, 454–455, 462, 474, 532
Royal Society of London, 7–8, 523
Rubin, Gerald, 246
Rule 41 motions, 392, 397–399, 403, 412–413
Rumsey, John, 134
Rusitzky, Louis, 179, 237
Rymer, William, 175–176, 177–178, 180

S

SAC. *See* Science Advisory Committee (SAC)
Sage, Nat, 29
Samson United Corporation, 60, 64
Sarnoff, David, 177
Schallhorn, Charles, 410–411, 472
Schumpeter, Joseph, 534
Schwan, Judith, 411
Schwartz, Herbert F.
Adelstein cross-examination, 415–419
application for pre-trial conference, 307–308
background, 192–193
Berkey Photo case, 240
on Carr's deposition program, 325
on conducting discovery prior to Christmas, 322–323
conveys his appreciation to Land, 441–442
decision of case, 465–478
delivery of lawsuit, 216
determination on amount of compensatory damages, 501
Driscoll examination, 390
on Kodak's patent misuse claims, 231
Fallon deposition, 316
Fallon's informal talks with Polaroid, 246
files brief in opposition to emergency motion, 492–493
files briefs with court of appeals, 488–489
files motion to limit cross-examination, 382–383
Land deposition, 295–297, 311–312
Land deposition preparation, 293
Land rebuttal presentation, 433–439
legal strategy discussions, 193–196, 199, 238
meeting to discuss negotiated resolution, 321–322
motion requesting court rule on protective order, 239
oral arguments on appeals, 491–492
Otto cross-examination, 414–415
persuading court not to stay the injunction, 475–476
response to Kodak motion to stay injunction, 485–487
review of litigation with Polaroid senior management, 288
review of Polaroid patents, 213–214
Rogers deposition, 266–273
Rogers examination, 393–397, 399–400
seeks status of Zobel's decision-making process, 460
settlement conference, 337–338
shares post-trial submissions with Land, 451
Sieg cross-examination, 408–410
Sieg deposition, 251–255

Schwartz, Herbert F. (*continued*)
status conference to consider form of judgment, 477–478
Thomas cross-examination, 405–407
Trautweiler cross-examination, 426–429
Young deposition, 266
Science Advisory Committee (SAC), 88, 92
Scientific American, 10, 45–46
SCM v. Xerox trial, 330
Sculley, John, 526, 528
Seiden & De Cuevas Inc., 278
Shamrock Holdings, 517
Shapiro, Irving, 324
Sherman Antitrust Act, 289
Shumard, Bob, 36–37
Sidewinder missile, 36
Sieg, Albert, 161, 206, 249, 252–253, 407–410
Silver, Julius, 11, 15, 19, 26, 112, 175, 236, 238, 314, 319–320, 326
Singer, Nina, 340, 460, 462, 463
Smith, Donald, 413
Smith, Edward, 490
Society of Automotive Engineers, 25
Society of Motion Picture Engineers, 25
Society of Photographic Engineers and Scientists (SPSE), 157, 161, 163
Solomonovitch, Avram. *See* Land, Abraham
Solomonovitch, Ella. *See* Land, Ella
Soluble Silver Complex Process, 47
Sonnenfeld, Jeffrey, 526, 529
Sony, 455, 515, 517, 518
South Africa, 526
Sparks, Robert, 21–22
Spearing, James, 22–23
SPSE. *See* Society of Photographic Engineers and Scientists (SPSE)
Sputnik, 92–93, 95
St. Louis Post-Dispatch, 23
Staud, Cyril "Sy," 47, 67, 108, 366
Stereopticons, 3
Stereoscopes, 3–4, 33
Stone, Harlan, 445
Strauss, Lewis, 27
Stubbins, Hugh, 455
Stuyvesant Asset Management Corporation, 230
Sullivan & Cromwell, 501
Summary judgment motions, 280–281, 287, 292, 304, 317, 319, 326–329
Sunglasses, 19–20, 29, 51
Svoboda, Karel, 10
SX-70 camera
battery issue, 179
camera design requirements, 115–117, 142
demand for, 172–173
film unit for, 113–115, 140–142, 154, 397, 418, 519–520
film unit opacification system, 117–119
as iconic symbol, 519–520
Kaprelian cross-examination, 430–431
Kodak reaction upon introduction, 407–410
Land demonstration in courtroom, 356
Land drops name of Aladdin, 167
lens requirements, 115, 142
light-piping problem, 148, 154
national distribution of, 171
new distribution network for, 520
price of, 197
prototypes for, 117, 140–144
public release, 167
Quillen observations on differences, 509
Time-Zero film, 280
threat to Kodak business, 251, 443
unveiling of Aladdin, 149–150, 154–157
viewfinder arrangement, 142–143
SX-70 project, 45, 49–52
Symmetrical Supports patent, 213, 361–363, 377–385, 403, 413, 415, 419–420, 436–438, 447, 465, 470–471, 511, 512
Symphony Hall, 310
Syre, Steven, 517

T

Taylor, Lloyd D., 215, 272, 392, 432, 465
Technological Capabilities Panel (TCP), 88–89, 90

Temporary National Economic Committee, 313
Tesla, Nicola, 220
35-millimeter cameras, 515–516
Thomas, Leo, 404–407
Thompson, Bob, 75
Thomson, Elihu, 523
Thomson-Houston, 176
Three-dimensional images, 33–34
Three-dimensional movies, 23, 25, 28, 39
Time, 131, 167
Timex, 60, 456
Toshiba, 517
Transargo, 47
Trap patent, 215, 320, 328, 345, 353
Trautweiler, Franz, 421, 443, 468–469
Truman administration, 88
Truman, Harry, 62, 65, 94
Tufts College, 61
Tufts University, 61
Tuttle, Fordyce, 17, 19

U

U-2 reconnaissance plane, 89–91
Ueberroth, Peter, 458
Union Pacific Railroad, 27
Uniroyal, 221
United Press International, 465, 466, 485
U.S. Air Force, 35, 36, 88, 90
U.S. Congress, 504
U.S. Court of Appeals for the Federal Circuit, 483, 497, 502, 504
U.S. Court of Appeals for the Second Circuit, 289
U.S. Department of Justice, 231, 242, 246
U.S. District Court for the District of Massachusetts, 214
U.S. Federal Trade Commission (FTC), 65, 230–231
U.S. Navy, 2, 32, 35, 36
U.S. Patent 1,918,848, 15–16 (*see also* Polarizers)
U.S. Patent 3,245,789, 205 (*see also* Negative Dye Developer patent)
U.S. Patent 3,362,821, 213 (*see also* L-Coat patent)
U.S. Patent 3,578,540, 213 (*see also* Symmetrical Supports patent)
U.S. Patent 3,594,165, 203 (*see also* Excedrin film unit patents)
U.S. Patent 3,689,262, 203 (*see also* Excedrin film unit patents)
U.S. Patent 3,753,392, 214 (*see also* Rear Pick patent)
U.S. Patent Board, 447
U.S. Patent Office, 11, 21, 26, 49, 60, 61, 193, 194, 225, 243, 293
U.S. Steel, 177, 221
U.S. Supreme Court, 176, 177, 212, 289, 443, 445, 493, 503, 511
U.S. Time, 60

V

Van DePoele, Charles, 220
Vaughn, William, 124
Vectograph, 33–34, 158
von Neumann, John, 88

W

Walker, Jimmy "Beau James," 7
Wall Street Journal, 106, 127, 149, 150, 166, 168, 187, 190, 229, 232, 242, 245, 297, 310, 391, 453, 496, 505, 518
Wallis, Brian, 521
Warburg, Felix, 25, 26
Warburg, James, 26
Wareham, Richard, 115–116, 140–141, 154, 215, 325–326, 446, 465
Warnke, Paul, 493
Washington Post, 494
Watson, Thomas, 57
Webster, William, 94
Wensberg, Peter, 172
Werle, Ira, 135, 137
Westinghouse Electric Company, 16
Weyde, Edith, 47
Wheatstone, Charles, 3

Wheelwright, George, III
American Optical and, 19
background, 12–13
departure from Polaroid, 29
formation of Polaroid, 26
three-dimensional movies, 23, 25
Whitmore, Kay, 458, 487, 492
Whitmore-Mader patent, 410–411
Whitney, Eli, 243
Wiesner, Jerome, 87
Willoughby's Camera, 483
Windshields, 16–17
Wolfarth, Eugene, 444
Wolff, Otto, 70, 390
Wood, Robert, 4, 20, 25
Woodward, Robert, 34, 67, 117, 530
Wright, Orville, 176, 243
Wright, Wilbur, 28, 176, 243
Wyman, Thomas, 185, 188

X

Xerox v. IBM, 230
Xerox, 221, 229–231, 242

Y

Yanes, Sam, 494
Young, Richard, 255, 266
Young, Thomas, 4
Yutzy, Henry, 58, 82, 90, 103, 108, 125, 360, 366

Z

Zausner, Sam, 192
Zobel, Rya
Adelstein examination, 415
application for pre-trial conference, 308
assignment of case, 306–307
conclusion of proceedings, 439–440
decision of case, 464, 465–478
Driscoll cross-examination, 390
dye developer process, 395
entry into courtroom, 348
expert witnesses at trial, 510–511
history of development of one-step photography, 512
inquiry to counsel on length of case, 415–416
issues permanent injunction against Kodak, 480–482
Kodak summary judgment motions, 319–320, 326–330, 336–337
Land cross-examination, 379–383
Land examination, 355–356, 359, 363–364, 372–373
on motion to limit cross-examination, 390–391
notice of appeal of judgment, 483–485
pre-trial conference, 316–317, 342
recusal of, 502
request for submission of post-trial briefs, 442
Rogers cross-examination, 400–402
Rule 41 motions, 412–413
settlement conference, 321–322, 337–338
settlement discussions, 338–344
status conference to consider form of judgment, 477
Trautweiler examination, 421–426
trial exhibits, 457–458
Zornow, Gerald B., 152, 163, 167, 173
Zucker, Harvey, 209